A Course in
Abstract Analysis

A Course in Abstract Analysis

John B. Conway

Graduate Studies
in Mathematics

Volume 141

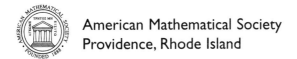

American Mathematical Society
Providence, Rhode Island

2010 *Mathematics Subject Classification.* Primary 28-01, 46-01.

For additional information and updates on this book, visit
www.ams.org/bookpages/gsm-141

Library of Congress Cataloging-in-Publication Data

Conway, John B., author.
  A course in abstract analysis / John B. Conway.
    pages ; cm. — (Graduate studies in mathematics ; volume 141)
  Includes bibliographical references and index.
  ISBN 978-0-8218-9083-7 (alk. paper)
  1. Measure theory. 2. Integration, Functional. 3. Functional analysis. I. Title.

QA312.C5785   2012
515—dc23
                                                      2012020947

For Ann,
The love of my life, the source of my happiness

Contents

Preface

I am an analyst. I use measure theory almost every day of my life. Yet for most of my career I have disliked it as a stand-alone subject and avoided teaching it. I taught a two-semester course on the subject during the second year after I earned my doctorate and never again until Fall 2010. Then I decided to teach our year long course that had a semester of measure theory followed by a semester of functional analysis, a course designed to prepare first-year graduate students for the PhD Qualifying exam. The spring before the course was to begin, I began to think about how I would present the material. In the process I discovered that with an approach different from what I was used to, there is a certain elegance in the subject.

It seems to me that the customary presentation of basic measure theory has changed little since I took it as a first-year graduate student. In addition, when I wrote my book on functional analysis [**8**], it was premised on students having completed a year long course in measure theory, something that seldom happens now. For these two reasons and because of my newly found appreciation of measure theory, I made the decision that I would write a book. For this project I resolved to look at this subject with fresh eyes, simplifying and streamlining the measure theory, and formulating the functional analysis so it depends only on the measure theory appearing in the same book. This would make for a self-contained treatment of these subjects at the level and depth appropriate for my audience. This book is the culmination of my effort.

What did I formerly find unpleasant about measure theory? It strikes me that most courses on measure theory place too much emphasis on topics I never again encountered as a working analyst. Some of these are natural enough within the framework of measure theory, but they just don't arise in the life of most mathematicians. An example is the question of the measurability of sets. To be sure we need to have our sets measurable, and this comes up in the present book; but when I studied measure theory I spent more time on this topic than I did in the more than 40 years that followed. Simply put, every set and every function I encountered after my first year in graduate school was obviously measurable. Part of my resolve when I wrote this book was to restrict such considerations to what was necessary and simplify wherever I could. Another point in the traditional approach is what strikes me as an overemphasis on pathology and subtleties. I think there are other things on which time is better spent when a student first encounters the subject.

In writing this book I continued to adhere to one of the principles I have tried to adopt in my approach to teaching over the last 20 years or so: start with the particular and work up to the general and, depending on the topic, avoid the most general form of a result unless there is a reason beyond the desire for generality. I believe students learn better this way. Starting with the most general result sometimes saves space and time in the development of the subject, but it does not facilitate learning. To compensate, in many places I provide references where the reader can access the most general form of a result.

Most of the emphasis in this book is on regular Borel measures on a locally compact metric space that is also σ-compact. Besides dealing with the setting encountered most frequently by those who use measure theory, it allows us to bypass a lot of issues. The idea is to start with a positive linear functional on $C(X)$ when X is a compact metric space and use this to generate a measure. The Riemann–Stieltjes integral furnishes a good source of examples. Needless to say, this approach calls for a great deal of care in the presentation. For example, it necessitates a discussion of linear functionals before we begin measure theory, but that is a topic we would encounter in a course like this no matter how we approached measure theory. There is also a bonus to this approach in that it gives students an opportunity to gain facility in manufacturing continuous functions with specified properties, a skill that I have found is frequently lacking after they finish studying topology and measure theory.

Chapter 1 contains the preliminary work. It starts with the Riemann–Stieltjes integral on a bounded interval. Then it visits metric spaces so that all have a common starting point, to provide some handy references, and to present results on manufacturing continuous functions, including partitions of unity, that are needed later. It then introduces topics on normed spaces needed to understand the approach to measure theory. Chapter 2 starts with a positive linear functional on $C(X)$ and shows how to generate a measure space. Then the properties of this measure space are abstracted and the theory of integration is developed for a general measure space, including the usual convergence theorems and the introduction of L^p spaces. Chapter 3 on Hilbert space covers just the basics. There is a later chapter on this subject, but here I just want to present what is needed in the following chapter so as to be sure to cover measure theory in a single semester. Chapter 4 starts by applying the Hilbert space results to obtain the Lebesgue–Radon–Nikodym Theorem. It then introduces complex-valued measures and completes the cycle by showing that when X is a σ-compact locally compact metric space, every bounded linear functional on $C_0(X)$ can be represented as integration with respect to a complex-valued Radon measure. The chapter then develops product measures and closes with a detailed examination of Lebesgue and other measures on Euclidean space, including the Fourier transform. That is the course on measure theory, and I had no difficulty covering it in a semester.

Chapter 5 begins the study of functional analysis by studying linear transformations, first on Banach spaces but quickly focusing on Hilbert space and reaching the diagonalization of a compact hermitian operator. This chapter and the subsequent ones are based on my existing book [8]. There are, however, significant differences. *A Course in Functional Analysis* was designed as a one-year course on the subject for students who had completed a year-long study of measure theory as well as having some knowledge of analytic functions. The second half of the present book only assumes the presentation on measures done in the first half and is meant to be covered in a semester. Needless to say, many topics in [8] are not touched here. Even when this book does a topic found in [8], it is usually treated with less generality and in a somewhat simpler form. I'd advise all readers to use [8] as a reference – as I did.

Chapter 6 looks at Banach spaces and presents the three pillars of functional analysis. The next chapter touches on locally convex spaces, but only to the extent needed to facilitate the presentation of duality. It does include, however, a discussion of the separation theorems that follow from the Hahn–Banach Theorem. Chapter 8 treats the relation between a Banach space and its dual space. It includes the Krein–Milman Theorem, which is applied to prove the Stone–Weierstrass Theorem. Chapter 9 returns to operator theory, but this time in the Banach space setting and gets to the Fredholm Alternative. Chapter 10 presents the basics of Banach algebras and lays the groundwork for the last chapter, which is an introduction to C*-algebras. This final chapter includes the functional calculus for normal operators and presents the characterization of their isomorphism classes.

When I taught my course I did not reach the end of the book, though I covered some topics in more detail and generality than they are covered here; I also presented some of the optional sections in this book – those that have a * in the title. Nevertheless, I wanted the readers to have access to the material on multiplicity theory for normal operators, which is one of the triumphs of mathematics. I suspect that with a good class like the one I had and avoiding the starred sections, the entire book could be covered in a year.

Biographies. I have included some biographical information whenever a mathematician's result is presented. (Pythagoras is the lone exception.) There is no scholarship on my part in this, as all the material is from secondary sources, principally what I could find on the web. In particular, I made heavy use of

http://www-history.mcs.st-andrews.ac.uk/history/BiogIndex.html

and Wikipedia. I did this as a convenience for the reader and from my experience that most people would rather have this in front of them than search it out. (A note about web addresses. There are a few others in this book and they were operational when I wrote the manuscript. We are all familiar with the fact that some web sites become moribund with time. If you experience this, just try a search for the subject at hand.)

I emphasize the personal aspects of the mathematicians we encounter along the way, rather than recite their achievements. This is especially so when I discover something unusual or endearing in their lives. I figure many students will see their achievements if they stick with the subject and most students at the start of their education won't know enough mathematics to fully appreciate the accomplishments. In addition I think the students will enjoy learning that these famous people were human beings.

Teaching. I think my job as an instructor in a graduate course is to guide the students as they learn the material, not necessarily to slog through every proof. In the book, however, I have given the details of the most tedious and technical proofs; but when I lecture I frequently tell my class, "Adults should not engage in this kind of activity in public." Students are usually amused at that, but they realize, albeit with my encouragement, that understanding a highly technical argument may be important. It certainly exposes them to a technique. Nevertheless, the least effective way to reach that understanding is to have someone stand in front of a student at a chalkboard and conscientiously go through all the details. The details should be digested by the student in the privacy of his/her office, away from public view.

I also believe in a gradual introduction of new material to the student. This is part of the reason for what I said earlier about going from the particular to the general. This belief is also reflected in making changes in some notation and terminology as we progress. A vivid example of this is the use of the term "measure." Initially it means a positive measure and then in the course of developing the material it migrates to meaning a complex-valued measure. I don't think this will cause problems; in fact, as I said, I think it facilitates learning.

Prerequisites. The reader is assumed to be familiar with the basic properties of metric spaces. In particular the concepts of compactness, connectedness, continuity, uniform continuity, and the surrounding results on these topics are assumed known. I also assume the student has had a good course in basic analysis on the real line. In particular, (s)he should know the Riemann integral and have control of the usual topics appearing in such a course. There are a few other things from undergraduate analysis that are assumed, though usually what appears here doesn't depend so heavily on their mastery.

For students. When I first studied the subject, I regarded it as very difficult. I found the break with ϵ-δ analysis dramatic, calling for a shift in thinking. A year later I wondered what all the fuss was about. So work hard at this, and I can guarantee that no matter how much trouble you have, it will eventually all clear up. Also I leave a lot of detail checking to the reader and frequently insert such things as (Why?) or (Verify!) in the text. I want you to delve into the details and answer these questions. It will check your understanding and give some perspective on the proof. I also strongly advise you to at least read all the exercises. With your schedule and taking other courses, you might not have the time to try to solve them all, but at least read them. They contain additional information. Learning mathematics is not a spectator sport.

Thanks. I have had a lot of help with this book. First my class was great, showing patience when a first draft of an argument was faulty, making comments, and pointing out typos. Specifically Brian Barg, Yosef Berman, Yeyao Hu, Tom Savistky, and David Shoup were helpful; Tanner Crowder was especially so, pointing out a number of typos and gaps. Also William J. Martin, who was an auditor, showed me a proof of Hölder's Inequality using Young's Inequality (though I decided not to use it in the book), and we had several enjoyable and useful discussions. A pair of friends helped significantly. Alejandro Rodríguez-Martínez did a reading of the penultimate draft as did William Ross. Bill, in addition to pointing out typos, made many pedagogical, stylistic, and mathematical comments which influenced the final product. I feel very fortunate to have such friends.

Needless to say, I am responsible for what you see before you.

Chapter 1

Setting the Stage

This chapter contains a mix of topics needed to study the remainder of this book. It has a little of metric space theory and some basics on normed vector spaces. Many readers will be familiar with some of these topics, but few will have seen them all. We start with a definition of the Riemann–Stieltjes integral, which I suspect is new to most.

1.1. Riemann–Stieltjes integrals

For a fixed closed, bounded interval $J = [a, b]$ in $\mathbb{R}$, the set of all real numbers, we want to define an extension of the usual Riemann integral from calculus. This extended integral will also assign a number to each continuous function on the interval, though later we will see how to extend it even further so we can integrate more general functions than the continuous ones. This more general integral, however, will be set in a far broader context than intervals in $\mathbb{R}$.

As in calculus, a *partition* of J is a finite, ordered subset $P = \{a = x_0 < x_1 < \cdots < x_n = b\}$. Say that P is a *refinement* of a partition Q if $Q \subseteq P$; so P adds additional points to Q.

1.1.1. Definition. A function $\alpha : J \to \mathbb{R}$ is of *bounded variation* if there is a constant M such that for every partition $P = \{a = x_0 < \cdots < x_n = b\}$ of J,

$$\sum_{j=1}^{n} |\alpha(x_j) - \alpha(x_{j-1})| \leq M.$$

The quantity

$$\mathrm{Var}(\alpha) = \mathrm{Var}(\alpha, J) = \sup\left\{\sum_{j=1}^{n} |\alpha(x_j) - \alpha(x_{j-1})| : P \text{ is a partition of } J\right\}$$

is called the *total variation* of α over J.

We'll see many examples of functions of bounded variation. The first below is specific and the second contains a collection of such functions.

1.1.2. Example. (a) The "mother of all" functions of bounded variation is $\alpha(x) = x$.

(b) Suppose $\alpha : [a, b] \to \mathbb{R}$ is a continuously differentiable function and M is a constant with $|\alpha'(t)| \leq M$ for all t in $[a, b]$. If $a = x_0 < x_1 < \cdots < x_n = b$, then for each j the Mean Value Theorem for derivatives says there is a point t_j in $[x_{j-1}, x_j]$ such that $\alpha(x_j) - \alpha(x_{j-1}) = \alpha'(t_j)(x_j - x_{j-1})$. Hence $\sum_j |\alpha(x_j) - \alpha(x_{j-1})| = \sum_j |\alpha'(t_j)|(x_j - x_{j-1}) \leq M(b - a)$, so that α is of bounded variation.

A function $\alpha : J \to \mathbb{R}$ is *increasing* if $\alpha(s) \leq \alpha(t)$ when $a \leq s \leq t \leq b$; the function is *decreasing* if $\alpha(s) \geq \alpha(t)$ when $a \leq s \leq t \leq b$. (We will usually avoid the terms non-decreasing and non-increasing as linguistically and psychologically awkward.) We sometimes use the terms *strictly increasing* and *strictly decreasing* when they are called for, though the reader will see this is not frequent. We note that α is a decreasing function if and only if the function $\beta : [-b, -a] \to \mathbb{R}$ defined by $\beta(s) = \alpha(-s)$ is increasing. It thus becomes possible to state and prove results for decreasing functions whenever we have a result for increasing functions. In what follows neither the separate statements for decreasing functions nor their proofs will be made explicit.

1.1.3. Example. (a) If f is a positive continuous function on $[a, b]$, then $\alpha(t) = \int_a^t f(x)dx$ is an increasing function. In fact any continuously differentiable function with a non-negative derivative is increasing.

(b) If α is a function of bounded variation and β is defined on J by $\beta(t) = \mathrm{Var}(\alpha, [a, t])$, then β is an increasing function.

If X is any set and $f, g : X \to \mathbb{R}$ are any functions, then

$$(f \vee g)(x) = \max\{f(x), g(x)\}$$

defines another function $f \vee g : X \to \mathbb{R}$. Similarly

$$(f \wedge g)(x) = \min\{f(x), g(x)\}$$

defines a function $f \wedge g : X \to \mathbb{R}$. We will see this notation frequently during the course of this book. The proofs of the next two results are routine and left to the reader.

1.1.4. Proposition. *If α, β are increasing functions on J, then so are $\alpha \vee \beta$ and $\alpha \wedge \beta$.*

For any interval $[a, b]$ let $BV[a, b]$ denote the set of all functions of bounded variation defined on $[a, b]$.

1.1.5. Proposition. (a) *Every increasing function on a bounded interval is of bounded variation.*

(b) *$BV[a, b]$ is a vector space over $\mathbb{R}$.*

Perhaps we might make explicit the definition of addition of two functions in $BV[a, b]$. If $\alpha, \beta \in BV[a, b]$, $\alpha + \beta$ is defined by $(\alpha + \beta)(t) = \alpha(t) + \beta(t)$ for all t in the interval. Similarly, if $a \in \mathbb{R}$ and $\alpha \in BV[a, b]$, $(a\alpha)(t) = a\alpha(t)$. This is referred to as defining the algebraic operations *pointwise* and will be seen repeatedly.

In light of the preceding proposition any linear combination of increasing functions is a function of bounded variation. The surprising thing is that the converse holds.

1.1.6. Proposition. *If $\alpha : [a, b] \to \mathbb{R}$ is a function of bounded variation, then we can write $\alpha = \alpha_+ - \alpha_-$, where $\alpha_\pm$ are increasing functions.*

Proof. Let $\alpha_+(t) = \frac{1}{2}[\mathrm{Var}(\alpha, [a, t]) + \alpha(t)]$ and $\alpha_-(t) = \frac{1}{2}[\mathrm{Var}(\alpha, [a, t]) - \alpha(t)]$. It is clear that $\alpha = \alpha_+ - \alpha_-$, so what we have to do is show that these functions are increasing. Let $t > s$, $\epsilon > 0$, and let $a = x_0 < \cdots < x_n = s$ such that $\sum_j |\alpha(x_j) - \alpha(x_{j-1})| > \mathrm{Var}(\alpha, [a, s]) - \epsilon$. Now it is easy to verify that
$$|\alpha(t) - \alpha(s)| \pm [\alpha(t) - \alpha(s)] \pm \alpha(s) \geq \pm \alpha(s)$$
(Note that we are not allowed to randomly make a choice of sign each time $\pm$ appears; it must be consistent.) Since $a = x_0 < \cdots < x_n < t$ is a partition of $[a, t]$, we get that

$$
\begin{aligned}
\mathrm{Var}(\alpha, [a, t]) \pm \alpha(t) &\geq \sum_{j=1}^{n} |\alpha(x_j) - \alpha(x_{j-1})| + |\alpha(t) - \alpha(s)| \\
&\quad \pm \big([\alpha(t) \quad \alpha(s)] + \alpha(s)\big) \\
&= \sum_{j=1}^{n} |\alpha(x_j) - \alpha(x_{j-1})| \\
&\quad + |\alpha(t) - \alpha(s)| \pm [\alpha(t) - \alpha(s)] \pm \alpha(s) \\
&\geq \mathrm{Var}(\alpha, [a, s]) - \epsilon \pm \alpha(s)
\end{aligned}
$$

Since ϵ is arbitrary, we have that $\mathrm{Var}(\alpha, [a, t]) \pm \alpha(t) \geq \mathrm{Var}(\alpha, [a, s]) \pm \alpha(s)$ and so the functions $\alpha_\pm$ are increasing.   ■

Now that we have discussed functions of bounded variation, gotten many examples, and discovered a structure of such functions (1.1.6), we might pose a question. Why the interest? The important thing for us is that we can define integrals or averaging processes for continuous functions by using a function of bounded variation. These integrals have geometric interpretations as well as applications to the study of various problems in analysis. Let's define the integrals, where the reader will notice a close similarity with the definition of the Riemann integral. Indeed if α is the increasing function $\alpha(t) = t$, then what we do below will result in the Riemann integral over J.

If α is a function in $BV(J)$, $f : J \to \mathbb{R}$ is some function, and P is a partition, define

$$S_\alpha(f, P) = \sum_{j=1}^{n} f(t_j)[\alpha(x_j) - \alpha(x_{j-1})]$$

where the points t_j are chosen in the subinterval $[x_{j-1}, x_j]$. Yes, the notation does not reflect the dependency of this sum on the choice of the points t_j, but I am afraid we'll just have to live with that; indicating such a dependency is more awkward than any gained benefit. When the function α is the special one, $\alpha(t) = t$, let $S_\alpha(f, P) = S(f, P)$. That is,

$$S(f, P) = \sum_{j=1}^{n} f(t_j)[x_j - x_{j-1}]$$

Define the *mesh* of the partition P to be the number $\|P\| = \max\{|x_j - x_{j-1}| : 1 \leq j \leq n\}$, and for any positive number δ let $\mathcal{P}_\delta$ denote the collection of all partitions P with $\|P\| < \delta$. Recall that for the Riemann integral the fundamental result on existence is that when $f : F \to \mathbb{R}$ is a continuous function, then there is a unique number I such that for every $\epsilon > 0$ there is a $\delta > 0$ with $|I - S(f, P)| < \epsilon$ whenever $P \in \mathcal{P}_\delta$. It is precisely this number I which is the Riemann integral of f and is denoted by $I = \int_a^b f(t)dt$. We now start the process of showing that a similar existence result holds if we replace the Riemann sum $S(f, P)$ by the sum $S_\alpha(f, P)$ for an arbitrary function of bounded variation α.

Here is another bit of notation that will simplify matters and is valid for any metric space (X, d). For a function $f : X \to \mathbb{R}$, the *modulus of continuity* of f for any $\delta > 0$ is the number $\omega(f, \delta) = \sup\{|f(x) - f(y)| : d(x, y) < \delta\}$. This will be infinite for some functions, but the main place we will use it is when X is compact and f is continuous. In that case f is uniformly continuous so that we have that for any $\epsilon > 0$ there is a δ such that $\omega(f, \delta) < \epsilon$.

Just as in the definition of the Riemann integral, we want to define the integral of a function with respect to a function of bounded variation α. Here is the crucial lemma to get us to that goal.

1.1.7. Lemma. *If α is a function of bounded variation on J and $f : J \to \mathbb{R}$ is a continuous function, then for any $\epsilon > 0$ there is a $\delta > 0$ such that $|S_\alpha(f, P) - S_\alpha(f, Q)| \leq \epsilon$ whenever $P, Q \in \mathcal{P}_\delta$.*

Proof. We start by observing that when P, Q are two partitions and $P \subseteq Q$, then $S_\alpha(f, P) \leq S_\alpha(f, Q)$; this is an easy consequence of the triangle inequality. For example, suppose $P = \{a = x_0 < \cdots < x_n = b\}$ and $Q = P \cup \{x_k^*\}$ with $x_{k-1} < x_k^* < x_k$. Writing out the definition of $S_\alpha(f, P)$ and applying the triangle inequality yields the desired relation. The proof of the general case is similar.

In light of the preceding observation, it suffices to prove that there is a δ such that when $P, Q \in \mathcal{P}_\delta$ and $P \subseteq Q$, then $|S_\alpha(f, P) - S_\alpha(f, Q)| \leq \epsilon/2$. In fact if this is done and we have that P, Q are arbitrary partitions in $\mathcal{P}_\delta$, then $P \cup Q \in \mathcal{P}_\delta$ and contains both P and Q. Hence we would have that $|S_\alpha(f, P) - S_\alpha(f, Q)| \leq |S_\alpha(f, P) - S_\alpha(f, P \cup Q)| + |S_\alpha(f, P \cup Q) - S_\alpha(f, Q)| \leq \epsilon$, completing the proof.

So we assume $P \subseteq Q$ and use the uniform continuity of f to find a δ such that $\omega(f, \delta) < \frac{\epsilon}{2} \mathrm{Var}(\alpha)$. To simplify matters we will assume that Q adds only one point to P and that this point lies between x_0 and x_1. That is we assume $P = \{a = x_0 < x_1 < \cdots < x_n = b\}$ and $Q = \{a = x_0 < x_0^* < x_1 < \cdots < x_n = b\}$. Now for $2 \leq j \leq n$, let $x_{j-1} \leq t_j, s_j \leq x_j$, $x_0 \leq t_1 \leq x_1$, $x_0 \leq s_0^* \leq x_0^*, x_0^* \leq s_1^* \leq x_1$. Note that

$$\left| f(t_1)[\alpha(x_1) - \alpha(x_0)] - \left\{ f(s_0^*)[\alpha(x_0^*) - \alpha(x_0)] + f(s_1^*)[\alpha(x_1) - \alpha(x_0^*)] \right\} \right|$$
$$= \left| f(t_1)[\alpha(x_0^*) - \alpha(x_0) + \alpha(x_1) - \alpha(x_0^*)] \right.$$
$$\left. - \left\{ f(s_0^*)[\alpha(x_0^*) - \alpha(x_0)] + f(s_1^*)[\alpha(x_1) - \alpha(x_0^*)] \right\} \right|$$
$$\leq |f(t_1) - f(s_0^*)| \, |\alpha(x_0^*) - \alpha(x_0)|$$
$$+ |f(t_1) - f(s_1^*)| \, |\alpha(x_1) - \alpha(x_0^*)|$$
$$\leq \omega(f, \delta) \left[|\alpha(x_0^*) - \alpha(x_0)| + |\alpha(x_1) - \alpha(x_0^*)| \right]$$

We therefore obtain

$$|S_\alpha(f, P) - S_\alpha(f, Q)| \leq \omega(f, \delta) \left[|\alpha(x_0^*) - \alpha(x_0)| + |\alpha(x_1) - \alpha(x_0^*)| \right]$$
$$+ \sum_{j=2}^n |f(t_j) - f(s_j)| \, |\alpha(x_j) - \alpha(x_{j-1})|$$
$$\leq \omega(f, \delta) \mathrm{Var}(\alpha)$$
$$< \epsilon/2$$

An inspection of the preceding argument shows that if Q had added more than a single point to P, then the same reasoning would prevail and yielded the same result.   ∎

It is important to emphasize that the value of the inequality obtained in this lemma is independent of the choices of transitory points t_j in $[x_{j-1}, x_j]$ that are used to define $S(\alpha, P)$. This amply justifies not incorporating them in the notation used to denote such a sum.

1.1.8. Theorem. *If α is a function of bounded variation on J and f : $J \to \mathbb{R}$ is a continuous function, then there is a unique number I with the property that for every $\epsilon > 0$ there is a $\delta > 0$ such that when $P \in \mathcal{P}_\delta$,*

$$|S_\alpha(f, P) - I| < \epsilon$$

The number I is called the Riemann[1]–Stieltjes[2] integral of f with respect to α, is denoted by

$$I = \int_a^b f d\alpha = \int f d\alpha$$

and satisfies

$$\left| \int f d\alpha \right| \leq \mathrm{Var}(\alpha) \max\{|f(t)| : t \in J\}$$

---

[1] Georg Friedrich Bernhard Riemann was born in 1826 in Breselenz, Germany. His early schooling was closely supervised by his father. When he entered the university at Göttingen in 1846, at his father's urging he began to study theology. Later, with his father's blessing, he switched to the Faculty of Philosophy so he could study mathematics. In 1847 he transferred to Berlin where he came under the influence of Dirichlet, which was permanent. He returned to Göttingen in 1849 where he completed his doctorate in 1851, working under the direction of Gauss. He took up a lecturer position there and in 1862 he married a friend of his sister. In the autumn of that same year he contracted tuberculosis. This began a period of ill health and he went between Göttingen and Italy, where he sought to recapture his health. He died in 1866 in Selasca, Italy on the shores of beautiful Lake Maggiore. Riemann is one of the giants of analysis and geometry. He made a series of discoveries and initiated theories. There is the Riemann zeta function, Riemann surfaces, the Riemann Mapping Theorem, and many other objects and concepts named after him besides this integral.

[2] Thomas Jan Stieltjes was born in 1856 in Zwolle, The Netherlands. He attended the university at Delft, spending most of his time in the library reading mathematics rather than attending lectures. This had the effect of causing him to fail his exams three years in a row and he left the university without a degree. (This phenomenon of talented people having trouble passing exams is not unique and examples exist in the author's personal experience.) The absence of a degree plagued the progress of his career, in spite of the recognition of his mathematical talent by some of the prominent mathematicians of the day. In 1885 he was awarded membership in the Royal Academy of Sciences in Amsterdam. He received his doctorate of science in 1886 for a thesis on asymptotic series. In the same year he secured a position at the University of Toulouse in France. He did fundamental work on continued fractions and is often called the father of that subject. He extended Riemann's integral to the present setting. He died in 1894 in Toulouse, where he is buried.

Proof. According to the preceding lemma, for every integer $n \geq 1$ there is a δ_n such that if $P, Q \in \mathcal{P}_{\delta_n}$, $|S_\alpha(f, P) - S_\alpha(f, Q)| < \frac{1}{n}$. We can choose the δ_n so that they are decreasing. Let K_n be the closure of the set of numbers $\{S_\alpha(f, P) : P \in \mathcal{P}_{\delta_n}\}$. If $|f(x)| \leq M$ for all x in J, then for any partition P, $|S_\alpha(f, P)| \leq M\mathrm{Var}(\alpha)$. So each set K_n is bounded and hence compact. Since the numbers δ_n are decreasing, for all $n \geq 1$ $\mathcal{P}_{\delta_n} \supseteq \mathcal{P}_{\delta_{n+1}}$ and so $K_n \supseteq K_{n+1}$. Finally by the choice of the δ_n, $\mathrm{diam}\, K_n \leq n^{-1} \to 0$. Therefore by Cantor's Theorem (Theorem 1.2.1 in the next section) $\bigcap_{n=1}^\infty K_n = \{I\}$ for a single number I. It is now routine to verify that I has the stated properties; its uniqueness is guaranteed by its construction. $\blacksquare$

A standard example comes, of course, when $\alpha(t) = t$ for all t in J and this is the Riemann integral. The proofs of the next two results are left to the reader as a way of fixing the ideas in his/her head. These results should not come as a surprise since their counterparts for the Riemann integral are well known.

1.1.9. Proposition. *Let α and β be functions of bounded variation on J, $f, g : J \to \mathbb{R}$ continuous functions, and $s, t \in \mathbb{R}$.*

(a) $\int_a^b (sf + tg)d\alpha = s\int_a^b f d\alpha + t\int_a^b g d\alpha$.

(b) *If $f(x) \geq 0$ for all x in J and α is increasing, then $\int_a^b f d\alpha \geq 0$.*

(c) $\int_a^b f d(s\alpha + t\beta) = s\int_a^b f d\alpha + t\int_a^b f d\alpha$.

1.1.10. Proposition. *If α is a function of bounded variation on J, f is a continuous function on J, and $a < c < b$, then $\int_a^b f d\alpha = \int_a^c f d\alpha + \int_c^b f d\alpha$.*

Here is an important result that enables us to compute some Riemann–Stieltjes integrals from what we know about the Riemann integral.

1.1.11. Theorem. *If α is a function on J that has a continuous derivative at every point of J, then*

$$\int_a^b f d\alpha = \int_a^b f(x)\alpha'(x)dx$$

for any continuous function f.

Proof. We already know from Example 1.1.2 that such a function is of bounded variation, so everything makes sense. Fix a continuous function f on J, let $\epsilon > 0$, and choose δ such that simultaneously $\left|\int f d\alpha - S_\alpha(f, P)\right| < \epsilon/2$ and $\left|\int f\alpha' dx - S(f\alpha', P)\right| < \epsilon/2$ whenever $P \in \mathcal{P}_\delta$. Momentarily fix $P = \{a = x_0 < \cdots < x_n = b\}$ in $\mathcal{P}_\delta$. For each subinterval $[x_{j-1}, x_j]$ defined by P, the Mean Value Theorem for derivatives implies there is a t_j in this subinterval with $\alpha(x_j) - \alpha(x_{j-1}) = \alpha'(t_j)(x_j - x_{j-1})$. Therefore

$$\left| \int_a^b f d\alpha - \int_a^b f(x)\alpha'(x)dx \right| \leq \left| \int_a^b f d\alpha - \sum_{j=1}^n f(t_j)[\alpha(x_j) - \alpha(x_{j-1})] \right|$$

$$+ \left| \sum_{j=1}^n f(t_j)\alpha'(t_j)(x_j - x_{j-1}) - \int_a^b f(x)\alpha'(x)dx \right| < \epsilon$$

Since ϵ was arbitrary, we have the desired equality. ∎

Extensions of the preceding theorem will occupy our attention later in §4.1 and beyond. Be warned, however, that there are increasing functions whose derivatives exist at many points but where nothing like the preceding result is true (2.4.9).

We introduce some special increasing functions whose roll in the general theory has significance.

1.1.12. Example. Fix s in J. When $a \leq s < b$, define

$$\alpha_s(t) = \begin{cases} 0 & \text{for } t \leq s \\ 1 & \text{for } t > s \end{cases}$$

and

$$\alpha_b(t) = \begin{cases} 0 & \text{for } t < b \\ 1 & \text{for } t = b \end{cases}$$

Each of these functions is increasing. Let $\epsilon > 0$ and choose $\delta > 0$ such that $\left| \int f d\alpha_s - S_{\alpha_s}(f, P) \right| < \epsilon/2$ whenever $P \in \mathcal{P}_\delta$ and also such that $|f(u) - f(t)| < \epsilon/2$ when $|u-t| < \delta$. Assume $s < b$. (A separate argument is required when $s = b$ and this is left to the reader. See Exercise 7.) Choose a P in $\mathcal{P}_\delta$ that contains s and let s_0 be the point in P that immediately follows s. Using s_0 as the point in $[s, s_0]$ at which to evaluate f, a moment's reflection reveals that $S_{\alpha_s}(f, P) = f(s_0)[\alpha(s_0) - \alpha(s)] = f(s_0)$. Thus $\left| f(s) - \int f d\alpha_s \right| \leq |f(s) - f(s_0)| + \left| S_{\alpha_s}(f, P) - \int f d\alpha_s \right| < \epsilon$. Since ϵ was arbitrary we have that

$$\int f d\alpha_s = f(s)$$

for every continuous function f on J. Similarly, $\int f d\alpha_b = f(b)$ for all such f.

Using the preceding example we can calculate the integrals with respect to many functions that only differ from a continuously differentiable one by having jump discontinuities. Consider the following.

1.1.13. Example. Define $\alpha : [0,1] \to \mathbb{R}$ by $\alpha(t) = t^2$ for $t \leq \frac{1}{2}$ and $\alpha(t) = t^2 + 1$ for $t > \frac{1}{2}$. What is $\int_0^1 f(t) d\alpha(t)$ for f in $C[0,1]$? Note that $\alpha(t) = t^2 + \alpha_{\frac{1}{2}}$, where $\alpha_{\frac{1}{2}}$ is defined in the preceding example. So using Proposition 1.1.9(c) and Theorem 1.1.11 we get

$$\int_0^1 f(t) d\alpha(t) = \int_0^1 f(t) d(t^2) + \int_0^1 f(t) d\alpha_{\frac{1}{2}}(t)$$
$$= 2 \int_0^1 t f(t) dt + f\left(\frac{1}{2}\right)$$

If α is increasing, then $\alpha(t) \leq \alpha(c)$ whenever $t < c$ and so $\alpha(c-) \equiv \lim_{t \uparrow c} \alpha(t)$ exists and $\alpha(c-) \leq \alpha(c)$; $\alpha(c-)$ is called the *left limit* of α at c. Similarly we define the *right limit* as $\alpha(c+) = \lim_{t \downarrow c} \alpha(t)$ and we have that $\alpha(c+) \geq \alpha(c)$. Note that α is continuous at c precisely when $\alpha(c-) = \alpha(c+)$. It is worthwhile pointing out that when α is discontinuous at c, then the discontinuity is called a jump discontinuity and the difference $\alpha(c+) - \alpha(c-)$ is exactly the size of the jump in values between the left and right of c. In other words, this difference is the size of the vertical jump in the graph of the function α at the point c. Another observation is that if $J = [a,b]$, then the definition of $\alpha(a-)$ is, in a sense, meaningless. We will declare, however, that $\alpha(a-) = \alpha(a)$. Similarly $\alpha(b+) = \alpha(b)$.

1.1.14. Proposition. *If $\alpha : J \to \mathbb{R}$ is a function of bounded variation, then α has at most a countable number of discontinuities.*

Proof. By Proposition 1.1.6 it suffices to show that the conclusion holds for increasing functions, so assume that α is increasing. If α is discontinuous at the points $c_n, n \geq 1$, then the sum of the jumps, $\sum_n [\alpha(c_n+) - \alpha(c_n-)]$, is at most $\alpha(b) - \alpha(a)$. If it were the case that α had an uncountable number of discontinuities, then we could find an $\epsilon > 0$ and an infinite sequence of discontinuities $\{c_n\}$ such that $\alpha(c_n+) - \alpha(c_n-) \geq \epsilon$. (Why?) In light of what we just pointed out, this would imply that $\alpha(b) - \alpha(a) = \infty$, which is nonsense. ∎

A function α is called *left-continuous* at c if the left limit $\alpha(c-)$ exists and is equal to $\alpha(c)$. The definition of *right-continuous* is analogous. So if $J = [a,b]$, every function α on J is left-continuous at a and right-continuous at b by default. Note that if $a \leq s < b$, α_s is left-continuous at s; however, α_b is not left-continuous at b.

1.1.15. Corollary. *If $\alpha : J \to \mathbb{R}$ is increasing and we define $\beta : J \to \mathbb{R}$ by $\beta(t) = \alpha(t-)$, then β is an increasing function that is left-continuous everywhere, has the same discontinuities as α on $[a,b)$, and agrees with α except possibly at its discontinuities.*

Proof. It is clear that β is increasing and left-continuous at each point of J. If c is a point of discontinuity of α, then $\beta(c) = \alpha(c-) < \alpha(c+)$. If $c < b$, then since there are points of continuity for α that approach c from the right, we have that $\beta(c+) = \alpha(c+) > \beta(c)$ and β is discontinuous at c. If $c = b$, then $\beta(b) = \alpha(b-) = \beta(b+)$, so that β is continuous at b irrespective of whether α is continuous at b. (See the function α_b defined in Example 1.1.12.)   ∎

Note that the integral of a continuous function with respect to a constant function is 0; hence $\int f d(\alpha+c) = \int f d\alpha$ for all continuous functions f. There are other ways that we can produce two functions of bounded variation that yield the same integral. (Compare the function defined in Exercise 4 with the functions in Example 1.1.12.) We want to examine when $\int f d\alpha = \int f d\beta$ for two fixed functions of bounded variation and for every continuous function f on J.

For a function of bounded variation α on J and $a \leq t \leq b$, define

1.1.16 $$\widetilde{\alpha}(t) = \alpha(t-) - \alpha(a) + [\alpha(b) - \alpha(b-)]\alpha_b$$

where α_b is defined in Example 1.1.12. Call $\widetilde{\alpha}$ the *normalization* of α. The first thing to observe is that $\alpha(t) - \widetilde{\alpha}(t) = \alpha(a)$ except at the points where α is discontinuous. Also if α is increasing, so is its normalization. Furthermore, if $\alpha = \alpha_+ - \alpha_-$, then $\widetilde{\alpha} = \widetilde{\alpha}_+ - \widetilde{\alpha}_-$. Therefore $\widetilde{\alpha}$ is also a function of bounded variation. Finally note that if α is continuous at b, then $\widetilde{\alpha}(t) = \alpha(t-) - \alpha(a)$.

1.1.17. Proposition. *If α is a function of bounded variation on J and $\widetilde{\alpha}$ is its normalization, then $\int f d\alpha = \int f d\widetilde{\alpha}$ for every continuous function f on J.*

Proof. We split this into two cases.

Case 1. α is continuous at b.

Let D be the set of points in J where α is discontinuous – a countable set. Here $\widetilde{\alpha}(t) = \alpha(t-) - \alpha(a)$; fix $\epsilon > 0$ and f in $C([a,b])$. Let $\delta > 0$ such that $\left|S_\alpha(f,P) - \int f d\alpha\right| < \epsilon/2$ and $\left|S_{\widetilde{\alpha}}(f,P) - \int f d\widetilde{\alpha}\right| < \epsilon/2$ whenever $P \in \mathcal{P}_\delta$. Since D is a countable set in $[a,b]$, we can choose $P = \{a = x_0 < \cdots < x_n = b\}$ in $\mathcal{P}_\delta$ such that $x_j \notin D$ for $0 < j \leq n$. We have by definition

$$S_\alpha(f,P) = f(a)[\alpha(x_1) - \alpha(a)] + \sum_{j=2}^{n} f(x_j)[\alpha(x_j) - \alpha(x_{j-1})]$$

and

$$S_{\widetilde{\alpha}}(f,P) = f(a)[\widetilde{\alpha}(x_1) - \widetilde{\alpha}(a)] + \sum_{j=2}^{n} f(x_j)[\widetilde{\alpha}(x_j) - \widetilde{\alpha}(x_{j-1})]$$

For $0 < j \leq n$, $x_j \notin D$ and so $\alpha(x_j) = \widetilde{\alpha}(x_j) + \alpha(a)$. Since $\widetilde{\alpha}(a) = 0$, we also have $\alpha(a) = \widetilde{\alpha}(a) + \alpha(a)$; thus $\alpha(x_j) - \alpha(x_{j-1}) = \widetilde{\alpha}(x_j) - \widetilde{\alpha}(x_{j-1})$ for $1 \leq j \leq n$. That is, $S_\alpha(f, P) = S_{\widetilde{\alpha}}(f, P)$. Therefore $\left| \int f d\alpha - \int f d\widetilde{\alpha} \right| < \epsilon$. Since ϵ was arbitrary, this proves Case 1.

Case 2. α is discontinuous at b.

Here $\widetilde{\alpha}(t) = \alpha(t-) + [\alpha(b) - \alpha(b-)]\alpha_b$. Consider the function $\beta = \alpha - [\alpha(b) - \alpha(b-)]\alpha_b$. It follows that β is continuous at b since $\beta(b-) = \alpha(b-) = \beta(b)$. By Case 1, $\int f d\beta = \int f d\widetilde{\beta}$. Moreover since $\alpha_b(t-) = 0$ for all t in J, including $t = b$, it follows that $\widetilde{\beta}(t) = \alpha(t-) = \widetilde{\alpha}(t) - [\alpha(b) - \alpha(b-)]\alpha_b(t)$. Now $\int f d\alpha - [\alpha(b) - \alpha(b-)]f(b) = \int f d\beta = \int f d\widetilde{\beta} = \int f d\widetilde{\alpha} - [\alpha(b) - \alpha(b-)]f(b)$. After canceling we get that $\int f d\alpha = \int f d\widetilde{\alpha}$. ∎

Later in this book we will prove the converse of the above result. To be precise, Proposition 4.5.3 shows that if α and β are two functions of bounded variation on the interval J, then $\int f d\alpha = \int f d\beta$ for every continuous function f on J if and only if $\widetilde{\alpha} = \widetilde{\beta}$.

Exercises. We continue to assume that $J = [a, b]$ unless the interval is otherwise specified.

(1) (This first exercise has rather little to do with this section and could have been assigned in your course on advanced calculus, but maybe it wasn't.) For $n \geq 0$ define $f_n : [0, 1] \to \mathbb{R}$ by $f_n(t) = t^n \sin\left(\frac{1}{t}\right)$ when $t \neq 0$ and $f_n(0) = 0$. (a) Show that f_0 is not continuous at 0. (b) Show that f_1 is continuous on $[0, 1]$ but is not differentiable at 0. (c) Show that f_2 is differentiable at every point of $[0, 1]$, but f_2' is not continuous at 0 and so f_2 is not twice differentiable at 0. (d) For $k \geq 1$ show that f_{2k} is $k - 1$ differentiable on $[0, 1]$, but $f_{2k}^{(k-1)}$ is not continuous at 0 and hence f_{2k} is not k-times differentiable on the unit interval. (e) Show that if $f(t) = e^{-t^{-2}} \sin\left(t^{-1}\right)$ for $t \neq 0$ and zero at $t = 0$, then f is infinitely differentiable on the unit interval.

(2) Show that a function of bounded variation is a bounded function; that is, there is a constant M with $|\alpha(t)| \leq M$ for all t in J.

(3) Show that the function f_1 in Exercise 1 is not of bounded variation even though it is continuous.

(4) Define α on the unit interval $[0, 1]$ by $\alpha(\frac{1}{2}) = 1$ and $\alpha(t) = 0$ when $t \neq \frac{1}{2}$. Observe that α is neither left nor right-continuous at $\frac{1}{2}$. Show that α is of bounded variation and find increasing functions $\alpha_\pm$ such that $\alpha = \alpha_+ - \alpha_-$. Compute $\int f d\alpha$ for an arbitrary continuous function f on the unit interval. Are you surprised?

(5) If α is an increasing function on J such that $\int f d\alpha = 0$ for every continuous function f on J, show that α is constant. Contrast this with Exercise 4. (Hint: First show that you can assume that $\alpha(a) = 0$, then show that α must be continuous at each point of J. Use Example 1.1.12. Now show that α is identically 0.)

(6) Give the details of the proof of Proposition 1.1.9

(7) If α_b is defined as in Example 1.1.12, show that $\int f d\alpha_b = f(b)$ for every continuous function f on J.

(8) Suppose a left-continuous increasing function $\alpha : J \to \mathbb{R}$ has a discontinuity at t_0 and $a_0 = \alpha(t_0+) - \alpha(t_0-)$. Let α_{t_0} be the increasing function defined as in Example 1.1.12: $\alpha_{t_0}(t) = 0$ for $t \le t_0$ and $\alpha_{t_0}(t) = 1$ for $t > t_0$. (Once again if t_0 is the right hand end point of J, a separate argument is required.) (a) Show that $\alpha - a_0\alpha_{t_0}$ is an increasing function that is continuous at t_0. (b) Show that any left-continuous increasing function α on J can be written as $\alpha = \beta + \gamma$, where both β and γ are increasing, β is continuous, and γ has the property that if γ is continuous on the open subinterval (c, d), then γ is constant there.

(9) If γ is an increasing function with the property of γ in Exercise 8, calculate $\int f d\gamma$ for any continuous function f on J.

(10) If A is any countable subset of J, show that there is an increasing function α on J such that A is precisely the set of discontinuities of α. (So, in particular, there is an increasing function on J with discontinuities at all the rational numbers in J.)

(11) Let $\{r_n\}$ denote the set of all rational numbers in the interval J and define $\alpha : J \to \mathbb{R}$ by $\alpha(t) = \sum \frac{1}{2^n}$ where the sum is taken over all n such that $r_n < t$. (a) Show that α is a strictly increasing function that is left-continuous and satisfies $\alpha(a) = 0$ and $\alpha(b) = 1$. (b) Show that α is continuous at each irrational number and discontinuous at all the rational numbers in J.

(12) If α is an increasing function, discuss the possibility of defining $\int f d\alpha$ when f has a discontinuity.

(13) Suppose $\alpha : [0, \infty) \to \mathbb{R}$ is an increasing function. Is it possible to define $\int_0^\infty f d\alpha$ for a continuous function $f : [0, \infty) \to \mathbb{R}$?

1.2. Metric spaces redux

As I stated in the Preface, the reader is expected to know the basics of metric spaces. My experience is that the students for whom this book is intended usually do, though there is an occasional student who seems to lack some of it. Such students, however, usually know the concepts as they apply to Euclidean spaces. So I am not going to review this material. I will include, however, two basic results we will use very often. The first is Cantor's Theorem, already used in the preceding section; the second is a list of equivalent formulations of compactness. The different statements of compactness will be used so often that I want to be sure that all readers are on the same page. Cantor's Theorem, besides already being used, will be used in the proof of the second result and can also be used to prove that a continuous function on a compact metric space is uniformly continuous (Exercise 2). Besides, its proof is short. Then in the remainder of this section we will see some results about continuous functions on a metric space that aren't encountered in all courses on the subject, and we'll explore a few other ideas that might not be so familiar to the reader.

The Baire Category Theorem is a topic from metric spaces that may not be familiar to all readers. This result will not be used in this section but will be seen in a significant way later in this book. In case a reader is not familiar with this, (s)he can see it together with a proof in §A.1.

Throughout this section (X, d) is a metric space. If $x \in X$ and $r > 0$, we let $B(x; r) = \{y \in X : d(x, y) < r\}$. For any subset A of X, cl A denotes the closure of A and int A denotes its interior. Also recall that a metric space is complete if every Cauchy sequence converges.

1.2.1. Theorem (Cantor's[3] Theorem). *A metric space (X, d) is complete if and only if for every sequence of closed subsets $\{F_n\}$ such that $F_{n+1} \subseteq F_n$ for each $n \geq 1$ and diam $F_n \equiv \sup\{d(x, y) : x, y \in F_n\} \to 0$, there is a point x_0 in X such that*

$$\bigcap_{n=1}^{\infty} F_n = \{x_0\}$$

Proof. Assume X is complete and $\{F_n\}$ is as stated in the theorem. If $x_n \in F_n$ and $m \geq n$, then $d(x_n, x_m) \leq$ diam F_n; hence $\{x_n\}$ is a Cauchy sequence. Since X is complete, there is an x_0 in X such that $x_n \to x_0$. It follows that $x_0 \in \bigcap_{n=1}^{\infty} F_n$. If there is another point x in this intersection, then $d(x, x_0) \leq$ diam F_n for every $n \geq 1$; hence $x = x_0$.

---

[3]Georg Cantor was the child of an international family. His father was born in Denmark and his mother was Russian; he himself was born in 1845 in St. Petersburg where his father was a successful merchant and stock broker. He is recognized as the father of set theory, having invented cardinal and ordinal numbers and proved that the irrational numbers are uncountable. He received his doctorate from the University of Berlin in 1867 and spent most of his career at the University of Halle. His work was a watershed event in mathematics but was condemned by many prominent mathematicians at the time. The work was simply too radical with counterintuitive results such as $\mathbb{R}$ and $\mathbb{R}^d$ having the same number of points. He began to suffer from depression around 1884. This progressed and plagued him the rest of his life. He died in a sanatorium in Halle in 1918.

Now assume that X satisfies the stated condition and let's prove that X is complete. If $\{x_n\}$ is a Cauchy sequence, put $F_n = \mathrm{cl}\{x_m : m \geq n\}$. So F_n is a closed set, $F_n \subseteq F_{n+1}$, and $\mathrm{diam}\, F_n \leq \sup\{d(x_n, x_m) : m \geq n\}$. Because we have a Cauchy sequence it follows that $\mathrm{diam}\, F_n \to 0$. Thus $\bigcap_{n=1}^{\infty} F_n = \{x_0\}$ for some point x_0. But then $d(x_0, x_n) \leq \mathrm{diam}\, F_n$, so $x_n \to x_0$.   ∎

The following is the basic result on compactness, whose definition we take as the property that every open cover has a finite subcover.

1.2.2. Theorem. *The following statements are equivalent for a closed subset K of a metric space (X, d).*

(a) *K is compact.*

(b) *If $\mathcal{F}$ is a collection of closed subsets of K having the property that every finite subcollection has non-empty intersection, then $\bigcap_{F \in \mathcal{F}} F \neq \emptyset$.*

(c) *Every infinite set in K has a limit point.*

(d) *Every sequence in K has a convergent subsequence.*

(e) *(K, d) is a complete metric space that is totally bounded. That is, if $r > 0$, then there are points $x_1, \ldots, x_n$ such that $X = \bigcup_{k=1}^{n} B(x_k; r)$.*

Proof. (a) *implies* (b). Let $\mathcal{F}$ be a collection of closed subsets of K having the property that every finite subcollection has non-empty intersection. Suppose $\bigcap\{F : F \in \mathcal{F}\} = \emptyset$. If $\mathcal{G} = \{X \backslash F : F \in \mathcal{F}\}$, then it follows that $\mathcal{G}$ is an open cover of X and therefore of K. By (a), there are $F_1, \ldots, F_n$ in $\mathcal{F}$ such that $K \subseteq \bigcup_{j=1}^{n}(X \backslash F_j) = X \backslash \left[\bigcap_{j=1}^{n} F_j\right]$. But since each F_j is a subset of K this implies $\bigcap_{j=1}^{n} F_j = \emptyset$, contradicting the assumption about $\mathcal{F}$.

(b) *implies* (c). Assume that (c) is false. So there is an infinite subset S of K with no limit point; it follows that there is an infinite sequence $\{x_n\}$ of distinct points in S with no limit point. Thus for each $n \geq 1$, $F_n = \{x_k : k \geq n\}$ is a closed set (it contains all its limit points) and $\bigcap_{n=1}^{\infty} F_n = \emptyset$. But each finite subcollection of $\{F_1, F_2, \ldots\}$ has non-empty intersection, contradicting (b).

(c) *implies* (d). Assume $\{x_n\}$ is a sequence of distinct points. By (c), $\{x_n\}$ has a limit point. We are tempted here to say that there is a sequence in the set $\{x_n\}$ that converges to x, but we have to manufacture an actual subsequence of the original sequence. This takes a little bit of effort that is left to the interested reader.

(d) *implies* (e). If $\{x_n\}$ is a Cauchy sequence in K, then (d) implies it is a convergent sequence; by Exercise 1 the original sequence converges.

Thus (K, d) is complete. Now fix an $r > 0$. Let $x_1 \in K$; if $K \subseteq B(x_1; r)$, we are done. If not, then there is a point x_2 in $K \backslash B(x_1; r)$. Once again, if $K \subseteq B(x_1; r) \cup B(x_2; r)$, we are done; otherwise pick an x_3 in $K \backslash [B(x_1; r) \cup B(x_2; r)]$. Continue. If this process does not stop after a finite number of steps, we produce an infinite sequence $\{x_n\}$ in K with $d(x_n, x_m) \geq r$ whenever $n \neq m$. But this implies that this sequence can have no convergent subsequence, contradicting (d).

(e) *implies* (d). Fix an infinite sequence $\{x_n\}$ in K and let $\{\epsilon_n\}$ be a decreasing sequence of positive numbers such that $\epsilon_n \to 0$. By (e) there is a covering of K by a finite number of balls of radius ϵ_1. Thus there is a ball $B(y_1; \epsilon_1)$ that contains an infinite number of points from $\{x_n\}$; let $\mathbb{N}_1 = \{n \in \mathbb{N} : d(x_n, y_1) < \epsilon_1\}$. Now consider the sequence $\{x_n : n \in \mathbb{N}_1\}$ and balls of radius ϵ_2. As we just did, there is a point y_2 in K such that $\mathbb{N}_2 = \{n \in \mathbb{N}_1 : d(y_2, x_n) < \epsilon_2\}$ is an infinite set. Using induction we can show that for each $k \geq 1$ we get a point y_k in K and an infinite set of positive integers $\mathbb{N}_k$ such that $\mathbb{N}_{k+1} \subseteq \mathbb{N}_k$ and $\{x_n : n \in \mathbb{N}_k\} \subseteq B(y_k; \epsilon_k)$. If $F_k = \mathrm{cl}\{x_n : n \in \mathbb{N}_k\}$, then $F_{k+1} \subseteq F_k$ and diam $F_k \leq 2\epsilon_k$. Since K is complete, Cantor's Theorem implies that $\bigcap_{k=1}^{\infty} F_k = \{x\}$ for some point x in X. Now pick integers n_k in $\mathbb{N}_k$ such that $n_k < n_{k+1}$. It follows that $x_{n_k} \to x$.

(e) *implies* (a). We first prove the following.

1.2.3. Claim. If X satisfies (d) and $\mathcal{G}$ be an open cover of X, then there is an $r > 0$ such that for each x in X there is a G in $\mathcal{G}$ such that $B(x; r) \subseteq G$.

Let $\mathcal{G}$ be an open cover of X and suppose the claim is false; so for every $n \geq 1$ there is an x_n in X such that $B(x_n; n^{-1})$ is not contained in any set G in $\mathcal{G}$. By (d) there is an x in X and a subsequence $\{x_{n_k}\}$ such that $x_{n_k} \to x$. Since $\mathcal{G}$ is a cover, there is a G in $\mathcal{G}$ such that $x \in G$; choose a positive ϵ such that $B(x; \epsilon) \subseteq G$. Let $n_k > 2\epsilon^{-1}$ such that $x_{n_k} \in B(x; \epsilon/2)$. If $y \in B(x_{n_k}; n_k^{-1})$, then $d(x, y) \leq d(x, x_{n_k}) + d(x_{n_k}, y) < \epsilon/2 + n_k^{-1} < \epsilon$, so that $y \in B(x; \epsilon) \subseteq G$. That is $B(x_{n_k}; n_k^{-1}) \subseteq G$, contradicting the restriction imposed on x_{n_k}. This establishes the claim.

From here it is easy to complete the proof. We know that (e) implies (d), so for an open cover $\mathcal{G}$ let $r > 0$ be the number guaranteed by (1.2.3). Now let $x_1, \ldots, x_n \in X$ such that $X = \bigcup_{k=1}^{n} B(x_k; r)$ and for $1 \leq k \leq n$ let $G_k \in \mathcal{G}$ such that $B(x_k; r) \subseteq G_k$. $\{G_1, \ldots, G_n\}$ is the sought after finite subcover. ∎

We might note that the statement in Claim 1.2.3 is sometimes called the Lebesgue Covering Lemma. It follows from part (d) of the last theorem that every compact metric space is separable. See Exercise 6.

We will in the course of this book look at functions that are real-valued as well as complex-valued. When we see a result that applies to both, we use $\mathbb{F}$ to denote that we are dealing either with $\mathbb{R}$ or $\mathbb{C}$. In the later chapters we will focus on complex-valued objects.

1.2.4. Definition. $C(X)$ denotes the space of all continuous functions $f : X \to \mathbb{F}$ and $C_b(X)$ denotes the space of all bounded continuous functions in $C(X)$. That is $f \in C_b(X)$ if and only if $f \in C(X)$ and there is a constant M such that $|f(x)| \leq M$ for all x in X. When we want to emphasize the underlying field of scalars, we will write $C(X, \mathbb{R}), C(X, \mathbb{C}), C_b(X, \mathbb{R})$, or $C_b(X, \mathbb{C})$. This, however, will not be frequent.

Of course when X is compact, $C(X) = C_b(X)$. $C(X)$ has a natural algebraic structure where the algebraic operations are defined pointwise as we did for $BV[a, b]$. If $f, g \in C(X), (f+g) : X \to \mathbb{F}$ is defined by $(f+g)(x) = f(x) + g(x)$ and we define $(fg) : X \to \mathbb{F}$ by $(fg)(x) = f(x)g(x)$. Similarly we can define af when $a \in \mathbb{F}$ and $f \in C(X)$. In these cases $f + g, fg$, and af belong to $C(X)$. With such definitions, $C(X)$ is an *algebra*. That is, it is a vector space over $\mathbb{F}$ that has a multiplicative structure and all the usual distributive laws are valid. In the same way $C_b(X)$ is an algebra. As vector spaces $C(X)$ and $C_b(X)$ are finite dimensional if and only if X is a finite metric space (Exercise 3). This will become apparent as soon as we see how to manufacture examples of continuous functions.

Consider $C(X) = C(X, \mathbb{R})$ as a vector space over $\mathbb{R}$ and impose an order structure on $C(X)$ pointwise as follows: if $f, g \in C(X)$, say $f \leq g$ if $f(x) \leq g(x)$ for all x in X. The notation $g \geq f$ means that $f \leq g$. Define the *positive cone* of $C(X)$ to be

$$C(X)_+ = \{f \in C(X) : f \geq 0\}$$

Similarly put this order structure on $C_b(X)$ and define the positive cone of $C_b(X)$ as $C_b(X)_+ = C(X)_+ \cap C_b(X)$. It is also possible to put an order structure on $C(X, \mathbb{C})$, though it is not as neat. Namely for f, g in $C(X, \mathbb{C})$ say that $f \leq g$ if both f and g are real-valued and $f(x) \leq g(x)$ for all x in X.

The proofs of the next two propositions are routine.

1.2.5. Proposition. *Let $f, g, h, k \in C(X)$.*

(a) *$f \leq g$ if and only if $g - f \geq 0$.*

(b) *If $f \leq g$ and $g \leq h$, then $f \leq h$.*

(c) *If $f \leq g$, then $f + h \leq g + h$.*

(d) *If $f \leq g$ and $h \leq k$, then $f + h \leq g + k$.*

(e) *If $f \leq g$ and $a \in [0, \infty)$, then $af \leq ag$; if $a \in (-\infty, 0]$, $af \geq ag$. Similar results hold for $C_b(X)$.*

Recall the definitions of $f \vee g$ and $f \wedge g$ from §1.1.

1.2.6. Proposition. *If $f, g \in C(X)$, then the following hold.*

(a) *$f \vee g$ belongs to $C(X)$.*

(b) *$f \wedge g$ belongs to $C(X)$.*

(c) *$|f|$, defined by $|f|(x) = |f(x)|$, belongs to $C(X)$.*

(d) *If $f_+ = f \vee 0$ and $f_- = -f \wedge 0$, then $f = f_+ - f_-$, $f_+ f_- = 0$, and $|f| = f_+ + f_-$.*
Similar results hold for $C_b(X)$.

Part (d) of this last proposition can be interpreted as saying that the positive cone $C(X)_+$ spans $C(X)$.

For any subset A of X, $\text{dist}(x, A) = \inf\{d(x, a) : a \in A\}$, the distance from x to A. Note that $\text{dist}(x, A) = \text{dist}(x, \text{cl}\, A)$.

1.2.7. Proposition. *If F is a closed subset of X, $|\text{dist}(x, F) - \text{dist}(y, F)| \leq d(x, y)$. Consequently the function $f : X \to \mathbb{R}$ defined by $f(x) = \text{dist}(x, F)$ is uniformly continuous.*

Proof. If $z \in F$, then $\text{dist}(x, F) \leq d(x, z) \leq d(x, y) + d(y, z)$. Taking the infimum over all z in F we get $\text{dist}(x, F) \leq d(x, y) + \text{dist}(y, F)$ or $\text{dist}(x, F) - \text{dist}(y, F) \leq d(x, y)$. Interchanging the roles of x and y we get $\text{dist}(y, F) - \text{dist}(x, F) \leq d(x, y)$, whence the result. ■

Urysohn's Lemma is usually proved in a first course on point set topology, but students don't seem to realize how much easier it is to prove in the metric space setting than in that of a normal topological space.

1.2.8. Theorem (Urysohn's[4] Lemma). *If A and B are two disjoint closed subsets of X, then there is a continuous function $f : X \to \mathbb{R}$ having the following properties:*

(a) *$0 \leq f(x) \leq 1$ for all x in X;*

---

[4]Pavel Samuilovich Urysohn was born in 1898 in Odessa, Ukraine. He was awarded his habilitation in June 1921 from the University of Moscow, where he remained as an instructor. He began his work in analysis but switched to topology where he made several important contributions, especially in developing a theory of dimension. His work attracted attention from the mathematicians of the day, and in 1924 he set out for a tour of the major universities in Germany, Holland, and France, meeting with Hausdorff, Hilbert, and others. That same year, while swimming off the coast of Brittany, France, he drowned. He is buried in Batz-sur-Mer in Brittany. In just three years he left his mark on mathematics.

(b) $f(x) = 0$ *for all* x *in* A;

(c) $f(x) = 1$ *for all* x *in* B.

Proof. Define $f : X \to \mathbb{R}$ by

$$f(x) = \frac{\operatorname{dist}(x, A)}{\operatorname{dist}(x, A) + \operatorname{dist}(x, B)}$$

which is well defined since the denominator never vanishes. It is easy to check that f has the desired properties. ∎

1.2.9. Corollary. *If* F *is a closed subset of* X *and* G *is an open set containing* F, *then there is a continuous function* $f : X \to \mathbb{R}$ *such that* $0 \le f(x) \le 1$ *for all* x *in* X, $f(x) = 1$ *when* $x \in F$, *and* $f(x) = 0$ *when* $x \notin G$.

Proof. In Urysohn's Lemma, take A to be the complement of G and $B = F$. ∎

This corollary can be thought of as the "local" version of Urysohn's Lemma, though this comment will only be seen to make sense once we have used it. Indeed, later in this book we will need to construct sequences of continuous functions that have various properties, and Urysohn's Lemma and the preceding corollary are the keys. Here is one such example but first a word about notation.

In this book the difference of two sets will be denoted by $A \backslash B$ rather than $A - B$. The reason for this is that often we will deal with vector spaces and if A and B are subsets of a vector space $\mathcal{X}$, the notation $A - B$ will mean $\{a - b : a \in A, b \in B\}$, the set of all differences of vectors from the two sets. So we use $A \backslash B$ to avoid confusion.

1.2.10. Proposition. *If* G *is an open subset and* F *is a closed subset of* X *such that* $F \subseteq G$, *then the following hold.*

(a) *There is a sequence of continuous functions* $\{f_n\}$ *such that for all* $n \ge 1$, $0 \le f_n(x) \le 1$, $f_n(x) = 1$ *when* $x \in F$, $f_n(x) = 0$ *when* $x \notin G$, $f_n \le f_{n+1}$, *and* $f_n(x) \nearrow 1$ *when* $x \in G$.

(b) *There is a sequence of continuous functions* $\{g_n\}$ *such that* $0 \le g_n(x) \le 1$, $g_n(x) = 1$ *when* $x \in F$, $g_n(x) = 0$ *when* $x \notin G$, $g_n \ge g_{n+1}$ *for all* $n \ge 1$, *and* $g_n(x) \searrow 0$ *when* $x \notin F$.

Proof. (a) Let $F_n = F \cup \{x : \operatorname{dist}(x, X \backslash G) \ge \frac{1}{n}\}$. So each F_n is closed, $F \subseteq F_n \subseteq F_{n+1}$, and $\bigcup_{n=1}^{\infty} F_n = G$. Use Urysohn's Lemma to find a continuous function $h_n : X \to [0, 1]$ such that $h_n(x) = 1$ when $x \in F_n$ and $h_n(x) = 0$ when $x \notin G$. Put $f_n = \max\{h_1, \dots, h_n\}$; so f_n is continuous by Proposition 1.2.6. Clearly this sequence of functions is increasing, $f_n(x) = 1$ when $x \in F_n$, which includes F, and $f_n(x) = 0$ when $x \notin G$. If $x \in G$, then there is an N with x in F_N; since $F_N \subseteq F_n$ when $n \geq N$, we have that $f_n(x) = 1$ for all $n \geq N$.

(b) Put $H = X \backslash F$ and $K = X \backslash G$. So K is closed, H is open, and $K \subseteq H$. By part (a) there is a sequence $\{f_n\}$ in $C(X)$ with $0 \leq f_n(x) \leq 1$ for all x in X, $f_n(x) = 1$ when $x \in K$, $f_n(x) = 0$ when $x \notin H$, $f_n \leq f_{n+1}$, and $f_n(x) \nearrow 1$ when $x \in H$. Let $g_n = 1 - f_n$ and verify that $\{g_n\}$ has the desired properties. ∎

It is worthwhile for what we will do later in this book to take the following point of view in the preceding proposition. For any set E in X, define the *characteristic function* of E, $\chi_E : X \to \mathbb{F}$, by

$$\chi_E(x) = \begin{cases} 1 & \text{if } x \in E \\ 0 & \text{if } x \notin E \end{cases}$$

The functions f_n in Proposition 1.2.10(a) can be said to satisfy $\chi_F \leq f_n \leq f_{n+1} \leq \chi_G$ and $f_n(x) \to \chi_G(x)$ for every x, even though the functions χ_F and χ_G may not be continuous. (See Exercise 10.) In fact, the conditions just given on $\{f_n\}$ capture all the properties given in Proposition 1.2.10. Similarly the functions g_n in part (b) of the preceding proposition satisfy $\chi_G \geq g_n \geq g_{n+1} \geq \chi_F$ and $g_n(x) \to \chi_F(x)$ for every x.

1.2.11. Theorem (Partition of unity). *If* $\{G_1, \ldots, G_n\}$ *is an open cover of* X, *then there are continuous functions* $\phi_1, \ldots, \phi_n$ *on* X *with the following properties:*

(a) $0 \leq \phi_k(x) \leq 1$ *for all* x *and* $1 \leq k \leq n$;

(b) $\phi_k(x) = 0$ *when* $x \notin G_k$ *and* $1 \leq k \leq n$;

(c) $\sum_{k=1}^{n} \phi_k(x) = 1$ *for all* x *in* X.

Proof. We begin by letting $g_k(x) = \text{dist}\,(x, X \backslash G_k)$. Note that when $x \in G_k$, $g_k(x) > 0$. Since the sets $\{G_k\}$ cover X, $g(x) = \sum_{k=1}^{n} g_k(x) > 0$ for all x in X. Put $\phi_k = g_k/g$. Clearly (a) and (c) hold. Since $g_k(x) = 0$ for $x \notin G_k$, the same holds for $\phi_k(x)$ and so (b) is also satisfied. ∎

For the open cover $\{G_1, \ldots, G_n\}$ and the functions $\phi_1, \ldots, \phi_n$ as in the preceding theorem, we say that these functions are a *partition of unity subordinate to the cover*. Using characteristic functions we can restate conditions (a) and (b) in this theorem as $0 \leq \phi_k \leq \chi_{G_k}$.

1.2.12. Corollary. *If* K *is a closed subset of* X *and* $\{G_1, \ldots, G_n\}$ *is an open cover of* K, *then there are continuous functions* $\phi_1, \ldots, \phi_n$ *on* X *with the following properties:*

(a) $0 \leq \phi_k \leq \chi_{G_k}$ *for* $1 \leq k \leq n$;

(b) $\sum_{k=1}^{n} \phi_k(x) = 1$ *for all x in K;*

(c) $\sum_{k=1}^{n} \phi_k(x) \leq 1$ *for all x in X.*

Proof. Note that if we put $G_{n+1} = X \backslash K$, then $\{G_1, \ldots, G_{n+1}\}$ is an open cover of X. Let $\{\phi_1, \ldots, \phi_{n+1}\}$ be a partition of unity subordinate to this cover. The reader can check that the functions $\phi_1, \ldots, \phi_n$ satisfy (a) and (b) since $\phi_{n+1}(x) = 0$ for x in K. For any x in X, $\sum_{k=1}^{n} \phi_k(x) = 1 - \phi_{n+1}(x) \leq 1$, giving (c). ∎

Partitions of unity are a way of putting together local results to get a global result. This comment will not make complete sense to the reader but will become evident later. The typical situation will be that we know how to do something in an open neighborhood like G_k, such as manufacturing a function f_k with certain properties. Using the partition of unity $\{\phi_k\}$ and writing $f = \sum_k f_k \phi_k$, we will obtain a function f defined on the entire space that reflects the local properties enjoyed by the functions f_k. Again, this is meant as a rough idea of how partitions of unity will be used, and the reader can be forgiven if the picture remains murky. We will use partitions of unity frequently in this book and the fog will then lift.

This section concludes with a result from topology, the Tietze Extension Theorem. This is valid for normal spaces, but only is stated here for metric spaces. No proof is given here since the author knows no proof for metric spaces that is simpler than the one for normal spaces. (Anyone who has such a proof is invited to communicate it to the author.)

1.2.13. Theorem (Tietze[5] Extension Theorem). *If X is a metric space, Y is a closed subset of X, and $f : Y \to \mathbb{F}$ is a continuous function with $|f(y)| \leq M$ for all y in Y, then there is a continuous function ϕ on X such that $\phi(y) = f(y)$ for y in Y and $|\phi(x)| \leq M$ for all x in X.*

Exercises. For these exercises (X, d) is always a metric space.

(1) Show that if $\{x_n\}$ is a Cauchy sequence in X and there is a subsequence $\{x_{n_k}\}$ that converges to a point x in X, then $x_n \to x$.

(2) Use (1.2.3) to show that if X is compact, Y is another metric space, and $f : X \to Y$ is a continuous function, then f is uniformly continuous.

---

[5] Heinrich Franz Friedrich Tietze was born in 1880 in Schleinz, Austria. In 1898 he entered the Technische Hochschule in Vienna. He continued his studies in Vienna and received his doctorate in 1904 and his habilitation in 1908, with a thesis in topology. His academic career was interrupted by service in the Austrian army in World War I; just before this he obtained the present theorem. After the war he had a position first at Erlangen and then at Munich, where he remained until his retirement. In addition to this theorem he made other contributions to topology and did significant work in combinatorial group theory, a field in which he was one of the pioneers. He had 12 PhD students, all at Munich, where he died in 1964.

(3) Show that $C_b(X)$ is a finite-dimensional vector space over $\mathbb{F}$ if and only if X is a finite metric space.

(4) If G is an open set and K is a compact set with $K \subseteq G$, then there is a $\delta > 0$ such that $\{x : \text{dist}(x, K) < \delta\} \subseteq G$. Find an example of an open set G in a metric space X and a closed, non-compact subset F of G such that there is no $\delta > 0$ with $\{x : \text{dist}(x, F) < \delta\} \subseteq G$.

(5) Recall that a metric space is *separable* if there is a countable dense subset. The following exercise will be used often in this book and so it is included here to be sure the reader is familiar with it. If (X, d) is separable, $E \subseteq X$, $r > 0$, and the balls $\{B(x; r) : x \in E\}$ are pairwise disjoint, then E is countable.

(6) Use Theorem 1.2.2(d) to show that a compact metric space is separable.

(7) Let I be any non-empty set and for each i in I, let X_i be a copy of $\mathbb{R}$ with the metric $d_i(x, y) = |x - y|$. Let X be the disjoint union of the sets X_i. That's a verbal description that can be used in any circumstance, but if you want precision you can say $X = \mathbb{R} \times I$, the Cartesian product where I has the discrete topology. Define a metric on X by letting d agree with d_i on X_i; when $x \in X_i$, $y \in X_j$, where $i \neq j$, then $d(x, y) = 1$. (a) Show that d is indeed a metric on X. (b) Show that $\{X_i : i \in I\}$ is the collection of components of X and each of these components is an open subset of X. (c) Show that (X, d) is separable if and only if I is a countable set.

(8) Show that a subset of $\mathbb{R}$ is connected if and only if it is an interval.

(9) For two subsets A and B of X, define the distance from A to B by $\text{dist}(A, B) = \inf\{d(a, b) : a \in A, b \in B\}$. (a) Show that $\text{dist}(A, B) = \text{dist}(B, A) = \text{dist}(\text{cl }A, \text{cl }B)$. (b) If A and B are two disjoint closed subsets of X such that B is compact, then $\text{dist}(A, B) > 0$. (c) Give an example of two disjoint closed subsets A and B of the plane $\mathbb{R}^2$ such that $\text{dist}(A, B) = 0$.

(10) If $E \subseteq X$, show that the characteristic function χ_E is a continuous function on X if and only if E is simultaneously an open and a closed set.

1.3. Normed spaces

The next concept plays a central role in analysis.

1.3.1. Definition. If $\mathcal{X}$ is a vector space over $\mathbb{F} = \mathbb{R}$ or $\mathbb{C}$, a *norm* on $\mathcal{X}$ is a function $\| \cdot \| : \mathcal{X} \to [0, \infty)$ having the following properties for a scalar a and vectors x, y in $\mathcal{X}$:

(a) $\|x\| = 0$ if and only if $x = 0$;

(b) $\|ax\| = |a|\|x\|$;

(c) $\|x + y\| \le \|x\| + \|y\|$ (triangle inequality).
A *normed space* is a pair $(\mathcal{X}, \|\cdot\|)$ consisting of a vector space $\mathcal{X}$ and a norm.

Most of the objects that are studied in analysis are connected to normed spaces or their cousins.

1.3.2. Example. (a) Let $\mathcal{X} = \mathbb{F}^d$ for some natural number d, and for $x = (x_1, \ldots, x_d)$ in $\mathbb{F}^d$ define $\|x\| = \left[\sum_{k=1}^{d} |x_k|^2\right]^{\frac{1}{2}}$. This is a norm but showing that the triangle inequality holds involves a bit of development. You may take this on faith until §3.1 below. Two other norms on $\mathbb{F}^d$ are $\|x\|_1 = \sum_{k=1}^{d} |x_k|$ and $\|x\|_\infty = \max\{|x_k| : 1 \le k \le d\}$.

(b) Let ℓ^1 denote the set of all sequences $\{a_n\}$ of numbers in $\mathbb{F}$ such that $\sum_{n=1}^{\infty} |a_n| < \infty$. Define addition and scalar multiplication for elements of ℓ^1 by: $\{a_n\} + \{b_n\} = \{a_n + b_n\}$ and $a\{a_n\} = \{aa_n\}$. The reader is asked to show that this transforms ℓ^1 into a vector space and that $\|\{a_n\}\| = \sum_{n=1}^{\infty} |a_n|$ defines a norm on ℓ^1.

(c) Let c_0 denote the set of all sequences $\{a_n\}$ of numbers from $\mathbb{F}$ such that $a_n \to 0$. If addition and scalar multiplication are defined entrywise on c_0 as they were on ℓ^1, then c_0 is a vector space. Moreover $\|\{a_n\}\| = \sup_n |a_n|$ defines a norm on c_0.

(d) Let ℓ^∞ denote the set of all bounded sequences of numbers $\{a_n\}$ from $\mathbb{F}$ and define addition and scalar multiplication entrywise as in the previous two examples; this makes ℓ^∞ into a vector space. If we set $\|\{a_n\}\| = \sup_n |a_n|$ for each $\{a_n\}$ in ℓ^∞, this is a norm.

(e) Let c_{00} denote the vector space of all finite sequences. There are many choices of a norm for c_{00}; for example, either of the norms used for c_0 or ℓ^1 can be used.

(f) Let $J = [a, b]$ be a bounded interval in the real line and let $BV(J)$ denote the functions of bounded variation on J. If we define $\|\alpha\| = |\alpha(a)| + \text{Var}(\alpha)$ for α in $BV(J)$, then this is a norm on $BV(J)$.

(g) If X is a metric space and $f \in C_b(X)$, $\|f\| = \sup\{|f(x)| : x \in X\}$ is a norm on $C_b(X)$. To show that the triangle inequality holds observe that $|f(x) + g(x)| \le |f(x)| + |g(x)| \le \|f\| + \|g\|$. Taking the supremum over all x gives the inequality. Verifying that the remaining properties of a norm are satisfied is straightforward. Actually if we consider $\mathbb{N}$ as a metric space, where the metric is the discrete or trivial metric ($d(x, y) = 1$ for $x \ne y$ and $d(x, x) = 0$), then ℓ^∞ in part (d) is seen to be identical with $C_b(\mathbb{N})$.

If $\mathcal{X}$ is a normed space, let

$$\text{ball}\,\mathcal{X} = \{X \in \mathcal{X} : \|x\| \leq 1\}$$

This closed unit ball will play a prominent role as our study of normed spaces progresses.

1.3.3. Proposition. *If $\mathcal{X}$ is a normed space and $x, y \in \mathcal{X}$, then $\big|\|x\| - \|y\|\big| \leq \|x - y\|$. Moreover if $d : \mathcal{X} \times \mathcal{X} \to [0, \infty)$ is defined by $d(x, y) = \|x - y\|$, then $(\mathcal{X}, d)$ is a metric space.*

Proof. We have $\|x\| = \|(x - y) + y\| \leq \|x - y\| + \|y\|$; hence $\|x\| - \|y\| \leq \|x - y\|$. If the rolls of x and y are interchanged in the preceding inequality and this new relation is combined with the preceding one, we get $\big|\|x\| - \|y\|\big| \leq \|x - y\|$. The proof that $d(x, y)$ is a metric is routine. ∎

When discussing a metric or any metric space concepts on a normed space it is always assumed that the above metric is the one under discussion.

1.3.4. Definition. A *Banach*[6] *space* is a normed space that is complete with respect to its metric. That is, every Cauchy sequence converges.

All the examples in (1.3.2) except for (e) are Banach spaces. The proof that c_0 is a Banach space takes a little work, but should be within the ability of the reader. The proof that ℓ^1 is complete takes a rather technical argument that will be encountered again and so we give this below.

1.3.5. Proposition. *ℓ^1 is a Banach space.*

Proof. Again the proof that ℓ^1 is a normed space is routine. For each $n > 1$, let $x^n = \{x_k^n\} \in \ell^1$ and assume that $\{x^n\}$ is a Cauchy sequence. If $k \geq 1$, then $|x_k^n - x_k^m| \leq \|x^n - x^m\|$, and so for each $k \geq 1$ we have that $\{x_k^n\}$ is a

---

[6]Stefan Banach was born in 1892 in Krakow, presently in Poland but then part of Austria-Hungary (Polish history is complicated). In 1922 the university in Lvov (a city that during its history has belonged to at least three countries and is presently in Ukraine) awarded Banach his habilitation for a thesis on measure theory. His earlier thesis on Integral Equations is sometimes said to mark the birth of functional analysis. He was one of the stars in a particularly brilliant constellation of Polish mathematicians during this period. Banach and his colleague Hugo Steinhaus (see Theorem 6.7.5 for a biographical note) in Lvov as well as other mathematicians in Warsaw began publishing a series of mathematical monographs. The first to appear in 1931 was Banach's *Théorie des Opérations Linéaires*, which had an enormous impact on analysis and continues to hold its place as a classic still worth reading. Banach was one of the founders of functional analysis, which had been brewing in academic circles for sometime. You will see Banach's name appear often as this book progresses. He died in Lvov in 1945 just after the end of World War II.

Cauchy sequence in $\mathbb{F}$; put $x_k = \lim_n x_k^n$. For any $K \geq 1$ and any $n \geq 1$,

$$\left(\sum_{k=1}^{K} |x_k|\right) \leq \left(\sum_{k=1}^{K} |x_k - x_k^n|\right) + \left(\sum_{k=1}^{K} |x_k^n|\right)$$

$$\leq \left(\sum_{k=1}^{K} |x_k - x_k^n|\right) + \|x^n\|$$

Choose n sufficiently large that $|x_k - x_k^n| < 1/K$ for $1 \leq k \leq K$. Since every Cauchy sequence is bounded (see Exercise 1), $\left(\sum_{k=1}^{K} |x_k|\right) \leq 1 + C$, where $C \geq \|x^n\|$ for all $n \geq 1$. Therefore $x = \{x_k\} \in \ell^1$. Now we will show that $x^n \to x$ in ℓ^1.

Let $\epsilon > 0$ and choose N_1 such that $\|x^n - x^m\| < \epsilon/3$ when $n, m \geq N_1$. Fix an $m \geq N_1$, and for this m, since $x - x^m \in \ell^1$, we can choose L such that $\sum_{k=L+1}^{\infty} |x_k - x_k^m| < \epsilon/3$. Now choose N_2 such that $|x_k - x_k^n| < \epsilon/3L$ for $1 \leq k \leq L$ when $n \geq N_2$. Put $N = \max\{N_1, N_2\}$. Adopt the notation that $T_L(x - x^n)$ is the truncation of the sequence $x - x^n$ after the L-th term; that is, $T_L(x - x^n)$ is the sequence that is $x_k - x_k^n$ for $1 \leq k \leq L$ and then is identically 0 for $k > L$. Observe that for $n \geq N_2$, $\|T_L(x - x^n)\| < L\frac{\epsilon}{3L} = \epsilon/3$. Let $R_L(x - x^n)$ be the remainder sequence, with a 0 in the first L places and then agreeing with $x - x^n$. If n is an arbitrary integer larger than N and m is that integer larger than N_1 that was fixed above, then

$$\|x - x^n\| \leq \|T_L(x - x^n)\| + \|R_L(x - x^n)\|$$

$$< \frac{\epsilon}{3} + \|R_L(x - x^m)\| + \|R_L(x^m - x^n)\|$$

$$< \frac{2\epsilon}{3} + \|x^m - x^n\|$$

$$< \epsilon \qquad \blacksquare$$

Here is another important example of a Banach space, one that will occupy a considerable amount of our attention.

1.3.6. Proposition. $C_b(X)$ *is a Banach space. Moreover a sequence* $\{f_n\}$ *in* $C_b(X)$ *converges to* f *if and only if* $f_n(x) \to f(x)$ *uniformly for* x *in* X.

Proof. We prove the last part first. Assume $\|f_n - f\| \to 0$ and let $\epsilon > 0$. So there is an integer N such that $\|f_n - f\| < \epsilon$ for $n \geq N$. Thus for any x in X and $n \geq N$, $|f_n(x) - f(x)| \leq \|f_n - f\| < \epsilon$. This says that $f_n(x) \to f(x)$ uniformly for x in X. Conversely, assume $f_n(x) \to f(x)$ uniformly for x in X. So if $\epsilon > 0$ there is an integer N such that for $n \geq N$, $|f_n(x) - f(x)| \leq \epsilon$ for all x in X. Thus $\|f_n - f\| = \sup\{|f_n(x) - f(x)| : x \in X\} \leq \epsilon$ when $n \geq N$.

To establish the completeness of $C_b(X)$, assume that $\{f_n\}$ is a Cauchy sequence in $C_b(X)$. Since $|f_n(x) - f_m(x)| \leq \|f_n - f_m\|$, we have, as in the preceding paragraph, that $\{f_n(x)\}$ is uniformly Cauchy in x. As is standard in metric space theory, this implies there is a continuous function f on X such that $f_n(x) \to f(x)$ uniformly in x. Choose N_1 such that $\|f_n - f_m\| < 1$ for $n, m \geq N_1$. Thus for $n \geq N_1$, $\|f_n\| \leq \|f_n - f_{N_1}\| + \|f_{N_1}\| \leq 1 + \|f_{N_1}\|$. So if $M = 1 + \max\{\|f_1\|, \ldots, \|f_{N_1}\|\}$, we have that $\|f_n\| \leq M$ for all $n \geq 1$. It follows that $|f(x)| \leq M$ for all x in X; that is, $f \in C_b(X)$. Since $f_n(x) \to f(x)$ uniformly on X, $f_n \to f$ in $C_b(X)$ from the first paragraph in this proof. ∎

The reader should now begin to undergo a change of point of view about functions. Namely, as indicated by the last result as well as the examples in (1.3.2), we want to start thinking of continuous functions as well as sequences as points in a larger space. The author remembers having some difficulty making this change when he was a graduate student, but it comes shortly with study and working with the functions as points. Indeed much of the power of modern analysis rests on adopting such a point of view and this should become apparent as we progress.

To prove one of the main results of this section we need the following.

1.3.7. Lemma. *If X is a metric space and S is a dense subset of X such that every Cauchy sequence of elements of S has a limit in X, then X is complete.*

Proof. Let $\{x_n\}$ be a Cauchy sequence in X and for every n let $s_n \in S$ such that $d(x_n, s_n) < n^{-1}$. We want to show that $\{s_n\}$ is a Cauchy sequence. If $\epsilon > 0$, choose N_1 such that $d(x_n, x_m) < \epsilon/3$ when $n, m \geq N_1$. Let $N \geq N_1$ such that $n^{-1} < \epsilon/3$ when $n \geq N$. Thus for $n, m \geq N$, $d(s_n, s_m) \leq d(s_n, x_n) + d(x_n, x_m) + d(x_m, s_m) < \epsilon$. Hence, by hypothesis, there is an x in X such that $s_n \to x$. Therefore $d(x_n, x) \leq n^{-1} + d(s_n, x) \to 0$. ∎

The next result concerns normed spaces. There is a similar result about metric spaces that some readers might have seen and will be stated after we prove this theorem, but the normed space version is really the one we want to concentrate on. One way for the reader to have a picture of what is happening in this proof is to keep in mind the process of defining the real numbers once we have the rational numbers. One way to do this is by means of a device know as a Dedekind cut, but another is by saying a real number is an equivalence class of Cauchy sequences of rational numbers. It is this last one that is analogous to what we undertake in the proof of the next theorem.

1.3.8. Theorem. *If $\mathcal{X}$ is a normed space, then there is a Banach space $\widehat{\mathcal{X}}$ and a linear isometry $U : \mathcal{X} \to \widehat{\mathcal{X}}$ such that $U(\mathcal{X})$ is dense in $\widehat{\mathcal{X}}$. Moreover $\widehat{\mathcal{X}}$ is unique in the sense that if $\mathcal{Y}$ is another Banach space with a linear isometry $W : \mathcal{X} \to \mathcal{Y}$ such that $W(\mathcal{X})$ is dense in $\mathcal{Y}$, then there is a surjective isometry $V : \widehat{\mathcal{X}} \to \mathcal{Y}$ such that $VU = W$.*

Proof. Let $\mathcal{X}_0$ denote the collection of all Cauchy sequences in $\mathcal{X}$. Note that $\mathcal{X}_0$ can be made into a vector space by defining $a\{x_n\} + \{y_n\} = \{ax_n + y_n\}$. Also observe that for $\{x_n\}$ in $\mathcal{X}_0$, $\{\|x_n\|\}$ is a Cauchy sequence in $\mathbb{R}$ by Proposition 1.3.3; so $\|\{x_n\}\| \equiv \lim \|x_n\|$ exists. It is routine to verify that this satisfies the properties of a norm on $\mathcal{X}_0$ except that it is possible that $\|\{x_n\}\| = 0$ without $\{x_n\}$ being the identically 0 sequence. (For example $\{x, 0, 0, \dots\} \in \mathcal{X}_0$ and $\|(x, 0, 0, \dots)\| = 0$.) We therefore introduce a relation on $\mathcal{X}_0$ by saying that $\{x_n\} \sim \{y_n\}$ if $\|x_n - y_n\| \to 0$; equivalently, if $\|\{x_n - y_n\}\| = 0$. It is left to the reader to check, using the various properties of the norm and Cauchy sequences, that $\sim$ is an equivalence relation on $\mathcal{X}_0$. Let $\widehat{\mathcal{X}}$ denote the set of equivalence classes for $\sim$. (This is a rather common event in mathematics. We have a set $\mathcal{X}$ and an equivalence relation. We examine the collection of equivalence classes $\widehat{\mathcal{X}}$ and want to impose a structure on $\widehat{\mathcal{X}}$ that reflects the structure of the original set $\mathcal{X}$. In this case we want to show that $\widehat{\mathcal{X}}$ is a normed space and, eventually, that it is a Banach space.)

To show that $\widehat{\mathcal{X}}$ is a vector space, let $\xi, \eta \in \widehat{\mathcal{X}}$ and let $\{x_n\}, \{y_n\}$ be representatives of ξ, η, respectively. We want to define $\xi + \eta$ as the equivalence class of the sequence $\{x_n + y_n\}$, which is easily seen to belong to $\mathcal{X}_0$. To be sure that $\xi + \eta$ is well defined, we must check that the definition is independent of the representatives chosen. That is, suppose $\{x_n'\} \sim \{x_n\}$ and $\{y_n'\} \sim \{y_n\}$. We must check that $\{x_n' + y_n'\} \sim \{x_n + y_n\}$. We leave the details to the reader. Similarly if $\xi \in \widehat{\mathcal{X}}$ with representative $\{x_n\}$ and $a \in \mathbb{F}$, we define $a\xi$ to be the equivalence class of the sequence $\{ax_n\}$. Once this is done it is easily checked that $\widehat{\mathcal{X}}$ satisfies the axioms of a vector space.

Now we put a norm on $\widehat{\mathcal{X}}$ by letting $\|\xi\| = \|\{x_n\}\|$ for any representative of ξ. Here Proposition 1.3.3 implies that if $\{x_n\} \sim \{y_n\}$, then $\|\{x_n\}\| = \|\{y_n\}\|$ so that this is well defined. Now note that if $\|\xi\| = 0$, then $\{x_n\} \sim \{0, 0, \dots\}$, the zero of $\widehat{\mathcal{X}}$, so that ξ is the zero of $\widehat{\mathcal{X}}$. Let $\{x_n\} \in \xi$ and $\{y_n\} \in \eta$. So $\|\xi + \eta\| = \lim \|x_n + y_n\| \leq \lim \|x_n\| + \lim \|y_n\| = \|\xi\| + \|\eta\|$. That is $\|\cdot\|$ satisfies the triangle inequality on $\widehat{\mathcal{X}}$. The fact that $\|a\xi\| = |a|\|\xi\|$ for every ξ in $\widehat{\mathcal{X}}$ is straightforward. Thus we have that $\widehat{\mathcal{X}}$ is a normed space.

If $x \in \mathcal{X}$, let ξ_x be the equivalence class of the sequence $\{x, x, \dots\}$. Clearly $\xi_x \in \widehat{\mathcal{X}}$. If we define $U : \mathcal{X} \to \widehat{\mathcal{X}}$ by $U(x) = \xi_x$, then it is easily verified that $U : \mathcal{X} \to \widehat{\mathcal{X}}$ is a linear isometry. We want to show that the range of U is dense in $\widehat{\mathcal{X}}$ and that every Cauchy sequence from $U(\mathcal{X})$ has a limit in $\widehat{\mathcal{X}}$. Once this is done, the preceding lemma implies that $\widehat{\mathcal{X}}$ is complete and we have finished the proof of the existence part of the theorem.

To show the density of $U(\mathcal{X})$ let $\{x_n\} \in \xi \in \widehat{\mathcal{X}}$ and let $\epsilon > 0$. So there is an integer N such that $\|x_n - x_m\| < \epsilon/2$ for $n, m \geq N$. Thus $\|U(x_N) - \xi\| = \lim_n \|x_N - x_n\| < \epsilon$. Therefore the range of U is dense in $\widehat{\mathcal{X}}$. Now let $\{U(x_n)\}$ be a Cauchy sequence in $U(\mathcal{X})$. Since U is an isometry, $\{x_n\}$ is a Cauchy sequence in $\mathcal{X}$; let ξ be its equivalence class in $\widehat{\mathcal{X}}$. We will show that $U(x_n) \to \xi$ in $\widehat{\mathcal{X}}$. In fact, if $\epsilon > 0$, there is an N such that $\|x_n - x_m\| < \epsilon/2$ for all $n, m \geq N$. Thus if $m \geq N$, $\|U(x_m) - \xi\| = \lim_n \|x_m - x_n\| < \epsilon$.

Now for the proof of uniqueness. Let $\mathcal{Y}$ and W be as in the statement of the theorem. If $\{x_n\} \in \xi \in \widehat{\mathcal{X}}$, the fact that W is an isometry implies that $\{W(x_n)\}$ is a Cauchy sequence in $\mathcal{Y}$. Thus there is a y in $\mathcal{Y}$ such that $W(x_n) \to y$. Let $V(\xi) = y$. The reader can check that V is well defined and linear. Also $\|y\| = \lim \|W(x_n)\| = \lim \|x_n\| = \|\xi\|$, so that V is an isometry. Moreover for any x in $\mathcal{X}$, $VU(x) = V(\xi_x) = W(x)$ by the definition of V. Thus V is a linear isometry with dense range and it must therefore be surjective. ∎

The unique Banach space $\widehat{\mathcal{X}}$ obtained in the preceding theorem is called the *completion* of $\mathcal{X}$. Using this same proof we can establish a similar result about general metric spaces, though we do not have to discuss any linearity conditions.

1.3.9. Theorem. *If X is a metric space then there is a complete metric space $\widehat{X}$ and an isometry $\tau : X \to \widehat{X}$ such that $\tau(X)$ is dense in $\widehat{X}$. Moreover $\widehat{X}$ is unique in the sense that if Y is a complete metric space such that there is an isometry $\phi : X \to Y$ with $\phi(X)$ dense in Y, then there is a surjective isometry $\psi : \widehat{X} \to Y$ with $\psi\tau = \phi$.*

As a practical point, it is easier to think of $\mathcal{X}$ as contained in $\widehat{\mathcal{X}}$ rather than work through the isometry U. When we actually complete a specific normed space, this is what happens. For example, if $\mathcal{X} = c_{00}$ as in Example 1.3.2(e) and it is given the supremum norm, then $\widehat{c_{00}} = c_0$.

Exercises.

(1) Show that if $\mathcal{X}$ is a normed space and $\{x_n\}$ is a Cauchy sequence in $\mathcal{X}$, then there is a constant C such that $\|x_n\| \leq C$ for all n.

(2) Let A be a subset of the normed space $\mathcal{X}$, and denote by $\bigvee A$ the intersection of all closed linear subspaces of $\mathcal{X}$ that contain A; this is called the *closed linear span of* A. Prove the following. (a) $\bigvee A$ is a closed linear subspace of $\mathcal{X}$; (b) $\bigvee A$ is the smallest closed linear

subspace of $\mathcal{X}$ that contains A. (c) $\bigvee A$ is the closure of

$$\left\{ \sum_{k=1}^{n} a_k x_k : n \geq 1, \text{ for } 1 \leq k \leq n, a_k \in \mathbb{F}, x_k \in A \right\}$$

(3) If $\mathcal{X}$ is a normed space, show that $\mathcal{X}$ is complete if and only if whenever $\{x_n\}$ is a sequence in $\mathcal{X}$ such that $\sum_{n=1}^{\infty} \|x_n\| < \infty$, then $\sum_{n=1}^{\infty} x_n$ converges in $\mathcal{X}$. (Note: To say that an infinite series $\sum_{n=1}^{\infty} x_n$ converges in a normed space means that the sequence of partial sums $\{\sum_{k=1}^{n} x_k\}$ converges. What else, right?)

(4) (a) Show that the space c_0 in Example 1.3.2(c) is a Banach space. (Hint: If $x^k = \{x_n^k\}$ is a Cauchy sequence in c_0, show that for every $\epsilon > 0$ there is an integer N such that $|x_n^k| < \epsilon$ for $n \geq N$ and all $k \geq 1$.) (b) Let c be all the sequences $\{x_n\}$ of numbers from $\mathbb{F}$ such that $\lim_n x_n$ exists, and show that c is a Banach space with the supremum norm $\|\{x_n\}\| = \sup_n |x_n|$. (c) If $e = \{1, 1, \dots\}$, show that $c = c_0 + \mathbb{F}e$.

(5) Show that c_{00} is dense in both c_0 and ℓ^1.

(6) Let $w = \{w_n\}$ be a sequence of strictly positive real numbers and define $\ell^1(w)$ to be all the sequences $\{x_n\}$ of numbers from $\mathbb{F}$ such that $\sum_{n=1}^{\infty} w_n |x_n| < \infty$. Define a norm on $\ell^1(w)$ by $\|\{x_n\}\| = \sum_{n=1}^{\infty} w_n |x_n|$ and show that $\ell^1(w)$ is a Banach space.

(7) Let w be as in the preceding exercise and define $c_0(w)$ to be all the sequences $\{x_n\}$ of numbers from $\mathbb{F}$ such that $\lim_n w_n x_n = 0$. If we define a norm on this space by $\|\{x_n\}\| = \sup_n w_n |x_n|$, show that $c_0(w)$ is a Banach space.

(8) Let J be a bounded interval in $\mathbb{R}$ and denote by $BV(J)$ the set of all functions of bounded variation on J. Show that with the norm defined in Example 1.3.2(f), $BV(J)$ is a Banach space.

(9) Let $\mathcal{X} = C([0, 1])$ but give it the norm $\|f\| = \int_0^1 |f(t)| dt$. Show that $\mathcal{X}$ with this norm is not complete.

(10) Find a sequence $\{f_n\}$ in $C([0, 1])$ such that when $n \neq m$, $\|f_n - f_m\| \geq 1$ and that this implies that ball $C([0, 1])$ is not compact. Can you generalize this to $C_b(X)$ for any infinite metric space X? What happens when X is a finite metric space?

(11) Show that the sequence $\{f_n\}$ found in Exercise 10 is a linearly independent set.

(12) (a) Show that if $\mathcal{X}$ is a normed space, then $\mathcal{X}$ is separable if and only if ball $\mathcal{X}$ is separable. (b) Show that c_0 is a separable Banach space. (c) Show that ℓ^∞ is not separable. (Hint: You know that the collection of all subsets of $\mathbb{N}$ is uncountable. Use these sets to

define an uncountable set of functions that will enable you to use Exercise 1.2.5.)

(13) This continues the preceding exercise. Recall that a compact metric space is separable. If (X, d) is a compact metric space, fix a countable dense subset D in X. Let $\mathcal{G}$ be the collection of all finite open covers of X by balls of the form $B(x; \frac{1}{n})$, where $x \in D$ and $n \geq 1$. (a) Show that $\mathcal{G}$ is countable. (b) For each $\gamma = \{G_1, \ldots, G_n\}$ in $\mathcal{G}$ fix a partition of unity $\Phi_\gamma = \{\phi_1, \ldots, \phi_n\}$ subordinate to γ and let $\mathcal{X}_\gamma$ denote the collection of all linear combinations with rational coefficients of the functions in Φ_γ. Show that $\bigcup_{\gamma \in \mathcal{G}} \mathcal{X}_\gamma$ is a countable dense subset in $C(X)$ and, hence, that $C(X)$ is separable.

(14) Let $\mathcal{X}$ be a normed space and let G be an open subset of $\mathcal{X}$. Show that G is a connected set if and only if for any two points a and b in G there are points $a = x_0, x_1, \ldots, x_n = b$ in G such that for $1 \leq j \leq n$ the line segment $[x_{j-1}, x_j] \equiv \{tx_j + (1-t)x_{j-1} : 0 \leq t \leq 1\} \subseteq G$.

1.4. Locally compact spaces

We want to extend the concept of a compact metric space. It is likely that at least some of the readers have seen the following definition, which can be made in the setting of an arbitrary topological space, not just a metric space as is done here.

1.4.1. Definition. A metric space (X, d) is *locally compact* if for every point x in X there is a radius $r > 0$ such that $B(x; r)$ has compact closure.

1.4.2. Example. (a) Euclidean space is a locally compact space.

(b) If X is any set and d is the discrete metric ($d(x, y) = 1$ when $x \neq y$ and $d(x, x) = 0$), then (X, d) is locally compact.

(c) Every compact metric space is locally compact, and here is a way to get additional examples of locally compact spaces once we are given a compact metric space. Let (Y, d) be a compact metric space, fix a point y_0 in Y, and let $X = Y \backslash \{y_0\}$. (X, d) is locally compact. Can it be compact? Concrete examples occur by letting $Y = [0, 1]$ and $y_0 = \{1\}$ or letting $Y = [0, 1] \sqcup \{2\}$ and $y_0 = \{2\}$.

(d) It is not necessarily the case that all locally compact metric spaces have the property that the closure of every ball of finite radius is compact, as is the case with $\mathbb{F}^d$. For example, let Y, X, and y_0 be as in the previous example. If y is a point of Y distinct from y_0 and $r > d(y, y_0)$, then the ball in X of radius r and centered at y may not have a compact closure in X.

The main interest we have in locally compact spaces is to do analysis on them. This is possible even for non-metric spaces. However in this book we limit ourselves to the metric space case for the sake of easier exposition. Note that with Euclidean spaces being locally compact, we do not lack examples where analysis on a locally compact space might be profitable. One of our principal tools will be to embed the locally compact metric space inside a compact metric space. This is not always possible, but let's set the stage. We need to introduce a special space of continuous functions on a locally compact space.

1.4.3. Definition. If X is a locally compact metric space, say that a continuous function $f : X \to \mathbb{F}$ *vanishes at infinity* if for every $\epsilon > 0$ the set $\{x \in X : |f(x)| \geq \epsilon\}$ is compact. Let $C_0(X)$ denote the set of all continuous functions on X that vanish at infinity. For any continuous function $f : X \to \mathbb{F}$, define the *support* of f to be the set cl $\{x : f(x) \neq 0\}$ and denote this by spt (f). Say that f has *compact support* if spt (f) is a compact subset of X. Let $C_c(X)$ denote the set of all continuous functions $f : X \to \mathbb{F}$ having compact support.

1.4.4. Example. (a) If $\mathbb{N}$ has the discrete metric, then $C_b(\mathbb{N}) = \ell^\infty$, the space of bounded sequences; $C_0(\mathbb{N}) = c_0$, the space of sequences that converge to 0; $C_c(\mathbb{N}) = c_{00}$, the space of all finitely non-zero sequences.

(b) Consider Example 1.4.2(c), where $X = Y \backslash \{y_0\}$. Here we can identify $C_0(X)$ with those functions in $C(Y)$ that vanish at y_0.

Say that a locally compact space X is *σ-compact* if $X = \bigcup_{n=1}^{\infty} K_n$ where each K_n is a compact subset. It is left as an exercise for the reader to show that if X is σ-compact, then we can write X as the union of compact sets K_n such that $K_n \subseteq$ int K_{n+1}. (See Exercise 2.) Note that Euclidean space is σ-compact as is the locally compact space given in Example 1.4.2(c). In this last case we can take $K_n = \{y \in Y : d(y, y_0) \geq \frac{1}{n}\}$, which is a compact subset of X.

1.4.5. Proposition. *Let X be a locally compact space.*

(a) *Both $C_c(X)$ and $C_0(X)$ are subalgebras of $C_b(X)$ and $C_c(X) \subseteq C_0(X)$. Moreover $C_0(X)$ is closed in $C_b(X)$. Hence $C_0(X)$ is a Banach space.*

(b) *$C_c(X)$ is dense in $C_0(X)$.*

(c) *If $f \in C_0(X)$, then f is uniformly continuous.*

(d) *If X is σ-compact, there is a sequence of functions $\{\phi_n\}$ in $C_c(X)$ such that $0 \leq \phi_n \leq \phi_{n+1} \leq 1$ for all $n \geq 1$ and for every f in $C_0(X)$, $\|f\phi_n - f\| \to 0$ as $n \to \infty$.*

Proof. (a) The fact that $C_c(X)$ and $C_0(X)$ are subalgebras of $C_b(X)$ and $C_c(X) \subseteq C_0(X)$ is routinely proven. Let $\{f_n\}$ be a sequence in $C_0(X)$ and assume that $f_n \to f$ in $C_b(X)$. That is $\|f_n - f\| = \sup\{|f_n(x) - f(x)| : x \in X\} \to 0$. Let $\epsilon > 0$ and put $K = \{x : |f(x)| \geq \epsilon\}$; we want to show that K is compact. Clearly K is closed. Let $N \geq 1$ such that $\|f_n - f\| < \epsilon/2$ for all $n \geq N$; fix an $n \geq N$. If $y \in K$, then $\epsilon \leq |f(y) - f_n(y)| + |f_n(y)| \leq \epsilon/2 + |f_n(y)|$, and so $y \in \{x : |f_n(x)| \geq \epsilon/2\}$. That is, K is a closed subset of $\{x : |f_n(x)| \geq \epsilon/2\}$, a compact set. Thus K is compact and so $f \in C_0(X)$.

(d) As in Exercise 2 we write $K = \bigcup_{n=1}^{\infty} K_n$ where each K_n is compact and $K_n \subseteq \operatorname{int} K_{n+1}$ for all $n \geq 1$. By Urysohn's Lemma we can find a ϕ_n in $C(X)$ such that $0 \leq \phi_n \leq 1$, $\phi_n(x) = 1$ for all x in K_n, and $\phi_n(x) = 0$ for all x in $X\backslash \operatorname{int} K_{n+1}$. It follows that $\phi_n C_c(X)$ and $0 \leq \phi_n \leq \phi_{n+1} \leq 1$. If $f \in C_0(X)$ and $\epsilon > 0$, let $K = \{x : |f(x)| \geq \epsilon\}$. Since K is compact, there is an integer N such that $K \subseteq \operatorname{int} K_n$ for all $n \geq N$. It is easy to see that $|f(x) - f(x)\phi_n(x)| = 0$ when $x \in K_n$. Hence for $n \geq N$, $\|f - f\phi_n\| < \epsilon$.

(b) If $f \in C_0(X)$ and $\{\phi_n\}$ is a sequence in $C_c(X)$ as in (d), then $f\phi_n \in C_c(X)$ for all $n \geq 1$ and $f\phi_n \to f$.

(c) Let $\epsilon > 0$ and put $L = \{x : |f(x)| \geq \epsilon/2\}$; so L is compact. Since X is locally compact, for every x in L there is an $r_x > 0$ such that $\operatorname{cl} B(x; r_x)$ is compact. Let $x_1, \ldots, x_m \in L$ such that $L \subseteq \bigcup_{j=1}^{m} B(x_j; r_{x_j})$. Choose $\gamma > 0$ such that $\operatorname{dist}(x, L) \leq \gamma$ implies $x \in \bigcup_{j=1}^{m} B(x_j; r_{x_j})$ and set $K = \{x : \operatorname{dist}(x, L) \leq \gamma\}$. Note that K is compact since it is contained in $\bigcup_{j=1}^{m} \operatorname{cl} B(x_j; r_{x_j})$. If we only consider f as a function on K, it is uniformly continuous there; so there exists δ such that $0 < \delta < \gamma$ and if $x, y \in K$ and $d(x, y) < \delta$, then $|f(x) - f(y)| < \epsilon$. Let $x, y \in X$ such that $d(x, y) < \delta$. If $x, y \in L$, then $|f(x) - f(y)| < \epsilon$. If $x \in L$ but $y \notin L$, then the fact that $d(x, y) < \delta < \gamma$ implies that $x, y \in K$; hence $|f(x) - f(y)| < \epsilon$. If neither point belongs to L, then $|f(x) - f(y)| \leq |f(x)| + |f(y)| < \epsilon/2 + \epsilon/2 = \epsilon$. ∎

Fix a locally compact metric space (X, d). The objective that will occupy us for the remainder of the section is to embed X in a compact metric space (X_∞, ρ) in a "congenial" way. In fact we want to have X_∞ differ from X by a single point that we denote by ∞, the point at infinity. For those who have studied abstract topological spaces this is called the *one-point compactification* of X. We will not assume that the reader has this background and will give a self contained exposition. It is possible that even the reader who has seen this topic has not addressed the question of when X_∞ is metrizable.

As a preview, consider the locally compact space X obtained from the compact space Y as in Example 1.4.2(c). Clearly the one-point compactification we seek is the space Y, where the point y_0 plays the role of the point

∞. The reader might keep this example in mind when we prove the next theorem.

Note that if Z is a set and d_1 and d_2 are two metrics on Z, then we say that the two metrics are *equivalent metrics* provided that the open sets defined by one metric coincide with those defined by the other. This is equivalent to the requirement that for a sequence $\{z_n\}$ in Z and a point z, $d_1(z_n, z) \to 0$ if and only if $d_2(z_n, z) \to 0$. See Exercise 6.

1.4.6. Theorem. *If (X, d) is a locally compact metric space, the following statements are equivalent.*

(a)   *There is a metric space (X_∞, ρ) such that $X_\infty = X \cup \{\infty\}$, X_∞ is compact, the metric d on X is equivalent to the restriction of ρ to X, and a sequence $\{x_n\}$ in X satisfies $\rho(x_n, \infty) \to 0$ if and only if for every compact subset K of X, there is an integer N such that $x_n \notin K$ for all $n \geq N$.*

(b)   *X is σ-compact.*

(c)   *The metric space $C_0(X)$ is separable.*

Before we begin the proof, let's observe a few things so as to better understand the theorem. Note that the metric ρ obtained for X_∞ is not required to agree with the original metric for X but only to be equivalent to it. In fact if $X = \mathbb{R}$, there is no way that the metric on the compactification will be the same as the metric on $\mathbb{R}$ since one is bounded and the other is not. Second, again if $X = \mathbb{R}$, then the space X_∞ differs from $\mathbb{R}$ by a single point; we have not added the two points $\pm\infty$. Finally note that if $f \in C_0(X)$ and we extend f to be defined on X_∞ by letting $f(\infty) = 0$, then this extension is continuous. In fact if $x_n \to \infty$ and $\epsilon > 0$, let $K = \{x \in X : |f(x)| \geq \epsilon\}$, which is compact since $f \in C_0(X)$. Thus there is an integer N such that $x_n \notin K$ for all $n \geq N$. Thus $|f(x_n)| < \epsilon$ for $n \geq N$. Finally let us observe that the last requirement of the metric ρ contained in (a) is equivalent to the requirement that for every $\epsilon > 0$, $\{x \in X : \rho(x, \infty) \geq \epsilon\}$ is compact. In fact this last statement is that the function $f : X \to [0, \infty)$ defined by $f(x) = \rho(x, \infty)$ vanishes at infinity. So the condition is seen to be the same as the condition that $f(x_n) \to 0$ whenever $\{x_n\}$ escapes every compact subset of X.

Proof. (a) *implies* (b). If ρ is the metric on X_∞ and $K_n = \{x \in X : \rho(x, \infty) \geq \frac{1}{n}\}$, then each K_n is compact and their union is all of X.

(b) *implies* (c). (The reader may want to first look at Exercise 1.3.13, which outlines a proof that $C(X)$ is separable when X is a compact metric space. The proof given here is basically the same but technically more complex.) We write $X = \bigcup_{n=1}^\infty K_n$, where each K_n is compact and $K_n \subseteq$ int K_{n+1} for all $n \geq 1$ (see Exercise 2). Find a decreasing sequence of positive numbers $\{\delta_n\}$ such that $\{x : \mathrm{dist}\,(x, K_n) < \delta_n\} \subseteq \mathrm{int}\, K_{n+1}$. (How?) For each n and each $k \geq n$, let $\{B(a_{nk}^j; \delta_k) : 1 \leq j \leq m_{nk}\}$ be open disks with a_{nk}^j in K_n such that $K_n \subseteq \bigcup_{j=1}^{m_{nk}} B(a_{nk}^j; \delta_k)$. Note that this union of open disks is contained in $\mathrm{int}\, K_{n+1}$. As in (1.2.12) for each $n \geq 1$ and $k \geq n$, let $\{\phi_{nk}^j : 1 \leq j \leq m_{nk}\}$ be continuous functions with $0 \leq \phi_{nk}^j \leq 1$, $\phi_{nk}^j(x) = 0$ for $x \notin B(a_{nk}^j; \delta_k)$, $\sum_{j=1}^{m_{nk}} \phi_{nk}^j(x) = 1$ when $x \in K_n$, and $\sum_{j=1}^{m_{nk}} \phi_{nk}^j \leq 1$. Let $\mathcal{M}$ be the linear span of the functions $\{\phi_{nk}^j : n \geq 1, k \geq n, \text{ and } 1 \leq j \leq m_{nk}\}$ with coefficients from the rational numbers $\mathbb{Q}$. Note that $\mathcal{M}$ is a countable subset of $C_c(X)$. We will show that $\mathcal{M}$ is dense in $C_c(X)$ and hence in $C_0(X)$ (1.4.5).

Fix f in $C_c(X)$ and let $\epsilon > 0$; so there is an integer n such that $f(x) = 0$ when $x \notin K_n$. Since f is uniformly continuous there is a $\delta > 0$ such that $|f(x) - f(y)| < \epsilon/2$ whenever $d(x, y) < \delta$. Pick $k \geq n$ such that $\delta_k < \delta$. For $1 \leq j \leq m_{nk}$ let $q_{nk}^j \in \mathbb{Q}$ such that $|q_{nk}^j - f(a_{nk}^j)| < \epsilon/2$. Hence $g = \sum_{j=1}^{m_{nk}} q_{nk}^j \phi_{nk}^j \in \mathcal{M}$. Now fix x in K_n. Thus

$$
\begin{aligned}
|f(x) - g(x)| &= \left| \sum_{j=1}^{m_{nk}} [f(x) - q_{nk}^j] \phi_{nk}^j(x) \right| \\
&\leq \sum_{j=1}^{m_{nk}} \left| f(x) - f(a_{nk}^j) \right| \phi_{nk}^j(x) + \sum_{j=1}^{m_{nk}} \left| f(a_{nk}^j) - q_{nk}^j \right| \phi_{nk}^j(x) \\
&\leq \sum_{j=1}^{m_{nk}} \left| f(x) - f(a_{nk}^j) \right| \phi_{nk}^j(x) + \epsilon/2
\end{aligned}
$$

Now when $\phi_{nk}^j(x) \neq 0$, $x \in B(a_{nk}^j; \delta_k)$ and so $\left| f(x) - f(a_{nk}^j) \right| < \epsilon/2$. Putting this inequality into the preceding one yields that $|f(x) - g(x)| < \epsilon$ when $x \in K_n$. Suppose now that $x \in \left[\bigcup_{j=1}^{m_{nk}} B(a_{nk}^j; \delta_k) \right] \backslash K_n$. So $f(x) = 0$. If $x \in B(a_{nk}^j; \delta_k)$, then $\left| f(a_{nk}^j) \right| = \left| f(a_{nk}^j) - f(x) \right| < \epsilon/2$. Therefore

$$
\begin{aligned}
|f(x) - g(x)| &= |g(x)| \\
&\leq \sum_{j=1}^{m_{nk}} \left| q_{nk}^j \right| \phi_{nk}^j(x) \\
&\leq \sum_{j=1}^{m_{nk}} \left[\left| q_{nk}^j - f(a_{nk}^j) \right| + \left| f(a_{nk}^j) \right| \right] \phi_{nk}^j(x) \\
&< (\epsilon/2 + \epsilon/2) \sum_{j=1}^{m_{nk}} \phi_{nk}^j(x) \\
&\leq \epsilon
\end{aligned}
$$

Finally if $x \notin \bigcup_{j=1}^{m_{nk}} B(a_{nk}^j; \delta_k)$, then $f(x) = 0 = g(x)$. Therefore we have that $\|f - g\| < \epsilon$ and so $\mathcal{M}$ is dense in $C_0(X)$.

(c) *implies* (a). Here is an outline of the proof. First note that when $f \in C_0(X)$, $\rho_f(x, y) = |f(x) - f(y)|$ is symmetric ($\rho_f(x, y) = \rho_f(y, x)$) and satisfies the triangle inequality. ρ_f is called a semimetric. Extending f to X_∞ by setting $f(\infty) = 0$ enables us to see that ρ_f defines a semimetric on X_∞. Now using the fact that $C_0(X)$ is separable we can use a countable dense subset of ball $C_0(X)$ to generate a sequence of such semimetrics and sum them up to get a true metric on X_∞. The details follow.

Let $\{f_n\}$ be a countable dense sequence in the unit ball of $C_0(X)$ and define $\rho(x, y) = \sum_{n=1}^\infty \frac{1}{2^n} |f_n(x) - f_n(y)|$ for all x, y in X_∞. Note that for x in X, $\rho(x, \infty) = \sum_{n=1}^\infty \frac{1}{2^n} |f_n(x)|$ since $f_n(\infty) = 0$ for all n. Clearly $\rho(x, y) = \rho(y, x)$. If $\rho(x, y) = 0$, then $f_n(x) = f_n(y)$ for all $n \geq 1$. Now if $x \neq y$ there is a function f in $C_c(X)$ such that $f(x) = 1$ and $f(y) = 0$, assuming $x \neq \infty$. (Why?) Let $n \geq 1$ such that $\|f - f_n\| < \frac{1}{2}$. It follows that $|f_n(x)| > \frac{1}{2}$ and $|f_n(y)| < \frac{1}{2}$. Thus it cannot be that $\rho(x, y) = 0$. We leave it to the reader to verify that the triangle inequality holds for ρ so that it is a metric on X.

It remains to prove that the inclusion map $(X, d) \to (X_\infty, \rho)$ is a homeomorphism onto its image and that $\{x : \rho(x, \infty) \geq \epsilon\}$ is compact for every $\epsilon > 0$.

Claim. A sequence $\{x_n\}$ converges to x in (X_∞, ρ) if and only if $f_k(x_n) \to f_k(x)$ for all $k \geq 1$.

If $x_n \to x$ in (X_∞, ρ), then the fact that $2^{-k} |f_k(x_n) - f_k(x)| \leq \rho(x_n, x)$ implies $f_k(x_n) \to f_k(x)$. Now assume that $f_k(x_n) \to f_k(x)$ for all $k \geq 1$ and let $\epsilon > 0$. Choose m such that $\sum_{k=m}^\infty \frac{1}{2^k} < \epsilon/2$ and choose N such that for $n \geq N$ and $1 \leq k \leq m$, $|f_k(x_n) - f_k(x)| < \epsilon/2m$. Thus for $n \geq N$,

$$\rho(x_n, x) < \epsilon/2 + \sum_{k=1}^m \frac{1}{2^k} \frac{\epsilon}{2m} < \epsilon$$

proving the claim.

Claim. The inclusion map $(X, d) \to (X_\infty, \rho)$ is a homeomorphism.

The first claim proves that if $x_n \to x$ in X, then $\rho(x_n, x) \to 0$; that is, the inclusion map $(X, d) \to (X_\infty, \rho)$ is continuous. For the converse, suppose that $\rho(x_n, x) \to 0$, where x and x_n belong to X for all $n \geq 1$. Suppose that $\{d(x_n, x)\}$ does not converge to 0; then there is an $\epsilon > 0$ and a subsequence $\{x_{n_j}\}$ such that $d(x_{n_j}, x) \geq \epsilon$ for all n_j. We can assume that ϵ is small, say $\epsilon < \frac{1}{2}$. Using Urysohn's Lemma there is a function f in $C_c(X)$ such that $0 \leq f \leq 1$, $f(x) = 1$, and $f(y) = 0$ when $d(y, x) \geq \epsilon/2$.

Since $\{f_k\}$ is dense in ball $C_0(X)$, there is an f_k such that $\|f_k - f\| < \epsilon/2$. Hence $|f_k(x_{n_j})| = |f_k(x_{n_j}) - f(x_{n_j})| < \epsilon/2$. On the other hand, $1 = f(x) \leq |f(x) - f_k(x)| + |f_k(x)| < \epsilon/2 + |f_k(x)|$, and so $|f_k(x)| > 1 - \epsilon/2 > \epsilon/2$. This contradicts the assumption that $f_k(x_{n_j}) \to f_k(x)$, which, by the first claim, contradicts the assumption that $\rho(x_n, x) \to 0$.

Claim. $\{x : \rho(x, \infty) \geq \epsilon\}$ is compact for every $\epsilon > 0$.

Let $\epsilon > 0$ and put $K = \{x \in X : \rho(x, \infty) \geq \epsilon\}$. Suppose K is not compact. For each $n \geq 1$, put $K_n = \{x \in X : |f_k(x)| \geq \frac{1}{n}$ for $1 \leq k \leq n\}$. Since K is closed and each K_n is compact, it cannot be that $K \subseteq K_n$. Therefore there is a point x_n in K such that $x_n \notin K_n$. That is $|f_k(x_n)| < \frac{1}{n}$ for $1 \leq k \leq n$. This says that for every $k \geq 1$, $\lim_n f_k(x_n) = 0$. According to the first claim this implies that $x_n \to \infty$ or that $\rho(x_n, \infty) \to 0$. Since each $x_n \in K$, this is a contradiction. Therefore K must be compact.

It remains to show that (X_∞, ρ) is compact. So let $\{x_n\}$ be a sequence in X_∞ and let's show it has a convergent subsequence. One of two things holds; either there is a compact set K such that $x_n \in K$ for infinitely many values of n, or every compact subset of X contains an x_n for only a finite number of values of n. In the first of these cases there is automatically a convergent subsequence. So assume the latter case is valid. Write $X = \bigcup_{j=1}^\infty K_j$, where each K_j is compact and $K_j \subseteq \text{int } K_{j+1}$ for all $j \geq 1$. Thus we can construct a subsequence $\{x_{n_j}\}$ such that for all $j \geq 1$, $x_{n_j} \notin K_j$. But if K is any compact subset of X, there is a j_0 such that $K \subseteq K_{j_0}$. Therefore $x_{n_j} \notin K$ whenever $j \geq j_0$. From the preceding paragraph we know that this implies that $x_{n_j} \to \infty$. ■

Here is a sample of the way we can use the one-point compactification to parlay results about compact spaces into results about σ-compact locally compact spaces.

1.4.7. Proposition. *Let X be σ-compact locally compact metric space with a compact subset K. If $K \subseteq \bigcup_{k=1}^n U_k$ where each U_k is open, then there are functions $\phi_1, \ldots, \phi_n$ in $C_c(X)$ such that the following hold.*

(a) *For $1 \leq k \leq n$, $0 \leq \phi_k \leq \chi_{U_k}$.*

(b) *$\sum_{k=1}^n \phi_k(x) = 1$ for all x in K.*

(c) *$\sum_{k=1}^n \phi_k(x) \leq 1$ for all x in X.*

Proof. Start by finding open sets $G_1, \ldots, G_n$ such that for $1 \leq k \leq n$, $\text{cl } G_k \subseteq U_k$ and $\text{cl } G_k$ is compact. Now consider X_∞ and apply Corollary 1.2.12. ■

It should be pointed out that we do not need Corollary 1.2.12 to prove
the preceding proposition and, in fact, the conclusion is valid without the
assumption that X is σ-compact. However under those circumstances the
proof requires more effort.

Exercises.

(1) Is $\mathbb{Q}$ locally compact?

(2) If X is a locally compact metric space that is σ-compact, show
that we can write X as the union of compact sets K_n such that
$K_n \subseteq \operatorname{int} K_{n+1}$.

(3) If (X_k, d_k) is a locally compact metric space for $1 \leq k \leq n$, show
that $X = \prod_{k=1}^{n} X_k$ with the metric $d(\{x_k\}, \{y_k\}) = \max_k d_k(x_k, y_k)$
is a locally compact metric space.

(4) It is known and usually proved in an elementary course on metric
spaces that if (X_k, d_k) is a metric space for all $k \geq 1$, then $X = \prod_{k=1}^{\infty} X_k$ is a metric space. Give a necessary and sufficient condition
that X is locally compact. (See the preceding exercise.)

(5) Show that every σ-compact metric space is separable.

(6) If Z is a set and d_1 and d_2 are two metrics on Z, then prove the
following statements are equivalent. (a) If $\{z_n\}$ is a sequence in Z
and $z \in Z$, then $d_1(z_n, z) \to 0$ if and only if $d_2(z_n, z) \to 0$. (b) A
set G is open in (Z, d_1) if and only if it is open in (Z, d_2). (c) A set
F is closed in (Z, d_1) if and only if it is closed in (Z, d_2).

(7) Show that every open subset of Euclidean space is σ-compact. More
generally, show that if X is a σ-compact metric space and if G is
an open subset, then G is σ-compact.

(8) Show that the one-point compactification of $\mathbb{R}$ is homeomorphic to
the circle.

(9) If X is a locally compact metric space and $f, g \in C_0(X)$, show that
$f \vee g$ and $f \wedge g \in C_0(X)$. Similarly, if $f, g \in C_c(X)$, show that $f \vee g$
and $f \wedge g \in C_c(X)$.

(10) Consider the metric space (X, d) defined in Exercise 1.2.7. (a)
Show that (X, d) is locally compact. (b) Show that (X, d) is not
σ-compact and hence $\widehat{X}$ is non-metrizable. (c) Find an infinite
subset A of X such that no sequence of points converges to ∞ in
the one-point compactification.

(11) If X is locally compact and σ-compact and $\phi \in C_0(X)$ such that
$\{x : \phi(x) = 0\} = \emptyset$, show that $\phi C_0(X)$ is dense in $C_0(X)$. (Hint:
use Proposition 1.4.5(b).) If X is not assumed to be σ-compact
and ϕ is any function in $C_0(X)$, what is the closure of $\phi C_0(X)$?

(12) If X is a locally compact metric space, is the condition that X is σ-compact equivalent to the condition that X is separable? (See Exercise 5.)

1.5. Linear functionals

1.5.1. Definition. If $\mathcal{X}$ is a vector space, a *linear functional* on $\mathcal{X}$ is a function $L : \mathcal{X} \to \mathbb{F}$ satisfying $L(ax + by) = aL(x) + bL(y)$ for all x, y in $\mathcal{X}$ and all scalars a and b. If $\mathcal{X}$ is a normed space, say that a linear functional L is *bounded* if there is a constant M such that $|L(x)| \leq M\|x\|$ for all x in $\mathcal{X}$.

The concept of a linear functional applies to any vector space, so it is unlikely to be of much value in the study of Banach space. The idea of a bounded linear functional, however, connects the concept to the norm and makes it more relevant.

1.5.2. Example. (a) Give $\mathcal{X} = \mathbb{F}^d$ the norm $\|x\| = \sum_{k=1}^d |x_k|$; if $y_1, \ldots, y_d \in \mathbb{F}$ and we define $L : \mathcal{X} \to \mathbb{F}$ by $L(x) = \sum_{k=1}^d x_k y_k$, then L is a bounded linear functional. To see that it is bounded, let $M = \max\{|y_k| : 1 \leq k \leq d\}$ and note that $|L(x)| \leq M\|x\|$.

(b) If X is a compact metric space, $x \in X$, and $L : C_b(X) \to \mathbb{F}$ is defined by $L(f) = f(x)$, then L is a bounded linear functional, where the constant M can be taken to be 1.

(c) Let $\mathcal{X} = \ell^1$ as in Example 1.3.2(b). If $\{b_n\} \in \ell^\infty$ as in Example 1.3.2(d), define $L : \ell^1 \to \mathbb{F}$ by

$$L(\{a_n\}) = \sum_{n=1}^{\infty} a_n b_n$$

It follows that L is a bounded linear functional on ℓ^1 with

$$|L(\{a_n\})| \leq \|\{b_n\}\|_\infty \|\{a_n\}\|$$

for all $\{a_n\}$ in ℓ^1.

(d) If J is a bounded interval in $\mathbb{R}$ and α is a function of bounded variation on J, then Theorem 1.1.8 says that $L : C(J) \to \mathbb{R}$ defined by $L(f) = \int f d\alpha$ is a bounded linear functional with $|L(f)| \leq \mathrm{Var}(\alpha)\|f\|$.

The reader can consult Exercises 3 and 4 for examples of unbounded linear functionals, though it might be underlined that in both these exercises the normed spaces are not Banach spaces. Getting an example of a Banach space and an unbounded linear functional on it requires the Axiom of Choice. See Exercise 5. The next result demonstrates that the boundedness of a linear functional is intimately connected to the topology defined on a normed space.

1.5.3. Proposition. *If $\mathcal{X}$ is a normed space and $L : \mathcal{X} \to \mathbb{F}$ is a linear functional, then the following statements are equivalent.*

(a) *L is bounded.*

(b) *L is a continuous function.*

(c) *L is a continuous function at 0.*

(d) *L is continuous at some point of $\mathcal{X}$.*

Proof. (a) *implies* (b). Let M be a constant such that $|L(x)| \leq M\|x\|$ for all x in $\mathcal{X}$. In fact $|L(x) - L(y)| = |L(x - y)| \leq M\|x - y\|$, from which it follows that not only is L continuous but it is uniformly continuous.

It is trivial that (b) implies (c) and that (c) implies (d).

(d) *implies* (a). Suppose L is continuous at x_0; so there is a $\delta > 0$ with $|L(x - x_0)| = |L(x) - L(x_0)| < 1$ whenever $\|x - x_0\| < \delta$. If x is an arbitrary non-zero vector in $\mathcal{X}$, then $\left\| \left[\frac{\delta}{2\|x\|} x + x_0 \right] - x_0 \right\| = \frac{\delta}{2} < \delta$. Hence $\left| L\left(\frac{\delta}{2\|x\|} x \right) \right| < 1$, and this implies that $|L(x)| \leq \frac{2}{\delta}\|x\|$. Since x was an arbitrary non-zero vector and the preceding inequality also holds for $x = 0$, we have that L is bounded where the constant M can be taken to be $\frac{2}{\delta}$. ∎

If L is a bounded linear functional on $\mathcal{X}$, we define

$$\|L\| = \sup\{|L(x)| : x \in \text{ball } \mathcal{X}\}$$

Clearly $\|L\| < \infty$; this is called the *norm* of L.

1.5.4. Proposition. *If L is a bounded linear functional on $\mathcal{X}$, then $|L(x)| \leq \|L\|\|x\|$ for all x in $\mathcal{X}$ and*

$$\|L\| = \sup\{|L(x)|/\|x\| : x \neq 0\}$$
$$= \inf\{M : |L(x)| \leq M\|x\| \text{ for all } x \in \mathcal{X}\}$$

Proof. If $x \in \mathcal{X}$, then $x/\|x\| \in \text{ball } \mathcal{X}$ and so $|L(x/\|x\|)| \leq M$, demonstrating the inequality. Since the first equation in the display is trivial, we will only prove the second, which is not a lot more difficult. Let α denote the infimum in question. We just saw that $\|L\|$ is a possible such constant, so $\alpha \leq \|L\|$. On the other hand if M is any of the constants in question, then the definition of $\|L\|$ shows that $\|L\| \leq M$. Hence $\|L\| \leq \alpha$. ∎

1.5.5. Definition. If $\mathcal{X}$ is a normed space and $\mathcal{X}^*$ denotes the collection of all bounded linear functionals on $\mathcal{X}$, then $\mathcal{X}^*$ is called the *dual space* of $\mathcal{X}$.

For a variety of reasons it is often convenient to denote the elements of $\mathcal{X}^*$ by x^* rather than something like L.

1.5.6. Proposition. *If $\mathcal{X}$ is a normed space, then $\mathcal{X}^*$ with its norm is a Banach space, where the operations of addition and scalar multiplication are defined pointwise: $(x^* + y^*)(x) = x^*(x) + y^*(x)$ and $(ax^*)(x) = ax^*(x)$.*

Proof. It is routine to show that $\mathcal{X}^*$ is a vector space and that its norm satisfies all the axioms of a norm on this vector space. We establish the completeness of $\mathcal{X}^*$. Let $\{x_n^*\}$ be a Cauchy sequence in $\mathcal{X}^*$. It follows that for every x in $\mathcal{X}$, $|x_n^*(x) - x_m^*(x)| = |(x_n^* - x_m^*)(x)| \leq \|x_n^* - x_m^*\| \|x\|$. Thus $\{x_n^*(x)\}$ is a Cauchy sequence in $\mathbb{F}$ and so $L(x) = \lim_n x_n^*(x)$ exists. The reader can show that this defines a linear functional $L : \mathcal{X} \to \mathbb{F}$; we want to show that L is bounded and that $\|x_n^* - L\| \to 0$. Since any Cauchy sequence in a normed space is uniformly bounded (Exercise 1.3.1.), $M = \sup_n \|x_n^*\| < \infty$. Therefore for any x in $\mathcal{X}$ we have that $|x_n^*(x)| \leq M\|x\|$ for all $n \geq 1$; it follows that $|L(x)| \leq M\|x\|$ and so $L \in \mathcal{X}^*$. Also if $\epsilon > 0$ and $x \in \text{ball } \mathcal{X}$, then for all $n, m \geq 1$, $|L(x) - x_n^*(x)| \leq |L(x) - x_m^*(x)| + |x_m^*(x) - x_n^*(x)| \leq |L(x) - x_m^*(x)| + \|x_m^* - x_n^*\|$. There is an integer N such that for $n, m \geq N$, $\|x_m^* - x_n^*\| < \epsilon$ and hence $|L(x) - x_n^*(x)| \leq |L(x) - x_m^*(x)| + \epsilon$ for any $m \geq N$. Letting $m \to \infty$ shows that $|L(x) - x_n^*(x)| \leq \epsilon$ for all $n \geq N$ and all x in ball $\mathcal{X}$; that is $\|L - x_n^*\| \leq \epsilon$ for $n \geq N$. ∎

It is worth emphasizing that the preceding proposition shows that even though $\mathcal{X}$ is not assumed to be a Banach space, its dual space is.

We will spend considerable time and effort in this book determining $\mathcal{X}^*$ for a variety of normed spaces. Why? As it turns out there is a rich theory about dual spaces that says that under certain assumptions on a subset A of a normed space $\mathcal{X}$, if we prove results about the action of every bounded linear functional x^* on A, then we can conclude that the results hold uniformly over A – that is, in relation to the topology defined by the norm. That is a powerful technique, analogous to a statement that if something holds pointwise it holds uniformly; though in this abstract formulation, such a principle surely seems quite distant.

Exercises.

(1) If X is a compact metric space and $L : C_b(X) \to \mathbb{F}$ is defined as in Example 1.5.2(b), show that $\|L\| = 1$.

(2) If $L : \ell^1 \to \mathbb{F}$ is defined as in Example 1.5.2(c), show that $\|L\| = \sup_n |b_n|$.

(3) Let $\mathcal{X} = C([0,1])$ but give it the norm $\|f\| = \int_0^1 |f(t)|dt$. (See Exercise 1.3.9.) Define $L : \mathcal{X} \to \mathbb{R}$ by $L(f) = f(\frac{1}{2})$ and show that L is an unbounded linear functional.

(4) Let $\mathcal{P}$ be the vector space of all polynomials with real coefficients and define $\|p\| = \max\{|p(x)| : 0 \le x \le 1\}$. (a) Show that $\|\cdot\|$ is a norm on $\mathcal{P}$. (b) If $L : \mathcal{P} \to \mathbb{R}$ is defined by $L(p) = p'(0)$, show that L is a linear functional on $\mathcal{P}$ that is not bounded.

(5) This exercise outlines the construction of an unbounded linear functional on a Banach space. Consider the Banach space c_0 and for each $n \ge 1$ let e_n denote the sequence with 1 in the n-th place and zeros elsewhere. Let $\{s_j : j \in J\}$ be vectors in c_0 such that $\mathcal{S} = \{e_n : n \ge 1\} \cup \{s_j : j \in J\}$ is an algebraic basis for c_0. That is, every element of c_0 is the linear combination of a finite number of elements from $\mathcal{S}$. Define $L : c_0 \to \mathbb{F}$ by

$$L \left(\sum_n a_n e_n + \sum_{j \in J} b_j s_j \right) = \sum_n n a_n e_n$$

for all collections of finitely non-zero scalars $\{a_n\} \cup \{b_j\}$. Show that L is an unbounded linear functional on c_0.

(6) Let $\mathcal{X}$ be a normed space and let $\mathcal{M}$ be a dense vector subspace of $\mathcal{X}$. If $L : \mathcal{M} \to \mathbb{F}$ is a bounded linear functional, show that there is a unique bounded linear functional $\widetilde{L} : \mathcal{X} \to \mathbb{F}$ such that $\widetilde{L}(x) = L(x)$ for all x in $\mathcal{M}$ and show that $\|\widetilde{L}\| = \|L\|$.

Chapter 2

Elements of Measure Theory

Measure theory is a vast subject, much larger than what will be covered in this book. There are mathematicians engaged in research on the topic, though their numbers are decreasing. It is, nevertheless, an incomparable tool and an engaging subject. We will see in this chapter the basic theory and later, in Chapter 4, we'll develop some additional topics. This should prepare the reader for work in analysis and probability, the two areas of mathematics that most use this subject. We begin with a particular type of linear functional on $C(X)$, and in the following section we'll see how such linear functionals give rise to measures. At that point we will start down the road of developing integration with respect to a measure.

2.1. Positive linear functionals on $C(X)$

Throughout this section (X, d) is a compact metric space. Recall from §1.2 that $C(X)$ has an order structure and we know what it means for two functions f and g in $C(X)$ to satisfy $f \leq g$.

2.1.1. Definition. A *positive linear functional* on $C(X)$ is a linear functional $L : C(X) \to \mathbb{R}$ such that $L(f) \geq 0$ whenever $f \geq 0$.

When L is a positive linear functional, $L(f) \leq L(g)$ whenever $f < g$, since $L(g) - L(f) = L(g - f) \geq 0$.

2.1.2. Example. (a) Let $X = [a, b] \subseteq \mathbb{R}$. If α is an increasing function on $[a, b]$, then the Riemann–Stieltjes integral $L(f) = \int_a^b f(x) d\alpha(x)$ defines a positive linear functional on $C[a, b]$ (1.1.9(b)). In particular, when $\alpha(x) = x$ for all x, the Riemann integral defines a positive linear functional.

(b) If X is arbitrary, $x_0 \in X$, and $L : C(X) \to \mathbb{R}$ is defined by $L(f) = f(x_0)$, then L is a positive linear functional. When $X = [a, b]$ this is just a special case of the example in part (a) where the increasing function is defined as in Example 1.1.12.

The proof of the next proposition is routine.

2.1.3. Proposition. *The set of all positive linear functionals forms a cone in the dual space of $C(X)$. That is, if L_1, L_2 are positive linear functionals on $C(X)$, then $(L_1 + L_2)(f) = L_1(f) + L_2(x)$ defines a positive linear functional on $C(X)$; if also a_1 is a positive real number, then $(a_1 L_1)(f) = a_1 L_1(f)$ defines a positive linear functional on $C(X)$.*

Also see Exercise 1 for an additional fact about this cone.

2.1.4. Proposition. *If $L : C(X) \to \mathbb{R}$ is a positive linear functional and $f \in C(X)$, then $|L(f)| \leq L(|f|)$ and L is bounded on the normed space $C(X)$. In fact, $\|L\| = L(1)$.*

Proof. First note that for any f in $C(X)$, $-|f| \leq f \leq |f|$; hence $-L(|f|) \leq L(f) \leq L(|f|)$. Thus $|L(f)| \leq L(|f|)$. Also if we put $a = \|f\|$ and consider a as a constant function, then $-a \leq f \leq a$. Therefore putting $M = L(1)$ we get $-aM = L(-a) \leq L(f) \leq L(a) = aM$. Thus $|L(f)| \leq aM = M\|f\|$; that is, $\|L\| \leq L(1)$. On the other hand $1 \in \text{ball } C(X)$, so the definition of the norm gives that $\|L\| \geq L(1)$. $\blacksquare$

Exercises.

(1) Show that the set of positive linear functionals on $C(X)$ is closed in the dual space $C(X)^*$.

(2) If g is a non-negative continuous function on X, L is a positive linear functional on $C(X)$, and we define $L_1 : C(X) \to \mathbb{R}$ by $L_1(f) = L(gf)$, show that L_1 is a positive linear functional. What is $\|L_1\|$?

(3) (a) If L is a positive linear functional on $C(X)$ and $f, g \in C(X)$, show that $L(f \vee g) \leq L(f) \vee L(g)$. (b) Define $L : C([0, 1]) \to \mathbb{R}$ by $L(f) = \int_0^1 f(t)dt$. Let $f(t) = t$ when $0 \leq t \leq \frac{1}{2}$ and $f(t) = \frac{1}{2}$ when $\frac{1}{2} \leq t \leq 1$; let $g(t) = \frac{1}{2}$ for $0 \leq t \leq \frac{1}{2}$ and $g(t) = 1 - t$ when $\frac{1}{2} \leq t \leq 1$. Compute $L(f \vee g)$ and $L(f) \vee L(g)$.

2.2. The Radon measure space

In this section X is again a compact metric space and L is a positive linear functional on $C(X)$. The goal is to associate with L what we will call a measure μ defined on a certain collection of subsets of X. This will conform to the concept we might associate intuitively with the word measure in that it will give us an idea of the size or content of these subsets relative to the action of L on $C(X)$. More importantly the properties of this measure will enable us to eventually (§2.4) define an integral with respect to μ, which will be a type of weighted average, and we will gain a representation of L as this integral. That is, we will show that $L(f) = \int f d\mu$ for all f in $C(X)$; but this lies in our future.

Recall that in §1.2, just before (1.2.11), we defined the characteristic function of a set E. Another term used in the literature in place of a characteristic function is the *indicator function*.

2.2.1. Definition. Fix a positive linear functional L on $C(X)$. If G is an open subset of X, define the *outer measure* of G to be the quantity

$$\mu^*(G) = \sup\{L(f) : f \in C(X) \text{ and } 0 \leq f \leq \chi_G\}$$

Note that for every f in $C(X)$ with $0 \leq f \leq \chi_G \leq 1$, we have that $L(f) \leq L(1) = \|L\|$, so that $\mu^*(G) \leq \|L\| < \infty$.

2.2.2. Example. Let $X = [a, b]$ and let L be the positive linear functional defined on $C[a, b]$ by the Riemann integral: $L(f) = \int_a^b f(t)dt$. If (c, d) is an open interval in $[a, b]$, then

$$\mu^*((c, d)) = d - c$$

the length of the interval. To see this let f_n be the continuous function with $f_n(x) = 1$ for $c + \frac{1}{n} \leq x \leq d - \frac{1}{n}$, $f(x) = 0$ for $a \leq x \leq c$ and $d \leq x \leq b$, and the graph of f_n forming straight line segments from $(c, 0)$ to $(c + \frac{1}{n}, 1)$ and from $(d - \frac{1}{n}, 1)$ to $(d, 0)$. It is easy to see that $f_n(x) \nearrow \chi_{(c,d)}(x)$ for all x in $[a, b]$ and so, by the next proposition, $L(f_n) \to \mu^*((c, d))$. Also $L(f_n) \to d - c$ as a computation will show, hence the conclusion stated above. Also see Exercise 4.

In the subsequent material, unless we want to state a result that actually uses the quantity $\|L\|$ we will always assume that

$$\|L\| = L(1) = 1$$

This will make many of the arguments cleaner.

The method used in exploring the example above can be abstracted (and justified).

2.2.3. Proposition. *If G is a non-empty open subset of X, then there is an increasing sequence of functions $\{f_n\}$ in $C(X)_+$ such that $0 \leq f_n \leq \chi_G$ for all $n \geq 1$ and $f_n(x) \to \chi_G(x)$ for every x in X. For any sequence $\{f_n\}$ having these properties, $L(f_n) \to \mu^*(G)$.*

Proof. The existence of such a sequence $\{f_n\}$ is furnished by Proposition 1.2.10, where we can take for the set F any closed subset of G.

Now assume $\{f_n\}$ is an increasing sequence in $C(X)_+$ having the stated properties. Since this sequence of functions increases and $0 \le f_n \le \chi_G$ for every $n \ge 1$, we have that the numbers $\{L(f_n)\}$ are increasing and bounded above by $\mu^*(G)$; so there is a number $\gamma \le \mu^*(G)$ such that $L(f_n) \to \gamma$. Now let f be an arbitrary function in $C(X)_+$ such that $0 \le f \le \chi_G$, let $0 < s < 1$, and let $\epsilon \ge 0$. Put $K = \{x : f(x) \ge \epsilon/2\}$; clearly K is compact and contained in G. Since $f_n(x) \to 1$ for each x in K, there is an integer N_x such that $f_n(x) - sf(x) > 0$ when $n \ge N_x$. Thus there is a radius $r_x > 0$ such that $f_{N_x}(y) - sf(y) > 0$ when $y \in B(x; r_x)$. Since the sequence of functions is increasing, this implies that $f_n(y) > sf(y)$ whenever $d(x, y) < r_x$ and $n \ge N_x$. But $\{B(x; r_x) : x \in K\}$ is an open cover of K and so there are points $x_1, \ldots, x_m$ in K with $K \subseteq \bigcup_{j=1}^m B(x_j; r_{x_j})$. If $N = \max\{N_{x_1}, \ldots, N_{x_m}\}$ and $n \ge N$, this implies that $f_n(x) > sf(x)$ for every x in K. Hence $(f_n \wedge sf)(x) = sf(x)$ for $x \in K$ and $n \ge N$. Also for $x \in G \backslash K$, $(f_n \wedge sf)(x) \le sf(x) \le s\epsilon/2$. Combining these estimates we have that $|(f_n \wedge sf)(x) - sf(x)| \le s\epsilon < \epsilon$ for all x in X and $n \ge N$. That is, $\|f_n \wedge sf - sf\| < \epsilon$ for $n \ge N$. Remembering that we are assuming that $\|L\| = 1$, this implies that $|L(f_n \wedge sf) - L(sf)| < \epsilon$ for $n \ge N$; since ϵ was arbitrary, we have that $L(f_n \wedge sf) \to L(sf) = sL(f)$. On the other hand $L(f_n \wedge sf) \le L(f_n) \to \gamma$. Thus $sL(f) \le \gamma$ whenever $0 < s < 1$; since s was arbitrary, we have that $L(f) \le \gamma$. Therefore we have shown that $L(f) \le \gamma$ when $f \in C(X)_+$ and $0 \le f \le \chi_G$. By definition, $\mu^*(G) \le \gamma$. ∎

The preceding proposition facilitates calculating $\mu^*(G)$. Here is an example in addition to Example 2.2.2 showing this; the reader can also look at Exercise 4.

2.2.4. Example. Consider the interval $J = [a, b]$, let $a < s < b$, and let $L : C(J) \to \mathbb{R}$ be defined by $L(f) = \int f d\alpha_s$, where α_s is the increasing function defined in Example 1.1.12. If G is any open set in J that contains s, then $\mu^*(G) = 1$; otherwise $\mu^*(G) = 0$. To see this suppose $s \in G$. Construct the increasing functions $\{f_n\}$ required in Proposition 2.2.3 with $f_n(s) = 1$ for all n. (How do you do this?) Using Example 1.1.12 we get that $\int f d\alpha_s = 1$ for all n, so the preceding proposition says that $\mu^*(G) = 1$. If $s \notin G$, then no matter how the functions f_n are chosen we have that $\int f_n d\alpha_s = 0$.

Now we begin to derive certain properties of this outer measure.

2.2.5. Proposition. *If $G_1, G_2, \ldots$ is a sequence of open subsets of X and $G = \bigcup_{n=1}^\infty G_n$, then $\mu^*(G) \le \sum_{n=1}^\infty \mu^*(G_n)$.*

Proof. Let $0 \leq f \leq \chi_G$ such that $\mu^*(G) \leq L(f) + \epsilon/2$ and put $K = \{x : f(x) \geq \epsilon/4\}$. Since K is compact, there is an N such that $K \subseteq \bigcup_{n=1}^{N} G_n$. By Corollary 1.2.12 there are continuous functions $\phi_1, \ldots, \phi_N$ such that for $1 \leq n \leq N$, $0 \leq \phi_n \leq \chi_{G_n}$, $\sum_{n=1}^{N} \phi_n(x) = 1$ when $x \in K$, and $\sum_{n=1}^{N} \phi_n \leq 1$. Note that $f\phi_n \leq \chi_{G_n}$ so that $L(f\phi_n) \leq \mu^*(G_n)$. Also when $x \in K$, $f(x) - \sum_{n=1}^{N} f(x)\phi_n(x) = 0$. On the other hand when $x \notin K$,

$$\left| f(x) - \sum_{n=1}^{N} f(x)\phi_n(x) \right| \leq \frac{\epsilon}{4} + \frac{\epsilon}{4} \sum_{n=1}^{N} \phi_n(x) \leq \epsilon/2$$

Therefore $\| f - \sum_{n=1}^{N} f\phi_n \| \leq \epsilon/2$ and so $|L(f - \sum_{n=1}^{N} f\phi_n)| \leq \epsilon/2$. This gives that

$$\mu^*(G) \leq \epsilon/2 + L(f)$$

$$= \epsilon/2 + L\left(f - \sum_{n=1}^{N} f\phi_n \right) + \sum_{n=1}^{N} L(f\phi_n)$$

$$\leq \epsilon + \sum_{n=1}^{N} \mu^*(G_n)$$

$$\leq \epsilon + \sum_{n=1}^{\infty} \mu^*(G_n)$$

Since ϵ was arbitrary, we have the desired inequality. ∎

Now we extend the definition of μ^* to arbitrary sets. Remember that always lurking in the shadows is the fixed positive linear functional L used to define the outer measure, and we continue to assume that $\|L\| = 1 = L(1)$.

2.2.6. Definition. If E is any subset of X, define

$$\mu^*(E) = \inf\{\mu^*(G) : G \text{ is open and } E \subseteq G\}$$

We also call this the *outer measure* of E. Note that when G is an open set, $\mu^*(G) = \inf\{\mu^*(H) : H \text{ is open and } G \subseteq H\}$, so this new definition is indeed an extension of μ^* to a larger class of sets. This justifies our continued use of the notation μ^* and the term outer measure. Here are some elementary facts about the outer measure.

2.2.7. Proposition. (a) $\mu^*(\emptyset) = 0$ and $\mu^*(X) = \|L\|$.

(b) If D and E are any sets with $D \subseteq E$, then $\mu^*(D) \leq \mu^*(E)$.

(c) If $\{E_n\}$ is any sequence of sets, $\mu^*\left(\bigcup_{n=1}^{\infty} E_n\right) \leq \sum_{n=1}^{\infty} \mu^*(E_n)$.

Proof. Since $\emptyset$ is open and there are no functions $0 \leq f \leq \chi_\emptyset$, the first part of (a) follows. Since X is open and $1 \leq \chi_X$, the second part of (a) is also immediate. Part (b) is clear from the definition.

To prove (c) let $E = \bigcup_{n=1}^\infty E_n$ and fix $\epsilon > 0$. For each $n \geq 0$ there is an open set G_n containing E_n such that $\mu^*(E_n) + \epsilon/2^n > \mu^*(G_n) \geq \mu^*(E_n)$. If $G = \bigcup_{n=1}^\infty G_n$, then G is open and contains E; so, using (2.2.5), $\mu^*(E) \leq \mu^*(G) \leq \sum_{n=1}^\infty \mu^*(G_n) \leq \sum_{n=1}^\infty [\mu^*(E_n) + \epsilon/2^n] = \epsilon + \sum_{n=1}^\infty \mu^*(E_n)$. Since ϵ was arbitrary, this proves the result. $\blacksquare$

2.2.8. Proposition. (a) *If K is compact, $\{\delta_n\}$ is a decreasing sequence of positive numbers that converges to 0, and $G_n = \{x : \mathrm{dist}\,(x, K) < \delta_n\}$, then $\mu^*(G_n) \to \mu^*(K)$.*

(b) *If G is open and $\{K_n\}$ is a sequence of compact subsets of G such that $K_n \subseteq K_{n+1}$ and $G = \bigcup_{n=1}^\infty K_n$, then $\mu^*(K_n) \to \mu^*(G)$.*

Proof. (a) Since $\{\delta_n\}$ is decreasing, the sets $\{G_n\}$ are decreasing and therefore $c = \lim_n \mu^*(G_n)$ exists. Since $K \subseteq G_n$ for all n, $\mu^*(K) \leq \mu^*(G_n)$ for all n and so $\mu^*(K) \leq c$. On the other hand if G is any open set containing K, the fact that K is compact implies there is an integer N such that $G_n \subseteq G$ for all $n \geq N$. (See Exercise 1.2.4.) This implies $c \leq \mu^*(G)$ for every choice of G, so that $\mu^*(K) = c$.

(b) For each n let G_n be an open set such that $K_n \subseteq G_n \subseteq G$ and $\mu^*(K_n) \leq \mu^*(G_n) \leq \mu^*(K_n) + \frac{1}{n}$. Let $g_n \in C(X)_+$ such that $\chi_{K_n} \leq g_n \leq \chi_{G_n}$. Put $f_n = \max\{g_1, \ldots, g_n\}$. So $\{f_n\}$ is an increasing sequence in $C(X)_+$ and $\chi_{K_n} \leq f_n \leq \chi_G$. Since $K_n \subseteq K_{n+1}$ for all n and $G = \bigcup_{n=1}^\infty K_n$, for any x in G there is an integer N such that $x \in K_n$ for all $n \geq N$. Thus $f_n(x) \to \chi_G(x)$ for all x in X. Therefore by (2.2.3), $L(f_n) \to \mu^*(G)$. But $L(f_n) \leq \mu^*(G_n) \leq \mu^*(K_n) + \frac{1}{n} \leq \mu^*(G) + \frac{1}{n}$, whence part (b). $\blacksquare$

Exercise 7 has an extension of part (a) of the preceding proposition.

2.2.9. Definition. Say that a set E is *measurable* if for every $\epsilon > 0$ there is a compact set K and an open set G such that $K \subseteq E \subseteq G$ and $\mu^*(G \backslash K) < \epsilon$. Of course this depends on the positive linear functional L; if there is the possibility of an ambiguity, we say the set E is *L-measurable* or *μ-measurable*. Denote the collection of measurable sets by $\mathcal{A} = \mathcal{A}_L = \mathcal{A}_\mu$. If $E \in \mathcal{A}$, we set $\mu(E) = \mu^*(E)$. The triple $(X, \mathcal{A}, \mu)$ is called the *Radon*[1] *measure space* associated with the positive linear functional L.

---

[1] Johann Radon was born in 1887 in what is today Decin, The Czech Republic. He received his doctorate from the University of Vienna in 1910, working on the calculus of variations. Somewhat unusual for his day, he held positions at several universities over the years, including Vienna, Göttingen, Hamburg, Griefswald, Erlangen, Breslau, and Innsbruck. He seemed a joyful man. He and his wife Maria had four children and at each of his appointments they organized an active social circle, with fancy dress parties and musical events. Tragedy struck them later and they lost

In light of the preceding proposition, every compact set and every open set is measurable. Note that if E is any measurable set, $\epsilon > 0$, and G and K are as in the definition, then $\mu(E) = \mu^*(E) = \mu^*(K \cup [E \backslash K]) \leq \mu^*(K) + \mu^*(E \backslash K) \leq \mu(K) + \mu^*(G \backslash K) < \mu(K) + \epsilon$. Similarly $\mu(E) \leq \mu(G) < \mu(E) + \epsilon$. It is worth pointing out that shortly we will define and explore an abstract version of a Radon measure space (2.4.1).

The reason for introducing the concept of a measurable set is that μ behaves itself when viewed as a function from the collection $\mathcal{A}$ of measurable sets into $[0, \|L\|]$, while it is erratic if considered as a function defined on 2^X. This good behavior becomes apparent in the next theorem, whose proof is the main objective of the rest of this section.

2.2.10. Theorem. *If $(X, \mathcal{A}, \mu)$ is the Radon measure space associate with the positive linear functional L on $C(X)$, the following hold.*

(a) *Every open set and every compact set belongs to $\mathcal{A}$ and, when $D, E \in \mathcal{A}$ and $D \subseteq E$, $\mu(D) \leq \mu(E) \leq L(1)$.*

(b) *If $E_1, E_2 \in \mathcal{A}$, then $E_1 \backslash E_2 \in \mathcal{A}$.*

(c) *If $\{E_1, E_2, \dots\}$ are measurable sets, then $E = \bigcup_{n=1}^\infty E_n$ is measurable and $\mu(E) \leq \sum_{n=1}^\infty \mu(E_n)$. If in addition the sets $\{E_n\}$ are pairwise disjoint, then $\mu(E) = \sum_{n=1}^\infty \mu(E_n)$.*

(d) *If $\{E_1, E_2, \dots\}$ are measurable sets, then $E = \bigcap_{n=1}^\infty E_n$ is measurable.*

(e) *If $\{E_1, E_2, \dots\}$ are measurable sets with $E_n \subseteq E_{n+1}$ for all $n \geq 1$ and $E = \bigcup_{n=1}^\infty E_n$, then $\mu(E) = \lim_n \mu(E_n)$.*

(f) *If $\{E_1, E_2, \dots\}$ are measurable sets with $E_n \supseteq E_{n+1}$ for all $n \geq 1$ and $E = \bigcap_{n=1}^\infty E_n$, then $\mu(E) = \lim_n \mu(E_n)$.*

Observe that many of these properties of a measure are in conformity with what we would intuitively associate with the concept of measuring sets. For example in part (a) we have that bigger sets have a bigger measure. In (c) we have that the measure of a set does not exceed the sum of the measures of its parts.

Proof. Continue to assume that $\|L\| = 1 = \mu(X)$. We have already pointed out that every open set and every compact set belong to $\mathcal{A}$. The rest of (a) is straightforward. To prove (b), fix $\epsilon > 0$ and let K_1, K_2 be compact sets and let G_1, G_2 be open sets with $K_j \subseteq E_j \subseteq G_j$ and $\mu(G_j \backslash K_j) < \epsilon/2$ for $j = 1, 2$. It is easy to check that $E_1 \backslash E_2 \subseteq G_1 \backslash K_2 = G$, an open set; and $E_1 \backslash E_2 \supseteq K_1 \backslash G_2 = K$, a compact set. Moreover $\mu^*[(E_1 \backslash E_2) \backslash K] \leq \mu^*(E_1 \backslash K_1) < \epsilon/2$; and $\mu^*[G \backslash (E_1 \backslash E_2)] \leq \mu^*(G_1 \backslash E_1) < \epsilon/2$. By Proposition

---

two sons, one to disease and another killed in World War II. In addition to his work in analysis Radon made contributions to differential geometry. He died in Vienna in 1956.

2.2.7(c), $\mu(G\backslash K) = \mu^*([G\backslash(E_1\backslash E_2)] \cup [(E_1\backslash E_2)\backslash K]) \leq \mu^*[G\backslash(E_1\backslash E_2)] + \mu^*[(E_1\backslash E_2)\backslash K] < \epsilon$.

Once we establish (c), the remaining parts will follow. For example, to see that part (d) follows, assume $\{E_n\}$ is a sequence in $\mathcal{A}$. By (b), $\{X\backslash E_n\} \subseteq \mathcal{A}$. By (c), $\bigcup_{n=1}^{\infty}(X\backslash E_n) \in \mathcal{A}$. But then part (b) again shows that $\bigcap_{n=1}^{\infty} E_n = X\backslash\bigcup_{n=1}^{\infty}(X\backslash E_n) \in \mathcal{A}$.

It is also true that (e) follows from (c). In fact if $F_1 = E_1$ and $F_n = E_n\backslash E_{n-1}$ for $n \geq 2$, then (c) implies $\mu(E) = \mu(\bigcup_{n=1}^{\infty} F_n) = \sum_{n=1}^{\infty} \mu(F_n)$. But $E_n = F_n \cup E_{n-1}$ and F_n and E_{n-1} are disjoint; so once again (c) implies $\mu(F_n) = \mu(E_n) - \mu(E_{n-1})$ when $n \geq 2$. Thus $\mu(E) = \mu(E_1) + \sum_{n=2}^{\infty}[\mu(E_n) - \mu(E_{n-1})] = \lim_n \mu(E_n)$.

We have moreover that (f) follows from (c) and (e). In fact if $\{E_n\}$ and E are as in (f), part (e) implies $\mu(X\backslash E) = \lim_n \mu(X\backslash E_n)$. But (c) implies that $\mu(X\backslash E) = \mu(X) - \mu(E)$ and $\mu(X\backslash E_n) = \mu(X) - \mu(E_n)$ for all n.

Therefore it only remains to prove (c). We do this by establishing a sequence of claims.

2.2.11. Claim. If $G_1, \ldots, G_n$ are pairwise disjoint open sets, then

$$\mu\left(\bigcup_{j=1}^{n} G_j\right) = \sum_{j=1}^{n} \mu(G_k)$$

Let $G = \bigcup_{j=1}^{n} G_j$; we already know that $\mu(G) \leq \sum_{j=1}^{n} \mu(G_j)$. If $\epsilon > 0$ and $1 \leq j \leq n$, let $0 \leq f_j \leq \chi_{G_j}$ such that $L(f_j) + \epsilon/n > \mu(G_j)$. Put $f = \sum_{j=1}^{n} f_j$. Note that because the sets G_j are pairwise disjoint we have that $0 \leq f \leq \chi_G$. Thus using Proposition 2.2.5, $\sum_{j=1}^{n} \mu(G_j) \geq \mu(G) \geq L(f) = \sum_{j=1}^{n} L(f_j) > -\epsilon + \sum_{j=1}^{n} \mu(G_j)$. Since ϵ was arbitrary, this proves the claim.

2.2.12. Claim. If $K_1, \ldots, K_n$ are pairwise disjoint compact sets,

$$\mu\left(\bigcup_{j=1}^{n} K_j\right) = \sum_{j=1}^{n} \mu(K_j)$$

Put $K = \bigcup_{j=1}^{n} K_j$ and for $\epsilon > 0$ let G be an open set containing K with $\mu(K) \leq \mu(G) < \mu(K) + \epsilon$. Find pairwise disjoint open sets $G_1, \ldots, G_n$ such that $K_j \subseteq G_j \subseteq G$ for $1 \leq j \leq n$. (How is this done?) Using Proposition 2.2.7(c) and the preceding claim we have that $\mu(K) \leq \sum_{j=1}^{n} \mu(K_j) \leq \sum_{j=1}^{n} \mu(G_j) = \mu\left(\bigcup_{j=1}^{n} G_j\right) \leq \mu(G) < \mu(K) + \epsilon$, establishing the claim.

2.2.13. Claim. If $E_1, \ldots, E_n$ are measurable sets, then $\bigcup_{j=1}^n E_j \in \mathcal{A}$ and $\mu\left(\bigcup_{j=1}^n E_j\right) \leq \sum_{j=1}^n \mu(E_j)$. If the sets $E_1, \ldots, E_n$ are pairwise disjoint, then $\mu\left(\bigcup_{j=1}^n E_n\right) = \sum_{j=1}^n \mu(E_j)$.

For $1 \leq j \leq n$, let K_j be a compact set and G_j an open set such that $K_j \subseteq E_j \subseteq G_j$ and $\mu(G_j \backslash K_j) < \epsilon/2n$. Put $G = \bigcup_{j=1}^n G_j$ and $K = \bigcup_{j=1}^n K_j$. So G is open, K is compact, and $K \subseteq \bigcup_{j=1}^n E_j \subseteq G$. Moreover $G \backslash K = \bigcup_{j=1}^n G_j \backslash K \subseteq \bigcup_{j=1}^n G_j \backslash K_j$. Since each $G_j \backslash K_j$ is open, we have that $\mu(G \backslash K) \leq \sum_{j=1}^n \mu(G_j \backslash K_j) < \epsilon/2$. Thus $\bigcup_{j=1}^n E_j \in \mathcal{A}$. Also $\mu\left(\bigcup_{j=1}^n E_j\right) \leq \mu(G) \leq \epsilon/2 + \mu(K) \leq \epsilon/2 + \sum_{j=1}^n \mu(K_j) \leq \epsilon + \sum_{j=1}^n \mu(E_j)$. Thus we have that $\mu\left(\bigcup_{j=1}^n E_j\right) \leq \sum_{j=1}^n \mu(E_j)$.

If $E_1, \ldots, E_n$ are pairwise disjoint, then so are the sets $\{K_1, \ldots, K_n\}$. Using the previous claim this implies $\sum_{j=1}^n \mu(E_j) \leq \epsilon/2 + \sum_{j=1}^n \mu(G_j) \leq \epsilon + \sum_{j=1}^n \mu(K_j) = \epsilon + \mu(K) \leq \epsilon + \mu\left(\bigcup_{j=1}^n E_j\right)$. The arbitrariness of ϵ therefore implies that $\mu\left(\bigcup_{j=1}^n E_j\right) = \sum_{j=1}^n \mu(E_j)$.

2.2.14. Claim. If $\{E_n\}$ is an infinite sequence of pairwise disjoint measurable sets and $E = \bigcup_{n=1}^\infty E_n$, then E is measurable and $\mu(E) = \sum_{n=1}^\infty \mu(E_n)$.

The first step is to show that $\sum_{n=1}^\infty \mu(E_n) < \infty$. Suppose this is not the case; thus there is an integer N such that $\sum_{n=1}^N \mu(E_n) > 2\mu(X)$. Let $\epsilon > 0$ and for $1 \leq n \leq N$ choose a compact set C_n contained in E_n such that $\mu(E_n) < \mu(C_n) + \epsilon/N$. Since the sets C_n are pairwise disjoint, Proposition 2.2.7(b) and Claim 2.2.12 imply $\mu(X) \geq \mu\left(\bigcup_{n=1}^N C_n\right) = \sum_{n=1}^N \mu(C_n) > \epsilon + \sum_{n=1}^N \mu(E_n) > 2\mu(X)$, a contradiction. Hence $\sum_{n=1}^\infty \mu(E_n) < \infty$.

Now let $\epsilon > 0$ and for each $n \geq 1$ let K_n be a compact set and G_n an open set such that $K_n \subseteq E_n \subseteq G_n$ and $\mu(G_n \backslash K_n) < \epsilon/2^{n+2}$. Also choose a positive integer N such that $\sum_{n=N+1}^\infty \mu(E_n) < \epsilon/2$. Put $K = \bigcup_{n=1}^N K_n$ and $G = \bigcup_{n=1}^\infty G_n$. So K is compact, G is open, and $K \subseteq E \subseteq G$. Moreover $G \backslash K \subseteq \bigcup_{n=N+1}^\infty G_n \cup \left[\left(\bigcup_{n=1}^N G_n\right) \backslash K\right] \subseteq \bigcup_{n=N+1}^\infty G_n \cup \bigcup_{n=1}^\infty (G_n \backslash K_n)$. Therefore by Proposition 2.2.7,

$$\mu(G \backslash K) \leq \mu\left(\bigcup_{n=N+1}^\infty G_n\right) + \mu\left(\bigcup_{n=1}^N G_n \backslash K_n\right)$$

$$\leq \sum_{n=N+1}^\infty \mu(G_n) + \sum_{n=1}^N \mu(G_n \backslash K_n)$$

$$\leq \sum_{n=N+1}^{\infty} [\mu(E_n) + \epsilon/2^{n+2}] + \sum_{n=1}^{N} \epsilon/2^{n+2}$$
$$< \epsilon/2 + \epsilon/4 + \epsilon/4 = \epsilon$$

Therefore $E \in \mathcal{A}$.

Again using (2.2.7) and Claim 2.2.13 we have $\mu(E) \leq \sum_{n=1}^{\infty} \mu(E_n) \leq \epsilon/2 + \sum_{n=1}^{N} \mu(E_n) = \epsilon/2 + \mu\left(\bigcup_{n=1}^{N} E_n\right) \leq \epsilon/2 + \mu(E)$. Again the arbitrariness of ϵ shows that $\mu(E) = \sum_{n=1}^{\infty} \mu(E_n)$.

We have one more claim before the proof of (c), and therefore of the theorem, is complete.

2.2.15. Claim. If $\{E_n\}$ is an infinite sequence of measurable sets, then $E = \bigcup_{n=1}^{\infty} E_n$ is measurable.

In fact put $F_1 = E_1$ and for $n \geq 2$ let $F_n = E_n \backslash (E_1 \cup \cdots \cup E_{n-1})$. By part (b) and Claim 2.2.13 each set F_n is measurable. Moreover the sets $\{F_n\}$ are pairwise disjoint so by Claim 2.2.14, $\bigcup_{n=1}^{\infty} F_n \in \mathcal{A}$. But this union is precisely the set E. $\blacksquare$

2.2.16. Example. If $X = [a,b]$ and $L : C[a,b] \to \mathbb{R}$ is defined by $L(f) = \int_a^b f \, dx$, the Riemann integral, the resulting Radon measure space

$$([a,b], \mathcal{A}, \lambda)$$

is called the *Lebesgue[2] measure space* on the interval and λ is called *Lebesgue measure*.

It is possible to show that non-measurable sets exist, even for Lebesgue measure. However this requires the use of the Axiom of Choice. The interested reader can see [**18**], Theorem 10.28.

---

[2]Henri Léon Lebesgue was born in 1875 in Beauvais, which is in the Picardie region of France just north of Paris. He began his education in Beauvais and entered the École Normale Supérieure in Paris in 1894, where he was awarded the teaching diploma in mathematics in 1897. He remained there for two years, reading mathematics in the library. In 1899 he went to Nancy as a professor at a lycée where he remained for another 2 years. In 1901 he formulated the theory of measure in a ground breaking paper that defined what we now know as the Lebesgue integral. This formed the basis of his doctoral dissertation and earned him the title of the father of measure theory. In 1902 he joined the faculty at the University of Rennes in Brittany. He married in 1903 and he and his wife had two children. Unfortunately, the union ended in divorce in 1916. He published two fundamental monographs, *Leçons sur l'intégration et la recherche des fonctions primitives* in 1904 and *Leçons sur les séries trigonométriques* in 1906. These books were unjustly criticized at the time by the classicists. Nevertheless he joined the faculty at the University of Poitiers and finally overcame his critics before joining the faculty at the Sorbonne in 1910. In 1921 he became Professor of Mathematics at the Collège de France and held this position until he died in 1941. Lebesgue received many honors throughout his life and made serious contributions in many areas, but he will forever be associated with measure theory and its impact on analysis. We will see his name frequently as this book progresses.

Lebesgue measure has particular significance and the reader would do well to see what every result we obtain says about it.

Exercises. We continue to assume that X is a compact metric space, L is a positive linear functional on $C(X)$, and μ^* is the associated outer measure.

(1) Show that if E is a subset of X and $\mu^*(E) = 0$, then E and every subset of E is a measurable set.

(2) Fix a point x_0 in X and define $L : C(X) \to \mathbb{R}$ by $L(f) = f(x_0)$ as in Example 2.1.2. Show that in this case every subset E of X is measurable and $\mu(E) = 1$ when $x_0 \in E$ while $\mu(E) = 0$ when $x_0 \notin E$.

(3) Show that every countable subset E of $[a, b]$ is Lebesgue measurable and that $\lambda(E) = 0$. As a consequence, if E is the set of irrational numbers in this interval, then $\lambda(E) = b - a$. That is, the set of irrational numbers in the interval has full measure while $\lambda(\mathbb{Q} \cap [a, b]) = 0$.

(4) If J is a bounded interval in $\mathbb{R}$ and α is an increasing function on J, let L be defined on $C(J)$ as integration with respect to α. (a) Show that if α is continuous at the points c and d in J, where $c < d$, then $\mu^*((c, d)) = \alpha(d) - \alpha(c)$. See Example 2.2.2. (b) What happens if α has a discontinuity at either c or d or both?

(5) Let J and L be as in the preceding exercise and show that α is continuous at a point x_0 if and only if $\mu^*(\{x_0\}) = 0$. If α has a discontinuity at x_0, what is $\mu^*(\{x_0\})$?

(6) Let $\{x_n\}$ be a sequence in X and let $\{a_n\}$ be a sequence of positive real numbers such that $\sum_{n=1}^{\infty} a_n < \infty$. (a) Show that $L(f) = \sum_{n=1}^{\infty} a_n f(x_n)$ defines a positive linear functional L on $C(X)$. Let μ be the measure defined by this positive linear functional. (b) If G is an open set, show that $\mu(G) = \sum \{a_n : x_n \in G\}$. (c) Identify the measurable sets. (Hint: Exercise 2 may be helpful.) (d) If E is a measurable set, what is $\mu(E)$?

(7) Extend part (a) of Proposition 2.2.8. If $\{G_n\}$ is any decreasing sequence of open sets with $K = \bigcap_{n=1}^{\infty} G_n$ compact, show that $\mu^*(G_n) \to \mu^*(K)$. By finding an example, show that the requirement that $\{G_n\}$ be a decreasing sequence is necessary.

2.3. Measurable functions

In the last section the measure μ and the collection of measurable sets were intimately connected to the topology of the underlying compact metric space. For a variety of reasons, some involving probability and its increasing role in today's mathematics and applications, we must now seek a more general setting that is divorced from a topological space. For any set X denote the collection of all subsets of X by 2^X. (This notation arises because we identify subsets with their characteristic functions. For any two sets A and B, A^B denotes the collection of all functions $f : B \to A$. Thus 2^X denotes the collection of functions $\chi : X \to \{0, 1\}$.)

2.3.1. Definition. If X is any set, then a subset $\mathcal{A}$ of 2^X is called a σ-*algebra* if it has the following properties:

(a) $\emptyset$ and X belong to $\mathcal{A}$;

(b) if $E, F \in \mathcal{A}$, then $E \backslash F \in \mathcal{A}$;

(c) If $\{E_1, E_2, \dots\} \subseteq \mathcal{A}$, then $\bigcup_{n=1}^{\infty} E_n \in \mathcal{A}$.

By an application of De Morgan's Law the following holds:

(d) If $\{E_1, E_2, \dots\} \subseteq \mathcal{A}$, then $\bigcap_{n=1}^{\infty} E_n \in \mathcal{A}$.

2.3.2. Example. (a) If X is a compact metric space and L is a positive linear functional on $C(X)$, then the collection of all measurable subsets of X for the Radon measure μ associated with L constitutes a σ-algebra. In fact this is immediate from Theorem 2.2.10.

(b) For any set X, 2^X is a σ-algebra.

(c) For any set X, $\mathcal{A} = \{\emptyset, X\}$ is a σ-algebra.

It follows that if $\{\mathcal{A}_i : i \in I\}$ is a collection of σ-algebras of subsets of X, then $\bigcap_i \mathcal{A}_i$ is a σ-algebra. So when $\mathcal{S}$ is a collection of subsets of X, the σ-*algebra generated by* $\mathcal{S}$ is defined to be $\bigcap\{\mathcal{A} : \mathcal{A}$ is a σ-algebra containing $\mathcal{S}\}$. Note that the σ-algebra generated by $\mathcal{S}$ is the smallest σ-algebra containing $\mathcal{S}$.

2.3.3. Definition. If X is a metric space, then the σ-algebra of *Borel*[3] *sets* is the σ-algebra generated by the collection of open subsets of X.

Clearly the σ-algebra of Borel sets is also generated by the closed subsets. The next result is immediate from Theorem 2.2.10.

---

[3]Emile Borel was born in Saint Affrique in the south of France in 1871. He published his first two papers in 1890, two years before receiving his doctorate in Paris and joining the faculty at Lille. He returned to Paris in 1897. In 1909 a special chair in the Theory of Functions was created for him at the Sorbonne. During World War I he was very supportive of his country and was put in charge of a central department of research. He also spent time at the front and in 1918 he was awarded the Croix de Guerre. In 1928 he set up the Institute Henri Poincaré. He was one of the founders of the modern theory of functions along with Baire and Lebesgue and he also worked on divergent series, complex variables, probability, and game theory. He continued to be very active in the French government, serving in the French Chamber of Deputies (1924–36) and as Minister of the Navy (1925–40). He died in 1956 in Paris.

2.3.4. Proposition. *If X is a compact metric space and L is a positive linear functional on $C(X)$, then the collection of measurable subsets for L includes the Borel sets.*

Usually the measurable sets for the Radon measure associated with L on $C(X)$ include more than the Borel sets. For example, every set having measure zero is measurable (see Exercise 2.2.1). We aren't going to explore this too deeply except to point out that Exercise 4 states that a set is measurable in this setting if and only if it is the union of a Borel set and a set of measure zero. It is worth underlining, however, that for every Radon measure the Borel sets form a σ-algebra that is contained in its collection of measurable sets, while clearly sets of measure zero for one such measure may not be measurable for another.

Let us denote by $\widehat{\mathbb{R}}$ the set $\mathbb{R} \cup \{\pm\infty\} = [-\infty, \infty]$. $\widehat{\mathbb{R}}$ is referred to as the *extended real numbers*. We need to sometimes think of $\widehat{\mathbb{R}}$ as a metric space. One way to do this is to find a "continuous", one-to-one function from $\widehat{\mathbb{R}}$ onto a closed interval $[a, b]$ in $\mathbb{R}$, $\phi : \widehat{\mathbb{R}} \to [a, b]$, and define $d(x, y) = |\phi(x) - \phi(y)|$ for all x, y in $\widehat{\mathbb{R}}$. One such function is $\phi(x) = \arctan x$, which maps $[-\infty, \infty]$ monotonically onto $[-\frac{\pi}{2}, \frac{\pi}{2}]$ so that the order structure of $\widehat{\mathbb{R}}$ is also preserved. We are not going to refer specifically to such a metric, but let's emphasize that a sequence in $\mathbb{R}$ converges to a point in $\mathbb{R}$ if and only if it also converges to this same point for any such metric on $\widehat{\mathbb{R}}$. See Exercise 1.

2.3.5. Definition. If X is a set and $\mathcal{A}$ is a σ-algebra of subsets of X, then a function $f : X \to \widehat{\mathbb{R}}$ is $\mathcal{A}$-*measurable*, or just *measurable* when the σ-algebra is understood, provided $f^{-1}(B) \in \mathcal{A}$ for every Borel subset B of $\widehat{\mathbb{R}}$. When X is a metric space and $\mathcal{A}$ is the collection of Borel subsets of X, we call the $\mathcal{A}$-measurable functions *Borel functions*. In general let $\mathcal{M}(X, \mathcal{A})$ denote the collection of all $\mathcal{A}$-measurable functions defined on X. We put $\mathcal{M} = \mathcal{M}(X) = \mathcal{M}(X, \mathcal{A})$ when one or both of X and $\mathcal{A}$ are understood.

Note that for any set E, $\chi_E \in \mathcal{M}(X, \mathcal{A})$ if and only if $E \in \mathcal{A}$ (Exercise 2).

2.3.6. Proposition. *Let X be a set and $\mathcal{A}$ a σ-algebra of subsets of X.*

(a) *If Y is another set and $f : X \to Y$, then $\{B \in 2^Y : f^{-1}(B) \in \mathcal{A}\}$ is a σ-algebra of subsets of Y.*

(b) *If $f : X \to \widehat{\mathbb{R}}$, then the following statements are equivalent.*

  (i) *f is measurable.*

  (ii) *$f^{-1}(G) \in \mathcal{A}$ whenever G is open in $\widehat{\mathbb{R}}$.*

  (iii) *For every a in $\widehat{\mathbb{R}}$, $\{x \in X : f(x) < a\} \in \mathcal{A}$.*

(iv) *For every a in $\widehat{\mathbb{R}}$, $\{x \in X : f(x) > a\} \in \mathcal{A}$.*

(v) *For every a in $\widehat{\mathbb{R}}$, $\{x \in X : f(x) \leq a\} \in \mathcal{A}$.*

(vi) *For every a in $\widehat{\mathbb{R}}$, $\{x \in X : f(x) \geq a\} \in \mathcal{A}$.*

Proof. The proof of (a) is an easy consequence of the fact that taking the inverse image under a function treats unions and intersections of sets very agreeably. We prove (b). In light of part (a) it follows that (i) and (ii) are equivalent. Since $[-\infty, a)$ is an open subset of $\widehat{\mathbb{R}}$, it is immediate that (ii) implies (iii).

(iii) implies (v): $[-\infty, a] = \bigcap_{n=1}^{\infty} [-\infty, a + \frac{1}{n})$.

(v) implies (iv): $(a, \infty] = \widehat{\mathbb{R}} \backslash [-\infty, a]$.

(iv) implies (vi): $[a, \infty] = \bigcap_{n=1}^{\infty} (a - \frac{1}{n}, \infty]$.

(vi) implies (iii): $[-\infty, a) = \widehat{\mathbb{R}} \backslash [a, \infty]$.

It remains to prove that conditions (iii) through (vi) imply (ii). If $-\infty \leq a < b \leq \infty$, then $(a, b) = (a, \infty] \cap [-\infty, b)$, so we have that $f^{-1}((a, b)) \in \mathcal{A}$. But every open subset of $\widehat{\mathbb{R}}$ is the union of a countable number of such open intervals, though we have to worry a bit more when the open set contains $\pm\infty$; so we get $f^{-1}(G) \in \mathcal{A}$ for every open set G in $\widehat{\mathbb{R}}$. ∎

We will frequently use the preceding proposition without reference.

2.3.7. Proposition. *Let X and Y be metric spaces and let $f : X \to Y$ be a continuous function. For every Borel subset A of Y, $f^{-1}(A)$ is a Borel subset of X. In particular, every continuous function $f : X \to \widehat{\mathbb{R}}$ is a Borel function.*

Proof. Let $\mathcal{B}$ denote the Borel subsets of X. First note that if U is an open subset of Y, then $f^{-1}(U)$ is open in X and so $f^{-1}(U) \in \mathcal{B}$. By part (a) of the preceding proposition, $\mathcal{A} = \{A \in 2^Y : f^{-1}(A) \in \mathcal{B}\}$ is a σ-algebra. Since this contains the open subsets of Y, it contains all the Borel subsets of Y. ∎

For X a set and $\mathcal{A}$ a σ-algebra of subsets of X, we would like to give $\mathcal{M} = \mathcal{M}(X, \mathcal{A})$ an algebraic structure; that is, we would like to add and multiply the functions in $\mathcal{M}$. Unfortunately we will have to be disappointed in this. Here is the difficulty. If $f, g \in \mathcal{M}$ and both are finite-valued, then there is no problem since we can define $(f + g) : X \to \mathbb{R}$ as usual by $(f + g)(x) = f(x) + g(x)$. It is impossible, however, to give a valid definition of $-\infty + \infty$; any attempt leads to ambiguities and no semblance of the laws of arithmetic would be valid. So if $f(x) = -\infty$ and $g(x) = \infty$, then $(f + g)(x)$ makes no sense. If we assume that the functions take values in $(-\infty, \infty]$ or both take values in $[-\infty, \infty)$, then we can make sense of $f + g$. In the next proposition and later this is what we mean by "take on appropriate values." For products, on the other hand, there is no fundamental difficulty since we can unambiguously define $(-\infty)(\infty) = -\infty$, $\infty\infty = \infty$, and $(-\infty)(-\infty) = \infty$ and produce a consistent arithmetic, though without any cancellation laws.

2.3.8. Proposition. *If $\mathcal{A}$ is a σ-algebra of subsets of X and if f and g in $\mathcal{M}(X, \mathcal{A})$ are appropriately valued* (see the note in the preceding paragraph), *then $f + g$ and fg are $\mathcal{A}$-measurable. Also if $a \in \mathbb{R}$, then $af \in \mathcal{M}(X, \mathcal{A})$.*

Proof. If $a \in \widehat{\mathbb{R}}$, then $\bigcup_{r \in \mathbb{Q}} (\{x : f(x) < r\} \cap \{x : g(x) < a - r\}) = \{x \in X : f(x) + g(x) < a\}$; it follows that $f + g \in \mathcal{M}$. To prove that the product of two functions in $\mathcal{M}$ belongs to $\mathcal{M}$, we first use the identity $fg = \frac{1}{2}[(f + g)^2 - f^2 - g^2]$ to reduce the problem to showing that $f^2 \in \mathcal{M}$ whenever $f \in \mathcal{M}$. For this we have that when $a \geq 0$, $\{x : f(x)^2 > a\} = \{x : f(x) < -\sqrt{a}\} \cap \{x : f(x) > \sqrt{a}\}$. Since $\{x : f(x)^2 > a\} = X$ when $a < 0$, this shows that $f^2 \in \mathcal{M}$. The proof of the last statement in the proposition is routine. ∎

$\mathcal{M}$ has a natural ordering: $f \leq g$ means $f(x) \leq g(x)$ for all x in X, and we can define $f \vee g(x)$ and $f \wedge g(x)$ as usual. (See §1.1.) Let $\mathcal{M}_+(X, \mathcal{A})$ be the non-negative functions in $\mathcal{M}(X, \mathcal{A})$; similarly we define $\mathcal{M}_+(X)$ and $\mathcal{M}_+$.

2.3.9. Proposition. *If $\mathcal{A}$ is a σ-algebra of subsets of X and f and g are in $\mathcal{M}(X, \mathcal{A})$, then $f \vee g$ and $f \wedge g$ are $\mathcal{A}$-measurable.*

Proof. For any a in $\widehat{\mathbb{R}}$, $\{x : f \vee g(x) > a\} = \{x : f(x) > a\} \cup \{x : g(x) > a\}$ and $\{x : f \wedge g(x) > a\} = \{x : f(x) > a\} \cap \{x : g(x) > a\}$. ∎

2.3.10. Corollary. *If $\mathcal{A}$ is a σ-algebra of subsets of X and $f \in \mathcal{M}(X, \mathcal{A})$, then $f_+ = f \wedge 0$ and $f_- = f \vee 0 \in \mathcal{M}_+(X, \mathcal{A})$.*

We note that when $f \in \mathcal{M}(X, \mathcal{A})$, then, just as we saw with continuous functions, $f = f_+ - f_-$ and $|f| = f_+ + f_-$ so that $|f| \in \mathcal{M}_+(X, \mathcal{A})$. Moreover $f_+ f_- = 0$ if f is finite-valued.

Recall that for any sequence $\{a_n\}$ of extended real numbers,

$$\limsup_n a_n = \lim_n [\sup\{a_n, a_{n+1}, \dots\}]$$

This is guaranteed to exist since $\sup\{a_n, a_{n+1}, \dots\}$ is a decreasing sequence in $\widehat{\mathbb{R}}$, though it is possibly infinite. Similarly we define $\liminf_n a_n$.

2.3.11. Proposition. *If $\{f_n\}$ is a sequence of $\mathcal{A}$-measurable functions, then so are $g_1(x) = \sup_n f_n(x)$, $g_2(x) = \inf_n f_n(x)$, $g_3(x) = \limsup_n f_n(x)$, and $g_4(x) = \liminf_n f_n(x)$.*

Proof. Let $a \in \widehat{\mathbb{R}}$. $\{x : g_1(x) > a\} = \bigcup_{n=1}^{\infty}\{x : f_n(x) > a\}$ and $\{x : g_2(x) \geq a\} = \bigcap_{n=1}^{\infty}\{x : f_n(x) \geq a\}$; hence g_1 and g_2 are measurable. Also note that $g_3^n(x) \equiv \sup\{f_k(x) : k \geq n\}$ defines a decreasing sequence of functions $\{g_3^n\}$, each of which is measurable by what was just proved. Moreover $g_3(x) = \inf_n g_3^n(x)$, and hence g_3 is measurable. Similarly if $g_4^n(x) \equiv \inf\{f_k(x) : k \geq n\}$, we see that $g_4(x) = \sup_n g_4^n(x)$ is measurable. ■

The astute reader may have noticed the lack of an example of a non-measurable function. If the σ-algebra involved is simple, this can easily be furnished. For example, let $\mathcal{A} = \{\emptyset, \mathbb{R}\}$ and consider $\chi_{(0,1)}$. (See Exercise 2.) However when the σ-algebra is robust, finding such a function can be difficult. For example if we start with a positive linear functional on $C(X)$, obtain the measure μ, and let $\mathcal{A}$ be the σ-algebra of sets measurable for μ, finding a non-measurable set or function is a challenge. At the end of §2.2 reference was made to a non-measurable set for Lebesgue measure. The characteristic function of that non-measurable set is a non-measurable function. (See Exercise 2.)

Exercises.

(1) Let ϕ be a one-to-one function of $\widehat{\mathbb{R}}$ onto the closed interval $[a, b]$ in $\mathbb{R}$. (a) Show that ϕ is monotonic and continuous from $\mathbb{R}$ onto the open interval (a, b). (b) Show that $\phi : \mathbb{R} \to (a, b)$ is a homeomorphism.

(2) If X is a set and $\mathcal{A}$ a σ-algebra of subsets, show that for any subset E of X, χ_E is an $\mathcal{A}$-measurable function if and only if $E \in \mathcal{A}$.

(3) If X is a set, $\mathcal{A}$ is a σ-algebra of subsets, and f is an $\mathcal{A}$-measurable function, show that for every real number a, $f^{-1}(\{a\}) \in \mathcal{A}$.

(4) If X is a compact metric space, L is a positive linear functional on $C(X)$, and $(X, \mathcal{A}, \mu)$ the associated Radon measure space, show that a subset E of X is in $\mathcal{A}$ if and only if $E = B \cup Z$, where B is a Borel set and Z is a set of measure 0. (See Exercise 2.2.1.)

(5) If X is a bounded interval and f is a function defined on X that has a derivative at each point of X, show that $f' : X \to \mathbb{R}$ is a Borel function.

2.4. Integration with respect to a measure

We have arrived at a watershed in the development of measure theory. Let's take a look at our progress so far. We started (§1.1) by developing the integral of a continuous function on a compact interval J in $\mathbb{R}$ with respect to a function of bounded variation. In particular, when that function is increasing we showed that the integral defines a positive linear functional on the Banach space $C(J)$. Then we studied in §2.1 the idea of a positive linear functional on $C(X)$ for any compact metric space X and showed there was a Radon measure space associated with each such linear functional (§2.2). Now we will abstract the properties of a Radon measure space, and then we will show how to define the integral of certain functions with respect to such an abstract measure. Moreover using this concept of an integral we'll see how for a positive linear functional $L : C(X) \to \mathbb{R}$ and its associated Radon measure μ, we have that $L(f) = \int f d\mu$ for every f in $C(X)$ (Theorem 2.4.24).

2.4.1. Definition. If X is a set and $\mathcal{A}$ a σ-algebra of subsets of X, then a *measure* is a function $\mu : \mathcal{A} \to [0, \infty]$ satisfying the following:

(a) $\mu(\emptyset) = 0$;

(b) if $\{E_1, E_2, \dots\} \subseteq \mathcal{A}$, then $\mu\left(\bigcup_{n=1}^{\infty} E_n\right) \leq \sum_{n=1}^{\infty} \mu(E_n)$;

(c) if $\{E_1, E_2, \dots\}$ are pairwise disjoint sets in $\mathcal{A}$, then

$$\mu\left(\bigcup_{n=1}^{\infty} E_n\right) = \sum_{n=1}^{\infty} \mu(E_n)$$

The triple $(X, \mathcal{A}, \mu)$ is called a *measure space*. If $\mu(X) < \infty$, then it is called a *finite measure space*.

The property of a measure given in part (c) of the definition is called *countable additivity*. There is a theory of measures that are only assumed to be finitely additive, but we will avoid this. We also want to emphasize that in this definition, unlike what happened in the Radon measure, the abstract measure is allowed to take infinite values.

2.4.2. Example. (a) If X is a compact metric space and L is a positive linear functional on $C(X)$, then the Radon measure space (2.2.9) associated with L is a finite measure space.

(b) If X is any set and we define $\mu : 2^X \to [0, \infty]$ by setting $\mu(E)$ equal to the number of points in E when the set is finite and letting $\mu(E) = \infty$ otherwise, then $(X, 2^X, \mu)$ is a measure space. This measure is called *counting measure*.

(c) For any interval $I = [a, b]$ in $\mathbb{R}$ we have already defined the Lebesgue measure λ (Example 2.2.16); and we have seen that for any open interval (c, d) in I, $\lambda((c, d)) = d - c$ (Example 2.2.2). Now write $\mathbb{R} = \bigcup_{n=-\infty}^{\infty} [n, n+1]$ and let λ_n be Lebesgue measure on $[n, n + 1]$ as defined in (2.2.16). Let X be a subset of $\mathbb{R}$ such that for every n in $\mathbb{Z}$, $X \cap [n, n + 1]$ is Lebesgue measurable; that is, it is measurable with respect to λ_n. Let $\mathcal{A}$ be the collection of all subsets E of X such that for every n in $\mathbb{Z}$, $E \cap [n, n+1]$ is Lebesgue measurable. It is easy to check that $\mathcal{A}$ is a σ-algebra of subsets of

X. For each E in $\mathcal{A}$ let $\lambda(E) = \sum_{n=-\infty}^{\infty} \lambda_n(E \cap [n, n+1])$. It follows that $(X, \mathcal{A}, \lambda)$ is a measure space. $\lambda = \lambda_X$ is called *Lebesgue measure* on X. It is left to the reader to verify that if E is any interval contained in X, then its Lebesgue measure is the length of E. It is clear that Lebesgue measure on $\mathbb{R}$ assumes the value $+\infty$.

For the remainder of this section we will always assume that $(X, \mathcal{A}, \mu)$ is a measure space. As we progress with this abstract concept we will want to return to the setting of a Radon measure space to see what the results we derive say for that situation. It is important that we maintain this generality, however, as there are many useful and important measure spaces that are not Radon. Lebesgue measure on Euclidean space, which is an infinite version of a Radon measure space, is one such example. But there are also examples in probability theory. The reader should be aware that later we will go a bit further and introduce the concept of measures that are allowed to assume negative values (§2.6).

2.4.3. Proposition. *Let $(X, \mathcal{A}, \mu)$ be a measure space.*

(a) *If $E \in \mathcal{A}$, $\mathcal{A}_E = \{E \cap F : F \in \mathcal{A}\}$, and $\mu_E(F) = \mu(E \cap F)$ for every F in $\mathcal{A}_E$, then $(E, \mathcal{A}_E, \mu_E)$ is a measure space.*

(b) *If $t \in [0, \infty)$, then $(X, \mathcal{A}, t\mu)$ is a measure space.*

(c) *If $\{E_n\}$ is a sequence in $\mathcal{A}$ such that $E_n \subseteq E_{n+1}$ for all n and $E = \bigcup_{n=1}^{\infty} E_n$, then $\mu(E_n) \to \mu(E)$.*

(d) *If $\{F_n\}$ is a decreasing sequence of sets in $\mathcal{A}$ all of which have finite measure and $F = \bigcap_{n=1}^{\infty} F_n$, then $\mu(F_n) \to \mu(F)$.*

Proof. The proofs of (a) and (b) are routine. To prove (c) let $F_1 = E_1$ and $F_n = E_n \backslash E_{n-1}$ for $n \geq 2$. It follows that $\{F_n\}$ is a sequence of pairwise disjoint sets in $\mathcal{A}$ and $\bigcup_{n=1}^{\infty} F_n = E$. Thus $\mu(E) = \sum_{n=1}^{\infty} \mu(F_n)$. But $\sum_{k=1}^{n} \mu(F_k) = \mu(\bigcup_{k=1}^{n} F_k) = \mu(E_n)$, so that $\mu(E_n) \to \mu(E)$. To prove (d) put $A = F_1$ and let $E_n = A \backslash F_n$ so that $\{E_n\}$ is an increasing sequence in $\mathcal{A}$ with union $E = A \backslash F$. By the preceding part and the fact that $\mu(A) < \infty$, $\mu(A) - \mu(F) = \mu(E) = \lim_n \mu(E_n) = \mu(A) - \lim_n \mu(F_n)$. Canceling $\mu(A)$ gives the result. ∎

2.4.4. Example. (a) In part (d) of the preceding proposition we need the requirement that all the sets F_n have finite measure. (Where was it used in the proof?) For example if we let λ be Lebesgue measure on $\mathbb{R}$ and $F_n = [n, \infty)$, then $\bigcap_{n=1}^{\infty} F_n = \emptyset$ while $\lambda(F_n) = \infty$ for all n.

(b) If λ is Lebesgue measure on $\mathbb{R}$ and E is any countable subset of $\mathbb{R}$, then $\lambda(E) = 0$. In fact, observe that since λ is countably additive, we need only show that $\lambda(E) = 0$ when E consists of a single point. But if $E = \{x\}$

and $n \geq 1$, then $\lambda(E) \leq \lambda((x - \frac{1}{n}, x + \frac{1}{n})) = 2/n$ and the result follows from part (d) of the last proposition.

No discussion of Lebesgue measure is complete without a discussion of the Cantor set and the Cantor function. Consider the unit interval $I = [0, 1]$, let $I_{1,1} = (\frac{1}{3}, \frac{2}{3})$, and put $F_1 = I \backslash I_{1,1}$. Note that F_1 is compact and consists of the two disjoint closed intervals $J_{1,1} = [0, \frac{1}{3}], J_{1,2} = [\frac{2}{3}, 1]$; each has length $\lambda(J_{1,j}) = \frac{1}{3}$. Now consider the open intervals that constitute the middle thirds of each $J_{1,j}$: $I_{2,1} = (\frac{1}{9}, \frac{2}{9}), I_{2,2} = (\frac{7}{9}, \frac{8}{9})$. Let

$$F_2 = F_1 \backslash (I_{2,1} \cup I_{2,2})$$

$$= \bigcup_{j=1}^{2^2} J_{2,j}$$

$$= \left[0, \frac{1}{3^2}\right] \cup \left[\frac{2}{3^2}, \frac{3}{3^2}\right] \cup \left[\frac{6}{3^2}, \frac{7}{3^2}\right] \cup \left[\frac{8}{3^2}, \frac{9}{3^2}\right]$$

Continue this process. At the n-th stage we have a closed set

$$F_n = \bigcup_{j=1}^{2^n} J_{n,j}$$

where each $J_{n,j}$ is a closed interval of length $\lambda(J_{n,j}) = 3^{-n}$. Now let $I_{n+1,j}$ be the open interval that constitutes the middle third of $J_{n,j}$. Write $J_{n,j} \backslash I_{n+1,j} = J_{n+1,2j-1} \cup J_{n+1,2j}$. Let

$$F_{n+1} = F_n \backslash \bigcup_{j=1}^{2^n} I_{n+1,j} = \bigcup_{j=1}^{2^{n+1}} J_{n+1,j}$$

and we have that $\lambda(J_{n+1,j}) = 3^{-(n+1)}$. So by induction we have a well defined process for obtaining the sequence of closed sets $\{F_n\}$.

By construction this sequence of closed sets is decreasing; hence

$$C = \bigcap_{n=1}^{\infty} F_n$$

is a non-empty compact set, called the *Cantor middle-third set* or the *Cantor ternary set*. Since F_n has 2^n pairwise disjoint closed intervals each of which has length 3^{-n}, the Lebesgue measure of this set is $\lambda(F_n) = (\frac{2}{3})^n$. Using (2.4.3(d)) we have that $\lambda(C) = \lim_n \lambda(F_n) = 0$. One consequence of this is that it has empty interior. (Why?) Since the only connected subsets of $\mathbb{R}$ are intervals, the fact that F has no interior implies that every component of the Cantor set is a singleton; that is, it is *totally disconnected*.

We are going to need some information about base-3 expansions of numbers in the unit interval. Some may have seen this, but many may not have. The next result gathers this in one place.

2.4.5. Proposition. *Each number x in $[0,1]$ has an expansion of the form*

2.4.6
$$x = \sum_{n=1}^{\infty} \frac{a_n}{3^n}$$

where $a_n = 0, 1, 2$ for all $n \geq 1$.

(a) *If in addition to the expansion for x in (2.4.6) we have another such expansion $x = \sum_{n=1}^{\infty} b_n 3^{-n}$, where the b_n are not all equal to the a_n, then there is an integer m such that $a_n = b_n$ when $n < m$, $|a_m - b_m| = 1$, and for $n > m$, $a_n = 0$ and $b_n = 2$ or vice versa. In fact if $a_m < b_m$, then one of the following two situations hold:*

(a.1) $a_m = 0, b_m = 1$, *and* $a_n = 2, b_n = 0$ *for* $n > m$;

(a.2) $a_m = 1, b_m = 2$, *and* $a_n = 2, b_n = 0$ *for* $n > m$.

(b) *A necessary and sufficient condition that x have two different representations of the form (2.4.6) is that $x = p3^{-m}$ for some integer $p > 0$.*

(c) *If C is the Cantor ternary set, then $x \in C$ if and only if x has an expansion (2.4.6) with $a_n \neq 1$ for every $n \geq 1$. Moreover when $a_n \neq 1$ for every n, the representation (2.4.6) is unique.*

Proof. The proof that every number x in the unit interval has such an expansion proceeds as follows. Locate x in one of the equal thirds of $[0,1]$: $[0, \frac{1}{3}], [\frac{1}{3}, \frac{2}{3}]$, or $[\frac{2}{3}, 1]$. Take $a_1 = 0, 1$, or 2 depending on whether x belongs to the first, second, or third interval. Now divide the subinterval that x belongs to into equal thirds, and let a_2 be $0, 1$, or 2 depending on whether x belongs to the first, second, or third subinterval. Continue. Note how this process introduces an ambiguity in case it coincides with the end point of one of the middle third intervals. We address this in part (a).

(a) Let m be the first positive integer such that $a_m \neq b_m$; so the first part of (a) is automatic. Without loss of generality we can assume that $a_m < b_m$. We have two cases to examine, so first suppose that $a_m = 0$; so $b_m = 1$ or 2. Thus $x = \sum_{n=1}^{m-1} a_n 3^{-n} + \sum_{n=m+1}^{\infty} a_n 3^{-n} \leq \sum_{n=1}^{m-1} a_n 3^{-n} + 2 \sum_{n=m+1}^{\infty} 3^{-n} = \sum_{n=1}^{m-1} a_n 3^{-n} + 3^{-m}$. On the other hand $x = \sum_n b_n 3^{-n} = \sum_{n=1}^{m-1} a_n 3^{-n} + b_m 3^{-m} + \sum_{n=m+1}^{\infty} b_n 3^{-n} \geq \sum_{n=1}^{m-1} a_n 3^{-n} + 3^{-m}$. Therefore we are forced to conclude that we have exactly the situation (a.1). Now suppose $a_m = 1$; so it must be that $b_m = 2$. Hence $x \leq \sum_{n=1}^{m-1} a_n 3^{-n} + 2 \cdot 3^{-m} \leq x$ and we conclude that (a.2) holds.

(b) The necessity that x has this form follows from (a). In fact if m is as in (a) and $a_m < b_m$, then $x = \sum_{n=1}^{m-1} a_n 3^{-n} + b_m 3^{-m}$ and we see that 3^{-m} is the common denominator. The proof of sufficiency is straightforward.

(c) Suppose that $x \in C$. Examine the proof that each x in the unit interval has the expansion (2.4.6). The first part of the construction of C is to delete the middle third interval in $[0, 1]$; this means we can choose $a_1 \neq 1$. Similarly at the n-th step in the construction of C we exclude the middle third so that we can have have $a_n \neq 1$. Similarly for the converse. Since $a_1 \neq 1$, $x \notin I_{1,1}$. Since $a_2 \neq 1$, $x \notin I_{2,1} \cup I_{2,2}$. Continue and we arrive at the conclusion that when the ternary expansion of x contains no $a_n = 1$, $x \in C$. Finally, (a) says that the only way we can have two ternary expansions coincide is when a 1 is present in the expansion. So not having a 1 implies the expression (2.4.6) is unique.   ∎

2.4.7. Example. The Cantor ternary set is perfect, compact, totally disconnected, uncountable, and has Lebesgue measure 0. In fact all these were established prior to the statement of Proposition 2.4.5 except for the fact that C is perfect and uncountable. The definition of a *perfect set* can be found in Exercise 2, where the reader is asked to establish this fact. To prove that C is uncountable we use part (c) of the preceding proposition, whose uniqueness statement establishes a one-to-one correspondence between C and the set $\{0, 2\}^{\mathbb{N}}$ consisting of all sequences $\{a_n\}$ of 0's and 2's. This latter set is uncountable.

By a variation on this process of defining the Cantor set, we can construct a perfect, totally disconnected, compact set having no interior but which has positive measure. The idea for this is that at each stage of the construction instead of deleting the middle third of each interval we delete a smaller size open subinterval. See Exercise 4.

We define the *Cantor function*, $c : C \to [0, 1]$, also referred to as the *Cantor–Lebesgue function*, as follows: for $x = \sum_{n=1}^{\infty} a_n 3^{-n}$ in C with $a_n \neq 1$ for all $n \geq 1$, let

2.4.8
$$c(x) = \frac{1}{2} \sum_{n=1}^{\infty} \frac{a_n}{2^n}$$

First we observe that c is an increasing function on C. In fact, if $x = \sum_{n=1}^{\infty} a_n 3^{-n} < y = \sum_{n=1}^{\infty} b_n 3^{-n}$ with $a_n, b_n \neq 1$ for all n, let m be the first integer where $a_m \neq b_m$. It must be that $a_m = 0$ and $b_m = 2$. In fact if this is not the case, then, since neither is 1, it has to be that $a_m = 2$ and $b_m = 0$.

Hence

$$0 < y - x = -\frac{2}{3^m} + \sum_{n=m+1}^{\infty} \frac{b_n - a_n}{3^n}$$

$$\leq -\frac{2}{3^m} + \sum_{n=m+1}^{\infty} \frac{2}{3^n}$$

$$= -\frac{2}{3^m} + \frac{1}{3^m}$$

a contradiction. Thus

$$c(y) - c(x) = \frac{1}{2}\frac{2}{2^m} + \frac{1}{2}\sum_{n=m+1}^{\infty} \frac{b_n - a_n}{2^n}$$

Again $|b_n - a_n| \leq 2$ for all $n \geq m + 1$, so the absolute value of the infinite sum in this formula is at most 2^{-m}. Hence it must be that $c(y) - c(x) \geq 0$.

When is $c(x) = c(y)$? Maintaining the notation of the last paragraph and examining the argument we see that the only way that $c(x) = c(y)$ is for $x = \sum_{n=1}^{m-1} a_n 3^{-n} + \sum_{n=m+1}^{\infty} 2 \cdot 3^{-n} = \sum_{n=1}^{m-1} a_n 3^{-n} + 3^{-m}$ and $y = \sum_{n=1}^{m-1} a_n 3^{-n} + 2 \cdot 3^{-m}$. We note that these values of x and y are the end points of one of the deleted open intervals in the m-th stage of the construction of C. Therefore we can extend c to be a continuous function

$$c : [0, 1] \to [0, 1]$$

by letting it be constant on each open interval in $[0, 1] \backslash C$. Observe that $c(x) = \frac{1}{2}$ if $x \in \left[\frac{1}{3}, \frac{2}{3}\right]$, $c(x) = \frac{1}{4}$ if $x \in \left[\frac{1}{9}, \frac{2}{9}\right]$, $c(x) = \frac{3}{4}$ if $x \in \left[\frac{7}{9}, \frac{8}{9}\right]$. We will also refer to this extended function as the *Cantor function*. An illustration of the graph of the Cantor function can be found at

http://www.math.uri.edu/~bkaskosz/flashmo/nfcantor.html

We gather together some properties of the Cantor function.

2.4.9. Example. (a) The Cantor function $c : [0, 1] \to [0, 1]$ is a continuous, increasing function that maps C, a set of Lebesgue measure 0, onto $[0, 1]$. We have already seen that c is increasing. Since each number in the unit interval has a binary expansion $\sum_{n=1}^{\infty} b_n 2^{-n}$, where $b_n = 0, 1$, it easily follows that $c(C) = [0, 1]$. Now since c is increasing on the entire interval, we know from Proposition 1.1.14 that c has only a countable number of jump discontinuities. But since c is surjective onto the unit interval, it can have no jumps and must therefore be continuous. (So we have an example showing that the continuous image of a set of measure 0 need not have measure 0.)

(b) It also has the property that it is differentiable off C with $c'(x) = 0$ for all $x \notin C$. Verifying this last statement is straightforward.

Now we return to the general theory. Fix a measure space $(X, \mathcal{A}, \mu)$; we want to find a class of functions f for which we can define the integral of f with respect to μ. In fact this definition of an integral must have the properties we normally associate with an integral, namely linearity: $\int (af + bg) d\mu = a \int f d\mu + b \int g d\mu$ for scalars a, b and functions f, g; and positivity: $\int f d\mu \geq 0$ when $f \geq 0$. Therefore the class of functions we will consider must have analogous properties. We should also have that for any set E in $\mathcal{A}$, $\int \chi_E d\mu = \mu(E)$ and so the considered class of functions will have to contain the characteristic functions. We will achieve our definition in stages and the first stage starts one step beyond characteristic functions.

2.4.10. Definition. A *simple measurable function* is an $\mathcal{A}$-measurable function $u : X \to \mathbb{R}$ that takes on only a finite number of values. Equivalently u is a simple measurable function if there are pairwise disjoint sets $E_1, \ldots, E_n$ in $\mathcal{A}$ and real numbers $a_1, \ldots, a_n$ such that $u = \sum_{k=1}^{n} a_k \chi_{E_k}$. If in addition the sets E_k have finite measure, we say that u is *integrable* and define

$$\int u d\mu = \sum_{k=1}^{n} a_k \mu(E_k)$$

Note that a simple function is $\mathcal{A}$-measurable if and only if $u^{-1}(a) \in \mathcal{A}$ for every $a \in \mathbb{R}$. Let $L_s^1(\mu)$ denote the set of all integrable simple functions. The next proposition is easily verified.

2.4.11. Proposition. $L_s^1(\mu)$ *is a vector space over* $\mathbb{R}$ *when we define addition and scalar multiplication of these functions pointwise. Moreover for* $u, v \in L_s^1(\mu)$ *and* $a, b \in \mathbb{R}$, $\int (au + bv) d\mu = a \int u d\mu + b \int v d\mu$; *and* $\int u d\mu \geq 0$ *when* $u(x) \geq 0$ *for all* x *in* X.

Also observe that if $u = \sum_{k=1}^{n} a_k \chi_{E_k}$, then $|u(x)| = \sum_{k=1}^{n} |a_k| \chi_{E_k}$. Thus when $u \in L_s^1(\mu)$, and $|u| : X \to \mathbb{R}$ is defined by $|u|(x) = |u(x)|$, then $|u| \in L_s^1(\mu)$. Thus we can almost define a norm on $L_s^1(\mu)$ by setting

2.4.12 $$\|u\| = \int |u| d\mu$$

The use of the word "almost" in reference to this norm is due to the fact that it lacks one feature of the definition. The difficulty, however, is more technical than actual. It is easy to show that the triangle inequality is satisfied and $\|au\| = |a| \|u\|$, but there can be non-zero functions u for which $\|u\| = 0$. For example, if $E \in \mathcal{A}$ with $E \neq \emptyset$ and $\mu(E) = 0$, then $\|\chi_E\| = 0$ but $\chi_E \neq 0$. We overcome this difficulty by identifying two functions that agree off a set of measure zero. That is, two functions u, v in $L_s^1(\mu)$ are identified if $\mu(\{x : u(x) \neq v(x)\}) = 0$; equivalently if $\|u - v\| = 0$. Once we make this identification, (2.4.12) does indeed define a norm on $L_s^1(\mu)$.

We formalize the preceding discussion not just for the simple functions but for all measurable functions.

2.4.13. Definition. Two $\mathcal{A}$-measurable functions f and g on X agree μ-*almost everywhere* if $\mu(\{x : f(x) \neq g(x)\}) = 0$. In symbols we write $f = g$ a.e. $[\mu]$.

Technically what we are doing is looking at the subspace $\mathcal{N}$ of the vector space $L_s^1(\mu)$ defined by $\mathcal{N} = \{u \in L_s^1(\mu) : \|u\| = 0\}$ and considering the quotient vector space $L_s^1(\mu)/\mathcal{N}$. This is precisely what is meant by saying that we identify u and v when $u - v \in \mathcal{N}$. Then we define a norm on this quotient space by $\|u + \mathcal{N}\| = \|u\|$. We have to verify that this is well defined (it is) and that on the quotient space this is indeed a norm (it is). This is getting too complicated! If we stay with such an approach, it would mean that every time we want to prove something about the quotient space we would have to check that it is independent of the representative we choose from the coset $u + \mathcal{N}$. This becomes cumbersome. In fact while technically precise, it's a pain in the neck. Hence we adopt the less technical approach mentioned above, which is precisely what is happening in the more formal approach of taking quotients but without the paraphernalia. The reader should be aware that this approach will be used several times as we progress. In general if there is some property P associated with the points of X, we will say that P holds a.e. $[\mu]$ if $\mu(\{x : P(x) \text{ does not hold}\}) = 0$. As a first such instance, we introduce a natural ordering on $L_s^1(\mu)$: $u \leq v$ means $u(x) \leq v(x)$ a.e. $[\mu]$.

Now that we have a normed space $L_s^1(\mu)$, we want to complete it as in Theorem 1.3.8. Of course it is not enough to simply take the abstract completion of $L_s^1(\mu)$. We need to find a concrete representation of this completion. We will accomplish this by showing the completion to consist of a collection of measurable functions f for which we can make sense of the expression $\int f d\mu$. That is the future. We begin by obtaining some functions that naturally belong to the completion. Let's start by making a small observation that follows immediately from Proposition 2.4.3.

2.4.14. Lemma. *If $u \in L_s^1(\mu)$ with $u \geq 0$ and $\nu(E) = \int \chi_E u d\mu$, then ν is also a measure on $(X, \mathcal{A})$.*

Proof. If $u(x) = \sum_{k=1}^n a_k \chi_{E_k}$, where the sets $E_1, \ldots, E_n$ are pairwise disjoint and the constants $a_1, \ldots, a_n$ are all positive, then $\chi_E u$ is given by $\sum_{k=1}^n a_k \chi_{E \cap E_k}$. Thus $\nu(E) = \sum_{k=1}^n a_k \mu_{E_k}(E)$. By Proposition 2.4.3 each $a_k \mu_{E_k}$ is a measure and clearly the sum of measures is a measure. ∎

Recall that $\mathcal{M}_+$ denotes all the non-negative measurable functions. Let $L_s^1(\mu)_+$ denote the non-negative elements in $L_s^1(\mu)$.

2.4.15. Definition. If f is a function in $\mathcal{M}_+$, define

$$\int f d\mu = \sup \left\{ \int u d\mu : u \in L^1_s(\mu)_+ \text{ and } u \leq f \text{ a.e. } [\mu] \right\}$$

$$= \sup \left\{ \int u d\mu : u \in L^1_s(\mu)_+ \text{ and } u \leq f \right\}$$

If for such a function f we have that $\int f d\mu < \infty$, say that f is *integrable*.

Establishing the equality in the definition is left as Exercise 5. It is easy to see that if $f, g \in \mathcal{M}_+$ and $f \leq g$ a.e. $[\mu]$, then $\int f d\mu \leq \int g d\mu$.

2.4.16. Theorem (Monotone Convergence Theorem). *If $\{f_n\}$ is a sequence in $\mathcal{M}_+$ such that $f_n(x) \leq f_{n+1}(x)$ a.e. $[\mu]$ for all n, then $f(x) = \lim_n f_n(x)$ exists a.e. $[\mu]$, this limit defines a function in $\mathcal{M}_+$, and $\int f_n d\mu \to \int f d\mu$.*

Proof. Let $E_n = \{x : f_n(x) > f_{n+1}(x)\}$. By hypothesis $\mu(E_n) = 0$; thus if $E = \bigcup_{n=1}^{\infty} E_n$, then $\mu(E) \leq \sum_{n=1}^{\infty} \mu(E_n) = 0$. Redefine each f_n by changing its values at points of E to be 0; note that it remains true that $f_n \in \mathcal{M}_+$. For these new functions we have $f_n(x) \leq f_{n+1}(x)$ for all x. Similarly by redefining the functions again on another set of measure 0, we can assume that $f(x) = \lim_n f_n(x)$ exists for all x in X. (**Reader be aware**! Without being explicit about it, we will use this type of argument several times in the future to phrase a result with a hypothesis involving statements a.e. $[\mu]$ but prove it under the assumption that the conditions hold everywhere.)

The fact that $f \in \mathcal{M}_+$ is immediate from (2.3.11). It is also transparent from the definition that $\{\int f_n d\mu\}$ is an increasing sequence of extended real numbers, so that $\alpha = \lim_n \int f_n d\mu$ exists, though it may be that $\alpha = \infty$. Since $f_n \leq f$ for each n, we have that $\alpha \leq \int f d\mu$. We must therefore show that $\alpha \geq \int f d\mu$. Now let $u \in L^1_s(\mu)_+$ such that $u \leq f$ and let $0 < s < 1$. Put $F_n = \{x : su(x) \leq f_n(x)\}$. It follows that each $F_n \in \mathcal{A}$. Since the sequence of functions $\{f_n\}$ is increasing, $F_n \subseteq F_{n+1}$ for all n; since $f_n(x) \to f(x)$ for all x, $\bigcup_{n=1}^{\infty} F_n = X$. On the other hand $\int f_n d\mu \geq \int_{F_n} f_n d\mu \geq s \int_{F_n} u d\mu$. Now $A \mapsto s \int_A u d\mu$ is a measure, so Proposition 2.4.3 implies $s \int_{F_n} u d\mu \to s \int u d\mu$. Therefore we have that $\alpha \geq s \int u d\mu$ for all $0 < s < 1$. Letting $s \to 1$, we have that $\alpha \geq \int u d\mu$ for all non-negative u in $L^1_s(\mu)$ with $u \leq f$. Thus, by definition, $\alpha \geq \int f d\mu$. ∎

Denote the completion of $L^1_s(\mu)$ by

$$L^1(\mu)$$

We now start the process of representing $L^1(\mu)$ as a space of functions.

2.4.17. Proposition. *If $f \in \mathcal{M}_+$, then there is a sequence $\{u_n\}$ of non-negative measurable simple functions such that for every x in X, $\{u_n(x)\}$*

is increasing and converges to $f(x)$. If in addition f is integrable, then the sequence $\{u_n\}$ can be chosen in $L_s^1(\mu)_+$.

Proof. The idea is simple for the proof of the existence of the sequence $\{u_n\}$. For each integer n, divide the interval $[0, n]$ into 2^n subintervals and look at the set of points in X where f takes values in these subintervals. To be specific, for any $n \geq 1$ and for $1 \leq k \leq n2^n$ let $E_{nk} = \{x : \frac{k-1}{2^n} \leq f(x) < \frac{k}{2^n}\}$; put

$$u_n = \sum_{k=1}^{n2^n} \frac{k-1}{2^n} \chi_{E_{nk}}$$

Observe that $E_{nk} = E_{n+1,2k-1} \cup E_{n+1,2k}$; using this it follows that $u_n(x) \leq u_{n+1}(x)$ for all x. It is left to the reader to show that $u_n(x) \to f(x)$ for every x. Also note that if f is integrable, then $\mu(E_{nk}) < \infty$ for every choice of n and k (see Exercise 6), so that $u_n \in L_s^1(\mu)_+$. ∎

2.4.18. Proposition. *Let $f, g \in \mathcal{M}_+$.*

(a) *$\int (f+g)d\mu = \int f d\mu + \int g d\mu$ and $\int (af)d\mu = a \int f d\mu$.*

(b) *If $f \geq g$ and f and g are integrable, then $\int (f-g)d\mu = \int f d\mu - \int g d\mu$.*

(c) *Each integrable function in $\mathcal{M}_+$ corresponds to an element of $L^1(\mu)$.*

Proof. (a) By the preceding proposition there are increasing sequences $\{u_n\}$ and $\{v_n\}$ in $L_s^1(\mu)_+$ such that for all x in X, $f(x) = \lim_n u_n(x), g(x) = \lim_n v_n(x)$. Thus $(f+g)(x) = \lim_n[u_n(x) + v_n(x)]$ on X. By the MCT and Proposition 2.4.11, $\int (f+g)d\mu = \lim_n \int (u_n + v_n)d\mu = \lim_n \int u_n d\mu + \lim_n \int v_n d\mu = \int f d\mu + \int g d\mu$. The proof of the other statement is similar.

(b) $f = (f-g) + g$, so (a) implies $\int f d\mu = \int (f-g)d\mu + \int g d\mu$.

(c) Use Proposition 2.4.17 to find a sequence $\{u_n\}$ in $L_s^1(\mu)_+$ that is increasing and converges to f a.e. $[\mu]$.

Claim. $\{u_n\}$ *is a Cauchy sequence in* $L_s^1(\mu)$.

We know that $\int u_n d\mu \nearrow \int f d\mu$ by the MCT. Let $\epsilon > 0$ and choose N such that for $n \geq N$, $0 \leq \int f d\mu - \int u_n d\mu < \epsilon/2$. If $m \geq n \geq N$, then $\|u_m - u_n\| = \int (u_m - u_n)d\mu = \int u_m d\mu - \int u_n d\mu = \int u_m d\mu - \int f d\mu + \int f d\mu - \int u_n d\mu \leq |\int u_m d\mu - \int f d\mu| + |\int f d\mu - \int u_n d\mu| < \epsilon$. That is, $\{u_n\}$ is a Cauchy sequence in $L^1(\mu)$ and we have established the claim.

Since $u_n \leq f$, part (b) implies that $\|f - u_n\| = \int (f - u_n)d\mu = \int f d\mu - \int u_n d\mu \to 0$. Suppose $\{v_n\}$ is another Cauchy sequence in $L_s^1(\mu)$ that is equivalent to $\{u_n\}$; hence $0 = \lim_n \|u_n - v_n\| = \int |u_n - v_n|d\mu$. Thus $\|v_n - f\| \leq \|v_n - u_n\| + \|u_n - f\| \to 0$. Hence $f \in L^1(\mu)$. ∎

2.4.19. Corollary. *If $\{f_n\}$ is a sequence in $\mathcal{M}_+$, then $\sum_{n=1}^{\infty} f_n \in \mathcal{M}_+$ and $\int \left(\sum_{n=1}^{\infty} f_n\right) d\mu = \sum_{n=1}^{\infty} \int f_n d\mu$.*

Proof. In light of the Monotone Convergence Theorem, it suffices to prove the conclusion for finite sums, since the partial sums of an infinite series of non-negative functions constitute an increasing sequence of non-negative functions in $\mathcal{M}$. But the proposition gives the proof for two such functions and the general case follows by induction. ∎

Now assume that $f \in \mathcal{M}$ and put $f_+ = f \vee 0$ and $f_- = -f \wedge 0$. We know that $f_+, f_- \in \mathcal{M}_+$ and that $f = f_+ - f_-$. Refer to this decomposition as the decomposition of f into its *positive and negative parts*. As usual we have that $|f| = f_+ + f_-$.

2.4.20. Definition. A function $f : X \to \widehat{\mathbb{R}}$ is *integrable* if it is measurable and both its positive and negative parts, f_+ and f_-, are integrable. In this case we define

$$\int f d\mu = \int f_+ d\mu - \int f_- d\mu$$

It follows immediately that a measurable function f is integrable if and only if $|f|$ is integrable. This will be used repeatedly. As a consequence of Proposition 2.4.18 we have the following.

2.4.21. Corollary. *If f is an integrable function, then f corresponds to an element of $L^1(\mu)$. Moreover $\|f\| = \int |f| d\mu = \int f_+ d\mu + \int f_- d\mu$.*

We will see below that the entirety of $L^1(\mu)$ consists of the integrable functions (Theorem 2.5.4). Perhaps it is more accurate to say that elements of $L^1(\mu)$ are equivalence classes of integrable functions, where two integrable functions are identified if they agree a.e. $[\mu]$. But to do this we first need to prove in the next section some basic results on convergence of sequences and series of integrable functions.

The reader was probably expecting at least part of the next result.

2.4.22. Proposition. (a) *The set of integrable functions is a vector space over $\mathbb{R}$. Moreover if f and g are integrable and $a, b \in \mathbb{R}$, then $\int (af + bg) d\mu = a \int f d\mu + b \int g d\mu$.*

(b) *If f and g are integrable functions and $f \leq g$, then $\int f d\mu \leq \int g d\mu$ and $\left| \int f d\mu \right| \leq \int |f| d\mu$.*

(c) *If f is an integrable function and $\phi : X \to \mathbb{R}$ is a measurable function such that there is a constant M with $|\phi(x)| \leq M$ a.e. $[\mu]$, then ϕf is integrable and $\int |\phi f| d\mu \leq M \int |f| d\mu$.*

(d) *If f is an integrable function, then $\mu \left(\{x : |f(x)| = \infty\}\right) = 0$.*

Proof. (a) The fact that the integrable functions form a vector space over $\mathbb{R}$ is rather straightforward: $|af + bg| \leq |a||f| + |b||g|$, so an application of Proposition 2.4.18 shows that $\int |af + bg| d\mu \leq |a| \int |f| d\mu + |b| \int |g| d\mu$. Thus $af + bg$ is integrable.

Now to show that the integral is linear. First we show that if g, h are two non-negative integrable functions and $f = g - h$, then f is integrable and $\int f d\mu = \int g d\mu - \int h d\mu$. Write $E_+ = \{x : f(x) \geq 0\}$ and $E_- = \{x : f(x) < 0\}$. So $f_+ = f\chi_{E_+} = g\chi_{E_+} - h\chi_{E_+}$ and $f_- = -f\chi_{E_-} = h\chi_{E_-} - g\chi_{E_-}$. Thus $g\chi_{E_+} = f_+ + h\chi_{E_+}$, $h\chi_{E_-} = f_- + g\chi_{E_-}$, and each of these functions is positive. By Proposition 2.4.18, $\int g\chi_{E_+} d\mu = \int f_+ d\mu + \int h\chi_{E_+} d\mu$ and $\int h\chi_{E_-} d\mu = \int f_- d\mu + \int g\chi_{E_-} d\mu$. Thus

$$\int f d\mu = \int f_+ d\mu - \int f_- d\mu$$

$$= \left(\int g\chi_{E_+} d\mu - \int h\chi_{E_+} d\mu \right) - \left(\int h\chi_{E_-} d\mu - \int g\chi_{E_-} d\mu \right)$$

$$= \left(\int g\chi_{E_+} d\mu + \int g\chi_{E_-} d\mu \right) - \left(\int h\chi_{E_+} d\mu + \int h\chi_{E_-} d\mu \right)$$

$$= \int g d\mu - \int h d\mu$$

From this it is easy to see that for any integrable functions f, g, $\int (f + g) d\mu = \int [(f_+ - f_-) + (g_+ - g_-)] d\mu = \int [(f_+ + g_+) - (f_- + g_-)] d\mu = \int (f_+ + g_+) d\mu - \int (f_- + g_-) d\mu = \int f_+ d\mu + \int g_+ d\mu - \int f_- d\mu - \int g_- d\mu = \int f d\mu + \int g d\mu$. The proof where we interject a scalar is routine.

(b) Since $f \leq g$, $g - f \geq 0$ and so $0 \leq \int (g - f) d\mu = \int g d\mu - \int f d\mu$. Next observe that $-|f| \leq f \leq |f|$, so that $-\int |f| d\mu \leq |\int f d\mu| \leq \int |f| d\mu$.

(c) We know that ϕf is measurable, but also $-M|f| \leq |\phi f| \leq M|f|$, so part (c) follows from (b).

(d) If $E = \{x : |f(x)| = \infty\}$, then for every integer n, $|f| \geq n\chi_E$, so that $\infty > \int |f| d\mu \geq \int n\chi_E d\mu = n\mu(E)$. Since n was arbitrary, the only way this can happen is if $\mu(E) = 0$. ∎

Note that part (d) of this last result says that when we deal with an integrable function, we can always assume that it is finite-valued everywhere since it is equivalent to such a function. Also note that part (c) implies that if the measure space is finite, $f \in \mathcal{M}$, and $|f(x)| \leq M$ for all x, then $|\int f d\mu| \leq M\mu(X)$ and so f is integrable.

The next theorem contains the purpose of associating the Radon measure with a positive linear functional on $C(X)$, but first we establish a lemma.

2.4.23. Lemma. *Let X be a compact metric space, $L : C(X) \to \mathbb{R}$ a positive linear functional, and μ the Radon measure associated with L. If K is compact, G is open, $K \subseteq G$, and $f \in C(X)$ such that $\chi_K \leq f \leq \chi_G$, then $\mu(K) \leq L(f) \leq \mu(G)$.*

Proof. The fact that $L(f) \leq \mu(G)$ is direct from the definition of the outer measure of an open set (2.2.1). The fact that $\chi_K \leq f \leq \chi_G \leq 1$, implies that $0 \leq 1 - f \leq \chi_{X \setminus K}$. So by the first part, $\mu(X) - \mu(K) \geq L(1 - f) = L(1) - L(f)$. Since $L(1) = \mu(X)$, we are done. ∎

2.4.24. Theorem. *If X is a compact metric space, $L : C(X) \to \mathbb{R}$ is a positive linear functional, and μ is the Radon measure associated with L, then every f in $C(X)$ is integrable with respect to μ and*

$$L(f) = \int f d\mu$$

Proof. Without loss of generality we can assume that $L(1) = 1 = \mu(X)$. Since continuous functions are Borel functions, they are measurable. Each f in $C(X)$ is also bounded and, since μ is a finite measure, f must therefore be integrable. Let $f \in C(X)$ and let $\epsilon > 0$; without loss of generality we can assume that $f \geq 0$. Since f is uniformly continuous, there is a $\delta > 0$ such that $d(x, y) < \delta$ implies $|f(x) - f(y)| < \epsilon/4$. Let $B_1, \ldots, B_n$ be balls of radius less that $\delta/2$ that cover X. Construct a Borel partition $\{A_1, \ldots, A_n\}$ of X such that $A_j \subseteq B_j$ for $1 \leq j \leq n$. (Details?) For $1 \leq j \leq n$ let $K_j \subseteq A_j \subseteq G'_j$, where K_j is compact, G'_j is open and contained in the ball B_j, and $\mu(G'_j \setminus K_j) < \epsilon/4$. Since $\bigcup_{i \neq j} K_i \subseteq \bigcup_{i \neq j} A_i = X \setminus A_j$, we can replace each G'_j by $G_j = G'_j \cap \left(X \setminus \bigcup_{i \neq j} K_i \right)$ to get that $K_i \cap G_j = \emptyset$ when $i \neq j$ but still $A_j \subseteq G_j \subseteq B_j$. Thus $\{G_1, \ldots, G_n\}$ is an open cover of X; let $\{\phi_1, \ldots, \phi_n\}$ be a partition of unity on X subordinate to this cover (1.2.11). Pick a point z_j in each K_j and put $a_j = f(z_j)$; we consider the continuous function $\sum_{j=1}^n a_j \phi_j$ and compare it to f.

For any x in X,

$$\left| f(x) - \sum_{j=1}^n a_j \phi_j(x) \right| = \left| \sum_{j=1}^n [f(x) - a_j] \phi_j(x) \right| \leq \sum_{j=1}^n |f(x) - a_j| \phi_j(x)$$

If $\phi_j(x) \neq 0$, then $x \in G_j \subseteq B_j$, so $d(x, z_j) < \delta = \operatorname{diam}(B_j)$; hence $|f(x) - a_j| < \epsilon/4$. Therefore $\|f - \sum_{j=1}^n a_j \phi_j\| < \epsilon/4$ and so

2.4.25 $$-\frac{\epsilon}{4} + \sum_j a_j L(\phi_j) < L(f) < \frac{\epsilon}{4} + \sum_j a_j L(\phi_j)$$

Note that since $K_i \cap G_j = \emptyset$ for $i \neq j$, we have that when $x \in K_j$ and $i \neq j$, $\phi_i(x) = 0$; hence $\phi_j(x) = 1$. Thus $\chi_{K_j} \leq \phi_j \leq \chi_{G_j}$ for $1 \leq j \leq n$. By the preceding lemma, $\mu(K_j) \le L(\phi_j) \le \mu(G_j)$. Also note that when $x \in A_j$, $d(x, z_j) < \delta$ so that $|f(x) - a_j| < \epsilon/4$. Thus

$$\sup\left\{\left|f(x) - \sum_j a_j \chi_{A_j}(x)\right| : x \in X\right\} < \epsilon/4$$

and so $\left|\int f d\mu - \sum_{j=1}^n a_j \mu(A_j)\right| = \left|\int \left(f - \sum_{j=1}^n a_j \chi_{A_j}\right) d\mu\right| < \epsilon/4$. This leads to

$$\int f d\mu < \epsilon/4 + \sum_{j=1}^n a_j \mu(A_j)$$

$$< \epsilon/2 + \sum_{j=1}^n a_j \mu(K_j)$$

$$\le \epsilon/2 + \sum_{j=1}^n a_j L(\phi_j)$$

Similarly

$$\int f d\mu > -\epsilon/4 + \sum_{j=1}^n a_j \mu(A_j)$$

$$> -\epsilon/2 + \sum_{j=1}^n a_j \mu(G_j)$$

$$\ge -\epsilon/2 + \sum_{j=1}^n a_j L(\phi_j)$$

If $\alpha = \sum_{j=1}^n a_j L(\phi_j)$, then combined with (2.4.25) these inequalities say that both $L(f)$ and $\int f d\mu$ belong to the open interval $(\alpha - \epsilon/2, \alpha + \epsilon/2)$. Thus $|L(f) - \int f d\mu| < \epsilon$. Since ϵ was arbitrary, we have that $L(f) = \int f d\mu$. ∎

We conclude this section with another result concerning Radon measures.

2.4.26. Theorem. *If X is a compact metric space and μ is a positive Radon measure on X, then $C(X)$ is dense in $L^1(\mu)$.*

Proof. Let $\mathcal{L}$ denote the closure of $C(X)$ in $L^1(\mu)$. Since we know, by definition, that $L_s^1(\mu)$ is dense in $L^1(\mu)$ and since $\mathcal{L}$ must be a linear subspace of $L^1(\mu)$, it suffices to show that $\chi_E \in \mathcal{L}$ for every measurable set E. If $\epsilon > 0$ the definition of measurable set implies there is a compact set K contained in E such that $\mu(E \backslash K) < \epsilon$. Since $\|\chi_E - \chi_K\| = \mu(E \backslash K) < \epsilon$, it suffices to show that $\chi_K \in \mathcal{L}$ for every compact set K. But in this case there is a sequence of continuous functions $\{f_n\}$ such that $\chi_K \leq f_n \leq 1$ and $f_n(x) \searrow \chi_K(x)$ for all x in X. Thus $\|\chi_K - f_n\| = \int (f_n - \chi_K) d\mu$; but $\int (\chi_K - f_n) d\mu \to 0$ by the MCT. ∎

Exercises. For these exercises X is a given set, $\mathcal{A}$ is a σ-algebra of subsets of X, and μ is a measure defined on $\mathcal{A}$.

(1) Verify the statements made in Example 2.4.2(c).

(2) Prove that the Cantor set is a perfect set. That is, if C is the Cantor set, $x \in C$, and $\epsilon > 0$, then there is a point y in C with $y \neq x$ and $|y - x| < \epsilon$.

(3) If $x \in [0, 1]$ has the ternary expansion (2.4.6) and it is impossible to write this without having some $a_n = 1$, show that if m is the first integer with $a_m = 1$, then $c(x) = \frac{1}{2} \sum_{n=1}^{m-1} 2^{-(m+1)}$.

(4) Show that there is a compact subset K of the unit interval that has no interior and has positive Lebesgue measure. (Imitate the construction of the Cantor ternary set but delete intervals less than a third of the length of the intervals in the n-th stage.) Show that K is totally disconnected and perfect (see Exercise 2). How large can you make the Lebesgue measure of K? Can you have $\lambda(K) = 1$?

(5) Prove the equality in Definition 2.4.15.

(6) If f is a positive integrable function and $a, b \in \mathbb{R}$, show that $\mu(\{x : a < f(x) < b\}) < \infty$. Is the word positive needed in this statement?

(7) Let $(X, \mathcal{A}, \mu)$ be a finite measure space and for sets E, F in $\mathcal{A}$ define $E \Delta F = (E \backslash F) \cup (F \backslash E)$. This is called the *symmetric difference* of E and F. Show that $\mu(E \Delta F) = \|\chi_E - \chi_F\|$, where the norm here is in $L^1(\mu)$. If two sets in $\mathcal{A}$ are identified when the measure of their symmetric difference is 0, then $d(E, F) = \mu(E \Delta F)$ defines a metric on $\mathcal{A}$ and with this metric, $(\mathcal{A}, d)$ is complete.

(8) Show that a function $f : X \to [-\infty, \infty]$ is $\mathcal{A}$-measurable if and only if $f^{-1}(G) \in \mathcal{A}$ for every open set G.

(9) Show that $\nu : \mathcal{A} \to [0, \infty]$ is a measure if it is finitely additive, $\nu(\emptyset) = 0$, and has the property that when $\{F_n\}$ is an increasing sequence in $\mathcal{A}$ with union F it follows that $\nu(F_n) \to \nu(F)$.

(10) If f is an integrable function and n is any positive integer, let $E_n - \{x : |f(x)| \geq n\}$. Show that $\lim_{n \to \infty} \mu(E_n) = 0$.

2.5. Convergence theorems

Many basic problems in analysis involve trying to interchange two infinite processes, something that is not always possible. For example, if $\{a_{nm}\}$

is a doubly indexed collection of real numbers, it is not always possible to conclude that $\lim_n \lim_m a_{nm} = \lim_m \lim_n a_{nm}$. (Construct an example where this does not happen.) The definition of an integral is an infinite process – we take the supremum of an infinite collection of numbers. So if we are presented with a sequence of measurable functions $\{f_n\}$, to conclude that $\int \lim f_n d\mu = \lim_n \int f_n d\mu$ requires justification; in fact, it too is not always true. We have already seen one such justification in the Monotone Convergence Theorem. Here we will see others and they will prove of great value.

We continue to assume we have a fixed measure space $(X, \mathcal{A}, \mu)$. The next result is very useful even though its proof is easy. Recall Proposition 2.3.11, where it was shown that the limit inferior of a sequence of measurable functions is also measurable.

2.5.1. Proposition (Fatou's[4] Lemma). *If $\{f_n\}$ is a sequence of positive measurable functions, then*

$$\int [\liminf f_n] d\mu \leq \liminf \int f_n d\mu$$

Proof. Put $F_n(x) = \inf\{f_k(x) : k \geq n\}$. It follows that $F_n \in \mathcal{M}_+$ and $F_n \leq f_k$ for all $k \geq n$; thus $\int F_n d\mu \leq \inf_{k \geq n} \int f_k d\mu$. Also $F_n(x) \nearrow f(x) = \liminf_k f_k(x)$, so the MCT implies that

$$\int f d\mu = \lim_n \int F_n d\mu \leq \lim_n \left[\inf_{k \geq n} \int f_k d\mu \right] = \liminf \int f_n d\mu. \qquad \blacksquare$$

One way the reader might remember the direction of this inequality is that whenever you hear "Fatou", think integral and put the integral sign first. The next result is one of the foundational stones of measure theory.

2.5.2. Theorem (Dominated Convergence Theorem). *If $\{f_n\}$ is a sequence of integrable functions, $f(x) = \lim_n f_n(x)$ exists a.e. $[\mu]$, and there is an integrable function F such that for each $n \geq 1$, $|f_n(x)| \leq F(x)$ a.e. $[\mu]$, then f is integrable and*

$$\int f_n d\mu \rightarrow \int f d\mu$$

---

[4]Pierre Fatou was born in 1878 in Lorient, France, a port city on the southern coast of Brittany. He obtained a degree in mathematics from the École Normale Supérieure in 1901. Because of the difficulty of obtaining a mathematics post, he got a job at the Paris Observatory where he continued to do mathematics. In 1906 he submitted a thesis on integration theory and analytic functions, some of the results of which are still studied today. He received the doctorate the following year. In 1926 he was given the title of "astronomer," in which area he would make several contributions. In particular, he used differential equations to prove results on planetary motion suggested but not substantiated by Gauss. He died in 1929 in Pornichet, France.

Proof. Again we note that by redefining the various functions on a set of measure 0, we may assume that $f(x) = \lim_n f_n(x)$ exists everywhere and $|f_n(x)| \leq F(x)$ for all x in X. (What are the details of this?) Clearly the fact that $|f_n| \leq F$ for all n implies that we have that $|f| \leq F$; since F is integrable, it follows that f is integrable. Also from the hypothesis we have that $f_n + F \geq 0$ for all n. Therefore Fatou's Lemma applies and we get

$$\int F d\mu + \int f d\mu = \int (F + f) d\mu$$
$$= \int \liminf (F + f_n) d\mu$$
$$\leq \liminf \int (F + f_n) d\mu$$
$$= \int F d\mu + \liminf \int f_n d\mu$$

Canceling, which we can do since F is integrable, we get

$$\int f d\mu \leq \liminf \int f_n d\mu$$

Now a fortunate thing is that we also have the $F - f_n \geq 0$, and so replacing each f_n in the preceding string of inequalities by $-f_n$ we get

$$\int F d\mu - \int f d\mu = \int (F - f) d\mu$$
$$\leq \liminf \int (F - f_n) d\mu$$
$$= \int F d\mu + \liminf \left(- \int f_n d\mu \right)$$
$$= \int F d\mu - \limsup \int f_n d\mu$$

Consequently $\limsup \int f_n d\mu \leq \int f d\mu$. Since the limit inferior is smaller than the limit superior, this give that $\int f_n d\mu \to \int f d\mu$. ∎

2.5.3. Example. Let λ denote Lebesgue measure on the unit interval and for $n \geq 2$ let $f_n = n\chi_{[1/n, 2/n]}$. It is easy to see that $f_n(x) \to 0$ for all x in $[0, 1]$, but $\int f_n d\lambda = 1$ for all $n \geq 2$. Thus the domination part of the hypothesis in the Dominated Convergence Theorem is required for the conclusion.

The preceding theorem will be referred to as the DCT. In the literature it is often called the Lebesgue Dominated Convergence Theorem after its discoverer. The power of the DCT is a prime example of the interchange of two infinite processes as mentioned at the start of this section.

Now we can finish off the proof that the elements of $L^1(\mu)$ are all represented as integrable functions.

2.5.4. Theorem. *Every element of $L^1(\mu)$ is represented by an integrable function.*

Proof. Let $\{u_n\}$ be a Cauchy sequence in $L_s^1(\mu)$. Passing to a subsequence if necessary, we can assume that $\|u_n - u_{n-1}\| < 2^{-n}$. Put $F_n(x) = \sum_{k=1}^{n} |u_k(x) - u_{k-1}(x)|$. Clearly $\{F_n\}$ is an increasing sequence of integrable functions and so $F(x) = \lim_n F_n(x) = \sum_{n=1}^{\infty} |u_n(x) - u_{n-1}(x)|$ is measurable. By the MCT, $\int F_n d\mu \to \int F d\mu$. Now $\int F_n d\mu = \sum_{k=1}^{n} \int |u_k - u_{k-1}| d\mu \le \sum_{k=1}^{n} 2^{-k} \le 1$, so F is integrable with $\int F d\mu \le 1$. Also $\|F - F_n\| = \int (F - F_n) d\mu \to 0$.

Now according to Proposition 2.4.22(d) there is a set E having $\mu(E) = 0$ such that $F(x) < \infty$ when $x \notin E$. So if we define $f : X \to \mathbb{R}$ by setting $f(x) = u_1(x) + \sum_{k=2}^{\infty} [u_k(x) - u_{k-1}(x)] = \lim_{n \to \infty} u_n(x)$ when $x \notin E$ and $f(x) = 0$ when $x \in E$, then we have a measurable function. Moreover $|u_n| \le F_n + |u_1| \le F + |u_1|$ for all $n \ge 1$. Therefore the DCT implies f is integrable and $\int u_n d\mu \to \int f d\mu$; in particular we know that f corresponds to an element of $L^1(\mu)$ (2.4.21). Moreover $|f - u_n| \le 2(F + |u_1|)$ so we can again use the DCT to conclude that $\|f - u_n\| = \int |f - u_n| d\mu \to 0$. Therefore f defines the element of $L^1(\mu)$ corresponding to the equivalence class of this Cauchy sequence. ∎

We have realized $L^1(\mu)$ as a rather concrete collection of objects, namely the equivalence classes of integrable functions. In fact the reader from now on should cease to think of $L^1(\mu)$ as the abstract completion of $L_s^1(\mu)$, and start to think of $L^1(\mu)$ solely as the space of functions integrable with respect to μ, where two functions are identified if they agree a.e. $[\mu]$. (In fact, a literature search will reveal that this is the traditional definition of $L^1(\mu)$.) This statement does not mean, however, that we forget about the fact that each integrable function is both the pointwise and norm limit of a sequence of functions from $L_s^1(\mu)$. Indeed this last fact is one of the principal tools we will use to prove results about $L^1(\mu)$ – first prove something about characteristic functions, extend it by linearity to simple functions, then take limits. We'll see this at work shortly in the proof of Proposition 2.5.6 below.

The proof of the preceding theorem can be used to show the following.

2.5.5. Proposition. *If $\{f_n\}$ is a sequence in $L^1(\mu)$ such that $\|f_n - f\| \to 0$, then there is a subsequence $\{f_{n_k}\}$ such that $f_{n_k}(x) \to f(x)$ a.e. $[\mu]$.*

Proof. Note that there is a subsequence $\{f_{n_k}\}$ such that $\|f_{n_k} - f_{n_{k-1}}\| < 2^{-k}$. Now proceed as in the previous proof. ∎

The next result constitutes a type of change of variables result.

2.5.6. Proposition. *If $\phi : X \to Y$ and $\mathcal{B} = \{B \subseteq Y : \phi^{-1}(B) \in \mathcal{A}\}$, then $\mathcal{B}$ is a σ-algebra of subsets of Y; if for each B in $\mathcal{B}$ we let $\nu(B) = \mu(\phi^{-1}(B))$, then $(Y, \mathcal{B}, \nu)$ is a measure space. Moreover whenever f is a ν-integrable function, $f \circ \phi$ is μ-integrable and*

2.5.7
$$\int f \, d\nu = \int f \circ \phi \, d\mu$$

Proof. The fact that $\mathcal{B}$ is a σ-algebra and ν is a measure on $\mathcal{B}$ is routine and left to the reader; we concentrate on establishing the part of the proposition that concerns integrable functions. First let $B \in \mathcal{B}$ with $\nu(B) < \infty$. Observe that $\chi_{\phi^{-1}(B)} = \chi_B \circ \phi$ and that $\int \chi_{\phi^{-1}(B)} d\mu = \mu(\phi^{-1}(B)) = \nu(B) = \int \chi_B \circ \phi \, d\nu < \infty$. That is, (2.5.7) is true when $f = \chi_B$ for a set B of finite measure. It is now routine to establish that the formula holds when $f \in L_s^1(\nu)$. If $f \in L^1(\nu)_+$, then Proposition 2.4.17 implies there is an increasing sequence $\{v_n\}$ in $L_s^1(\nu)_+$ such that $v_n(y) \nearrow f(y)$ for all y in Y; thus $\|f - v_n\| = \int (f - v_n) d\nu \to 0$. It is easy to check that $f \circ \phi$ is $\mathcal{A}$-measurable and $\{v_n \circ \phi\}$ is an increasing sequence in $L_s^1(\mu)_+$ that converges pointwise to $f \circ \phi$. Moreover $\|v_n \circ \phi - v_m \circ \phi\| = \int |v_n \circ \phi - v_m \circ \phi| d\mu = \int |v_n - v_m| d\nu$ since $|v_n - v_m|$ is a simple function. Therefore $\{v_n \circ \phi\}$ is a Cauchy sequence in $L_s^1(\mu)_+$. It follows that $v_n \circ \phi \to f \circ \phi$ in $L^1(\mu)$. When f is an arbitrary function in $L^1(\nu)$, we can write it as the difference of two positive functions in $L^1(\nu)_+$ and apply what was just established. ∎

In most cases where we will use this result we will be presented with a set Y and a σ-algebra of subsets $\mathcal{C}$ and the function ϕ will have the property that $\phi^{-1}(C) \in \mathcal{A}$ for every C in $\mathcal{C}$. That is, $\phi : (X, \mathcal{A}) \to (Y, \mathcal{C})$ is *measurable*. In this case we note that if $\mathcal{B}$ is defined as in the statement of the proposition, then $\mathcal{C} \subseteq \mathcal{B}$. So when we define ν on $\mathcal{B}$ it is defined on $\mathcal{C}$.

2.5.8. Definition. *If f and $\{f_n\}$ are measurable functions, say that $\{f_n\}$ converges in measure to f if whenever $c > 0$,*

$$\lim_n \mu \left(\{x : |f_n(x) - f(x)| \geq c\} \right) = 0$$

In a sense this is a measure-theoretic version of uniform convergence in that if the sequence of functions converges to f uniformly on X, then for every positive constant c, the set $\{x : |f_n(x) - f(x)| \geq c\}$ is empty for sufficiently large values of n. Here we just require that the measure of these sets becomes small.

2.5.9. Theorem. *Let f and $\{f_n\}$ be measurable functions.*

(a) *If f and $\{f_n\}$ belong to $L^1(\mu)$ and $\|f - f_n\| \to 0$, then $f_n \to f$ in measure.*

(b) *If $f_n \to f$ in measure, then there is a subsequence that converges to f a.e. $[\mu]$.*

Proof. (a) If $c > 0$, put $E_n = \{x : |f_n(x) - f(x)| \geq c\}$, and observe that $\|f_n - f\| \geq \int_{E_n} |f_n - f| d\mu \geq c\mu(E_n)$.

(b) From the definition of convergence in measure we can find a subsequence $\{f_{n_k}\}$ such that if $E_k = \{x : |f_{n_j}(x) - f(x)| \geq 2^{-k}$ for all $n_j \geq n_k\}$, then $\mu(E_k) < 2^{-k}$. If $F_k = \bigcup_{j=k}^{\infty} E_j$, then $\mu(F_k) \leq \sum_{j=k}^{\infty} \mu(E_j) \leq (\frac{1}{2})^{k-1}$. Note that if $x \notin F_k$, then for all $j \geq k$, $|f_{n_j}(x) - f(x)| < 2^{-k}$. If we put $F = \bigcap_{k=1}^{\infty} F_k$, then $\mu(F) = 0$; and when $x \notin F$ and $j \geq k$, we have $|f_{n_j}(x) - f(x)| < 2^{-k}$. That is $f_{n_k}(x) \to f(x)$ when $x \notin F$ and so $f_{n_k} \to f$ a.e. $[\mu]$. ∎

2.5.10. Corollary. *If $\{f_n\}$ is a sequence in $L^1(\mu)$ that converges in measure to a measurable function f and if there is an F in $L^1(\mu)$ with $|f_n| \leq F$ for all $n \geq 1$, then f is integrable and $\|f_n - f\| \to 0$.*

Proof. According to the preceding theorem, there is a subsequence $\{f_{n_k}\}$ that converges a.e. $[\mu]$ to f. But then the DCT implies $f \in L^1(\mu)$ and $\|f_{n_k} - f\| \to 0$. Why does the original sequence converge to f? If it does not, then there is an $\epsilon > 0$ such that for every n there is an $m \geq n$ with $\|f_m - f\| \geq \epsilon$. From here it follows that there is a subsequence $\{f_{n_k}\}$ such that $\|f_{n_k} - f\| \geq \epsilon$ for all n_k. But we still have that $\{f_{n_k}\}$ converges to f in measure and so there is a further subsequence $\{f_{n_{k_i}}\}$ that converges to f a.e. $[\mu]$. But then, as above, $\|f_{n_{k_i}} - f\| \to 0$, a contradiction. ∎

One might wonder if convergence in measure implies the original sequence converges almost everywhere. The next example is standard and shows this to not be the case.

2.5.11. Example. There is a sequence of integrable functions that converges in measure to 0 but does not converge a.e. $[\mu]$. In fact the measure is Lebesgue measure on $[0, 1]$ and the functions $\{f_n\}$ are all characteristic functions of intervals. It is rather cumbersome to write down formulas for the functions, but here is their description. Let $f_1 = \chi_{(0,1/2)}, f_2 = \chi_{(1/2,1)}, f_3 = \chi_{(0,1/3)}, f_4 = \chi_{(1/3,2/3)}, f_5 = \chi_{(2/3,1)}, \ldots$. In other words, the functions f_n are characteristic functions of intervals that are getting smaller but at any stage n there is an integer $N \geq n$ such that $\bigcup_{k=n}^{N}\{x \in [0, 1] : f_k(x) = 1\}$ includes all the irrational numbers in $[0, 1]$. Consequently for an irrational number x any tail end of the sequence $\{f_n(x)\}$ contains infinitely many 0's and 1's. Since the irrational numbers have full measure (see Exercise 2.2.3), the sequence of functions cannot converge to 0 off a set of Lebesgue measure zero. On the other hand, $\int |f_n| d\lambda \to 0$, so that it converges in measure.

2.5.12. Theorem (Egorov's[5] Theorem). *If the measure space is finite and $\{f_n\}$ is a sequence of measurable functions that converges to f a.e. $[\mu]$, then for every $\epsilon > 0$ there is a measurable set A such that $\mu(X \backslash A) < \epsilon$ and $f_n \to f$ uniformly on A.*

Proof. Begin by doing one of our customary tricks: by redefining each function on a set of measure 0, we can assume that $f_n(x) \to f(x)$ for all x in X.

Claim. For any pair of positive numbers δ and ϵ, there is a set D in $\mathcal{A}$ and an integer N such that $\mu(X \backslash D) < \epsilon$ and $|f_n(x) - f(x)| < \delta$ for all $n \geq N$ and all x in D.

In fact, put $E_m = \{x : |f_n(x) - f(x)| < \delta$ for all $n \geq m\}$. So the sets $\{E_m\}$ are increasing and $\bigcup_{m=1}^{\infty} E_m = X$. Therefore $\mu(X \backslash E_m) \to 0$ by Proposition 2.4.3(d) and the fact that $\mu(X) < \infty$. Choose N such that $\mu(X \backslash E_n) < \epsilon$ for $n \geq N$ and put $D = E_N$.

The claim falls short of proving Egorov's Theorem since the set D obtained there depends on the number δ as well as ϵ. It is now our task to remove this dependency on δ; we do this with a standard analyst's trick. According to the claim, for every positive integer k there is a set D_k in $\mathcal{A}$ and an integer N_k such that $\mu(X \backslash D_k) < \epsilon/2^k$ and $|f_n(x) - f(x)| < 1/k$ for all x in D_k and $n \geq N_k$. Put $A = \bigcap_{k=1}^{\infty} D_k$. If follows that $\mu(X \backslash A) \leq \sum_{k=1}^{\infty} \mu(X \backslash D_k) < \epsilon$. Moreover for x in A and $n \geq N_k$ we have that $|f_n(x) - f(x)| < 1/k$, so that we get that $f_n \to f$ uniformly on A. ∎

Exercises.

(1) Find a doubly indexed collection of real numbers $\{a_{nm}\}$ such that

$$\lim_n \lim_m a_{nm} \neq \lim_m \lim_n a_{nm}$$

---

[5]Dimitri Egorov (sometimes written Egoroff) was born in 1869 in Moscow and in 1887 he enrolled in Moscow University to study mathematics and physics, obtaining his doctorate in 1901. After a year abroad he returned to Moscow in 1903 as a professor. He made significant contributions to differential geometry and turned to the study of integral equations, during which time the present result came to be. With his first student Lusin (see 4.3.3) he established a group in real analysis. He rose in the hierarchy of the mathematics community and became president of the Moscow Mathematical Society in 1923. He was deeply religious and was distraught by the persecution of clerics under the communists. He began to use his position to shelter dismissed academics and tried to prevent the imposition of Marxist ideology on science. In 1929 this resulted in his being dismissed from his position as director of the Institute for Mechanics and Mathematics and publicly rebuked. Later he was arrested as a "religious sectarian" and thrown in prison. The Moscow Mathematical Society refused to expel him and those who participated in the next mathematics meeting were themselves expelled by an "Initiative group" that took over the society in 1930. In prison Egorov went on a hunger strike and, close to death, was taken to the prison hospital in Kazan. He died there in 1931.

(2) We can think of infinite series as a special case of the integral as follows. Consider the measure space $(\mathbb{N}, 2^{\mathbb{N}}, \mu)$, where, as usual, $2^{\mathbb{N}}$ is the σ-algebra of all subsets of $\mathbb{N}$ and μ is counting measure. Thus an integrable function f for this measure space corresponds to an absolutely summable series. State the DCT in this setting. (Note that conditionally convergent series do not fit into the theory of the integral.)

(3) Carry out the details of the proof of Proposition 2.5.5.

(4) Say that a sequence of measurable functions $\{f_n\}$ converges to f *almost uniformly* if for every $\delta > 0$ there is a measurable set E with $\mu(E) < \delta$ and such that $f_n \to f$ uniformly on $X \backslash E$. (So the conclusion of Egorov's Theorem is that $f_n \to f$ almost uniformly.) (a) Prove that if $f_n \to f$ almost uniformly, then it converges a.e. $[\mu]$ and in measure. (b) Prove that if $f_n \to f$ in measure, then it has a subsequence that converges almost uniformly. (Hint: Examine closely the proof of Theorem 2.5.9(b).)

2.6. Signed measures

In this section we continue our odyssey and extend the concept of a measure to allow it to take on negative values; then we define the integral of a function with respect to this generalization. This is similar to the idea of integrating with respect to a function of bounded variation rather than an increasing function. Indeed, as we'll see, there is more than an analogy here.

2.6.1. Definition. If X is a set and $\mathcal{A}$ is a σ-algebra of subsets of X, a *signed measure* on $(X, \mathcal{A})$ is a function $\mu : \mathcal{A} \to [-\infty, \infty]$ having the following properties:

(a)   there do not coexist sets E and F in $\mathcal{A}$ such that $\mu(E) = \infty$ and $\mu(F) = -\infty$;

(b) $\mu(\emptyset) = 0$;

(c)   if $\{E_n\}$ is a sequence of sets in $\mathcal{A}$ that are pairwise disjoint, then

$$\mu\left(\bigcup_{n=1}^{\infty} E_n\right) = \sum_{n=1}^{\infty} \mu(E_n)$$

Part (a) in this definition says that the function is allowed to take on one of the values ∞ or $-\infty$ but not both. Without such a restriction, condition (c) would be nonsensical since if there were disjoint sets E and F with $\mu(E) = \infty$ and $\mu(F) = -\infty$, it would be impossible to make sense of $\mu(E) + \mu(F)$.

We may sometimes say the phrase "positive measure." This stands for a measure as defined previously or, equivalently, a signed measure that takes its values in $[0, \infty]$. Also to say that a signed measure is finite is to say that it is finite-valued – the range of its values is in $(-\infty, \infty)$. For the remainder of this section X is a set, $\mathcal{A}$ is a σ-algebra of subsets of X, and μ is a signed measure defined on $\mathcal{A}$. Finally, to avoid the chance of confusion, when we discuss a measure space $(X, \mathcal{A}, \mu)$, we mean what we meant in the past – μ is a positive measure.

2.6.2. Example. (a) Every measure is a signed measure.

(b) Any linear combination of finite signed measures on $(X, \mathcal{A})$ is a signed measure.

(c) If μ_1, μ_2 are two positive measures on $(X, \mathcal{A})$ and one of them is finite, then $\mu = \mu_1 - \mu_2$ is a signed measure.

(d) Suppose μ is a signed measure on $(X, \mathcal{A})$ and $Y \in \mathcal{A}$. If we let $\mathcal{A}_Y = \{E \cap Y : E \in \mathcal{A}\} = \{F \in \mathcal{A} : F \subseteq Y\}$ and define $\mu_Y : \mathcal{A}_Y \to [-\infty, \infty]$ by $\mu_Y(F) = \mu(F)$ for all F in $\mathcal{A}_Y$, then μ_Y is a signed measure on $(Y, \mathcal{A}_Y)$.

(e) If $(X, \mathcal{A}, \mu)$ is a measure space and $f \in L^1(\mu)$, then $\nu(E) = \int_E f d\mu = \int f \chi_E d\mu$ is a finite signed measure.

The proof of the next result is similar to the proofs of the corresponding results for measures.

2.6.3. Proposition. *Let μ be a signed measure on $(X, \mathcal{A})$.*

(a) *If $E, F \in \mathcal{A}$ and $F \subseteq E$, then $\mu(E \backslash F) = \mu(E) - \mu(F)$.*

(b) *If $\{E_n\}$ is an increasing sequence in $\mathcal{A}$ with union E, then $\mu(E) = \lim_n \mu(E_n)$.*

(c) *If $\{E_n\}$ is a decreasing sequence in $\mathcal{A}$ with intersection E and $\mu(E_1)$ is finite, then $\mu(E) = \lim_n \mu(E_n)$.*

Perhaps this is a good place for a word of caution. It is not necessarily the case that when $\{E_n\}$ is a sequence of sets in $\mathcal{A}$, $\mu(\bigcup_{n=1}^\infty E_n) \leq \sum_{n=1}^\infty \mu(E_n)$. Indeed if ν is a positive measure and $\mu = -\nu$, this is false unless the sets E_n are pairwise disjoint. Also be aware that for a signed measure it may be that a set has measure 0 while some of its subsets have non-zero measure. (What's an example?)

Now the idea is to show that every signed measure can be obtained from positive measures as in Example 2.6.2(c).

2.6.4. Definition. If μ is a signed measure on $(X, \mathcal{A})$ and $E \in \mathcal{A}$, say that E is a *positive set* for μ if $\mu(A) \geq 0$ for every A in $\mathcal{A}$ that is contained in E. Similarly, say that F is a *negative set* for μ if $\mu(B) \leq 0$ for every set B in $\mathcal{A}$ that is contained in F.

Let's emphasize that for E to be a positive set it does not suffice that $\mu(E) \geq 0$. Let ν be the signed measure defined in Example 2.6.2(e). If $E = \{x \in X : f(x) \geq 0\}$, then E is a positive set for ν and so is $\{x \in X : f(x) > 0\}$.

2.6.5. Proposition. *Let μ be a signed measure on $(X, \mathcal{A})$.*

(a) *A set E in $\mathcal{A}$ is a positive set for μ if and only if it is a negative set for $-\mu$.*

(b) *If $\{E_n\}$ is a sequence of sets that are positive for μ, then $\bigcup_{n=1}^{\infty} E_n$ is positive for μ.*

Proof. The proof of (a) is routine. To see (b) let $A \in \mathcal{A}$ such that $A \subseteq \bigcup_{n=1}^{\infty} E_n$. Put $F_1 = E_1$ and, when $n \geq 2$, $F_n = E_n \backslash (F_1 \cup \cdots \cup F_{n-1})$; let $A_n = A \cap F_n$. Thus the sets $\{A_n\}$ are pairwise disjoint and $\mu(A_n) \geq 0$ since $A_n \subseteq E_n$; hence $\mu(A) = \mu(\bigcup_{n=1}^{\infty} A_n) = \sum_{n=1}^{\infty} \mu(A_n) \geq 0$. ∎

In light of part (a) in the preceding proposition, almost any result we obtain for positive sets of a signed measure can be phrased and proven for negative sets. For example, if $\{F_n\}$ is a sequence of sets that are negative for μ, then $\bigcup_{n=1}^{\infty} F_n$ is negative for μ.

2.6.6. Theorem (Hahn[6] Decomposition Theorem). *If μ is a signed measure on $(X, \mathcal{A})$, then there is a positive set P and a negative set N for μ such that $P \cap N = \emptyset$ and $P \cup N = X$. This pair of sets is unique in the following sense. If (P_1, N_1) is a second pair having the same properties, then $\mu(E) = 0$ for any set E in $\mathcal{A}$ that is contained in $[(P \backslash P_1) \cup (P_1 \backslash P)] \cup [(N \backslash N_1) \cup (N_1 \backslash N)]$.*

Proof. Note that since the empty set is a positive set for μ, the collection of positive sets for μ is non-void, and so we can define $\alpha = \sup\{\mu(E) : E \text{ is positive for } \mu\}$. Let $\{E_n\}$ be a sequence of positive sets for μ such that $\mu(E_n) \to \alpha$. If we let $P_1 = \bigcup_{n=1}^{\infty} E_n$, then according to the preceding proposition, P_1 is a positive set; also $\mu(P_1) = \alpha$ since $\mu(P_1) \geq \mu(E_n)$ for all n. (Why?) We will refer to such a positive set as a maximizing positive set. Now let N be a maximizing positive set for $-\mu$; thus N is a minimizing negative set for μ.

---

[6] Hans Hahn was born in Vienna in 1879. He received his doctorate in 1902 from the University of Vienna with a thesis in the calculus of variations. In 1909, after temporary positions in Vienna and Innsbruck, he was a regular faculty member in Czernowitz, Austria-Hungary, which was renamed Cernauti when it became part of Romania after World War I, and then Chernovtsy after it became part of the USSR in 1940. Presently it is Chernivtsi in Ukraine. He served in the Austrian Army in World War I and became a member of the Faculty in Bonn from 1916 until 1920, when he returned to Vienna to take up a chaired position. He had 10 doctoral students, among them Witold Hurewicz and Kurt Gödel. He continued to work on various parts of analysis, making several important contributions. His most famous is the Hahn–Banach Theorem, which appears later in this book. He died in Vienna in 1934.

If $A \subseteq P_1 \cap N$, then $\mu(A) \geq 0$ and $\mu(A) \leq 0$ so we have that $\mu(A) = 0$. Thus if we let $P_2 = P_1 \backslash N$, we still have a maximizing positive set and $P_2 \cap N = \emptyset$. Consider the set $\Delta = X \backslash (P_2 \cup N)$ in $\mathcal{A}$ and define the signed measure μ_Δ on $(\Delta, \mathcal{A}_\Delta)$ as in Example 2.6.2(d). By the first part of the proof, there is a maximizing positive set D for μ_Δ in $\mathcal{A}_\Delta$. But this implies that $P_2 \cup D$ is a positive set for μ and $\alpha \geq \mu(P_2 \cup D) = \alpha + \mu(D)$. Hence $\mu(D) = 0$. Similarly a maximizing positive set for $-\mu_\Delta$ has measure zero. Thus $\mu(E) = 0$ for every subset of Δ. Hence if we set $P = P_2 \cup \Delta$, then P is a positive set for μ and P and N satisfy the desired properties. The proof of uniqueness follows the same lines as the just completed argument. ■

Observe that the set that appears in the uniqueness statement in the preceding theorem is the union of the two symmetric differences (see Exercise 2.4.7) of the positive and negative sets of the two decompositions.

If ν is the signed measure defined in Example 2.6.2(e), then the Hahn decomposition (P, N) for ν is given by $P = \{x : f(x) \geq 0\}$ and $N = \{x : f(x) < 0\}$. We might take this opportunity to revisit a matter previously discussed, namely that the elements in $L^1(\mu)$ are not functions but equivalence classes of functions. Thus this example of the Hahn decomposition depends on fixing a representative function f. Of course another choice of representative will yield a different decomposition, but one that is equivalent as in the uniqueness statement of the theorem. See Exercise 2.

For the signed measure μ let (P, N) be the Hahn decomposition and put $\mu_+ = \mu_P, \mu_- = -\mu_N$. It follows that μ_+ and μ_- are positive measures, at least one of which is finite, and $\mu = \mu_+ - \mu_-$. This is called the *Jordan*[7] *decomposition* of μ; μ_+ and μ_- are called the *positive* and *negative* parts of μ. (OK. There is a bit of awkwardness here in that the negative part of μ is a positive measure. As long as we are cognizant of this there should be no confusion, just an awkwardness of language that sometimes arises.)

2.6.7. Definition. If μ is a signed measure on $(X, \mathcal{A})$, define

$$|\mu| = \mu_+ + \mu_-$$

---

[7]Camille Jordan was born in Lyon, France in 1838. He was initially trained as an engineer and worked in that profession before turning to mathematics. He received a doctorate in 1861 with a thesis that had two parts, one in algebra and the other in analysis. He returned to work as an engineer and did not become a mathematics professor until 1876. His mathematics was deep and ranged over a number of topics, with many results that bear his name: The Jordan Curve Theorem, Jordan composition series in group theory, Jordan canonical forms for matrices, and the present Jordan decomposition of measures. (Note that two other mathematical concepts, Jordan algebras and Gauss–Jordan elimination, are named after two different people who are not related to him.) In addition to being a prolific mathematician, he and his wife had eight children. In 1912 he retired. Sadness entered his life during World War I when he lost three of his six sons. (The three remaining sons became a government minister, a historian at the Sorbonne, and an engineer.) He died in Paris in 1922.

The measure $|\mu|$ is called the *variation* of μ.

It is easy to see that $|\mu|$ is a positive measure. Observe that μ is a finite signed measure precisely when $|\mu|(X) < \infty$. Finally we might return to the uniqueness part of the Hahn decomposition and rephrase that condition using $|\mu|$. In fact the condition there is precisely the statement that $\chi_P = \chi_{P_1}$ and $\chi_N = \chi_{N_1}$ a.e.$[|\mu|]$.

It is helpful to derive another formula for $|\mu|$, one that avoids reference to the Hahn and Jordan decompositions. To do this we introduce another term that has an obvious definition. If $E \in \mathcal{A}$, say that $\{A_1, \ldots, A_n\}$ is an $\mathcal{A}$-partition of E provided each set $A_k \in \mathcal{A}$, $A_j \cap A_k = \emptyset$ when $j \neq k$, and $E = A_1 \cup \cdots \cup A_n$.

2.6.8. Proposition. *If μ is a signed measure, then for every E in $\mathcal{A}$*

$$|\mu|(E) = \sup \left\{ \sum_{k=1}^n |\mu(A_k)| : \{A_1, \ldots, A_n\} \text{ is an } \mathcal{A}\text{-partition of } E \right\}$$

Proof. Fix the set E and let α denote the supremum on the right hand side of the above equation. If (P, N) is the Hahn decomposition for μ, then $\alpha \geq |\mu(P \cap E)| + |\mu(N \cap E)| = \mu_+(E) + \mu_-(E) = |\mu|(E)$. Now for any set B in $\mathcal{A}$ we have that $|\mu(B)| = |\mu_+(B) - \mu_-(B)| \leq \mu_+(B) + \mu_-(B) = |\mu|(B)$. If we apply this to each summand in the definition of α, we see that $\alpha \leq |\mu|(E)$, completing the proof. ∎

If $\mu = \mu_+ - \mu_-$ is a signed measure on $(X, \mathcal{A})$, we say an $\mathcal{A}$-measurable function f is *integrable* with respect to μ if it belongs to $L^1(|\mu|)$. It is easy to see that this is equivalent to the requirement that f is integrable with respect to both $\mu_\pm$. When f is integrable with respect to μ, we set $\int f d\mu = \int f d\mu_+ - \int f d\mu_-$.

Define $M(X, \mathcal{A})$ to be the set of all finite signed measures for $(X, \mathcal{A})$. The reader can quickly check the axioms to see that $M(X, \mathcal{A})$ is a vector space over the real numbers. In fact, this is the content of Example 2.6.2(b). For any μ in $M(X, \mathcal{A})$ define

2.6.9 $$\|\mu\| = |\mu|(X)$$

2.6.10. Proposition. *$\| \cdot \|$ as defined in (2.6.9) is a norm on the vector space $M(X, \mathcal{A})$, and with this norm $M(X, \mathcal{A})$ becomes a Banach space.*

Proof. It is trivial to check that $\|\mu\| = 0$ if and only if $\mu = 0$ and that $\|a\mu\| = |a|\|\mu\|$ whenever a is in $\mathbb{R}$. To prove the triangle inequality fix μ

and ν in $M(X, \mathcal{A})$ and let $\{A_1, \ldots, A_n\}$ be an $\mathcal{A}$-partition of X. Then

$$\sum_{k=1}^{n} |\mu(A_k) + \nu(A_k)| \leq \sum_{k=1}^{n} |\mu(A_k)| + \sum_{k=1}^{n} |\nu(A_k)|$$

$$\leq \|\mu\| + \|\nu\|$$

Taking the supremum over all partitions gives the triangle inequality by Proposition 2.6.8.

Now to show that $M(X, \mathcal{A})$ is a Banach space let $\{\mu_n\}$ be a Cauchy sequence in $M(X, \mathcal{A})$. If $A \in \mathcal{A}$, then $|\mu_n(A) - \mu_m(A)| \leq \|\mu_n - \mu_m\|$, so that $\{\mu_n(A)\}$ is a Cauchy sequence in $\mathbb{R}$; let $\lim_n \mu_n(A) = \mu(A)$. Clearly $\mu : \mathcal{A} \to \mathbb{R}$ defines a function and $\mu(\emptyset) = 0$. Also if $E, F \in \mathcal{A}$ and $E \cap F = \emptyset$, then $\mu(E \cup F) = \lim_n \mu_n(E \cup F) = \lim_n[\mu_n(E) + \mu_n(F)] = \mu(E) + \mu(F)$. So μ is finitely additive. To show that μ is a measure let $\{E_k\}$ be an infinite sequence of pairwise disjoint sets in $\mathcal{A}$ with $E = \bigcup_{k=1}^{\infty} E_k$ and let $\epsilon > 0$. Choose N such that for $n, m \geq N$, $\|\mu_n - \mu_m\| < \epsilon/5$. Fix $m \geq N$ and choose an integer p_0 such that $|\mu_m(E) - \sum_{k=1}^{p} \mu_m(E_k)| < \epsilon/5$ whenever $p \geq p_0$. Now fix an arbitrary $p \geq p_0$ and choose $n \geq N$ such that $|\mu(E) - \mu_n(E)| < \epsilon/5$ and $|\mu(\bigcup_{k=1}^{p} E_k) - \mu_n(\bigcup_{k=1}^{p} E_k)| < \epsilon/5$. Therefore

$$\left| \mu(E) - \sum_{k=1}^{p} \mu(E_k) \right| \leq |\mu(E) - \mu_n(E)| + |\mu_n(E) - \mu_m(E)|$$

$$+ \left| \mu_m(E) - \sum_{k=1}^{p} \mu_m(E_k) \right|$$

$$+ \left| \mu_m \left(\bigcup_{k=1}^{p} E_k \right) - \mu_n \left(\bigcup_{k=1}^{p} E_k \right) \right|$$

$$+ \left| \mu_n \left(\bigcup_{k=1}^{p} E_k \right) \quad \mu \left(\bigcup_{k=1}^{p} E_k \right) \right|$$

$$\leq |\mu(E) - \mu_n(E)| + \|\mu_n - \mu_m\|$$

$$+ \left| \mu_m(E) - \sum_{k=1}^{p} \mu_m(E_k) \right|$$

$$+ \|\mu_m - \mu_n\| + \left| \mu_n \left(\bigcup_{k=1}^{p} F_k \right) - \mu \left(\bigcup_{k=1}^{p} E_k \right) \right|$$

$$< \epsilon$$

Thus $\mu \in M(X, \mathcal{A})$. Now to show that $\|\mu - \mu_n\| \to 0$.

Let $\epsilon > 0$ and choose N such that $\|\mu_n - \mu_m\| < \epsilon/3$ for $m, n \geq N$. Fix $n \geq N$. Now choose an $\mathcal{A}$-partition of X, $\{A_1, \ldots, A_k\}$, such that

$\|\mu - \mu_n\| < \epsilon/3 + \sum_{j=1}^{k} |(\mu - \mu_n)(A_j)|$. Choose $m \geq N$ such that $|\mu(A_j) - \mu_m(A_j)| < \epsilon/(3k)$ for $1 \leq j \leq k$. Thus

$$\|\mu - \mu_n\| \leq \epsilon/3 + \sum_{j=1}^{k} |\mu(A_j) - \mu_m(A_j)| + \sum_{j=1}^{k} |\mu_m(A_j) - \mu_n(A_j)|$$
$$\leq \epsilon/3 + k(\epsilon/(3k)) + \|\mu_m - \mu_n\|$$
$$< \epsilon \qquad\qquad\qquad\qquad\qquad\qquad \blacksquare$$

Exercises.

(1) Let $\mu_1, \mu_2 \in M(X, \mathcal{A})$ and let (P_i, N_i), $i = 1, 2$, be their Hahn decompositions. Express in terms of P_1, P_2, N_1, N_2 the Hahn decomposition of $\mu = \mu_1 + \mu_2$.

(2) Let $(X, \mathcal{A}, \mu)$ be a measure space. If $f \in L^1(\mu)$ and $\nu(E) = \int_E f d\mu$ for every E in $\mathcal{A}$, show that ν is a signed measure and determine its Hahn decomposition, its Jordan decomposition, and its variation. Discuss the uniqueness criterion in this case.

(3) If μ is a signed measure on $(X, \mathcal{A})$ and ν is a positive measure such that $|\mu(E)| \leq \nu(E)$ for every set E in $\mathcal{A}$, show that $|\mu|(E) \leq \nu(E)$ for every E in $\mathcal{A}$.

(4) Let μ be a signed measure with Jordan decomposition $\mu = \mu_+ - \mu_-$. Show that if $\mu = \nu - \eta$ where ν and η are positive measures, then $\nu \geq \mu_+$ and $\eta \geq \mu_-$.

2.7. L^p-spaces

Throughout this section $(X, \mathcal{A}, \mu)$ is a measure space. We have introduced the space $L^1(\mu)$ and represented it as a collection of equivalence classes of functions. Now we introduce additional spaces of functions that are related to $L^1(\mu)$.

2.7.1. Definition. If $0 < p < \infty$, $L^p(\mu)$ consists of all equivalence classes of $\mathcal{A}$-measurable functions f such that $|f|^p \in L^1(\mu)$. When $p = \infty$ the space $L^\infty(\mu)$ consists of the equivalence classes of $\mathcal{A}$-measurable functions f such that there is a set Z in $\mathcal{A}$ and a constant M with $\mu(Z) = 0$ and $|f(x)| \leq M$ for all x in $X \backslash Z$. The functions in $L^\infty(\mu)$ are called *essentially bounded* functions.

There is a literature on the spaces $L^p(\mu)$ for $0 < p < 1$, and the curious reader can start with [**9**]. Many of these spaces exhibit pathological structure. In this book we will not treat this topic, but rather focus on those

L^p-spaces with $1 \leq p \leq \infty$. Also note that as with $L^1(\mu)$, we are considering equivalence classes of functions, with two functions being identified when they agree a.e. $[\mu]$.

We now put a norm on these spaces and show that they are all Banach spaces. In fact for $1 \leq p < \infty$ the norm is easy to define:

2.7.2
$$\|f\|_p = \left(\int |f|^p d\mu \right)^{\frac{1}{p}}$$

Showing that this is a norm, in particular that it satisfies the triangle inequality, takes some effort. Defining the norm on $L^\infty(\mu)$ is a bit more involved, but once defined it is easier to show it is a norm so that is where we start.

For f in $L^\infty(\mu)$ define

2.7.3
$$\|f\|_\infty = \inf \left\{ \sup\{|f(x)| : x \in X \backslash Z\} : Z \in \mathcal{A} \text{ and } \mu(Z) = 0 \right\}$$

There may be some measurable sets Z with $\mu(Z) = 0$ for which the supremum is infinite, but we know from the definition of $L^\infty(\mu)$ that there is at least one such that the supremum is finite. So there is at least one finite number in the set over which we take the infimum.

2.7.4. Proposition. *If $f \in L^\infty(\mu)$, then there is a set Z in $\mathcal{A}$ with $\mu(Z) = 0$ such that $\|f\|_\infty = \sup\{|f(x)| : x \in X \backslash Z\}$.*

Proof. According to the definition of the norm, there is a sequence $\{Z_n\}$ in $\mathcal{A}$ such that $\mu(Z_n) = 0$ for all n and $\sup\{|f(x)| : x \in X \backslash Z_n\} \to \|f\|_\infty$. If $Z = \bigcup_{n=1}^\infty Z_n$, then $\mu(Z) \leq \sum_{n=1}^\infty \mu(Z_n) = 0$. It follows that $\|f\|_\infty \leq \sup\{|f(x)| : x \in X \backslash Z\} \leq \sup\{|f(x)| : x \in X \backslash Z_n\} \to \|f\|_\infty$. ∎

The preceding proposition allows us to redefine a function f in $L^\infty(\mu)$ so that it agrees a.e. $[\mu]$ with the original one, and hence defines the same element of $L^\infty(\mu)$, and such that $\|f\|_\infty = \sup\{|f(x)| : x \in X\}$.

2.7.5. Proposition. *The expression (2.7.3) defines a norm on $L^\infty(\mu)$.*

Proof. Let $f, g \in L^\infty(\mu)$. As was just mentioned, we can assume that $\|f\|_\infty = \sup\{|f(x)| : x \in X\}$ and $\|g\|_\infty = \sup\{|g(x)| : x \in X\}$. Thus $\|f + g\|_\infty \leq \sup\{|f(x) + g(x)| : x \in X\} \leq \sup\{|f(x)| : x \in X\} + \sup\{|g(x)| : x \in X\} = \|f\|_\infty + \|g\|_\infty$. The proofs of the other requirements for being a norm are straightforward. ∎

To prove that we have a norm on the spaces $L^p(\mu)$ when p is finite requires an inequality, the proof of which starts with an inequality about numbers.

2.7.6. Lemma. *If* $1 < p, q < \infty$ *and* $\frac{1}{p} + \frac{1}{q} = 1$, *then for* $a, b > 0$ *we have*

$$ab \leq \frac{a^p}{p} + \frac{b^q}{q}$$

Proof. We introduce the function $f : (0, \infty) \to \mathbb{R}$ defined by $f(t) = \frac{a^p t^p}{p} + \frac{b^q t^{-q}}{q}$. Differentiation gives that $f'(t) = a^p t^{p-1} - b^q t^{-q-1}$. Using the relation between p and q to get that $p + q = pq$, we see that f has exactly one critical point at $t = a^{-\frac{1}{q}} b^{\frac{1}{p}}$. It is easy to see that $f''(t) > 0$ for all t, so this is a local minimum. Since $\lim_{t \to 0} f(t) = \infty = \lim_{t \to \infty} f(t)$, it is a global minimum and so we have that $\frac{a^p}{p} + \frac{b^q}{q} = f(1) \geq f(a^{-\frac{1}{q}} b^{\frac{1}{p}}) = ab$. ∎

2.7.7. Theorem (Hölder's[8] Inequality). *If* $1 < p, q < \infty$, $\frac{1}{p} + \frac{1}{q} = 1$, $f \in L^p(\mu)$, *and* $g \in L^q(\mu)$, *then* $fg \in L^1(\mu)$ *and*

$$\|fg\|_1 \leq \|f\|_p \|g\|_q$$

Proof. We will use the preceding lemma but still introduce a parameter t, $0 < t < \infty$ to get

$$\int |fg| d\mu = \int (t|f|)(t^{-1}|g|) d\mu$$
$$\leq \int \left[\frac{t^p |f|^p}{p} + \frac{t^{-q} |g|^q}{q} \right] d\mu$$
$$= \frac{t^p}{p} \|f\|_p^p + \frac{t^{-q}}{q} \|g\|_q^q$$
$$= F(t)$$

which is valid for all positive values of t. We again apply Lemma 2.7.6 with $a = \|f\|_p$ and $b = \|g\|_q$. The above inequality holds for all values of $t > 0$, so in particular it holds for the critical value for $F(t)$: $t = a^{-\frac{1}{q}} b^{\frac{1}{p}} = \|f\|_p^{-\frac{1}{q}} \|g\|_q^{\frac{1}{p}}$. That is

$$\int |fg| d\mu \leq F\left(\|f\|_p^{-\frac{1}{q}} \|g\|_q^{\frac{1}{p}} \right) = \|f\|_p \|g\|_q$$

giving Hölder's Inequality. ∎

There is another proof of Hölder's Inequality that can be found in [**18**]. We can extend this inequality to the case where $p = 1$ and therefore $q = \infty$ as follows.

---

[8]Otto Ludwig Hölder was born in Stuttgart in 1859 and died in Leipzig in 1937. He received his doctorate at the University of Tübingen and spent most of his career in Leipzig. His research began with Fourier series and potential theory but gravitated toward group theory, where he did most of his work. Besides this inequality he is known for the Jordan–Hölder Theorem in group theory. At Leipzig he had 43 doctoral students, including Emil Artin.

2.7.8. Proposition. *If $f \in L^1(\mu)$ and $g \in L^\infty(\mu)$, then $fg \in L^1(\mu)$ and*

$$\|fg\|_1 \le \|f\|_1 \|g\|_\infty$$

Proof. This is easy. We can assume that $|g(x)| \le \|g\|_\infty$ for all x in X. Hence $\int |fg| d\mu \le \int \|g\|_\infty |f| d\mu = \|f\|_1 \|g\|_\infty$. ∎

An application of Hölder's Inequality, which will henceforward also be used to refer to the case where $p = 1$ and $q = \infty$, yields the following important result that will lead to a proof that the L^p-norm satisfies the triangle inequality and is, therefore, a norm. It will also be used later in this book in a meaningful way.

2.7.9. Theorem. *If $1 \le p < \infty$ and $\frac{1}{p} + \frac{1}{q} = 1$, then for every f in $L^p(\mu)$*

$$\|f\|_p = \sup\left\{\left|\int fg d\mu\right| : g \in L^q(\mu) \text{ and } \|g\|_q \le 1\right\}$$
$$= \sup\{\|fg\|_1 : g \in L^q(\mu) \text{ and } \|g\|_q \le 1\}$$

The correct interpretation of this theorem when $p = 1$ is obtained by letting $q = \infty$.

Proof. Fix f in $L^p(\mu)$ and set β equal to the first supremum above and α equal to the second supremum. If $g \in L^q(\mu)$ with $\|g\|_q \le 1$, then $|\int fg d\mu| \le \int |fg| d\mu = \|fg\|_1$. So we clearly have that $\beta \le \alpha$. For the same type of function g, Hölder's Inequality shows that $\|fg\|_1 \le \|f\|_p$ so that $\alpha \le \|f\|_p$. Thus we have that $\beta \le \alpha \le \|f\|_p$; we must show that $\|f\|_p \le \beta$.

Assume that $1 < p < \infty$. Let's remark that the relation between p and q implies $|f|^{p-1} \in L^q(\mu)$ and $\||f|^{p-1}\|_q = (\|f\|_p)^{p/q}$. For every x in X, define $\text{sign}(f)(x) = 1$ when $f(x) \ge 0$ and $\text{sign}(f)(x) = -1$ when $f(x) < 0$. (We will use the notion of the $\text{sign}(f)$ in the future as well. It is handy.) It is easy to see that $\text{sign}(f)$ is a measurable function and $f(x)\text{sign}(f)(x) = |f(x)|$. Thus $g = \left[(\|f\|_p)^{p/q}\right]^{-1} |f|^{p-1}\text{sign}(f)$ has $\|g\|_q = 1$; and $\beta \ge |\int fg d\mu| = \int fg d\mu = \left[(\|f\|_p)^{p/q}\right]^{-1} \int |f|^p d\mu = \|f\|_p^{p-p/q} = \|f\|_p$.

Now assume that $p = 1, q = \infty$. Here let $E = \{x : f(x) \ne 0\}$ and put $g = \text{sign}(f)\chi_E$. Clearly $g \in \text{ball } L^\infty(\mu)$ and $\int fg d\mu = \|f\|_1$. ∎

Thus not only is the theorem valid, but the supremum is actually attained. In the preceding theorem, the case where $p = \infty$ is conspicuously absent. What's going on? The same result holds but only if we put a restriction on the measure space. Say that a set E in $\mathcal{A}$ is an *atom* if $\mu(E) \ne 0$ and for every subset F of E that belongs to $\mathcal{A}$, either $\mu(F) = 0$ or $\mu(F) = \mu(E)$. Getting an example is easy, but let's take the time to introduce a concept that will surface frequently in measure theory.

2.7.10. Definition. If X is a set and $x \in X$, define the *unit point mass* at x to be the measure δ_x defined on 2^X or any σ-algebra of subsets of X by

$$\delta_x(E) = \begin{cases} 1 & \text{if } x \in E \\ 0 & \text{if } x \notin E \end{cases}$$

If λ is Lebesgue measure on $[0,1]$ and δ_0 is the unit point mass at 0, then $\mu = \lambda + \delta_0$ has an atom at $\{0\}$ whereas λ has no atoms. (A measure with no atoms such as λ is said to be a *continuous measure*.) Note that $\mu = \lambda + \infty\delta_0$ has an *infinite atom* at 0. Infinite atoms play havoc with measure theory. For example if a measure μ has an infinite atom E, then every μ-integrable function must vanish identically on E; so that as far as integrable functions are concerned, the infinite atom might just as well not be there. If a measure only has finite atoms, then little damage is done as long as there aren't too many of them. For a function f in $L^\infty(\mu)$, the existence of an infinite atom is more serious. Here f can take on non-zero values on the infinite atom even though an integrable function must vanish there. So the relationship between $L^\infty(\mu)$ and $L^1(\mu)$ we discuss in the next proposition completely breaks down when there are infinite atoms.

2.7.11. Proposition. *If μ has no infinite atoms and $f \in L^\infty(\mu)$, then*

$$\|f\|_\infty = \sup\{\|fg\|_1 : g \in L^1(\mu) \text{ and } \|g\|_1 \leq 1\}$$

Proof. The fact that $\sup\{\|fg\|_1 : g \in L^1(\mu) \text{ and } \|g\|_1 \leq 1\} \leq \|f\|_\infty$ is easily seen by using Hölder's Inequality. For the reverse inequality take any $\epsilon > 0$ and let $F \subseteq \{x : |f(x)| \geq \|f\|_\infty - \epsilon\}$ such that $F \in \mathcal{A}$ and $0 < \mu(F) < \infty$. (Note that we have just used the fact that there are no infinite atoms.) Put $g = \mu(F)^{-1}\chi_F$. So $g \in L^1(\mu)$ and $\|g\|_1 = 1$. Also $\int |fg|d\mu = \mu(F)^{-1}\int_F |f|d\mu \geq \|f\|_\infty - \epsilon$. Since ϵ was arbitrary, we are done. ∎

2.7.12. Theorem. *If $1 \leq p \leq \infty$, then $L^p(\mu)$ is a Banach space. Moreover the simple measurable functions belonging to $L^p(\mu)$ are dense in $L^p(\mu)$.*

Proof. First we establish that these are normed spaces. When $p = \infty$ this was already done (2.7.5). If $p < \infty$ and $g \in L^q(\mu)$ with $\|g\|_q \leq 1$ and $f_1, f_2 \in L^p(\mu)$, then $\int |(f_1+f_2)g|d\mu \leq \int |f_1 g|d\mu + \int |f_2 g|d\mu \leq \|f_1\|_p + \|f_2\|_p$. By Theorem 2.7.9 the triangle inequality follows. The rest of the verification that $\|\cdot\|_p$ is a norm is straightforward.

We know that $L^1(\mu)$ is complete, so to show that the other $L^p(\mu)$ spaces are complete we first assume that $1 < p < \infty$. This proof is similar to the proof of Theorem 2.5.4. Let $\{f_n\}$ be a Cauchy sequence in $L^p(\mu)$. Passing to a subsequence if necessary, we can assume that $\|f_n - f_{n-1}\|_p < 2^{-n}$. Put

$F_n(x) = \sum_{k=1}^{n} |f_k(x) - f_{k-1}(x)|$. Clearly $\{F_n\}$ is an increasing sequence and hence by (2.3.11) $F(x) = \lim_n F_n(x) = \sum_{n=1}^{\infty} |f_n(x) - f_{n-1}(x)|$ is a measurable function. Without loss of generality we can assume that this infinite series converges for all x in X. (Why?) Since $\|F_n\|_p \le \sum_{k=1}^{n} \|f_k - f_{k-1}\|_p \le 1$, by the MCT we have that $F \in L^p(\mu)$ and $\|F\|_p \le 1$. We use F as a control. Let $f(x) = f_1(x) + \sum_{n=1}^{\infty} [f_n(x) - f_{n-1}(x)]$, a series dominated by the one that defines F; thus $f \in L^p(\mu)$. Also $\|f - f_n\|_p = \|\sum_{k=n+1}^{\infty} [f_k - f_{k-1}]\|_p \le \sum_{k=n+1}^{\infty} \|f_k - f_{k-1}\|_p \to 0$ as $n \to \infty$.

Now to show that $L^\infty(\mu)$ is complete let $\{f_n\}$ be a Cauchy sequence. Without loss of generality we can assume that for every $\epsilon > 0$ there is an N such that for $n, m \ge N$, $\sup\{|f_n(x) - f_m(x)| : x \in X\} < \epsilon$. (Why?) In particular for every x, $\{f_n(x)\}$ is a Cauchy sequence in $\mathbb{R}$; let $f(x) = \lim_n f_n(x)$. If follows that f is an $\mathcal{A}$-measurable function. Moreover if $M \ge \sup\{|f_n(x)| : x \in X\}$ for all n, then $|f(x)| \le M$ for all x; hence $f \in L^\infty(\mu)$. Finally if $\epsilon > 0$ and N is chosen so that $\sup\{|f_n(x) - f_m(x)| : x \in X\} < \epsilon/2$ for $n, m \ge N$, then for every x and $n, m \ge N$, $|f(x) - f_n(x)| \le |f(x) - f_m(x)| + \|f_n - f_m\|_\infty < \epsilon/2 + |f(x) - f_m(x)|$. Now choose m sufficiently large that $|f(x) - f_m(x)| < \epsilon/2$ and we have that $|f(x) - f_n(x)| < \epsilon$ for all $n \ge N$ and all x in X; that is, $\|f - f_n\|_\infty < \epsilon$ for $n \ge N$.

It remains to show that the measurable simple functions are dense in $L^p(\mu)$. Again we already have this for $p = 1$ by Theorem 2.5.4. Assume $1 < p < \infty$ and let $\epsilon > 0$. According to Exercise 1 there is an integer N such that the set $E = \{x : \frac{1}{N} \le |f(x)| \le N\}$ has $\int_{X \backslash E} |f|^p d\mu < \epsilon^p/2$ and $\mu(E) < \infty$. Now write $[-N, -\frac{1}{N}] \cup [\frac{1}{N}, N]$ as the union of pairwise disjoint subintervals $\{I_k : 1 \le k \le n\}$ each of which has length $|I_k| < \epsilon/[2\mu(X \backslash E)]^{\frac{1}{p}}$. Let $A_k = f^{-1}(I_k)$, pick t_k in the interval I_k, and set $u = \sum_{k=1}^{n} t_k \chi_{A_k}$. Observe that $\bigcup_{k=1}^{n} A_k = E$. Thus $\|f - u\|_p^p = \int |f - \sum_{k=1}^{n} t_k \chi_{A_k}|^p = \int_{X \backslash E} |f|^p d\mu + \sum_{k=1}^{n} \int_{A_k} |f(x) - t_k|^p d\mu < \epsilon^p/2 + \epsilon^p [2\mu(E)]^{-1} \sum_{k=1}^{n} \mu(A_k) = \epsilon^p$. That is, $\|f - u\|_p < \epsilon$.

Now for the case $p = \infty$, which is similar to the previous case but technically easier. Let $f \in L^\infty(\mu)$ and assume that $\|f\|_\infty = \sup\{|f(x)| : x \in X\}$. If $\epsilon > 0$, let $\{I_k : 1 \le k \le n\}$ be intervals all of which have length $|I_k| < \epsilon$ and whose union is the interval $[-\|f\|_\infty, \|f\|_\infty]$. Let $A_k = f^{-1}(I_k)$ and pick a t_k in each interval I_k; let $u = \sum_{k=1}^{n} t_k \chi_{A_k}$. So $\{A_k\}$ is an $\mathcal{A}$-partition of X, and if $x \in A_k$, $|f(x) - u(x)| = |f(x) - t_k| < \epsilon$. Hence $\|f - u\|_\infty < \infty$. ∎

We want to use the preceding theorem to specify when the space $L^p(\mu)$ is separable. Recall Exercise 2.4.7 where we introduced the symmetric difference between two sets E and F: $E \Delta F = (E \backslash F) \cup (F \backslash E)$. Define two sets E, F in $\mathcal{A}$ to be equivalent if $\mu(E \Delta F) = 0$, and let $\mathcal{A}_1$ denote the collection of all equivalence classes of sets in $\mathcal{A}$ having finite measure. It was noted in Exercise 2.4.7 that for sets of finite measure, $\mu(E \Delta F) = \int |\chi_E - \chi_F| d\mu = \|\chi_E - \chi_F\|_1$. Thus if we define $\rho(E, F) = \mu(E \Delta F)$, ρ defines a metric on $\mathcal{A}_1$.

2.7.13. Proposition. *If $(X, \mathcal{A}, \mu)$ is any measure space, the following statements are equivalent.*

(a) *For all p with $1 \le p < \infty$, $L^p(\mu)$ is separable.*

(b) *There is a p with $1 \le p < \infty$ such that $L^p(\mu)$ is separable.*

(c) *$L^1(\mu)$ is separable.*

(d) *If $\mathcal{A}_1$ denotes the collection of equivalence classes of sets in $\mathcal{A}$ having finite measure and ρ is the metric on $\mathcal{A}_1$ defined by $\rho(E, F) = \mu(E \Delta F)$, then $(\mathcal{A}_1, \rho)$ is a separable metric space.*

Proof. Trivially it follows that (a) implies (c) and this implies (b). Now assume that (b) holds and let's prove (d). Since every subset of a separable metric space is itself separable, this says that $\mathcal{A}_1$ with the metric $\rho_p(E, F) = \|\chi_E - \chi_F\|_p$ is a separable metric space; let $\mathcal{D}$ be a countable dense subset of this metric space. Notice that $\rho_p(E, F) = \rho(E, F)^{\frac{1}{p}}$. So if $E \in \mathcal{A}_1$, there is a D in $\mathcal{D}$ with $\rho(E, D) = \rho_p(E, D)^p < \epsilon$ and we see that (d) holds.

Finally assume that (d) is true and we prove (a). Let $\mathcal{D}$ be a countable dense subset of $(\mathcal{A}_1, \rho)$ and let $\epsilon > 0$. If $f \in L^p(\mu)$, Theorem 2.7.12 implies there is a $u = \sum_{k=1}^n a_k \chi_{E_k}$ in $L^p(\mu)$ such that $\|f - u\|_p < \epsilon/2$. For $1 \le k \le n$ let $q_k \in \mathbb{Q} \setminus \{0\}$ such that $|q_k - a_k| < \frac{\epsilon}{4n\mu(E_k)^{1/p}}$. Also pick D_k in $\mathcal{D}$ such that $\|\chi_{E_k} - \chi_{D_k}\|_p = \rho(E_k, D_k)^{\frac{1}{p}} < \frac{\epsilon}{4n|q_k|}$. Then

$$\left\| u - \sum_{k=1}^n q_k \chi_{D_k} \right\|_p \le \left\| u - \sum_{k=1}^n q_k \chi_{E_k} \right\|_p + \left\| \sum_{k=1}^n q_k [\chi_{E_k} - \chi_{D_k}] \right\|_p$$

$$\le \left\| \sum_{k=1}^n [a_k - q_k] \chi_{E_k} \right\|_p + \sum_{k=1}^n |q_k| \, \|\chi_{E_k} - \chi_{D_k}\|_p$$

$$\le \sum_{k=1}^n |a_k - q_k| \mu(E_k)^{\frac{1}{p}} + \sum_{k=1}^n |q_k| \frac{\epsilon}{4n|q_k|}$$

$$< \epsilon/4 + \epsilon/4 = \epsilon/2$$

Thus $\|f - \sum_{k=1}^n q_k \chi_{D_k}\|_p \le \|f - u\|_p + \|u - \sum_{k=1}^n q_k \chi_{D_k}\|_p < \epsilon$. This says that the linear span of the set of characteristic functions $\{\chi_D : D \in \mathcal{D}\}$ with rational coefficients is dense in $L^p(\mu)$; but such a linear span is countable and so $L^p(\mu)$ is separable. ∎

$L^\infty(\mu)$ is usually not separable. The reason for this is that if E and F are two measurable subsets with $\mu(E\Delta F) > 0$, then $\|\chi_E - \chi_F\|_\infty = 1$. So if we can find an uncountable collection of non-equivalent measurable sets, we have an uncountable number of functions in $L^\infty(\mu)$ that are uniformly separated and hence $L^\infty(\mu)$ cannot be separable. (Why? You might want to look at Exercise 1.2.5.) An example is the measure space $(\mathbb{N}, 2^{\mathbb{N}}, \mu)$, where μ is counting measure. The cardinality of $2^{\mathbb{N}}$ is c and $\mu(E\Delta F) > 0$ whenever E and F are distinct sets. Also see Exercise 10.

2.7.14. Proposition. *If X is a compact metric space and μ is the Radon measure on X defined by a positive linear functional on $C(X)$, then $L^p(\mu)$ is separable for all finite p.*

Proof. If $\mathcal{B}$ is the collection of equivalence classes of Borel sets and ρ is the metric defined on $\mathcal{B}$ as in Proposition 2.7.13, then we know that for every E in $\mathcal{B}$ there is an open set G containing E such that $\mu(G\backslash E) < \epsilon$. This says that the collection of open sets $\mathcal{G}$ is dense in $\mathcal{B}$. So we need only show that $(\mathcal{G}, \rho)$ is separable.

Recall that a compact metric space is separable; let $\{x_n\}$ be a countable dense subset of X, let $\{r_n\}$ be an enumeration of the rational numbers in $(0,1)$, and let $\mathcal{D}$ be the collection of all finite unions of balls of the form $B(x_n; r_k)$ with $n \geq 1, k \geq 1$; so $\mathcal{D}$ is a countable subset of $\mathcal{G}$. Now if $G \in \mathcal{G}$, there is a sequence $\{D_j\}$ in $\mathcal{D}$ such that $D_j \subseteq D_{j+1}$ for all $j \geq 1$ and $G = \bigcup_{j=1}^\infty D_j$. (Why?) Thus $\mu(D_j) \to \mu(G)$. But this implies $\rho(D_j, G) \to 0$ and so $\mathcal{D}$ is dense in $\mathcal{G}$. Hence $L^p(\mu)$ is separable by Theorem 2.7.13. ∎

The reader might look at Exercise 1.3.13, where (s)he is asked to show that $C(X)$ is separable when X is a compact metric space.

Exercises. Unless otherwise specified, $(X, \mathcal{A}, \mu)$ is an arbitrary measure space.

(1) Assume $1 \leq p < \infty$ and let $f \in L^p(\mu)$. (a) Show that $\mu(\{x : |f(x)| = \infty\}) = 0$. (b) If for each $n \geq 1$, $E_n = \{x : \frac{1}{n} \leq |f(x)| \leq n\}$, show that $\mu(E_n) < \infty$. (c) Show that $\lim_n \int_{E_n} |f|^p d\mu = \|f\|_p^p$. (d) Show that $\lim_n \int_{X\backslash E_n} |f|^p d\mu = 0$.

(2) If λ is Lebesgue measure on the unit interval and $1 \leq p \leq \infty$, for which real numbers s is x^s in $L^p(\lambda)$?

(3) If λ is Lebesgue measure on $[0,\infty)$ and $1 \leq p \leq \infty$, for which real numbers s is x^s in $L^p(\lambda)$?

(4) If X is a compact metric space and μ is a Radon measure on X, show that $C(X)$ is dense in $L^p(\mu)$ for $1 \leq p < \infty$. What goes wrong when $p = \infty$?

(5) If $1 \leq p < \infty$, λ is Lebesgue measure on $\mathbb{R}$, and $f \in L^p(\lambda)$ such that f is continuous, show that $f \in C_0(\mathbb{R})$.

(6) If $1 \leq p < \infty$, $p^{-1} + q^{-1} = 1$, and $f \in L^1(\mu)$, show that there are g in $L^p(\mu)$ and h in $L^q(\mu)$ such that $f = gh$.

(7) If μ is a finite measure, show that $L^p(\mu) \subseteq L^1(\mu)$ for $p \geq 1$.

(8) If $f_n, f \in L^p$, $1 \leq p < \infty$, and $f_n(x) \to f(x)$ a.e. $[\mu]$, show that $f_n \to f$ in L^p-norm if and only if $\|f_n\|_p \to \|f\|_p$.

(9) If μ is a finite measure and $f \in L^\infty(\mu)$, show that $f \in L^p(\mu)$ for $1 \leq p < \infty$ and prove that $\lim_{p\to\infty} \|f\|_p = \|f\|_\infty$. (Hint: Use $\limsup$ and $\liminf$.)

(10) If λ is Lebesgue measure on the unit interval, show that $L^\infty(\lambda)$ is not separable.

(11) If X is a compact metric space and μ is a Radon measure on X, give a necessary and sufficient condition that $L^\infty(\mu)$ be separable.

(12) (Vitali's[9] Convergence Theorem) Let $f_n, f \in L^p(\mu)$ with $f_n(x) \to f(x)$ a.e. $[\mu]$. Show that $\|f_n - f\|_p \to 0$ if and only if the following two conditions hold: (a) for every $\epsilon > 0$ there is a set A in $\mathcal{A}$ with $\mu(A) < \infty$ and $\sup_n \int_{X \setminus A} |f_n|^p d\mu < \epsilon$; (b) for every $\epsilon > 0$ there is a $\delta > 0$ such that when $E \in \mathcal{A}$ and $\mu(E) < \delta$, $\sup_n \int_E |f_n|^p d\mu < \epsilon$.

---

[9]Giuseppe Vitali was born in Ravenna, Italy in 1875. After starting work toward a doctorate, Vitali left school in 1901 for unknown reasons. From 1904 to 1923 he taught at the Liceo C. Colombo in Genoa (roughly equivalent to an American high school). With the advent of the fascists he returned to mathematics and obtained a chair in Modena. In 1930 he obtained a chair in Bologna. It seems he never received a doctorate but did important and interesting research and hence his success. He was the first to give an example of a non-measurable set for Lebesgue measure. Starting in 1926 he developed an illness that impaired his research but did not stop it. He died in Bologna in 1932.

Chapter 3

A Hilbert Space Interlude

Here we just peek behind the curtain of Hilbert space theory, though in this chapter's last section we will see a fundamental and crucial result of the subject, the Riesz Representation Theorem. In fact it is this theorem that necessitates the insertion of this chapter this early in the book. The idea here is to do just enough to facilitate the completion of measure theory that occurs in the following chapter. Later in Chapter 5 we'll draw back that curtain and get a more complete view of the subject.

3.1. Introduction to Hilbert space

Here we encounter the first place where we must discuss complex numbers. We will make a short excursion through this topic here, but we will increasingly be involved with the complex numbers as we go on in this book – complex measures, functions, and normed spaces. Be prepared. The reader should feel free to consult Wikipedia or the first section of my book on complex variables [**7**] for a discussion of $\mathbb{C}$ if (s)he is not conversant with the topic. What follows is just a short excursion.

We remind the reader that a complex number is $z = x + iy$, where x and y are real numbers, called the *real part* and *imaginary part* of z, respectively. When z is so expressed, we define the complex conjugate of z as $\bar{z} = x - iy$. It follows that $x = \frac{1}{2}(z + \bar{z})$ and $y = \frac{1}{2i}(z - \bar{z})$. Denote this by $x = \operatorname{Re} z$ and $y = \operatorname{Im} y$. We also know that if $z \in \mathbb{C}$, $z\bar{z} = |z|^2 = x^2 + y^2$. When $z \neq 0$, define $\operatorname{sign} z = z/|z|$; let $\operatorname{sign} 0 = 1$. (Note that this is consistent with the definition of $\text{sign}(f)$ in §2.7.) We have that

$$|\text{sign}\, z| = 1, \quad z = |z|\,\text{sign}\, z, \quad z\,\overline{\text{sign}\, z} = |z|$$

We will want to treat the scalar field $\mathbb{F}$, which could be the real or complex numbers. For a in $\mathbb{F}$ we define $\bar{a} = a$ when $\mathbb{F} = \mathbb{R}$ and $\bar{a}$ is the complex conjugate of a when $\mathbb{F} = \mathbb{C}$. In this section we will need to incorporate this "conjugate" into the concepts we introduce, and it is convenient to do this even when the underlying scalar field is the real numbers. Indeed, we will see later in §5.4 some results that hold when the scalars are complex but fail otherwise. This will start a process where we migrate to always having our scalars be complex.

3.1.1. Definition. For a vector space $\mathcal{H}$ over $\mathbb{F}$ an *inner product* is a function $\langle\cdot,\cdot\rangle : \mathcal{H} \times \mathcal{H} \to \mathbb{F}$ satisfying the following for x, y, z belonging to $\mathcal{H}$ and a, b in $\mathbb{F}$:

(a) $\langle ax + by, z\rangle = a\langle x, z\rangle + b\langle y, z\rangle$;

(b) $\langle x, ay + bz\rangle = \bar{a}\langle x, y\rangle + \bar{b}\langle x, z\rangle$;

(c) $\langle x, x\rangle \geq 0$ with equality if and only if $x = 0$;

(d) $\langle x, y\rangle = \overline{\langle y, x\rangle}$.

We might also make it clear that when $\mathbb{F} = \mathbb{C}$ and we say that an element a in $\mathbb{C}$ satisfies $a \geq 0$, we mean that $a \in \mathbb{R} \subseteq \mathbb{C}$ and $a \geq 0$. Note that by taking $a = b = 0$ in (b) we get that for all x, $\langle x, 0\rangle = 0$; similarly $\langle 0, x\rangle = 0$ for all x.

3.1.2. Example. (a) For $x, y \in \mathbb{F}^d$, if $\langle x, y\rangle = \sum_{k=1}^{d} x_k \overline{y_k}$, then this is an inner product on $\mathbb{F}^d$. We might note that if $a_1, \ldots, a_d$ are strictly positive real numbers, then $\langle x, y\rangle = \sum_{k=1}^{d} a_k x_k \overline{y_k}$ also defines an inner product on $\mathbb{F}^d$.

(b) Let $\mathcal{H}$ denote the collection of all finitely non-zero sequences of scalars in $\mathbb{F}$; that is, $\{x_n\} \in \mathcal{H}$ if $x_n \in \mathbb{F}$ for all n and $x_n = 0$ except for a finite number of coordinates n. We can make $\mathcal{H}$ into a vector space by defining addition and scalar multiplication coordinatewise. If we define $\langle x, y\rangle = \sum_{n=1}^{\infty} x_n \overline{y_n}$, then this is an inner product on $\mathcal{H}$.

(c) If $(X, \mathcal{A}, \mu)$ is a measure space and $\mathcal{H} = L^2(\mu)$, define $\langle f, g\rangle = \int f\bar{g}\, d\mu$ for f, g in $L^2(\mu)$. It is easy to check that this defines an inner product.

(d) Let X be a compact metric space, let $\mathcal{H} = C(X)$, and for a positive regular Borel measure on X define $\langle f, g\rangle = \int f(x)\overline{g(x)}d\mu(x)$; then this is an inner product on $C(X)$.

The next result will be referred to as the CBS Inequality.

3.1.3. Theorem (Cauchy–Bunyakovsky–Schwarz Inequality[1]). *If $\mathcal{H}$ is a vector space with an inner product, then*

$$|\langle x, y \rangle|^2 \leq \langle x, x \rangle \langle y, y \rangle$$

for all x, y in $\mathcal{H}$. Moreover equality occurs if and only if x, y are linearly dependent.

Proof. Notice that for x, y in $\mathcal{H}$ and a in $\mathbb{F}$ we have that

$$0 \leq \langle x - ay, x - ay \rangle$$
$$= \langle x, x \rangle - a\langle y, x \rangle - \bar{a}\langle x, y \rangle + |a|^2 \langle y, y \rangle$$

Consider the polar representation of the number $\langle y, x \rangle$: $\langle y, x \rangle = \beta e^{i\theta}$ with $\beta \geq 0$. If t is an arbitrary real number and we substitute $a = e^{-i\theta}t$ in the above inequality, then we get

$$0 \leq \langle x, x \rangle - e^{-i\theta}t\beta e^{i\theta} - e^{i\theta}t\beta e^{-i\theta} + t^2\langle y, y \rangle$$
$$= \langle x, x \rangle - 2\beta t + t^2\langle y, y \rangle$$
$$= \gamma - 2\beta t + \alpha t^2 \equiv q(t)$$

where $\gamma = \langle x, x \rangle$ and $\alpha = \langle y, y \rangle$. Thus $q(t)$ is a quadratic polynomial with real coefficients and $q(t) \geq 0$ for all $t \geq 0$. So the graph of $q(t)$ stays above the x-axis except that it might be tangent at a single point; that is,

---

[1]Augustin Louis Cauchy was born in Paris in August 1789, a month after the storming of the Bastille. He was educated in engineering and his first job was in 1810 working on the port facilities at Cherbourg in preparation for Napoleon's contemplated invasion of England. In 1812 he returned to Paris and his energies shifted toward mathematics. His contributions were monumental, with a plethora of results in analysis bearing his name. His collected works fill 27 published volumes. As a human being he left much to be desired. He was highly religious with a totally dogmatic personality, often treating others with dismissive rudeness. Two famous examples were his treatment of Abel and Galois, where he refused to consider their monumental works. Both suffered an early death. Perhaps better treatment by Cauchy would have given them the recognition that would have resulted in a longer life and a productive career to the betterment of mathematics; we'll never know. He had two doctoral students, one of which was Bunyakovsky. Cauchy died in 1857 in Sceaux near Paris.

Viktor Yakovlevich Bunyakovsky was born in 1804 in what is presently Vinnitsa in the Ukraine. In 1825 he received a doctorate in Paris working under Cauchy. The next year he went to St Petersburg, where he made his career. In 1859 this result appeared in his monograph on integral inequalities; he seems to be the first to have discovered this, 25 years before Schwarz. He made contributions to number theory, geometry, and applied mathematics. He died in St Petersburg in 1889.

Hermann Amandus Schwarz was born in 1843 in Hermsdorf, Silesia, presently in Poland. He began his studies at Berlin in Chemistry, but switched to mathematics and received his doctorate in 1864 under the direction of Weierstrass. He held positions at Halle, Zurich, Göttingen, and Berlin. His work centered on various geometry problems that were deeply connected to analysis. This included work on surfaces and conformal mappings in analytic function theory, any student of which will see his name in prominence. He died in Berlin in 1921.

$q(t) = 0$ has at most one real root. From the quadratic formula we get that $0 \geq 4\beta^2 - 4\alpha\gamma$. Therefore

$$0 \geq \beta^2 - \alpha\gamma = |\langle x, y \rangle|^2 - \langle x, x \rangle \langle y, y \rangle$$

proving the inequality.

The proof of the necessary and sufficient condition for equality consists of a careful analysis of the above inequalities to see the implications of such an equality. The reader is urged to carry out this argument. ∎

Observe that when we examine the space $L^2(\mu)$, the CBS Inequality is the same as Hölder's Inequality for $p = 2 = q$. Also note that in the proof of the CBS Inequality the property of an inner product that $\langle x, x \rangle = 0$ implies that $x = 0$ is used only in deriving the condition for equality. A function that has all the properties of an inner product save this one is called a *semi-inner product*. Therefore when we have a semi-inner product the CBS Inequality holds and a necessary and sufficient condition for equality is that there exist scalars a, b, not both 0, such that $\langle ax + by, ax + by \rangle = 0$.

3.1.4. Corollary. *If $\langle \cdot, \cdot \rangle$ is an inner product on $\mathcal{H}$, then $\|x\| = \sqrt{\langle x, x \rangle}$ defines a norm on $\mathcal{H}$.*

Proof. If $x, y \in \mathcal{H}$, then using the CBS Inequality we get

$$\begin{aligned}
\|x + y\|^2 &= \langle x + y, x + y \rangle \\
&= \|x\|^2 + \langle y, x \rangle + \langle x, y \rangle + \|y\|^2 \\
&= \|x\|^2 + 2\mathrm{Re}\,\langle x, y \rangle + \|y\|^2 \\
&\leq \|x\|^2 + 2|\langle x, y \rangle| + \|y\|^2 \\
&\leq \|x\|^2 + 2\|x\|\|y\| + \|y\|^2 \\
&= (\|x\| + \|y\|)^2
\end{aligned}$$

Therefore the triangle inequality holds. The remainder of the proof that $\|\cdot\|$ defines a norm is straightforward. ∎

Let's underline that in the course of the preceding proof we established that for all x, y in $\mathcal{H}$ we have

3.1.5 $$\|x + y\|^2 = \|x\|^2 + 2\mathrm{Re}\,\langle x, y \rangle + \|y\|^2$$

We will find this useful in the future; it is called the *polar identity*.

When we have that $\mathcal{H} = L^2(\mu)$ as in Example 3.1.2(c), the norm defined by the inner product is exactly $\|\cdot\|_2$ as defined in (2.7.2) for $p = 2$. Now that we have a normed space we can use what we know about that subject, though because of the way this norm arises we will see it will have extra structure.

3.1.6. Definition. A *Hilbert[2] space* is a vector space with an inner product such that with respect to the norm defined by this inner product it is a complete metric space.

So every Hilbert space is a Banach space. For the remainder of this chapter $\mathcal{H}$ is a Hilbert space.

3.1.7. Example. (a) $\mathbb{F}^d$ with any of the inner products defined in Example 3.1.2(a) is a Hilbert space.

(b) $L^2(\mu)$ is a Hilbert space.

(c) We define ℓ^2 as the space of all sequences $\{x_n\}$ such that $\sum_{n=1}^{\infty} |x_n|^2 < \infty$. If we consider the measure space $(\mathbb{N}, 2^{\mathbb{N}}, \mu)$, where μ is counting measure, then ℓ^2 is precisely $L^2(\mu)$, and so ℓ^2 is another example of a Hilbert space.

(d) The inner product spaces in parts (b) and (d) of (3.1.2) are not Hilbert spaces since they fail to be complete.

When we are presented with an incomplete inner product space, we can complete it just as we did for a normed space in Theorem 1.3.8. This result is stated here, though its proof will not be given as it follows routinely by mimicking the proof of (1.3.8).

3.1.8. Theorem. *If $\mathcal{X}$ is a vector space over $\mathbb{F}$ that has an inner product and $\mathcal{H}$ is its completion as a normed space, then there is an inner product on $\mathcal{H}$ such that $\langle x, y \rangle_{\mathcal{H}} = \langle x, y \rangle_{\mathcal{X}}$ for all x, y in $\mathcal{X}$ and the norm on $\mathcal{H}$ is defined by this inner product.*

What is the completion of the spaces in Examples 3.1.2(b) and 3.1.2(d)? The completion of the first of these is exactly ℓ^2. The completion of the space in Example 3.1.2(d) is $L^2(\mu)$.

---

[2]David Hilbert was born in 1862 in Königsberg, Prussia, now Kaliningrad, Russia. He received his doctorate from the University of Königsberg in 1885. He continued there on the faculty until 1895 when he was appointed to a chair in mathematics at Göttingen, where he remained for the rest of his life. His first major work at Göttingen was the proof of the basis theorem in a part of algebra called invariant theory. Previously the world's leader in this subject, Paul Albert Gordon, had proved a special case by horrendous computations. Gordon was highly critical of what he called Hilbert's formal approach. Needless to say, history smiled on this formal approach and Gordon receded into a historical note. (Except that Emmy Noether was his student.) This so-called formal approach turned out to characterize Hilbert's method in mathematics. He encountered a problem, stripped it down to its essence, recast it in his own framework, and proceeded to advance the entire subject in terms that resonate with the present generation of mathematicians. One of his most famous contributions was the presentation of 23 problems at the 1900 International Congress of Mathematicians in Paris. Many remain unsolved and those that were solved brought instant fame to the solver. They set the research direction for much of twentieth century mathematics and their influence continues. Hilbert contributed to many areas of mathematics besides algebra, including several parts of analysis and geometry. He was one of the subject's titans. In a 1930 address in Königsberg to accept being made an honorary citizen he said, "We must know, we shall know." A fitting epitaph. Hilbert died in 1943 in Göttingen.

Exercises.

(1) Carry out the details of the proof of the necessary and sufficient condition for equality in the CBS Inequality.

(2) Let w be a sequence of strictly positive real numbers and define $\ell^2(w)$ to be the set of all sequences $\{x_n\}$ of numbers from $\mathbb{F}$ such that $\sum_{n=1}^{\infty} w_n |x_n|^2 < \infty$. Show that $\langle \{x_n\}, \{y_n\} \rangle = \sum_{n=1}^{\infty} w_n x_n \overline{y_n}$ defines an inner product on $\ell^2(w)$ and that this is a Hilbert space. Is there a measure μ such that $\ell^2(w)$ is naturally identified with $L^2(\mu)$?

(3) Generalize the definition of ℓ^2 as follows. Let I be any set and for a function $x : I \to \mathbb{F}$ say that $\sum_{i \in I} x(i) = a$ if for every $\epsilon > 0$ there is a finite subset J_0 of I such that $|a - \sum_{i \in J} x(i)| < \epsilon$ for every finite set J containing J_0. (a) Show that if such an x converges, there are at most a countable number of i in I such that $x(i) \neq 0$. Define $\ell^2(I)$ to be the set of all functions $x : I \to \mathbb{F}$ such that $\sum_{i \in I} |x(i)|^2 < \infty$. For x, y in $\ell^2(I)$ let $\langle x, y \rangle = \sum_{i \in I} x(i) \overline{y(i)}$. (b) Show that this defines an inner product. (c) Find a measure space $(X, \mathcal{A}, \mu)$ such that $\ell^2(I)$ is naturally identified with $L^2(\mu)$.

3.2. Orthogonality

The subject of this section is the very thing that makes the structure of a Hilbert space so transparent; it has a concept of angle. We can define the angle between two non-zero vectors x and y as $\arccos[\langle x, y \rangle / \|x\| \|y\|]$. The ambiguity of the function arccos of course lends ambiguity to this process. However we will not pursue this concept of the angle and instead focus on the concept of two vectors forming a right angle.

3.2.1. Definition. If $x, y \in \mathcal{H}$, say that x and y are *orthogonal*, in symbols $x \perp y$, if $\langle x, y \rangle = 0$. If A and B are two subsets of $\mathcal{H}$, say that these sets are orthogonal if $x \perp y$ for any x in A and any y in B. If we have a set of vectors S in $\mathcal{H}$, we say that the vectors in S are *pairwise orthogonal* if $x \perp y$ for any two distinct elements of S.

Note that in Euclidean space (3.1.2(a)) this coincides with the traditional definition of orthogonality.

3.2.2. Example. Let $\mathcal{H} = L^2([0, 1], \lambda)$, where λ is Lebesgue measure. For n in $\mathbb{Z}$ let $e_n(t) = \exp(2\pi i n t)$. It follows that $e_n \perp e_m$ whenever $n \neq m$.

3.2.3. Theorem (Pythagorean Theorem). *If $x_1, \ldots, x_n$ are pairwise orthogonal vectors in $\mathcal{H}$, then*

$$\|x_1 + \cdots + x_n\|^2 = \|x_1\|^2 + \cdots + \|x_n\|^2$$

Proof. We prove the case where $n = 2$; the general case proceeds by induction. If $x \perp y$, then the polar identity (3.1.5) implies that $\|x + y\|^2 = \|x\|^2 + 2\text{Re}\,\langle x, y \rangle + \|y\|^2 = \|x\|^2 + \|y\|^2$. ∎

We note that if $x \perp y$, then $x \perp -y$ and so we also have that $\|x - y\|^2 = \|x\|^2 + \|y\|^2$.

3.2.4. Theorem (Parallelogram Law). *If $x, y \in \mathcal{H}$, then*

$$\|x + y\|^2 + \|x - y\|^2 = 2(\|x\|^2 + \|y\|^2)$$

Proof. The proof consists in looking at the polar identity in two different ways and then adding both:

$$\|x + y\|^2 = \|x\|^2 + 2\text{Re}\,\langle x, y \rangle + \|y\|^2$$
$$\|x - y\|^2 = \|x\|^2 - 2\text{Re}\,\langle x, y \rangle + \|y\|^2$$

Now add. ∎

3.2.5. Definition. If $\mathcal{X}$ is any vector space and $A \subseteq \mathcal{X}$, say that A is *convex* if $tx + (1 - t)y \in A$ whenever $x, y \in A$ and $0 \leq t \leq 1$.

Since $\{tx + (1 - t)y : 0 \leq t \leq 1\}$ is the straight line segment joining x and y, this means that A is convex when the line joining any two points in A is wholly contained in A. It is easy to draw pictures of subsets of the plane that are and are not convex. Also observe that the convex subsets of $\mathbb{R}$ are precisely the intervals.

Any open or closed ball in a normed space is convex. If $\mathcal{X}$ is any vector space, then any linear subspace is convex and the intersection of any collection of convex subsets is again convex. So if $A \subseteq \mathcal{X}$, we can define the *convex hull* of A as the intersection of all convex subsets of $\mathcal{X}$ that contain A; denote this by co(A). Similarly, when $\mathcal{X}$ is a normed space we define the *closed convex hull* of A, denoted by $\overline{\text{co}}(A)$, as the closure of co(A). See Exercise 7.

The next result is one the salient facts in the structure of Hilbert space. It says that the distance from a point to a closed convex subset of $\mathcal{H}$ is always attained at a unique point.

3.2.6. Theorem. *If $x \in \mathcal{H}$ and K is a closed convex subset of $\mathcal{H}$, then there is a unique x_0 in K such that*

$$\|x - x_0\| = \text{dist}\,(x, K) = \inf\{\|x - y\| : y \in K\}$$

Proof. Replacing K by $K - x$, we see that without loss of generality we may assume that $x = 0$. (Verify!) By definition there is a sequence $\{y_n\}$ in

K such that $\|y_n\| \to \mathrm{dist}\,(0, K) = d$. The Parallelogram Law implies that

$$\left\|\frac{y_n - y_m}{2}\right\|^2 = \frac{1}{2}(\|y_n\|^2 + \|y_m\|^2) - \left\|\frac{y_n + y_m}{2}\right\|^2$$

Now we use the convexity of K to conclude that $\frac{1}{2}(y_n + y_m) \in K$ and so that $\|\frac{1}{2}(y_n + y_m)\|^2 \geq d^2$. If $\epsilon > 0$ we can choose N such that for $n \geq N$, $\|y_n\|^2 < d^2 + \frac{1}{4}\epsilon^2$. Therefore for $n, m \geq N$

$$\left\|\frac{y_n - y_m}{2}\right\|^2 < \frac{1}{2}\left(2d^2 + \frac{1}{2}\epsilon^2\right) - d^2 = \frac{1}{4}\epsilon^2$$

That is, $\|y_n - y_m\| < \epsilon$ for $n, m \geq N$; thus $\{y_n\}$ is a Cauchy sequence. Since K is closed there is an x_0 in K such that $y_n \to x_0$. Also $d \leq \|x_0\| \leq \|x_0 - y_n\| + \|y_n\| \to d$. This proves existence.

To establish uniqueness of the point x_0, suppose there is another point y_0 in K with $\|y_0\| = d$. Since $\frac{1}{2}(x_0 + y_0) \in K$, $d \leq \|\frac{1}{2}(x_0 + y_0)\| \leq \frac{1}{2}(\|x_0\| + \|y_0\|) = d$, so that $d = \|\frac{1}{2}(x_0 + y_0)\|$. From here we get that the Parallelogram Law implies that $d^2 = \|\frac{1}{2}(x_0 + y_0)\|^2 = d^2 - \|\frac{1}{2}(x_0 - y_0)\|$, whence $x_0 = y_0$. ∎

Now let's see what happens when the convex set in the preceding theorem is a linear subspace.

3.2.7. Theorem. *If $\mathcal{M}$ is a closed linear subspace of $\mathcal{H}$, $x \in \mathcal{H}$, and x_0 is the unique point in $\mathcal{M}$ such that $\|x - x_0\| = \mathrm{dist}\,(x, \mathcal{M})$, then $x - x_0 \perp \mathcal{M}$. Conversely if $x_0 \in \mathcal{M}$ such that $x - x_0 \perp \mathcal{M}$, then x_0 is the unique point in $\mathcal{M}$ with $\|x - x_0\| = \mathrm{dist}\,(x, \mathcal{M})$.*

Proof. Let $x_0 \in \mathcal{M}$ such that $\|x - x_0\| = \mathrm{dist}\,(x, \mathcal{M})$. If y is an arbitrary vector in $\mathcal{M}$, then $x_0 + y \in \mathcal{M}$ and so $\|x - x_0\|^2 \leq \|x - (x_0 + y)\|^2 = \|(x - x_0) - y\|^2 = \|x - x_0\|^2 - 2\mathrm{Re}\,\langle x - x_0, y\rangle + \|y\|^2$. Thus

$$2\mathrm{Re}\,\langle x - x_0, y\rangle \leq \|y\|^2$$

for any y in $\mathcal{M}$. Now we perform a standard trick. Fix a y in $\mathcal{M}$ and suppose $\langle x - x_0, y\rangle = re^{i\theta}$ with $r \geq 0$. Substitute $te^{i\theta}y$ for y in the preceding inequality. This gives us $2\mathrm{Re}\,[te^{-i\theta}re^{i\theta}] \leq t^2\|y\|^2$ or $2tr \leq t^2\|y\|^2$; this is valid for all t. Letting $t \to 0$ shows that $r = 0$; that is, $\langle x - x_0, y\rangle = 0$ so that $x - x_0 \perp y$.

For the converse assume that $x_0 \in \mathcal{M}$ such that $x - x_0 \perp \mathcal{M}$. Thus for any y in $\mathcal{M}$ we have that $x - x_0 \perp x_0 - y$ and so $\|x - y\|^2 = \|(x - x_0) + (x_0 - y)\|^2 = \|x - x_0\|^2 + \|x_0 - y\|^2 \geq \|x - x_0\|^2$. Therefore $\|x - x_0\| = \mathrm{dist}\,(x, \mathcal{M})$. ∎

If $A \subseteq \mathcal{H}$ and $A \neq \emptyset$, let $A^\perp = \{x \in \mathcal{H} : x \perp A\}$. It is easy to see that no matter what properties the set A has, $A^\perp$ is always a closed linear subspace of $\mathcal{H}$. Also let's introduce a bit of notation that could have been introduced earlier in this book. If $\mathcal{X}$ is any normed space, the notation $\mathcal{M} \leq \mathcal{X}$ will mean that $\mathcal{M}$ is a closed linear subspace of $\mathcal{X}$. We need to make a distinction between vector subspaces of $\mathcal{X}$ that are closed and those that are not. A *linear manifold* or *submanifold* is a vector subspace of $\mathcal{X}$ that is not necessarily closed; a *subspace* is a closed linear manifold.

In light of Theorem 3.2.7, whenever $\mathcal{M} \leq \mathcal{H}$ we can define a function $P : \mathcal{H} \to \mathcal{H}$ as follows: for any x in $\mathcal{H}$, Px is the unique vector in $\mathcal{M}$ with $x - Px \perp \mathcal{M}$. Note that when $x \in \mathcal{M}$, $Px = x$. This function will be called the *orthogonal projection* of $\mathcal{H}$ onto $\mathcal{M}$, or simply the *projection* of $\mathcal{H}$ onto $\mathcal{M}$. Sometimes this is designated by $P = P_{\mathcal{M}}$.

Whenever $\mathcal{X}$ and $\mathcal{Y}$ are vector spaces and $T : \mathcal{X} \to \mathcal{Y}$ is a linear transformation, the *kernel* and *range* of T are defined by

$$\ker T = \{x \in \mathcal{X} : Tx = 0\}$$
$$\operatorname{ran} T = \{Tx : x \in \mathcal{X}\}$$

We note that $\ker T$ is a linear manifold in $\mathcal{X}$ and $\operatorname{ran} T$ is a linear manifold in $\mathcal{Y}$.

3.2.8. Theorem. *If $\mathcal{M} \leq \mathcal{H}$ and P is the orthogonal projection of $\mathcal{H}$ onto $\mathcal{M}$, then the following statements are true.*

(a) *P is a linear transformation.*

(b) *$\|Px\| \leq \|x\|$ for all x in $\mathcal{H}$.*

(c) *$P^2 = P$ (here $P^2 = P \circ P$, the composition of P with itself).*

(d) *$\ker P = \mathcal{M}^\perp$ and $\operatorname{ran} P = \mathcal{M}$.*

Proof. Keep in mind that for every x in $\mathcal{H}$, $x - Px \perp \mathcal{M}$ and $\|x - Px\| = \operatorname{dist}(x, \mathcal{M})$.

(a) Let $x_1, x_2 \in \mathcal{H}$ and $a_1, a_2 \in \mathbb{F}$. If $y \in \mathcal{M}$, then $\langle [a_1 x_1 + a_2 x_2] - [a_1 P x_1 + a_2 P x_2], y \rangle = a_1 \langle x_1 - P x_1, y \rangle + a_2 \langle x_2 - P x_2, y \rangle = 0$. Since $a_1 P x_1 + a_2 P x_2 \in \mathcal{M}$, the uniqueness statement in Theorem 3.2.7 implies $P(a_1 x_1 + a_2 x_2) = a_1 P x_1 + a_2 P x_2$ and so P is linear.

(b) If $x \in \mathcal{H}$, then $x - Px \perp Px$ since $x - Px \in \mathcal{M}^\perp$ while $Px \in \mathcal{M}$. Therefore $\|x\|^2 = \|x - Px\|^2 + \|Px\|^2 \geq \|Px\|^2$.

(c) Since $Py = y$ for any y in $\mathcal{M}$, for any x in $\mathcal{H}$ we have that $P(Px) = Px$.

(d) If $Px = 0$, then $x = x - Px \in \mathcal{M}^\perp$. Conversely if $x \in \mathcal{M}^\perp$, then 0 is the unique point in $\mathcal{M}$ such that $x - 0 = x \perp \mathcal{M}$; thus $Px = 0$. Thus $\ker P = \mathcal{M}^\perp$. The fact that $P\mathcal{H} = \mathcal{M}$ is clear. ∎

3.2.9. Corollary. *If $\mathcal{M} \leq \mathcal{H}$, then $\mathcal{M} = (\mathcal{M}^\perp)^\perp$.*

Proof. For this proof we use Exercise 4: if $P = P_{\mathcal{M}}$, then $I - P$ is the projection onto $\mathcal{M}^{\perp}$, where I denotes the identity linear transformation. By the preceding theorem, $(\mathcal{M}^{\perp})^{\perp} = \ker(I - P)$. But $0 = (I - P)x$ if and only if $Px = x$; that is, if and only if $x \in \operatorname{ran} P$. Thus $(\mathcal{M}^{\perp})^{\perp} = \operatorname{ran} P = \mathcal{M}$. ∎

3.2.10. Corollary. *If $A \subseteq \mathcal{H}$, then $(A^{\perp})^{\perp}$ is the closed linear span of A in $\mathcal{H}$.*

For a discussion of the closed linear span of a set, the reader should look at Exercise 1.3.2. The proof of this corollary as well as the next one is straightforward.

3.2.11. Corollary. *If $\mathcal{Y}$ is a linear submanifold of $\mathcal{H}$, then $\mathcal{Y}$ is dense in $\mathcal{H}$ if and only if $\mathcal{Y}^{\perp} = (0)$.*

The preceding corollaries are quite powerful tools. For example, this last corollary says something very deep: that a topological property, the density of $\mathcal{Y}$, is equivalent to an algebraic one, $\mathcal{Y}^{\perp} = (0)$. Whenever two such disparate conditions are equivalent, the force is with us.

Exercises.

(1) Let λ be normalized area measure on the closed unit disk $\operatorname{cl} \mathbb{D}$ in the complex plane. That is, λ is the Radon measure on $\operatorname{cl} \mathbb{D}$ associated with the positive linear functional $f \mapsto \pi^{-1} \int \int_{\mathbb{D}} f(x + iy)\,dx\,dy$ and $\lambda(\mathbb{D}) = 1$. For $n, m \geq 0$, let $f_{nm}(z) = z^n \bar{z}^m$. When is $f_{nm} \perp f_{jk}$? (Hint: Use polar coordinates.)

(2) Let x, y be two linearly independent vectors in $\mathcal{H}$ with $1 = \|x\| = \|y\|$. Show that for $0 < t < 1$, $\|ty + (1 - t)x\| < 1$. What does this say about the geometry of the boundary of ball $\mathcal{H}$? What happens if instead of the Hilbert space $\mathcal{H}$, we look at $\mathbb{F}^2$ with the norm $\|(x, y)\| = |x| + |y|$? (Also see the next exercise.)

(3) Consider $\mathbb{F}^2$ with the norm $\|(x, y)\| = |x| + |y|$. (a) Use a compactness argument to show that if K is any closed subset of $\mathbb{F}^2$ and $x \in \mathbb{F}^2$, then there is a point x_0 in K with $\|x - x_0\| = \operatorname{dist}(x, K)$. (b) Give an example of a closed convex subset K of $\mathbb{F}^2$ and a point x in $\mathbb{F}^2$ such that the distance from x to K is not attained at a unique point. (Also see the end of the preceding exercise.)

(4) If $\mathcal{M} \leq \mathcal{H}$ and P is the projection onto $\mathcal{M}$, show that $I - P$ is the projection onto $\mathcal{M}^{\perp}$.

(5) Let $\mathcal{M}, \mathcal{N} \leq \mathcal{H}$, $P = P_{\mathcal{M}}$, $Q = P_{\mathcal{N}}$. If $PQ = QP$, show that $PQ = P_{\mathcal{M} \cap \mathcal{N}}$. In this case, what is $\ker PQ$?

(6) Let X be a compact metric space and give $C(X)$ its usual supremum norm. Show that this norm is not given by an inner product

      (so that $C(X)$ is not a Hilbert space) by showing that the norm does not satisfy the Parallelogram Law.

(7) If $\mathcal{X}$ is a normed space, show that $\overline{\text{co}}(A)$ is the intersection of all closed convex subsets of $\mathcal{X}$ that contain A.

(8) If F is a closed subset of the normed space $\mathcal{X}$, show that F is convex if and only if for any x, y in F, $\frac{1}{2}(x+y) \in F$. Give a counterexample to this statement if F is not assumed to be closed.

(9) Show that if $\mathcal{H} = \mathbb{F}^d$ with the usual inner product and if K is any closed subset of $\mathcal{H}$, then for any x in $\mathcal{H}$ there is a point x_0 in K with $\|x - x_0\| = \text{dist}(x, K)$. (Later we'll show that this is true in any Hilbert space. Of course since we have not assumed that K is convex, the point x_0 is not necessarily unique.)

(10) Is ℓ^1 a Hilbert space? Why?

3.3. The Riesz Representation Theorem

For Hilbert spaces it is possible to give a simple representation of all the bounded linear functionals on it and the proof of this representation is rather simple. As we will see this is not true of all Banach spaces. For example, when X is a compact metric space, the representation of the elements of $C(X)^*$ will occupy a lot of our energy. In a sense it is our motivation for establishing measure theory. (This motivation is not historically true – measure theory preceded a consideration of bounded linear functionals. The original motivation was to be able to fully explore the Fourier transform and trigonometric series.)

    I am certain that at this point the student will not fully appreciate the importance of representing linear functionals on a Banach or Hilbert space. Why should (s)he? So I'll have to ask the reader to be patient until we see the power of this illustrated by applications. Perhaps the reader's patience can be increased if (s)he reflects on the name of one of the subjects that is partially covered in this book: Functional Analysis. In fact one application of the next theorem on representing linear functionals will appear in the first section of the next chapter. Another application of linear functionals in a Banach space setting will take place in §8.5 when we use them to prove the Stone–Weierstrass Theorem, which generalizes the result that the polynomials are uniformly dense in $C[0, 1]$.

    Start by realizing that if $x_0 \in \mathcal{H}$ and $L : \mathcal{H} \to \mathbb{F}$ is defined by $L(x) = \langle x, x_0 \rangle$, then L is a bounded linear functional on $\mathcal{H}$. The fact that it is linear is a direct consequence of the definition of an inner product; the fact that it is bounded follows from the CBS Inequality: $|L(x)| = |\langle x, x_0 \rangle| \leq \|x\| \|x_0\|$.

So $\|L\| \leq \|x_0\|$. In fact, if $x = \|x_0\|^{-1}x_0$, then $\|x\| = 1$ and $L(x) = \|x_0\|$; hence $\|L\| = \|x_0\|$.

3.3.1. Theorem (Riesz[3] Representation Theorem). *If $\mathcal{H}$ is a Hilbert space and $L : \mathcal{H} \to \mathbb{F}$ is a continuous linear functional, then there is a unique vector x_0 in $\mathcal{H}$ such that $L(x) = \langle x, x_0 \rangle$ for all x in $\mathcal{H}$ and $\|L\| = \|x_0\|$.*

Proof. The theorem is trivially true when $L \equiv 0$; so assume that L is not the zero linear functional and let $\mathcal{M} = \ker L$. Since L is continuous, its kernel is a closed linear subspace of $\mathcal{H}$; since $L \neq 0$, $\mathcal{M} \neq \mathcal{H}$. Thus $\mathcal{M}^{\perp} \neq (0)$ and we can find a vector y_0 in $\mathcal{M}^{\perp}$ such that $L(y_0) = 1$. If $x \in \mathcal{H}$ and $a = L(x)$, then $L(x - ay_0) = 0$; that is, $x - L(x)y_0 \in \mathcal{M}$. Therefore $0 = \langle x - L(x)y_0, y_0 \rangle = \langle x, y_0 \rangle - L(x)\|y_0\|^2$. Since x was arbitrary, we see that for $x_0 = \|y_0\|^{-2}y_0$, $L(x) = \langle x, x_0 \rangle$ for all x in $\mathcal{H}$. Since the proof of the norm equality was done before the statement of the theorem, we are done. ∎

3.3.2. Corollary. *If $(X, \mathcal{A}, \mu)$ is a measure space and $L : L^2(\mu) \to \mathbb{F}$ is a bounded linear functional, then there is a function h in $L^2(\mu)$ such that*

$$L(f) = \langle f, h \rangle = \int f\bar{h}\, d\mu$$

for all f in $L^2(\mu)$ and $\|h\|_2 = \|L\|$.

3.3.3. Corollary. *If $L : \ell^2 \to \mathbb{F}$ is a bounded linear functional, there is a unique sequence $\{b_n\}$ in ℓ^2 such that $L(\{a_n\}) = \sum_{n=1}^{\infty} a_n\bar{b}_n$ for all $\{a_n\}$ in ℓ^2 and $\|L\| = \left(\sum_{n=1}^{\infty} |b_n|^2\right)^{\frac{1}{2}}$.*

Exercises.

(1) Suppose $\mathcal{H}$ is a Hilbert space and $\mathcal{M}$ is a closed subspace. Show that if $L : \mathcal{M} \to \mathbb{F}$ is a bounded linear functional, then there is a bounded linear functional $\widetilde{L} : \mathcal{H} \to \mathbb{F}$ such that $\widetilde{L}(x) = L(x)$ for every x in $\mathcal{M}$ and $\|\widetilde{L}\| = \|L\|$. (Later we will prove this same result in any normed space. It is called the Hahn–Banach Theorem.)

(2) If $\mathcal{H}$ is a Hilbert space, $\mathcal{M}$ is a closed subspace of $\mathcal{H}$, and $x \in \mathcal{H}$ such that $x \notin \mathcal{M}$, show that there is a bounded linear functional

---

[3]Frigyes Riesz was born in 1880 in Győr in present day Hungary. His younger brother, Marcel Riesz, was also a famous mathematician; indeed there is an important theorem in complex/harmonic analysis called the F. and M. Riesz Theorem. F. Riesz obtained his doctorate from the University of Budapest in 1902. He is widely considered as one of the founders of functional analysis and has additional representation theorems named after him besides the one in this section. For example see Theorem 4.3.8. He made fundamental contributions to Fourier analysis and ergodic theory and was a driving force in establishing modern mathematics in Hungary, including the founding of the János Bolyai Mathematical Institute in Szeged in 1922. He brought a high degree of artistry and elegance to mathematics. In 1945 he accepted the chair of mathematics at the University of Budapest, where he remained until his death in 1956.

L on $\mathcal{H}$ such that $L(y) = 0$ for every y in $\mathcal{M}$, $L(x) = 1$, and $\|L\| = [\mathrm{dist}\,(x, \mathcal{M})]^{-1}$.

(3) If $\mathcal{H}$ is a Hilbert space and $\mathcal{M}$ is a closed linear subspace, show that a necessary and sufficient condition for there to be a bounded linear functional L on $\mathcal{H}$ with $\mathcal{M} = \ker L$ is that there is a single vector x in $\mathcal{H}$ with $\mathcal{H} = \mathcal{M} \vee \mathbb{C}x = \bigvee\{\mathcal{M} \cup \{x\}\}$. (This condition can be expressed by saying that $\mathcal{M}$ has codimension 1.)

Chapter 4

A Return to Measure Theory

In this chapter we'll use the results we just developed on Hilbert space to obtain a decomposition of one measure with respect to another and then use this to introduce and explore the concept of a complex-valued measure. Then we'll see some applications of what we have developed by representing bounded linear functionals on certain Banach spaces we have already seen – $C(X), C_0(X), L^p(\mu)$. This will be followed by the introduction of the idea of a product measure. That is, we have a measure μ on a space X and a measure ν on a space Y and we define a measure $\mu \times \nu$ on $X \times Y$. This is the extension of the idea of a multiple integral seen in calculus. Then we'll explore measure theory on Euclidean spaces, including the Fourier transform, and that will conclude the development of measure theory in this book.

4.1. The Lebesgue Radon–Nikodym Theorem

Again we fix a set X and $\mathcal{A}$, a σ-algebra of subsets of X. In this section we begin to explore the relationships between signed measures.

4.1.1. Definition. If μ and ν are signed measures on $(X, \mathcal{A})$, say that ν is *absolutely continuous* with respect to μ if $\nu(E) = 0$ whenever $|\mu|(E) = 0$. In symbols we denote this by $\nu \ll \mu$.

Let's underline three things. First, both measures are allowed to take on infinite values. Second, μ enters the definition only by using its variation, $|\mu|$; hence allowing μ to be a signed measure is a bit of a formality that will, however, allow us some flexibility in the future. Third, if $\nu = \nu_+ - \nu_-$ is the Jordan decomposition of ν, then it is easy to see that $\nu \ll \mu$ if and only if $\nu_\pm \ll \mu$, and this is equivalent to the condition that $|\nu| = \nu_+ + \nu_- \ll \mu$.

4.1.2. Example. If $f \in \mathcal{M}_+(X, \mathcal{A})$, $g \in L^1(\mu)$, and we define $\nu(E) = \int_E f d\mu - \int_E g d\mu$ for every E in $\mathcal{A}$, then ν is a signed measure and $\nu \ll \mu$.

The Radon–Nikodym Theorem below says that the only way to obtain a measure absolutely continuous with respect to μ is as in the preceding example, at least under a mild restriction on the measure space $(X, \mathcal{A}, \mu)$.

4.1.3. Definition. If ν and μ are signed measures on $(X, \mathcal{A})$, say that they are *mutually singular* if we can find a set E in $\mathcal{A}$ such that $|\nu|(E) = 0$ and $|\mu|(X\backslash E) = 0$. In symbols we write $\nu \perp \mu$.

In other words, the statement that ν and μ are mutually singular means that the two measures "live" on or are carried by disjoint sets.

4.1.4. Example. (a) Let $\mathcal{B}$ be the σ-algebra of Borel subsets of $\mathbb{R}$ and let λ be Lebesgue measure. If δ_0 is the unit point mass at 0, then λ and δ_0 are mutually singular. In fact these two measures are carried by the disjoint sets $\{0\}$ and $\mathbb{R}\backslash\{0\}$.

(b) Let c be the Cantor function on $[0, 1]$. Since c is increasing we can define the positive linear functional $L : C[0, 1] \to \mathbb{R}$ by $L(f) = \int_0^1 f dc$ and thus generate the Radon measure μ. If we consider both μ and Lebesgue measure λ on $[0, 1]$ as defined on the Borel sets, then μ and λ are mutually singular. In fact if C is the Cantor set, then, because c is constant on every subinterval of $[0, 1]\backslash C$, we have that $\mu([0, 1]\backslash C) = 0$ while $\lambda(C) = 0$.

(c) If ω is any signed measure and E and F are two disjoint sets in $\mathcal{A}$, then $\mu = \omega_E$ and $\nu = \omega_F$ are mutually singular. Every pair of mutually singular signed measures can be obtained in this way, for if ν, μ, and E are as in the definition and we put $F = X\backslash E$ and $\omega = \mu + \nu$, then $\mu = \omega_E$ and $\nu = \omega_F$.

4.1.5. Definition. Say that a signed measure μ on $(X, \mathcal{A})$ is *σ-finite* if there is a sequence $\{E_n\}$ in $\mathcal{A}$ such that $X = \bigcup_{n=1}^\infty E_n$ and $|\mu|(E_n) < \infty$ for all $n \geq 1$.

Clearly every finite measure space is σ-finite as is the measure space defined by Lebesgue measure on $\mathbb{R}$.

4.1.6. Theorem (Lebesgue–Radon–Nikodym[1] Theorem). *If μ and ν are σ-finite positive measures on $(X, \mathcal{A})$, then the following hold.*

---

[1]Otton Marcin Nikodym was born in 1887 in Zablotow, then in Austria-Hungary now in Ukraine. In 1911 he became a high school teacher in Krakow and stayed there until 1924, doing mathematical research but with little concern about an advanced degree. Under some prodding

(a) There are unique measures ν_a and ν_s on $(X, \mathcal{A})$ such that $\nu = \nu_a + \nu_s$, $\nu_a \perp \nu_s$, $\nu_a \ll \mu$, and $\nu_s \perp \mu$.

(b) There is a unique non-negative $\mathcal{A}$-measurable function f such that $\int \psi \, d\nu_a = \int \psi f d\mu$ for every function ψ in $\mathcal{M}_+(X, \mathcal{A})$. In particular, $\nu_a(E) = \int_E f d\mu$ for every E in $\mathcal{A}$.

Proof. We consider two cases.

Case 1. Both μ and ν are finite positive measures. Let $\omega = \mu + \nu$. If u is any simple $\mathcal{A}$-measurable function, then the CBS Inequality implies that $\left| \int u \, d\nu \right| \leq \int |u| \, d\nu \leq \int |u| \, d\omega \leq \sqrt{\omega(X)} \|u\|_2$, where the norm is for the space $L^2(\omega)$. Since the simple functions are dense in $L^2(\omega)$, this says that the linear functional L defined on the simple functions in $L^2(\omega)$ by $L(u) = \int u \, d\nu$ extends to a bounded linear functional $L : L^2(\omega) \to \mathbb{R}$. (See Exercise 1.5.6.) By the Riesz Representation Theorem there is an h_0 in $L^2(\omega)$ such that $L(g) = \langle g, h_0 \rangle = \int g h_0 \, d\omega = \int g h_0 d\mu + \int g h_0 d\nu$ for all g in $L^2(\omega)$. So for simple functions u we have $\int u d\nu = L(u) = \int u h_0 d\mu + \int u h_0 d\nu$. In particular if we take for u the characteristic function of some set, we see that $\nu(E) = \int_E h_0 \, d\omega$ for every E in $\mathcal{A}$.

Since all the measures in question are positive, it easily follows that $h_0(x) \geq 0$ a.e. $[\omega]$. On the other hand if $\epsilon > 0$ and $A_\epsilon = \{x \in X : h_0(x) \geq 1 + \epsilon\}$, then we have that $(1 + \epsilon)\omega(A_\epsilon) \leq \int_{A_\epsilon} h_0 \, d\omega = \nu(A_\epsilon) \leq \omega(A_\epsilon)$. Thus $\omega(A_\epsilon) = 0$ for every $\epsilon > 0$ and so $h_0(x) \leq 1$ a.e. $[\omega]$. So we can assume that $0 \leq h_0 \leq 1$ on X.

By using Proposition 2.4.17 and the MCT we have that for every ϕ in $\mathcal{M}_+(X, \mathcal{A})$, $\int \phi \, d\nu = \int \phi h_0 \, d\omega = \int \phi h_0 \, d\nu + \int \phi h_0 \, d\mu$. Hence

4.1.7
$$\int \phi(1 - h_0) \, d\nu = \int \phi h_0 d\mu$$

for all ϕ in $\mathcal{M}_+(X, \mathcal{A})$. Put $E = \{x \in X : 0 \leq h_0(x) < 1\}$ and $F = X \backslash E = \{x \in X : h_0(x) = 1\}$. For every set A in $\mathcal{A}$ define

$$\nu_a(A) = \nu(A \cap E)$$
$$\nu_s(A) = \nu(A \cap F)$$

Clearly $\nu = \nu_a + \nu_s$ and $\nu_a \perp \nu_s$ since they are carried by disjoint sets. In (4.1.7) take $\phi = \chi_F$ and we get $0 = \int_F (1 - h_0) \, d\nu = \mu(F)$; hence $\nu_s \perp \mu$, the

from Sierpinski, his former professor, he took the doctoral exams at the University of Warsaw. He began teaching at the university in Krakow and publishing his research. After a year at the Sorbonne, he received his habilitation from Warsaw in 1927. At this time he showed the existence of what has become known as a Nikodym set in the unit square S in the Euclidean plane. This is a subset E of S with area equal to 1 such that for every point (x, y) in E, there is a straight line through (x, y) that meets E only at (x, y). Just after World War II he left Poland for Belgium and France and in 1948 he joined the faculty at Kenyon College in the US, retired in 1966, and moved to Utica NY where he died in 1974. He made contributions to measure theory and functional analysis.

last part of (a). Also for $n \geq 1$ and ψ in $\mathcal{M}_+(X, \mathcal{A})$, substitute the function $\psi(1 + h_0 + \cdots + h_0^n)$ for ϕ in (4.1.7) to get

$$\int \psi(1 + h_0 + \cdots + h_0^n)h_0 d\mu = \int \psi(1 + h_0 + \cdots + h_0^n)(1 - h_0)\, d\nu$$

$$= \int \psi(1 - h_0^{n+1})\, d\nu$$

$$= \int_E \psi(1 - h_0^{n+1})\, d\nu$$

$$= \int \psi d\nu_a - \int_E \psi h_0^{n+1} d\nu$$

But on the set E, $0 \leq h_0(x) < 1$, so $\{\psi(x)h_0^{n+1}(x)\}$ is decreasing to 0. Therefore the MCT implies that the right hand side of this equation converges to $\int \psi\, d\nu_a$. On the other hand $\{\psi(1 + h_0 + \cdots + h_0^n)h_0\}$ is an increasing sequence of functions. Therefore there is a function f in $\mathcal{M}_+(X, \mathcal{A})$ such that $[1 + h_0(x) + \cdots + h_0^n(x)]h_0(x) \nearrow f(x)$ for all x, and so $\psi(x)[1 + h_0(x) + \cdots + h_0^n(x)]h_0(x) \nearrow \psi(x)f(x)$ for all x. Again the MCT implies that $\int \psi(1 + h_0 + \cdots + h_0^n)h_0 d\mu \nearrow \int \psi f d\mu$. Thus when we pass to the limit, the last displayed equation implies

$$\int \psi f\, d\mu = \int \psi\, d\nu_a$$

for every function ψ in $\mathcal{M}_+(X, \mathcal{A})$. In particular, we get $\int_A f\, d\mu = \nu_a(A)$ for every A in $\mathcal{A}$; thus $\nu_a \ll \mu$ and we also have (b).

It remains to establish the uniqueness part of the statement. Assume that we also have $\nu = \nu_a' + \nu_s'$ and $\int \psi d\nu_a' = \int \psi f' d\mu$ for every ψ in $\mathcal{M}_+(X, \mathcal{A})$, where ν_a', ν_s' are carried by disjoint sets E', F', respectfully, and $X = E' \cup F'$. Since $\nu_a \ll \mu$ we have that $\nu_s' \perp \nu_a$ and so $\nu_s'(E) = 0$. Thus for any set A in $\mathcal{A}$ we have that $\nu_a(A) = \nu(A \cap E) = \nu_a'(A \cap E) + \nu_s'(A \cap E) = \nu(A \cap E \cap E')$. Similarly $\nu_a'(A) = \nu(A \cap E \cap E')$. Therefore $\nu_a = \nu_a'$ and it follows that $\nu_s = \nu - \nu_a = \nu - \nu_a' = \nu_s'$.

Since $\nu_a = \nu_a'$, $\int (f - f')\psi d\mu = 0$ for every ψ in $\mathcal{M}_+(X, \mathcal{A})$. Set $A = \{x : f(x) > f'(x)\}$. Putting $\psi = \chi_A$ we get that $0 = \int_A (f - f')d\mu$ and the integrand is non-negative; hence $\mu(A) = 0$ so that $f(x) \leq f'(x)$ a.e. $[\mu]$. A similar argument with the set $\{x : f'(x) > f(x)\}$ shows we also have that $f(x) \geq f'(x)$ a.e. $[\mu]$.

This completes the proof of Case 1. We want to point out, however, that because ν is finite, the function f belongs to $L^1(\mu)$. (Take $\psi = 1$.)

Case 2. The measures μ and ν are σ-finite. Under this assumption we can find sets $\{X_n\}$ in $\mathcal{A}$ such that for every $n \geq 1$, $X_n \subseteq X_{n+1}$ and both $\nu(X_n)$ and $\mu(X_n)$ are finite. Consider the restrictions of ν and μ to $(X_n, \mathcal{A}_{X_n})$, denoted by ν_n and μ_n, and apply Case 1. So for each $n \geq 1$

we have $\nu_n = \nu_{na} + \nu_{ns}$, where $\nu_{na} \perp \nu_{ns}$ and there is a unique function h_n in $\mathcal{M}_+(X_n, \mathcal{A}_{X_n})$ such that $\int_{X_n} \psi \, d\nu_{na} = \int_{X_n} \psi h_n d\mu$ for all ψ in $\mathcal{M}_+$. By using the uniqueness part of Case 1 it follows that when $m \geq n$ and $E \in \mathcal{A}$ with $E \subseteq X_n$, $\nu_{ma}(E) = \nu_{na}(E)$, $\nu_{ms}(E) = \nu_{ns}(E)$, and $h_n(x) = h_m(x)$ a.e. $[\mu]$ on X_n.

Since the sets $\{X_n\}$ are increasing, for A in $\mathcal{A}$ we can define $\nu_a(A) = \lim_n \nu_{na}(A)$ and $\nu_s = \lim_n \nu_{ns}(A)$. It is easily verified that these are measures, $\nu = \nu_a + \nu_s$, $\nu_a \ll \mu$, and $\nu_a \perp \nu_s$. Define f_n on X by setting $f_n(x) = \sup\{h_1(x), \ldots, h_n(x)\}$ for x in X_n and $f_n(x) = 0$ otherwise. It follows that $f_n \in \mathcal{M}_+$, $\int_{X_n} \psi \, d\nu = \int \psi f_n d\mu$ for all ψ in $\mathcal{M}_+$, and that $\{f_n\}$ is an increasing sequence; put $f(x) = \lim_n f_n(x)$ for all x in X. So $f \in \mathcal{M}_+$. If ψ is any function in $\mathcal{M}_+$, $\{\psi f_n\}$ is also an increasing sequence converging pointwise to ψf. By the MCT we have that $\int \psi \, d\nu_a = \lim_n \int_{X_n} \psi \, d\nu_{na} = \lim_n \int \psi f_n d\mu = \int \psi f d\mu$. The uniqueness follows as in Case 1. ■

Historically the statement in part (a) is called the Lebesgue Decomposition Theorem and the statement in (b) is called the Radon–Nikodym Theorem. The decomposition of $\nu = \nu_a + \nu_s$ in part (a) is called the *Lebesgue decomposition* of ν with respect to μ. The function f in part (b) is called the *Radon–Nikodym derivative* of ν_a with respect to μ and is denoted by $f = \frac{d\nu_a}{d\mu}$. The proof given here of Theorem 4.1.6 is due, I believe, to John von Neumann.

4.1.8. Corollary. *If μ and ν are σ-finite positive measures on $(X, \mathcal{A})$ with $\nu \ll \mu$ and $f = \frac{d\nu}{d\mu}$, then whenever $g \in L^1(\nu)$, $fg \in L^1(\mu)$ and*

$$\int g \, d\nu = \int g f \, d\mu$$

Proof. Since $\nu \geq 0$, we have that $f \geq 0$ (Exercise 4). We know from (4.1.6) that $\int g \, d\nu = \int g f \, d\mu$ whenever $g \geq 0$. If $g \in L^1(\nu)$, write $g = g_+ - g_-$ and we have the proof. ■

The next result is also called the Lebesgue–Radon–Nikodym Theorem, but in this version we allow ν to be a signed measure but require it to be finite-valued.

4.1.9. Corollary. *If μ is a σ-finite positive measure on $(X, \mathcal{A})$ and ν is a finite signed measure, then the following hold.*

(a) *There are unique signed measures ν_a and ν_s on $(X, \mathcal{A})$ such that $\nu = \nu_a + \nu_s$, $\nu_a \perp \nu_s$, $\nu_a \ll \mu$, and $\nu_s \perp \mu$.*

(b) *There is a unique function f in $L^1(\mu)$ such that $\nu_a(E) = \int_E f d\mu$ for every E in $\mathcal{A}$, and for every ψ in $L^1(|\nu_a|)$ we have that $f\psi \in L^1(\mu)$ and $\int \psi\, d|\nu_a| = \int \psi f d\mu$.*

Proof. Let $\nu = \nu_+ - \nu_-$ be the Jordan decomposition of ν and apply Theorem 4.1.6 to both $\nu_\pm$. Let $f_\pm = d(\nu_\pm)_a/d\mu$ and put $f = f_+ - f_-$. Because ν is finite, it follows readily that $f \in L^1(\mu)$. The rest of the corollary follows from Theorem 4.1.6. (At the risk of beating a dead horse, let's underline that in this corollary we insist that the function ψ belong to $L^1(|\nu_a|)$ rather than only requiring that ψ is a non-negative measurable function as in (4.1.6). This is forced on us since we are allowing ν to be a signed measure.) ∎

4.1.10. Example. The assumption that μ is σ-finite is crucial for the Radon–Nikodym Theorem. For example consider the σ-algebra $\mathcal{A}$ of Borel subsets of $\mathbb{R}$ and let μ be counting measure defined on $\mathcal{B}$: $\mu(E)$ is the number of points in E when E is a finite set and $\mu(E) = \infty$ otherwise. Clearly μ is not σ-finite. Now let λ be Lebesgue measure. Clearly $\lambda \ll \mu$. However there is no positive Borel function f with $\lambda(E) = \int_E f d\mu$. In fact if such a function f existed and we let $E = [0,1]$, we would have that $1 = \int_{[0,1]} f d\mu$; but then f would have to vanish at all but a countable number of points of $[0,1]$ and this cannot be. (Why?)

Do we need to assume that the measure ν in the Radon–Nikodym Theorem is σ-finite? No; see Exercise 3. In the material below when we want to refer to Theorem 4.1.6 or any of the small extensions to more general measures we will cite the LRNT.

In a sense the next result might be how we would have defined the concept of an absolutely continuous measure, if left on our own. It certainly more closely matches and extends what we have for the definition of a continuous function. (We might also mention here that we will define the concept of an absolutely continuous function, and this will be related to the concept of an absolutely continuous measure. See §4.10.) The drawback of using this equivalent formulation of an absolutely continuous measure is that it requires one of the measures to be finite. Thus we retain the original definition and have this special case for finite measures.

4.1.11. Proposition. *If μ is a σ-finite measure and ν a finite signed measure on $(X, \mathcal{A})$, then $\nu \ll \mu$ if and only if for every $\epsilon > 0$ there is a $\delta > 0$ such that $|\nu|(E) < \epsilon$ whenever $\mu(E) < \delta$.*

Proof. Without loss of generality we may assume that ν is positive. It follows quickly that if for every $\epsilon > 0$ there is a $\delta > 0$ such that $\nu(E) < \epsilon$ whenever $\mu(E) < \delta$, then $\nu \ll \mu$. In fact if $\mu(A) = 0$, then $\mu(A) < \delta$ and so $\nu(A) < \epsilon$ for every ϵ; since ϵ is arbitrary, $\nu(A) = 0$.

Now assume that $\nu \ll \mu$ and let $f = \frac{d\nu}{d\mu}$. Since ν is a positive finite measure, f is a positive function in $L^1(\mu)$; we can assume that $0 \leq f(x) < \infty$ for all x in X. (Why?) Let $\epsilon > 0$. If $F_n = \{x : f(x) \geq n\}$, then $F_n \supseteq F_{n+1}$ and $\bigcap_{n=1}^{\infty} F_n = \emptyset$. Again the fact that ν is a finite measure implies $\nu(F_n) \to 0$. Pick m such that $\nu(F_m) < \epsilon/2$. For any set E in $\mathcal{A}$ we have that $\nu(E) = \nu(E \cap F_m) + \nu(E \backslash F_m) < \epsilon/2 + \int_{E \backslash F_m} f d\mu \leq \epsilon/2 + m\mu(E \backslash F_m) \leq \epsilon/2 + m\mu(E)$. Therefore if we let $\delta = \epsilon/2m$, we have that $\mu(E) < \delta$ implies that $\nu(E) < \epsilon$. ∎

4.1.12. Example. Let ν be Lebesgue measure on $X = [1, \infty)$ and let $\mu(E) = \int_E x^{-2} d\nu$ for every Borel set E in $[1, \infty)$. It follows that $\nu \ll \mu$. (It also follows that $\mu \ll \nu$, but this is not relevant to this example.) For any $\delta > 0$, the set $E_\delta = [\delta^{-1}, \infty)$ has $\mu(E_\delta) = \delta$ but $\nu(E_\delta) = \infty$. Hence the requirement that ν be finite in the preceding proposition is essential.

4.1.13. Proposition (Chain Rule). *Let ν be a σ-finite signed measure and let μ and η be σ-finite positive measures on $(X, \mathcal{A})$. If $\nu \ll \mu$ and $\mu \ll \eta$, then $\nu \ll \eta$ and*

$$\frac{d\nu}{d\eta} = \frac{d\nu}{d\mu}\frac{d\mu}{d\eta}$$

Proof. It is immediate that $\nu \ll \eta$. If $f = \frac{d\nu}{d\mu}$ and $g = \frac{d\mu}{d\eta}$, then we know that for every E in $\mathcal{A}$ and every h in $\mathcal{M}_+$, $\nu(E) = \int_E f d\mu$ and $\int h d\mu = \int h g d\eta$. Therefore $\int_E \frac{d\nu}{d\eta} d\eta = \nu(E) = \int \chi_E f d\mu = \int \chi_E f g d\eta = \int_E f g d\eta$ for every set E in $\mathcal{A}$. Thus the proposition follows by the uniqueness of the Radon–Nikodym derivative. ∎

For another proof of the Lebesgue Decomposition Theorem, see [**3**].

Exercises.

(1) When μ is a σ-finite measure, show that there can be no infinite atoms and there are at most a countable number of finite atoms $\{E_n\}$. Show that when the measure is σ-finite it can be written as the sum of a *continuous measure* – one with no atoms – and a *purely atomic measure* – a measure of the form $\eta = \sum_{n=1}^{\infty} a_n \delta_{x_n}$, where each a_n is a positive scalar, $\{x_n\}$ is a sequence of distinct points in X, and δ_{x_n} is the unit point mass at x_n.

(2) In Theorem 4.1.6, look at each of the statements that some pair of the three conditions $\nu_a \perp \nu_s$, $\nu_a \ll \mu$, and $\nu_s \perp \mu$ imply the third, and either prove it or give a counterexample.

(3) (a) Let μ be a finite measure and let ν be an arbitrary measure such that $\nu \ll \mu$. For each $n \geq 1$ let $\{P_n, N_n\}$ be the Hahn decomposition of the signed measure $\nu - n\mu$; put $P = \bigcap_{n=1}^{\infty} P_n$, $N = \bigcup_{n=1}^{\infty} N_n$.

Show that $\nu|N$ is a σ-finite measure and if $A \in \mathcal{A}$ such that $A \subseteq P$, then either $\nu(A) = 0$ or $\nu(A) = \infty$. (b) Use part (a) to show that the Radon–Nikodym Theorem holds when μ is σ-finite but ν is an arbitrary measure.

(4) In the Lebesgue–Radon–Nikodym Theorem show that if ν is a positive measure, then $f(x) \geq 0$ a.e. $[\mu]$.

4.2. Complex functions and measures

We will extend the concept of a measure to allow it to take on complex values. In the process we will also allow our functions to assume complex values, and then we will begin to integrate complex-valued functions with respect to complex-valued measures. To set the notation we refer the reader to the start of §3.1.

Note that for any set S, a function $Z : S \to \mathbb{C}$ can be written as $Z = X + iY$, where $X, Y : S \to \mathbb{R}$ are defined by $X(s) = \operatorname{Re} Z(s)$ and $Y(s) = \operatorname{Im} Z(s)$ and these functions are called the real and imaginary parts of Z. Also for a function $Z : S \to \mathbb{C}$, we define $\operatorname{sign} Z : S \to \mathbb{C}$ by $(\operatorname{sign} Z)(s) = \operatorname{sign}[Z(s)]$.

Essentially any property of a complex-valued function can be defined and examined by a consideration of its real and imaginary parts. We have already seen this, to an extent, when we examined continuous functions in Chapter 1. Note, however, that it is often preferable to have a definition and examination of a complex-valued function without resorting to its real and imaginary parts. This was certainly true for continuity where a function $f : X \to \mathbb{C}$, for a metric space X, is defined as continuous in the same way we define a function from X into any metric space to be continuous. Similarly, if $\mathcal{A}$ is a σ-algebra of subsets of an arbitrary set X and $f : X \to \mathbb{C}$, we say that f is a *measurable function* if $f^{-1}(G) \in \mathcal{A}$ for every open subset G of the complex plane. It then becomes an exercise to show that f is measurable if and only if $\operatorname{Re} f$ and $\operatorname{Im} f$ are measurable (Exercise 1). Also we have that $\operatorname{sign}(f)$ is measurable. Note that if $u = \operatorname{sign}(f)$, then $|u| = 1$ and $f = u|f|$. If μ is a positive measure defined on $(X, \mathcal{A})$, we define $L^p_{\mathbb{C}}(\mu)$ to be the equivalence classes of measurable functions f on X such that $|f| \in L^p(\mu)$. Again it is an exercise to show that $f \in L^p_{\mathbb{C}}(\mu)$ if and only if $\operatorname{Re} f, \operatorname{Im} f \in L^p(\mu)$. For f in $L^1_{\mathbb{C}}(\mu)$ we define $\int f d\mu = \int \operatorname{Re} f d\mu + i \int \operatorname{Im} f d\mu$. Needless to say, it must be verified that the usual properties of integrals continue to hold for complex-valued functions. The reader will be forgiven, however, if (s)he takes many of these statements on faith and verifies only the occasional one that tweaks his/her fancy. There are some facts that were established for real-valued functions whose proofs in the complex case need a bit of effort; that will be supplied here. The next proposition is the first such example.

4.2.1. Proposition. *If $(X, \mathcal{A}, \mu)$ is a measure space and $f : X \to \mathbb{C}$ is an integrable function, then $\left| \int f d\mu \right| \leq \int |f| d\mu$.*

Proof. If $a = \mathrm{sign} \left[\overline{\int f d\mu} \right]$, then

$$\left| \int f d\mu \right| = a \int f d\mu = \mathrm{Re} \int a f d\mu = \int \mathrm{Re}\,(af) d\mu \leq \int |f| d\mu$$

since for any complex number z, $\mathrm{Re}\, z \leq |z|$. ∎

We note that with the preceding proposition established, Hölder's Inequality is seen to remain valid in the complex case. In fact for f in $L_{\mathbb{C}}^p(\mu)$ and g in $L_{\mathbb{C}}^q(\mu)$ we have $|f| \in L^p(\mu)$ and $|g| \in L^q(\mu)$ so that $|fg| \in L^1(\mu)$. Thus fg is integrable with respect to μ and so $\left| \int fg d\mu \right| \leq \int |fg| d\mu \leq \|f\|_p \|g\|_q$. As in the real-valued case it now follows that $L_{\mathbb{C}}^p(\mu)$ is a Banach space.

The next result extends Theorem 2.7.9 to the complex case.

4.2.2. Theorem. *If $1 \leq p < \infty$ and $\frac{1}{p} + \frac{1}{q} = 1$, then for every f in $L_{\mathbb{C}}^p(\mu)$*

$$\|f\|_p = \sup \left\{ \left| \int fg d\mu \right| : g \in \mathrm{ball}\, L_{\mathbb{C}}^q(\mu) \right\}$$

The correct interpretation of this when $p = \infty$ is obtained by letting $q = 1$. If μ has no infinite atoms and $f \in L^\infty(\mu)$, then

$$\|f\|_\infty = \sup \left\{ \left| \int fg d\mu \right| : g \in \mathrm{ball}\, L_{\mathbb{C}}^1(\mu) \right\}$$

Proof. This is derived from (2.7.9) by the following device. For f in $L_{\mathbb{C}}^p(\mu)$, observe that $\|f\|_p = \|\,|f|\,\|_p = \sup \left\{ \left| \int |f| g d\mu \right| : g \in \mathrm{ball}\, L_{\mathbb{C}}^q(\mu) \right\}$. Now let $u = \mathrm{sign}(f)$; so $|u| = 1$ and $\{ug : g \in \mathrm{ball}\, L_{\mathbb{C}}^q(\mu)\} = \mathrm{ball}\, L_{\mathbb{C}}^q(\mu)$. The equalities now follow. ∎

All the convergence results from §2.5 have their complex counterparts, except, of course, for the MCT. In fact the MCT doesn't make any sense in the complex case. Since the results proven for $L_{\mathbb{R}}^p(\mu)$ carry over to $L_{\mathbb{C}}^p(\mu)$, we will now drop the subscript $\mathbb{C}$ in $L_{\mathbb{C}}^p(\mu)$. Henceforth

$$L^p(\mu)$$

will denote the space $L_{\mathbb{F}}^p(\mu)$, where $\mathbb{F}$ could be either $\mathbb{R}$ or $\mathbb{C}$.

Now for complex-valued measures.

4.2.3. Definition. A *complex measure* or *complex-valued measure* on $(X, \mathcal{A})$ is a function $\mu : \mathcal{A} \to \mathbb{C}$ such that $\mathrm{Re}\, \mu$ and $\mathrm{Im}\, \mu$ are finite signed measures.

A point of emphasis here is that we do not allow the measures $\operatorname{Re}\mu$ and $\operatorname{Im}\mu$ to assume any infinite values since this would introduce a multitude of arithmetic ambiguities; the point at infinity in the extended complex numbers or Riemann sphere $\mathbb{C}_\infty$ is fine as a geometric concept, but arithmetically it just does not behave itself.

A measurable function $f : X \to \mathbb{C}$ is *integrable* with respect to a complex measure μ if $|f|$ is integrable with respect to to $\operatorname{Re}\mu$ and $\operatorname{Im}\mu$ and we define $\int f d\mu \equiv \int f d\operatorname{Re}\mu + i \int f d\operatorname{Im}\mu$. Once again we leave it to the reader to satisfy his/her whims in verifying that the basic properties of an integral (for example, linearity) are satisfied.

Note that for a complex measure μ we can construct the Jordan decomposition of $\operatorname{Re}\mu$ and $\operatorname{Im}\mu$ to get a *Jordan decomposition* of μ: $\mu = (\mu_1 - \mu_2) + i(\mu_3 - \mu_4)$, where each of the measures μ_j, $1 \leq j \leq 4$, is positive. Though we have that the measures μ_1 and μ_2 are carried by disjoint sets in $\mathcal{A}$ and that the same applies to the pair of measures μ_3 and μ_4, we can make no such statement about all four measures as the reader can easily verify by writing down an almost random example.

The reader can check that if μ is a complex measure and η is a positive measure, then the definition that μ is absolutely continuous with respect to η makes complete sense. Similarly, to say that two complex measures are mutually singular means the same thing as when the measures are signed.

4.2.4. Theorem (Lebesgue–Radon–Nikodym Theorem). *If μ is a complex measure and η is a σ-finite positive measure on $(X, \mathcal{A})$, then the following hold.*

(a)   *There are unique complex measures μ_a and μ_s on $(X, \mathcal{A})$ such that $\mu = \mu_a + \mu_s$, $\mu_a \perp \mu_s$, $\mu_a \ll \eta$, and $\mu_s \perp \eta$.*

(b)   *There is a unique function f in $L^1(\eta)$ such that whenever g is integrable with respect to μ_a, then $gf \in L^1(\eta)$ and $\int g \, d\mu_a = \int gf d\eta$. In particular, $\mu_a(E) = \int_E f d\eta$ for every E in $\mathcal{A}$.*

The proof is easily obtained by applying Theorem 4.1.6 and Corollary 4.1.9 to the real and imaginary parts of μ.

There is an issue with complex measures, however, that presents technical problems: defining its variation. Recall in (2.6.7) that for a signed measure $\mu = \mu_+ - \mu_-$, we defined the variation of μ as the measure $|\mu| = \mu_+ + \mu_-$; then, in Proposition 2.6.8, we gave an alternative formula for $|\mu|(E)$ for any measurable set E. We use that alternative formula as the definition of the variation of a complex measure.

4.2.5. Theorem. *If μ is a complex measure on $(X, \mathcal{A})$ and for any set E in $\mathcal{A}$ we define*

$$|\mu|(E) = \sup\left\{\sum_{k=1}^{n} |\mu(A_k)| : \{A_1, \ldots, A_n\} \text{ is an } \mathcal{A}\text{-partition of } E\right\}$$

then $|\mu|$ is a finite positive measure. Moreover, if ν is any positive measure such that $|\mu(E)| \leq \nu(E)$ for all E in $\mathcal{A}$, then $|\mu| \leq \nu$.

Proof. Consider the Jordan decomposition $\mu = \mu_1 - \mu_2 + i(\mu_3 - \mu_4)$ and let $\eta = \mu_1 + \mu_2 + \mu_3 + \mu_4$. We note that η is a finite positive measure and $\mu \ll \eta$. According to Theorem 4.2.4 there is a function f in $L^1(\eta)$ such that whenever g is integrable with respect to μ, $gf \in L^1(\eta)$ and

$$\int g \, d\mu = \int g f \, d\eta$$

We will show that for any E in $\mathcal{A}$, the expression for $|\mu|(E)$ in the statement of the theorem equals $\int_E |f| \, d\eta$, which will show that $|\mu|$ is a measure. If $\{A_1, \ldots, A_n\}$ is an $\mathcal{A}$-partition of E, then it follows from the definition that $\sum_{k=1}^{n} |\mu(A_k)| = \sum_{k=1}^{n} \left|\int_{A_k} f \, d\eta\right| \leq \int_E |f| \, d\eta$; hence $|\mu|(E) \leq \int_E |f| \, d\eta$.

Now fix E, consider the measure space $(E, \mathcal{A}_E, \eta_E)$, and apply Theorem 4.2.2: $\int_E |f| \, d\eta = \sup\left\{\left|\int_E fg \, d\eta\right| : g \in \text{ball } L^\infty(\eta_E)\right\}$. Now the bounded simple functions are dense in $L^\infty(\eta_E)$ (2.7.12), so the simple functions in ball $L^\infty(\eta_E)$ are dense in ball $L^\infty(\eta_E)$. (Why?) Thus for any $\epsilon > 0$ there is a simple function u in ball $L^\infty(\eta_E)$ with $\int_E |f| \, d\eta < \epsilon + \left|\int_E fu \, d\eta\right|$; let $u = \sum_k a_k \chi_{A_k}$. So $\{A_1, \ldots, A_n\}$ is an $\mathcal{A}$-partition of E and, because $\|u\|_\infty \leq 1$, $|a_k| \leq 1$ for $1 \leq k \leq n$. Therefore

$$\int_E |f| \, d\eta < \epsilon + \left|\int_E fu \, d\eta\right|$$

$$= \epsilon + \left|\sum_{k=1}^{n} a_k \int_{A_k} f \, d\eta\right|$$

$$\leq \epsilon + \sum_{k=1}^{n} \left|\int_{A_k} f \, d\eta\right|$$

$$= \epsilon + \sum_{k=1}^{n} |\mu(A_k)|$$

$$\leq \epsilon + |\mu|(E)$$

Since ϵ was arbitrary, we obtain the other needed inequality and $|\mu|(E) = \int_E |f| \, d\eta$.

The fact that $|\mu|$ satisfies the minimizing property stated in the theorem is an easy manipulation. (Also see Exercise 2.6.3.) ■

Now we can extend Proposition 2.6.10 to the space $M(X, \mathcal{A})$ of complex-valued measures defined on $(X, \mathcal{A})$ and show that $\|\mu\| = |\mu|(X)$ defines a norm on $M(X, \mathcal{A})$ and with this norm $M(X, \mathcal{A})$ is a Banach space over the complex numbers. At this point it is instructive to call the reader's attention to Corollary 4.3.4 in the next section.

It is easy to see that a measurable function $f : X \to \mathbb{C}$ is integrable with respect to a complex measure μ if and only if it is integrable with respect to the positive measure $|\mu|$. It may have occurred to the reader that extending results from signed measures to complex measures presents few difficulties unless there is an inequality involved. That is the way it seems to the author as well.

Terminology Notice. *The term "positive measure" retains its original meaning as a measure that takes its values in the non-negative extended real numbers. Similarly we will continue to use the term "signed measure" to mean a measure that takes its values in the extended real numbers. Henceforward, however, the term "measure" with no modifier will mean a complex-valued measure.*

The only real change here is that the term "measure" previously was reserved for what we are now calling a positive measure, though for emphasis we may sometimes use the term "complex measure." This decree will necessitate some care. The point is we want to discuss both complex measures and signed measures, and so we have to accommodate the fact that complex measures cannot take on infinite values. Therefore a signed measure is not a complex measure if it attains an infinite value. Thus we will frequently have statements such as, "Suppose μ is either a measure or a signed measure." We note that a signed measure can be a positive measure and a measure can be a finite signed measure.

Exercises.

   (1) If $\mathcal{A}$ is a σ-algebra of subsets of X and $f : X \to \mathbb{C}$, show that $\operatorname{Re} f$ and $\operatorname{Im} f$ are $\mathcal{A}$-measurable if and only if $f^{-1}(G) \in \mathcal{A}$ for every open subset G of $\mathbb{C}$. Also show that if f is measurable, then so is $\operatorname{sign}(f)$.

   (2) (a) Fix a positive measure η. If f is a complex-valued function in $L^1(\eta)$ and $\mu(E) = \int_E f d\eta$, show that $|\mu|(E) = \int_E |f| d\eta$ for every E in $\mathcal{A}$. (b) Show that if μ is any complex-valued measure, then there is a measurable function u with $|u(x)| = 1$ a.e. $[\|\mu\|]$ and $\mu(E) = \int_E u d|\mu|$ for every E in $\mathcal{A}$. (c) If μ is a real-valued signed measure, how is this function u related to the Hahn decomposition of μ?

4.3. Linear functionals on $C(X)$

In this section we will use measures to represent all the bounded linear functionals on $C(X)$ when X is a compact metric space. This completes the cycle started in §1.1 with the introduction of the Riemann–Stieltjes integral and the definition of the Radon measure from the existence of a positive linear functional on $C(X)$ in Chapter 2. In the following section we will extend this to the representation of linear functionals on $C_0(X)$ when X is a locally compact, σ-compact, metric space. We start, however, with a concept that is couched in the context of a locally compact space as are several of the initial results.

4.3.1. Definition. When X is a locally compact metric space, we say that μ is a *regular Borel measure* on X if μ is either a (complex) measure or a signed measure defined on a σ-algebra $\mathcal{A}$ containing the Borel sets and the following hold:

(a) $|\mu|(K) < \infty$ for every compact set;

(b) For every E in $\mathcal{A}$, $|\mu|(E) = \inf\{|\mu|(G) : E \subseteq G \text{ and } G \text{ is open}\}$;

(c) For every E in $\mathcal{A}$, $|\mu|(E) = \sup\{|\mu|(K) : K \subseteq E \text{ and } K \text{ is compact}\}$.

Note that a regular Borel measure can be defined on a larger σ-algebra than the Borel sets; it's only required that the Borel sets are contained in its domain of definition. See Exercise 1. Also note that if the measure is a signed measure, there is no requirement that it be finite. This is important since what we develop will apply to Lebesgue measure on $\mathbb{R}$.

When X is compact, every Radon measure space arising from a positive linear functional on $C(X)$ is an example of a regular Borel measure. In fact the reader may have noticed that the conditions in this definition are precisely what was used to define a measurable set in that situation (2.2.9).

I have to caution the reader that there is a concept of a regular Borel measure for arbitrary locally compact topological spaces that differs from what is stated above. That concept coincides with what we have defined when X is a σ-compact metric space. We are trying to avoid these technical points, so, to simplify the discourse, we make the following agreement that will hold for the remainder of this section.

For the remainder of this section and the next X is assumed to be a σ-compact locally compact metric space unless otherwise specified.

4.3.2. Proposition. *Let μ be a regular Borel measure on X.*

(a) *Both $|\mu|$ and the parts of the Jordan decomposition of μ are regular Borel measures.*

(b) *If $f : X \to \mathbb{R}$ is a Borel function such that $\int_K |f| d|\mu| < \infty$ for every compact set and $\nu(E) = \int_E f d|\mu| > -\infty$ for every Borel set E, then ν is a regular Borel signed measure.*

(c) *If $f \in L^1(|\mu|)$, then $\nu(E) = \int_E f d|\mu|$ defines a regular Borel measure.*

Proof. The proof of (a) is routine as is the fact in (b) that ν is a signed measure. Note that $\nu_{\pm} = \int_E f_{\pm} d|\mu|$ are the two parts of the Jordan decomposition of ν. If we can show that both $\nu_{\pm}$ are regular, it follows that ν is regular. That is, without loss of generality we can assume that $f \geq 0$ on X and ν is a positive measure.

The hypothesis on f assures that $\nu(K)$ is finite for every compact set K. If E is any Borel set, let $\{K_n\}$ be an increasing sequence of compact sets such that $|\mu|(E) = \lim_n |\mu|(K_n)$ and $K_n \subseteq E$ for every n. It follows from the MCT that $\nu(E) = \lim_n \nu(K_n)$. That is, (c) in the definition is satisfied by ν. To establish that (b) holds we divide the proof into two parts. First assume that there is an open set G containing E with $\nu(G) < \infty$. Now let $\{G_n\}$ be a decreasing sequence of open sets such that $E \subseteq G_n \subseteq G$ and $|\mu|(G_n) \searrow |\mu|(E)$. It follows that $\{f[\chi_G - \chi_{G_n}]\}$ is an increasing sequence of functions converging pointwise a.e.$[|\mu|]$ to $f[\chi_G - \chi_E]$ so that the MCT implies that $\nu(G) - \nu(G_n) = \int f[\chi_G - \chi_{G_n}] d\mu \to \nu(G) - \nu(E)$. Since $\nu(G) < \infty$ it follows that $\nu(E) = \lim_n \nu(G_n)$. Now assume that E is arbitrary and write $X = \bigcup_{n=1}^{\infty} K_n$, where each K_n is compact and $K_n \subseteq \operatorname{int} K_{n+1}$. Let $\epsilon > 0$. Since $|\mu|(\operatorname{int} K_{n+1}) < \infty$, the first case implies we can find an open set $G_n \supseteq E \cap K_n$ such that $\nu(G_n \backslash (E \cap K_n)) < \epsilon/2^n$. Put $G = \bigcup_{n=1}^{\infty} G_n$. So $E \subseteq G$ and $G \backslash E = \bigcup_{n=1}^{\infty}(G_n \backslash E) \subseteq \bigcup_{n=1}^{\infty}[G_n \backslash (E \cap K_n)]$. Thus $\nu(G \backslash E) \leq \sum_{n=1}^{\infty} \nu(G_n \backslash (E \cap K_n)) < \epsilon$.

(c) Here we observe that by considering the two functions $\operatorname{Re} f$ and $\operatorname{Im} f$, we can apply part (b) to both $\operatorname{Re} \nu$ and $\operatorname{Im} \nu$. ∎

4.3.3. Theorem (Lusin's[2] Theorem). *Let X be a locally compact space and let μ be a positive regular Borel measure on X. If $f : X \to \mathbb{C}$ is a measurable*

---

[2]Nikolai Lusin (or Luzin) was born in 1883 in Irkutsk, Siberia. His father was a Russian–Buryat businessman who soon moved his family to Tomsk so that Lusin could attend the gymnasium. Apparently his teachers considered mathematics as a collection of facts to be memorized, and he responded poorly. (Who wouldn't?) His father hired a tutor who had ability, recognized talent in the young Lusin, and made mathematics interesting for him. The father then moved his family to Moscow where Lusin entered the university in 1901. Unfortunately the father lost all his money on the stock exchange and Lusin had to move into a room owned by a widow – whose daughter Lusin was later to marry. He remained rather indifferent to mathematics until he came under the influence of Egorov (see 2.5.12). Nevertheless he was deeply affected by the poverty he observed and was torn between the study of mathematics and medicine, which he hoped to use to relieve the suffering he saw about him. Finally he settled into mathematics in 1909, though religious and moral questions continued to dog him. He was awarded the doctorate in 1915. In 1917 he was appointed a professor of pure mathematics at Moscow, where he and Egorov formed an impressive group in analysis. They attracted students who later became leaders of the subject in the Soviet Union. He made significant contributions to function theory, number theory, and set

function that vanishes off some measurable set E with $\mu(E) < \infty$, then for every $\epsilon > 0$ there is a continuous function ϕ with compact support such that $\mu(\{x : \phi(x) \neq f(x)\}) < \epsilon$. If, in addition, $f \in L^\infty(\mu)$, then ϕ can be chosen such that $\|\phi\|_\infty \leq \|f\|_\infty$.

Proof. First we treat the case where f is bounded on X; assume $|f(x)| \leq M$ for all x. Choose a compact set L contained in E and an open set U containing E such that $\mu(U \backslash L) < \epsilon/4$. Since f is bounded and $\mu(L) < \infty$, $f|L \in L^1(\mu|L)$. By Theorem 2.4.26 there is a sequence $\{h_n\}$ in $C(L)$ such that $\int_L |h_n - f| d\mu \to 0$. By passing to a subsequence, we may assume that $h_n(x) \to f(x)$ a.e. $[\mu|L]$ (2.5.9). By Egorov's Theorem (2.5.12) there is a subset F of L such that $\mu(L \backslash F) < \epsilon/4$ and $h_n(x) \to f(x)$ uniformly on F. Let K be compact such that $K \subseteq F$ and $\mu(F \backslash K) < \epsilon/4$. Thus the restriction of f to K is continuous. Let G be an open subset of X such that $K \subseteq G \subseteq U$ and $\mu(G \backslash K) < \epsilon/4$. Now take a continuous extension ϕ of $f|K$ to all of X with $\phi(x) = 0$ when $x \notin G$ and $\|\phi\| \leq \|f|K\| \leq M$ (1.2.13). Hence $\{x : \phi(x) \neq f(x)\} \subseteq U \backslash K \subseteq (U \backslash E) \cup (E \backslash L) \cup (L \backslash F) \cup (F \backslash K)$, so that we have completed the proof when f is bounded.

Now assume that f is any measurable function that vanishes off E, a set of finite measure. Put $E_n = \{x : 0 < |f(x)| < n\}$; so $E_n \subseteq E_{n+1}$ for all $n \geq 1$ and $\bigcup_{n=1}^\infty E_n = E$. Hence $\mu(E_n) \to \mu(E) < \infty$. Fix $n \geq 1$ such that $\mu(E \backslash E_n) < \epsilon/2$. According to the first part of the proof there is a ϕ in $C_c(X)$ such that $\mu(\{x : \phi(x) \neq \chi_{E_n}(x) f(x)\} < \epsilon/2$. The same function satisfies the conclusion of the theorem. ∎

We note in passing that in Lusin's Theorem we had no need for X to be σ-compact.

4.3.4. Corollary. *If X is a locally compact space and μ is a regular Borel measure on X, then*

$$\|\mu\| = |\mu|(X) = \sup\left\{\left|\int \phi d\mu\right| : \phi \in C_c(X) \text{ and } \|\phi\| \leq 1\right\}$$

Proof. Let γ denote the supremum in the statement of the corollary. If $\phi \in \text{ball } C_c(X)$, then $|\int \phi d\mu| \leq \int |\phi| d|\mu| \leq \|\mu\|$; thus $\gamma \leq \|\mu\|$. If $\epsilon > 0$, Proposition 4.2.5 implies there is a Borel partition $\{A_1, \ldots, A_n\}$ of X such that $\|\mu\| < \epsilon/2 + \sum_{k=1}^n |\mu(A_k)|$. For $1 \leq k \leq n$ let $a_k \in \mathbb{C}$ such that $|a_k| = 1$ and $|\mu(A_k)| = u_k \mu(A_k)$; put $f = \sum_{k=1}^n a_k \chi_{A_k}$. Thus $f \in L^\infty(|\mu|)$, $\|f\|_\infty = 1$, and $\int f d\mu = \sum_{k=1}^n |\mu(A_k)|$. Since μ is a finite measure, Lusin's Theorem implies there is a function ϕ in $C_c(X)$ with $\|\phi\| \leq 1$ and $|\mu|(\{x :$

theory. He had 17 doctoral students, including Pavel Aleksandrov, Andrei Kolmogorov, Mikhail Lavrentiev, and Pavel Urysohn. He died in 1950 in Moscow.

$\phi(x) \neq f(x)\}) < \epsilon/4$. This implies

$$\|\mu\| < \epsilon/2 + \int f d\mu$$

$$\leq \epsilon/2 + \left| \int (f - \phi) d\mu \right| + \left| \int \phi d\mu \right|$$

$$\leq \epsilon/2 + 2|\mu| (\{x : \phi(x) - f(x) \neq 0\}) + \gamma$$

$$< \epsilon + \gamma$$

Since ϵ was arbitrary, we have that $\|\mu\| \leq \gamma$, completing the proof. ∎

4.3.5. Corollary. *If X is a locally compact space and μ and ν are two regular Borel measures on X such that $\int \phi \, d\mu = \int \phi \, d\nu$ for every ϕ in $C_c(X)$, then $\mu = \nu$.*

Proof. By looking at the difference of the two measures, it is equivalent to show that if $\int \phi d\mu = 0$ for every ϕ in $C_c(X)$, then $\mu = 0$. This is immediate from the preceding corollary. ∎

Though the focus in this section is on linear functionals on $C(X)$ for a compact space X, we continue to assume that X is locally compact and prove a lemma that will also be used in the next section when we explore linear functionals on $C_0(X)$. In this lemma, however, we only assume that we are dealing with real-valued functions.

4.3.6. Lemma. *If X is locally compact and $L : C_0(X) \to \mathbb{R}$ is a bounded linear functional, then L can be written as $L = L_+ - L_-$, where $L_\pm$ are two positive linear functionals on $C_0(X)$.*

Proof. For f in $C_0(X)$ and $f \geq 0$, define

$$L_+(f) = \sup\{L(g) : g \in C_0(X), 0 \leq g \leq f\}$$

Since the zero function is a candidate for a function g in the definition of $L_+(f)$, $L_+(f) \geq 0$. It is also easy to see that when a is a positive scalar, $L_+(af) = aL_+(f)$. Finally if f_1, f_2 are two functions in $C_0(X)_+$ and $0 \leq g \leq f_1 + f_2$, then put $g_1 = g \wedge f_1$ and $g_2 = g - g_1$. It follows that $g_1, g_2 \in C_0(X)$, $g = g_1 + g_2$, and $0 \leq g_j \leq f_j$, $j = 1, 2$. (To show this inequality for g_2, for example, consider the case where, for some x in X, $g_1(x) = f_1(x)$ and then the case that $g_1(x) = g(x)$.) From here we see that the supremum $L_+(f_1 + f_2) = \sup\{L(g) : g \in C_0(X), 0 \leq g \leq f_1 + f_2\}$ splits and we have that

4.3.7 $$L_+(f_1 + f_2) = L_+(f_1) + L_+(f_2)$$

Now for any f in $C_0(X)$ we write $f = f_+ - f_-$ and define $L_+(f) = L_+(f_+) - L_+(f_-)$. If g, h are also positive continuous functions on X that vanish at infinity with $f = g - h = f_+ - f_-$, then $g + f_- = f_+ + h$ and so (4.3.7) implies that $L_+(g) + L_+(f_-) = L_+(f_+) + L_+(h)$; that is $L_+(f_+) - L_+(f_-) = L_+(g) - L_+(h)$. From here we get that if $f, g \in C_0(X)$, $L_+(f+g) = L_+([f_+ + g_+] - [f_- + g_-]) = L_+(f_+) + L_+(g_+) - L_+(f_-) - L_+(g_-) = L_+(f) + L_+(g)$. So L_+ is a linear functional. Finally we show that L_+ is bounded. In fact suppose $M = \|L\|$. If $f \in C_0(X)_+$ and $0 \le g \le f$, then $\|g\| \le \|f\|$; so $L(g) \le |L(g)| \le M\|f\|$. From the definition of L_+ we have that $0 \le L_+(f) \le M\|f\|$ when $f \in C_0(X)_+$. If f is any function in $C_0(X)$ and we write $f = f_+ - f_-$, then $|L_+(f)| \le |L_+(f_+)| + |L_+(f_-)| \le M(\|f_+\| + \|f_-\|) \le 2M\|f\|$.

Let $L_- = L_+ - L$. So L_- is a bounded linear functional on $C_0(X)$. If $f \in C_0(X)$ and $f \ge 0$, then the definition of $L_+(f)$ shows that $L_+(f) \ge L(f)$, so that $L_-(f) \ge 0$; that is, L_- is also a positive linear functional on $C_0(X)$. ∎

Note that the preceding lemma is a Jordan decomposition for bounded linear functionals. The main goal of this section is to prove the following.

4.3.8. Theorem (Riesz Representation Theorem). *If X is a compact metric space and $L : C(X) \to \mathbb{C}$ is a bounded linear functional, then there is a unique regular Borel measure μ such that*

$$L(f) = \int f \, d\mu$$

for all f in $C(X)$. Moreover $\|L\| = \|\mu\| = |\mu|(X)$.

Proof. This proof proceeds in stages. First we consider the case where the underlying scalars are real; so assume $L : C(X) \to \mathbb{R}$ is a bounded linear functional. By Lemma 4.3.6, $L = L_+ - L_-$, where $L_\pm$ are positive. Now apply Theorem 2.4.24 to $L_\pm$ to obtain two positive Radon measures $\mu_\pm$ such that $L_\pm(f) = \int f \, d\mu_\pm$ for all f in $C(X)$; let $\mu = \mu_+ - \mu_-$ and we have the existence of a finite regular Borel signed measure μ such that $L(f) = \int f \, d\mu$. The uniqueness of μ follows immediately from Corollary 4.3.5, and the equality $\|L\| = \|\mu\|$ is Corollary 4.3.4.

Now assume that $L : C(X) \to \mathbb{C}$ is a bounded linear functional. Define L_1 and L_2 from $C_{\mathbb{R}}(X)$ into $\mathbb{R}$ by $L_1(f) = \operatorname{Re} L(f), L_2(f) = \operatorname{Im} L(f)$. It follows that both are real linear functionals with norms at most $\|L\|$. By the first part of the proof there are finite signed measures μ_1, μ_2 such that $L_j(f) = \int f \, d\mu_j$ for all f in $C_{\mathbb{R}}(X)$. For a complex-valued f in $C(X)$, $L(f) = L(\operatorname{Re} f) + iL(\operatorname{Im} f) = [L_1(\operatorname{Re} f) + iL_2(\operatorname{Re} f)] + i[L_1(\operatorname{Im} f) + iL_2(\operatorname{Im} f)] = [\int \operatorname{Re} f \, d\mu_1 + i \int \operatorname{Re} f \, d\mu_2] + i [\int \operatorname{Im} f \, d\mu_1 + i \int \operatorname{Im} f \, d\mu_2]$. If we let $\mu = \mu_1 + i\mu_2$, a little algebra shows that $L(f) = \int f \, d\mu$. Also μ is a measure and, again, (4.3.5) and (4.3.4) show that this measure is unique and $\|L\| = \|\mu\|$. This completes the proof. ∎

It is worth spending a moment to rephrase the preceding theorem using the language of the dual space initiated in §1.5. See Exercise 5 below.

Exercises.

(1) If X is a compact metric space and μ is a signed measure defined on the Borel sets with $|\mu|(K) < \infty$ for all compact sets, prove that the following statements are equivalent. (a) μ is regular. (b) For every Borel set E and every $\epsilon > 0$ there is a compact set K contained in E such that $|\mu|(E \backslash K) < \epsilon$. (c) For every Borel set E and every $\epsilon > 0$ there is an open set G containing E such that $|\mu|(G \backslash E) < \epsilon$.

(2) If μ and ν are regular Borel measures and $\nu = \nu_a + \nu_s$ is the Lebesgue decomposition of ν with respect to μ, show that ν_a and ν_s are regular Borel measures.

(3) If X is a compact metric space, let $L : C(X) \to \mathbb{R}$ be a non-zero bounded linear functional. Show that L has the property that $L(fg) = L(f)L(g)$ for all f, g in $C(X)$ if and only if there is a unique point x in X such that $L(f) = f(x)$ for all f in $C(X)$.

(4) If X is a compact metric space and $L : C(X) \to \mathbb{C}$ bounded linear functional, say that L is positive if $L(f) \geq 0$ whenever f is real-valued and $f \geq 0$. Show that L is positive if and only if the measure μ obtained in Theorem 4.3.8 is positive.

(5) Recall the definition of the dual space $\mathcal{X}^*$ of a normed space $\mathcal{X}$ (1.5.5). For a compact metric space X let $M(X)$ denote the set of all regular Borel measures on X and give it the norm $\|\mu\| = |\mu|(X)$. Show that if $\mu \in M(X)$ and $L_\mu : C(X) \to \mathbb{F}$ is defined by $L_\mu(f) = \int f d\mu$ for every f in $C(X)$, then the map $\mu \mapsto L_\mu$ defines a linear isometry of $M(X)$ onto $C(X)^*$.

(6) Can you use Lusin's Theorem to give another proof of Theorem 2.4.26?

4.4. Linear functionals on $C_0(X)$

There are various ways in which Theorem 4.3.8 can be extended. Much of the work needed to prove the first of these extensions was done in the preceding section.

4.4.1. Theorem (Riesz Representation Theorem). *If X is a locally compact, σ-compact metric space and $L : C_0(X) \to \mathbb{C}$ is a bounded linear functional, then there is a unique finite regular Borel measure μ such that $L(f) = \int f d\mu$ for all f in $C_0(X)$ and $\|L\| = \|\mu\|$.*

Proof. Here we use the one-point compactification of X, $X_\infty = X \cup \{\infty\}$, and the fact that when X is σ-compact, X_∞ is a compact metric space (1.4.6). We also use the fact that $C_0(X)$ can be identified with $\{f \in C(X_\infty) : f(\infty) = 0\}$. Note that if $f \in C(X_\infty)$ and $f(\infty)$ is the function constantly equal to $f(\infty)$, then $f - f(\infty) \in C_0(X)$. If $L : C_0(X) \to \mathbb{C}$ is a bounded linear functional on $C_0(X)$, define $\widehat{L} : C(X_\infty) \to \mathbb{C}$ by $\widehat{L}(f) = L(f - f(\infty))$. Since $|\widehat{L}(f)| \le \|L\|\|f - f(\infty)\| \le 2\|L\|\|f\|$, $\widehat{L}$ is a bounded linear functional on $C(X_\infty)$. By Theorem 4.3.8 there is a regular Borel measure $\widehat{\mu}$ on X_∞ such that $\widehat{L}(f) = \int_{X_\infty} f d\widehat{\mu}$ for all f in $C(X_\infty)$.

Any measure $\widehat{\mu}$ on X_∞ can be written as $\widehat{\mu} = \mu + a\delta_\infty$, where μ is a regular Borel measure on X, $a \in \mathbb{C}$, and δ_∞ is the unit point mass at ∞. Thus if $f \in C_0(X)$, then $f \in C(X_\infty)$ and $L(f) = \widehat{L}(f) = \int_{X_\infty} f d(\mu + a\delta_\infty) = \int_X f d\mu$. (In fact the constant a must be 0; see Exercise 1.) The equality of the norms follows from Corollary 4.3.4. ∎

Let us return to Theorem 2.4.26 and Proposition 2.7.14 and extend these to this more general setting.

4.4.2. Proposition. *If X is a locally compact, σ-compact metric space and μ is a positive regular Borel measure on X, then $C_c(X)$ is dense in $L^p(\mu)$ for all finite p. Consequently, $L^p(\mu)$ is separable.*

Proof. We know (2.7.12) that the simple functions belonging to $L^p(\mu)$ are dense. Now proceed as in the proof of (2.4.26), where this result was stated and proved when X is compact. Namely, we need only show that when E is a measurable set and $\mu(E) < \infty$, then there is a function f in $C_c(X)$ with $\|f - \chi_E\|_p$ arbitrarily small. But if $\epsilon > 0$, we use the regularity of μ to find a compact set K with $K \subseteq E$ such that $\mu(E \backslash K) < (\epsilon/2)^p$. Pick an open set G containing K with $\text{cl } G$ compact and $\mu(G \backslash K) < (\epsilon/2)^p$. If f is continuous with $\chi_K \le f \le \chi_G$, then $f \in C_c(X)$; note that $|f(x) - \chi_K(x)|^p \le 1$ and $|\chi_E(x) - \chi_K(x)|^p \le 1$ for all x. Hence $\int |f - \chi_K|^p d\mu = \int_{G \backslash K} |f - \chi_K|^p d\mu \le \mu(G \backslash K)$, so that $\|f - \chi_K\|_p < \epsilon/2$. Also $\int |\chi_E - \chi_K|^p d\mu \le \mu(E \backslash K)$ so that $\|\chi_E - \chi_K\|_p < \epsilon/2$. By the triangle inequality this establishes that $\|\chi_E - f\|_p < \epsilon$.

By Theorem 1.4.6, $C_0(X)$ is separable; hence $C_c(X)$ is separable. It follows that $L^p(\mu)$ is separable. ∎

The restriction to metric spaces can also be dispensed with. Theorem 4.3.8 can be proved for any compact Hausdorff space; the assumption of metrizability makes the arguments easier. Similarly Theorem 4.4.1 can be proved for any locally compact space, even those that are not σ-compact.

The assumption that X be σ-compact was made just so that X_∞ is metrizable and we could apply Theorem 4.3.8. The reader can consult [**18**]. The following can also be found in [**18**] as the Riesz Representation Theorem.

4.4.3. Theorem (Riesz Representation Theorem). *If X is a locally compact space and $L : C_c(X) \to \mathbb{R}$ is a positive linear functional, then there is a unique positive regular Borel measure μ on X such that*

$$L(f) = \int f \, d\mu$$

for all f in $C_c(X)$.

Note that the linear functional in this theorem is defined only on the continuous functions with compact support and there is no assumption of boundedness. It is however, restricted by assuming it is positive. From this result it is easy to deduce the following theorem, applicable to all locally compact spaces, not only those that are metrizable and σ-compact.

4.4.4. Theorem. *If X is locally compact and for each complex regular Borel measure μ we define $L_\mu : C_0(X) \to \mathbb{C}$ by $L_\mu(f) = \int f \, d\mu$, then L_μ is a bounded linear functional on $C_0(X)$ with $\|L_\mu\| = \|\mu\|$. Conversely, given a bounded linear functional L on $C_0(X)$, there is a unique regular Borel measure μ such that $L = L_\mu$. Thus the map $\mu \mapsto L_\mu$ defines an isometric isomorphism of $M(X)$ onto $C_0(X)^*$.*

Exercises.

(1) Show that the constant a found in the first half of the proof of Theorem 4.4.1 is 0.

(2) Let μ be a positive regular Borel measure on the locally compact metric space X. If $f \in L^1(\mu)$, show that there is a σ-compact open set G such that $f(x) = 0$ a.e. $[\mu]$ off G.

(3) This exercise requires the reader to be familiar with the Stone–Čech compactification. Namely for a completely regular space X, which includes the case where X is a metric space, there is a compact space βX containing X as a dense subset such that every bounded continuous function $f : X \to \mathbb{C}$ has a continuous extension $f^\beta : \beta X \to \mathbb{C}$ with $\sup\{|f(x)| : x \in X\} = \sup\{|f^\beta(y)| : y \in \beta X\}$. Using the version of Theorem 4.3.8 for arbitrary compact spaces, show that $L : C_b(X) \to \mathbb{C}$ is a bounded linear functional if and only if there is a regular Borel measure μ on βX such that $L(f) = \int_{\beta X} f^\beta d\mu$ for all f in $C_b(X)$, in which case $\|L\| = \|\mu\|$.

4.5. Functions of bounded variation

In this section we will tidy up the results on functions of bounded variation by giving a necessary and sufficient condition for two such functions to determine the same Riemann–Stieltjes integral, thus closing that topic. Recall (§1.1) that for $J = [a, b]$, $BV(J)$ denotes the space of all functions of bounded variation $\alpha : J \to \mathbb{R}$. For every such function we defined $\widetilde{\alpha}$, the normalization of α (1.1.16). Say that a function of bounded variation α is *normalized* if $\alpha = \widetilde{\alpha}$. Equivalently, α is normalized if (a) $\alpha(a) = 0$ and (b) α is left-continuous at every point of the open interval (a, b). Note that for $a \leq s \leq b$ the functions α_s defined in Example 1.1.12 are normalized. In Proposition 1.1.17 we showed that $\int f d\alpha = \int f d\widetilde{\alpha}$ for every f in $C(J)$. Here we prove a type of converse.

Denote the set of all normalized functions of bounded variation on J by $NBV(J)$. In Example 1.3.2(f) we placed a norm on $BV(J)$. We modify this; for α in $NBV(J)$ define the norm $\|\alpha\| = \mathrm{Var}(\alpha)$. In fact this is the original norm on $BV(J)$ restricted to $NBV(J)$. It follows that $NBV(J)$ is a Banach space.

Since for any function of bounded variation α, $L(f) = \int f d\alpha$ defines a bounded linear functional from $C(J)$ into $\mathbb{R}$, Theorem 4.3.8 implies there is a regular Borel signed measure μ_α such that $\int f d\alpha = \int f d\mu_\alpha$ for every continuous function f on J. Now assume that we are given a regular Borel signed measure μ on J. Define $\alpha_\mu : J \to \mathbb{R}$ by

4.5.1
$$\alpha_\mu(t) = \begin{cases} \mu([a, t)) & \text{for } t < b \\ \mu([a, b]) & \text{when } t = b \end{cases}$$

We claim that α_μ is a normalized function of bounded variation. To show it is a function of bounded variation, let $P = \{a = x_0 < x_1 < \cdots < x_n = b\}$. For $j < n$, $\alpha_\mu(x_j) - \alpha_\mu(x_{j-1}) = \mu([x_{j-1}, x_j))$ while $\alpha_\mu(b) - \alpha_\mu(x_{n-1}) = \mu([x_{n-1}, b])$. Thus

$$\sum_{j=1}^{n} |\alpha_\mu(x_j) - \alpha_\mu(x_{j-1})| = \sum_{j=1}^{n-1} |\mu([x_{j-1}, x_j))| + |\mu([x_{n-1}, b])| \leq \|\mu\|$$

and $\mathrm{Var}(\alpha) \leq \|\mu\|$. Clearly $\alpha_\mu(a) = 0$. If $t < b$ and $t_n \nearrow t$, then $\{[a, t_n)\}$ is an increasing sequence of sets whose union is $[a, t)$, and so $\alpha(t_n) = \mu([a, t_n)) \to \mu([a, t)) = \alpha(t)$. Therefore α_μ is left-continuous on the open interval and, consequently, normalized.

4.5.2. Theorem. *If α is a function of bounded variation on the interval J, then there is a regular Borel signed measure μ_α such that $\int f d\mu_\alpha = \int f d\alpha$ for every continuous function f on J. Conversely, if μ is a regular Borel signed measure on J, then (4.5.1) defines a normalized function of bounded* *variation α_μ on J such that $\int f d\alpha_\mu = \int f d\mu$ for every continuous function f on J.*

Proof. As was pointed out before, the first part of the theorem follows from the Riesz Representation Theorem. For the converse we have already seen that for a given μ, α_μ is a normalized function of bounded variation. We must show that α_μ and μ integrate continuous functions with the same result. We begin by showing that α_μ is continuous at a point t of J if and only if $\mu(\{t\}) = 0$. In fact if α_μ is continuous at $t < b$, then $\mu([t - \frac{1}{n}, t + \frac{1}{n})) = \alpha_\mu(t + \frac{1}{n}) - \alpha_\mu(t - \frac{1}{n}) \to 0$. Since $\{t\} = \bigcap_{n=1}^\infty [t - \frac{1}{n}, t + \frac{1}{n})$, we have that $\mu(\{t\}) = 0$. When $t = b$, write $\{b\} = \bigcap_{n=1}^\infty [b - \frac{1}{n}, b]$ and make a similar argument. Now assume that $\mu(\{t\}) = 0$. Here $\alpha_\mu(t) = \mu([a, t)) = \mu([a, t])$. So if $t < b$ and $t_n \downarrow t$, $[a, t] = \bigcap_{n=1}^\infty [a, t_n)$ and therefore $\alpha_\mu(t_n) = \mu([a, t_n)) \to \mu([a, t]) = \alpha_\mu(t)$; hence α_μ is continuous at t. If $t = b$, observe that $\alpha_\mu(b) = \mu([a, b))$. If $t_n \nearrow b$, then $\alpha_\mu(t_n) = \mu([a, t_n)) \to \alpha_\mu(b)$. This proves that α_μ is continuous at b.

Now fix a continuous function f on J and an $\epsilon > 0$. Choose $\delta > 0$ such that both of the following are satisfied: $|f(s) - f(t)| < (\epsilon/2)\|\mu\|$ when $|s - t| < \delta$, and $\left| \int f d\alpha_\mu - S_{\alpha_\mu}(f, P) \right| < \epsilon/2$ when $P \in \mathcal{P}_\delta$. Choose $P = \{a = x_0 < \cdots < x_n = b\}$ in $\mathcal{P}_\delta$ so that for $1 \leq j < n$, α_μ is continuous at x_j. It follows from this continuity condition and the definition of α_μ that $S_{\alpha_\mu}(f, P) = \sum_{j=1}^n f(x_j)\mu([x_{j-1}, x_j])$. Therefore

$$\left| \int f d\alpha_\mu - \int f d\mu \right| \leq \left| \int f d\alpha_\mu - S_{\alpha_\mu}(f, P) \right|$$

$$+ \left| \sum_{j=1}^n f(x_j)\mu([x_{j-1}, x_j]) - \int f d\mu \right|$$

In this inequality the first summand on the right hand side is dominated by $\epsilon/2$ since $P \in \mathcal{P}_\delta$. The second summand is equal to $\left| \int (g - f) d\mu \right|$, where $g = \sum_{j=1}^n f(x_j)\chi_{[x_{j-1}, x_j]}$. But when $t \in [x_{j-1}, x_j]$, $|f(x_j) - f(t)| < (\epsilon/2)\|\mu\|$ so that $|g(t) - f(t)| < (\epsilon/2)\|\mu\|$ a.e. $[\mu]$ on J. By (2.4.22)(b), $\left| \int (g - f) d\mu \right| < \epsilon/2$. Thus $\left| \int f d\alpha_\mu - \int f d\mu \right| < \epsilon$ for an arbitrary ϵ, and the theorem is proven. ∎

There remains the question of the uniqueness of the Riemann–Stieltjes integral. That is, if J is a closed interval and α and β are two functions of bounded variation such that $\int f d\alpha = \int f d\beta$ for every f in $C(J)$, what is the relation between α and β? Proposition 1.1.17 shows that a sufficient condition occurs when $\beta = \widetilde{\alpha}$. Now we prove the converse. We have done enough that the proof becomes trivial.

4.5.3. Proposition. *If α and β are two functions of bounded variation on the interval J, then $\int f d\alpha = \int f d\beta$ for every continuous function f on J if and only if they have the same normalization.*

Proof. As mentioned, Proposition 1.1.17 furnishes half this proof. So assume $\int f d\alpha = \int f d\beta$ for every f in $C(J)$. By replacing these functions by their normalizations, we may assume that α and β are normalized. If μ_α, μ_β are the corresponding regular Borel signed measures as in (4.5.1), then we have that $\int f d\mu_\alpha = \int f d\mu_\beta$ for every f in $C(J)$. By Corollary 4.3.5, $\mu_\alpha = \mu_\beta$. But then (4.5.1) implies $\alpha = \beta$. ∎

4.5.4. Example. Let c be the Cantor function (2.4.8). This is a normalized continuous increasing function, so there is a positive regular Borel measure μ_c on $[0,1]$ with $\int f dc = \int f d\mu_c$ for every continuous function on the unit interval. On any of the deleted intervals in the complement of the Cantor set C, the function c is constant. This says that the measure μ_c is carried by C. On the other hand if λ is Lebesgue measure on the unit interval, $\lambda(C) = 0$, so that λ is carried by $[0,1]\backslash C$. Thus $\mu_c \perp \lambda$.

Exercise.

(1) This is more a project than an exercise. Develop the Riemann–Stieltjes integral with respect to a function of bounded variation on the real line.

4.6. Linear functionals on L^p-spaces

Here we continue to find representations of bounded linear functionals, this time on the spaces $L^p(\mu)$ for $1 \le p < \infty$. Be aware that the proof of the next theorem is long. There are two cases to consider: when the measure μ is finite and when it is infinite. In each of these cases there are two parts: when p is not equal to 1 and when it is. All this is a consequence of having one theorem cover a lot of territory. So be patient as you work your way through this; it's worth it.

4.6.1. Theorem. *Let $(X, \mathcal{A}, \mu)$ be a measure space, let $1 \le p < \infty$, and let $1/p + 1/q = 1$. If $g \in L^q(\mu)$ and we define $L_g : L^p(\mu) \to \mathbb{C}$ by*

$$L_g(f) = \int fg d\mu$$

for all f in $L^p(\mu)$, then L_g is a bounded linear functional on $L^p(\mu)$ and $\|L_g\| = \|g\|_q$. Conversely, if $1 < p < \infty$ and $L : L^p(\mu) \to \mathbb{C}$ is a bounded linear functional, then there is a unique g in $L^q(\mu)$ such that $L = L_g$. If $p = 1$, $(X, \mathcal{A}, \mu)$ is σ-finite, and $L : L^1(\mu) \to \mathbb{C}$ is a bounded linear functional, then there is a unique g in $L^\infty(\mu)$ such that $L = L_g$.

Proof. We start by pointing out that when $g \in L^q(\mu)$, one of the consequences of Hölder's Inequality (2.7.9) is that L_g, as defined in the statement of this theorem, is a bounded linear functional on $L^p(\mu)$ and $\|L_g\| \leq \|g\|_q$. The reverse inequality will surface in the course of the remainder of the proof.

Assume that L is a bounded linear functional on $L^p(\mu)$, $1 \leq p < \infty$.

Case 1. $\mu(X) < \infty$. In this case $\chi_E \in L^p(\mu)$ for every set E belonging to $\mathcal{A}$. Define $\nu(E) = L(\chi_E)$; we want to show that ν is a measure and that $\nu \ll \mu$. Actually, this last statement is clear once we know that ν is a measure since when $\mu(E) = 0$, $\chi_E = 0$ in $L^p(\mu)$ and so $\nu(E) = L(\chi_E) = 0$. It is immediate that $\nu(\emptyset) = 0$ and that $\nu(E) \in \mathbb{C}$ for every measurable set E. It is trivial that ν is finitely additive and we must show that it is countably additive. To do this let $E_1, E_2, \ldots \in \mathcal{A}$ such that $E_1 \supseteq E_2 \supseteq \cdots$ and $\bigcap_{n=1}^{\infty} E_n = \emptyset$; we want to show that $\nu(E_n) \to 0$. (See Exercise 1.) But $\|\chi_{E_n}\|_p = \left[\int |\chi_{E_n}|^p d\mu\right]^{\frac{1}{p}} = \mu(E_n)^{\frac{1}{p}} \to 0$ (2.4.3(d)) as $n \to \infty$, and hence $\nu(E_n) = L(\chi_{E_n}) \to 0$ since L is continuous. Thus ν is a measure. By the Radon–Nikodym Theorem there is an $\mathcal{A}$-measurable function g on X such that $\int h d\nu = \int h g d\mu$ for every function h that is integrable with respect to $|\nu|$. In particular we have that for every simple function f,

4.6.2
$$\int f g d\mu = \int f d\nu = L(f)$$

Claim. $g \in L^q(\mu)$ and $\|g\|_q \leq \|L\|$.

Let $t > 0$ and put $E_t = \{x : |g(x)| \leq t\}$. Suppose $f \in L^p(\mu)$ such that $f(x) = 0$ when $x \notin E_t$. There is a sequence $\{f_n\}$ of simple functions such that each f_n vanishes off E_t, $|f_n| \leq |f|$, and $f_n(x) \to f(x)$ for all x in X. Note that $|f_n g| \leq |fg| \leq t|f|$ and that $\int |f| d\mu = \int |f| \cdot 1 d\mu \leq \|f\|_p \mu(X)^{\frac{1}{q}}$, so f is integrable. By the DCT, $L(f_n) = \int f_n g d\mu \to \int f g d\mu$. Also $|f_n - f|^p \leq 2^p |f|^p$ so that again the DCT implies $\|f_n - f\|_p \to 0$ and thus $L(f_n) \to L(f)$. Therefore we have shown that for any $t > 0$ and any f in $L^p(\mu)$ that vanishes off E_t, (4.6.2) holds. Now, unfortunately for the flow of exposition, we must break up the proof of the claim into two subcases depending on the value of p.

Case 1a. $1 < p < \infty$. With E_t as above, define a function f by setting $f = \chi_{E_t} |g|^q / g$ on the set $A = \{x : g(x) \neq 0\}$ and $f(x) = 0$ elsewhere. Note that $|g|^q \chi_{E_t} = fg$. So, using the fact that $pq - p = q$, we get that

$$\int |f|^p d\mu = \int_{E_t \cap A} \frac{|g|^{pq}}{|g|^p} d\mu = \int_{E_t} |g|^q d\mu$$

Hence

$$\int_{E_t} |g|^q d\mu = \int f g d\mu = L(f) \leq \|L\| \|f\|_p = \|L\| \left[\int_{E_t} |g|^q d\mu \right]^{\frac{1}{p}}$$

and so

$$\|L\| \geq \left[\int_{E_t} |g|^q d\mu \right]^{1 - \frac{1}{p}} = \left[\int_{E_t} |g|^q d\mu \right]^{\frac{1}{q}}$$

Letting $t \to \infty$ shows that $\|L\| \geq \|g\|_q$ and we have that $g \in L^q(\mu)$. Since we already know that $\|L_g\| \leq \|g\|_q$, we get that $\|L\| = \|L_g\| = \|g\|_q$. Also we have that $L(f) = L_g(f)$ for every f in $L^p(\mu)$ such that $f = 0$ off E_t, that is, for every f in $\chi_{E_t} L^p(\mu)$. But $\bigcup \{\chi_{E_t} L^p(\mu) : t > 0\}$ is a dense linear manifold in $L^p(\mu)$ and both L and L_g are bounded linear functionals, so $L = L_g$ on $L^p(\mu)$. This concludes the proof of Case 1a.

Case 1b. $p = 1$, so that $q = \infty$. We continue to use the set E_t defined above. This is easier than Case 1a. If $\epsilon > 0$, put $A = \{x : |g(x)| > \|L\| + \epsilon\}$. If we let $f = \chi_{E_t \cap A}(|g|/g)$, then $\|f\|_1 = \mu(A \cap E_t)$ and so

$$\|L\| \mu(A \cap E_t) \geq |L(f)| = \int f g d\mu = \int_{A \cap E_t} |g| d\mu \geq (\|L\| + \epsilon) \mu(A \cap E_t)$$

If we let $t \to \infty$ we get that $\|L\| \mu(A) \geq (\|L\| + \epsilon) \mu(A)$. The only way this can happen is if $\mu(A) = 0$. Since ϵ was arbitrary, $\|g\|_\infty \leq \|L\|$ and $g \in L^\infty(\mu)$, concluding the proof of Case 1 where $\mu(X)$ is finite.

Case 2. $(X, \mathcal{A}, \mu)$ is arbitrary. For any set E in $\mathcal{A}$ we consider the measure space $(E, \mathcal{A}_E, \mu_E)$ as in Proposition 2.4.3. Note that $L^p(\mu_E)$ can be identified with $\{f \in L^p(\mu) : f = 0 \text{ off } E\}$, which is a closed linear subspace of $L^p(\mu)$; in fact this subspace is exactly $\chi_E L^p(\mu)$. Define $L_E : L^p(\mu_E) \to \mathbb{C}$ by setting $L_E(f) = L(f)$ for all f in $L^p(\mu_E)$. In other words, L_E is the restriction of L to the corresponding subspace of $L^p(\mu)$. It is easy to see that each L_E is a bounded linear functional and, in fact, $\|L_E\| \leq \|L\|$ for all E in $\mathcal{A}$.

Let $\mathcal{E} = \{E \in \mathcal{A} : \mu(E) < \infty\}$. By Case 1 for each E in $\mathcal{E}$ there is a function g_E in $L^q(\mu_E)$ such that $\|g_E\|_q = \|L_E\| \leq \|L\|$ and

4.6.3
$$\int_E f g_E d\mu = L_E(f)$$

for every f in $L^p(\mu_E)$. We want to show now that for any D, E in $\mathcal{E}$, $g_D = g_E$ a.e. $[\mu]$ on $D \cap E$. But note that $L^p(\mu_{D \cap E}) \subseteq L^p(\mu_D) \cap L^p(\mu_E)$; and, moreover, on $L^p(\mu_{D \cap E})$, $L_D = L_E = L_{D \cap E}$. Hence $g_E = g_D = g_{D \cap E}$ a.e. $[\mu]$ on $D \cap E$. But we have a problem here. The functions g_E and g_D

agree a.e. $[\mu]$, not everywhere, and there are (possibly) uncountably many sets of finite measure. We cannot do our usual thing and redefine each g_E so that two such functions agree everywhere. Even though we have that any two of the functions agree on their common domains, how can we put them together to get a measurable function? Once again we bifurcate.

Case 2a. $1 < p < \infty$. Let $\sigma = \sup\{\|g_E\|_q : E \in \mathcal{E}\}$. Note two things: $\sigma \le \|L\|$ and if $D, E \in \mathcal{E}$ with $D \subseteq E$, then $\|g_D\|_q \le \|g_E\|_q$. Therefore we can take sets $\{E_n\}$ from $\mathcal{E}$ such that $E_n \subseteq E_{n+1}$ and $\|g_{E_n}\|_q \to \sigma$; put $F = \bigcup_{n=1}^{\infty} E_n$. We can redefine the functions g_{E_n} on a set of measure zero so that $g_{E_n}(x) = g_{E_{n+1}}(x)$ when $x \in E_n$. Also if $E \in \mathcal{E}$ and $E \cap F = \emptyset$, then $\|g_{E \cup E_n}\|_q = \|g_E\|_q + \|g_{E_n}\|_q \to \|g_E\|_q + \sigma$. Hence it must be that $g_E = 0$ a.e. $[\mu]$ whenever $E \cap F = \emptyset$. Define a measurable function g on X by setting $g(x) = g_{E_n}(x)$ when $x \in E_n$ and $g(x) = 0$ when $x \notin F$. It follows that $\|g\|_q = \sigma$ (Why?) so that $g \in L^q(\mu)$. Moreover for any set E in $\mathcal{E}$, $g_E = g$ a.e. $[\mu]$ on E. Now if $f \in L^p(\mu)$, then $\{x : f(x) \ne 0\} = \bigcup_{n=1}^{\infty} D_n$, where for every $n \ge 1$, $D_n \in \mathcal{E}$ and $D_n \subseteq D_{n+1}$; for example, we could let $D_n = \{x : |f(x)| > \frac{1}{n}\}$. (This is precisely where we are using the fact that $p < \infty$. If $f \in L^\infty(\mu)$ it is not necessarily true that $\{x : f(x) \ne 0\}$ is σ-finite.) We have that $f\chi_{D_n} \to f$ in $L^p(\mu)$ and so, by (4.6.3),

$$L(f) = \lim_n L(f\chi_{D_n}) = \lim_n L_{D_n}(f) = \lim_n \int_{D_n} fg \, d\mu = \int fg \, d\mu$$

Case 2b. $p = \infty$. Here we are assuming that the measure space is σ-finite, so let $X = \bigcup_{n=1}^{\infty} E_n$, where each $E_n \in \mathcal{A}$, $\mu(E_n) < \infty$, and $E_n \subseteq E_{n+1}$. We proceed just as in the preceding case to define a function g on all of X that belongs to $L^\infty(\mu)$. Just as in that case we also have that $L(f) = \int fg \, d\mu$ for all f in $L^1(\mu)$. ∎

The dual of $L^\infty(\mu)$ is not $L^1(\mu)$. To show this, consider the measure space $(\mathbb{N}, 2^{\mathbb{N}}, \mu)$, where μ is counting measure. So in fact this measure is σ-finite. As in Exercise 4.4.3, we identify ℓ^∞ with $C_b(\mathbb{N})$, which is in turn identified with $C(\beta\mathbb{N})$. Thus the dual of ℓ^∞ is identified with $M(\beta\mathbb{N})$ while ℓ^1 is identified with the regular Borel measures on $\mathbb{N}$.

We mentioned before that there is a theory of the spaces $L^p(\mu)$ when $0 < p < 1$ and referred to the book [9] as a starting place for those who are curious. In fact for $p < 1$, $L^p(\mu)$ is not a normed space but it has a metric

$$d(f, g) = \int |f - g|^p d\mu$$

Note this does not have a p-th root. (If you introduce a p-th root, the triangle inequality is invalid.) If $\mu = \lambda$, Lebesgue measure on $[0, 1]$, then when $0 < p < 1$ there are no linear functionals on the L^p space that are continuous for this metric. It is precisely because of this phenomenon that we are not pursuing the study of these spaces. As the title of the subject "functional analysis" implies, the power of its methods relies on having a rich supply of continuous linear functionals.

Exercises.

(1) If $\mathcal{A}$ is a σ-algebra of subsets of X and $\nu : \mathcal{A} \to \mathbb{C}$ is a function satisfying: (i) $\nu(\emptyset) = 0$; (ii) ν is finitely additive; (iii) when $\{E_n\}$ is a decreasing sequence in $\mathcal{A}$ with $\bigcap_{n=1}^{\infty} E_n = \emptyset$, $\nu(E_n) \to 0$, then ν is a measure.

(2) (a) If $\mathcal{H}$ is a Hilbert space and $\{x_n\}$ is an orthonormal sequence in $\mathcal{H}$, show that $\langle x_n, y \rangle \to 0$. (b) Show that if λ is Lebesgue measure on the unit interval, then $\int g(t) \cos 2\pi nt \, d\lambda(t) \to 0$ as $n \to \infty$ for every g in $L^2(\lambda)$.

(3) If $1 \leq p < \infty$ and $f \in L^p(\mu)$, let $E_t = \{x \in X : |f(x)| > t\}$ whenever $t > 0$. (a) Prove that $\mu(E_t) \leq t^{-p}\|f\|_p^p$. (This is called Chebyshev's Inequality.) (b) Show that if $f_n \to 0$ in $L^p(\mu)$, then $f_n \to 0$ in measure.

(4) Here is another proof that the simple functions are dense in $L^2(\mu)$, one that illustrates the power of using continuous linear functionals. (a) If $1 \leq p < \infty$, $f \in L^p(\mu)$, and $\int_E f d\mu = 0$ for every measurable set E with $\mu(E) < \infty$, show that $f = 0$. (b) Show that the set of simple functions that belong to $L^2(\mu)$ are dense in $L^2(\mu)$. Hint: Use Corollary 3.2.11. (This same proof will show that the simple functions are dense in $L^p(\mu)$, but we will need a bit more functional analysis for this.)

4.7. Product measures

In this section we develop the concept of multiple or iterated integrals for a finite number of measures. Actually, we only do this for two measures, but the generalization to a finite number is routine. We again restrict ourselves to σ-compact locally compact metric spaces, though at the end we'll state a more abstract version with a reference to a proof.

If X and Y are given sets and $f : X \times Y \to \mathbb{C}$, then for x in X and y in Y define functions $f_x : Y \to \mathbb{C}$ and $f^y : X \to \mathbb{C}$ by

$$f_x(y_1) = f(x, y_1), \qquad f^y(x_1) = f(x_1, y)$$

for all y_1 in Y and x_1 in X.

4.7.1. Theorem (Fubini's[3] Theorem). *Let X and Y be σ-compact locally compact metric spaces. If μ and ν are finite positive regular Borel measures on X and Y, respectively, then there is a unique finite positive regular Borel measure η on $X \times Y$ with the following properties.*

(a) *$\eta(A \times B) = \mu(A)\nu(B)$ for all Borel sets A in X and B in Y.*

(b) *If $f \in L^1(\eta)$, then $f_x \in L^1(\nu)$ for a.e. $[\mu]$ value of x in X and $f^y \in L^1(\mu)$ for a.e. $[\nu]$ value of y in Y. Moreover*

4.7.2
$$\int f(x,y)d\eta(x,y) = \int \left[\int f(x,y)d\mu(x)\right] d\nu(y)$$
$$= \int \left[\int f(x,y)d\nu(y)\right] d\mu(x)$$

The measure η that appears in Fubini's Theorem is usually written

$$\eta = \mu \times \nu$$

and called the *product measure* of μ and ν. So another way to write (4.7.2) is

$$\int_{X \times Y} f d\mu \times \nu = \int_Y \left[\int_X f^y d\mu\right] d\nu(y) = \int_X \left[\int_Y f_x d\nu\right] d\mu(x)$$

To prove this theorem we will use the Riesz Representation Theorem (4.4.1) for representing linear functionals on $C_0(X)$ as an integral with respect to a finite regular Borel measure. To do this we first need to prove a preliminary result that has independent interest.

When X and Y are metric spaces, let's agree that the metric we have on $X \times Y$ is

$$d[(x,y),(x_1,y_1)] = d_X(x,x_1) + d_Y(y,y_1)$$

There are many other choices for the metric on $X \times Y$ that can be made, but all reasonable choices give the same collection of open sets and convergent sequences as the one just defined. (Can you give another such metric?) We choose this one to make it specific and to somewhat simplify some of the arguments. If $f : X \to \mathbb{C}$ and $g : Y \to \mathbb{C}$ are two functions, define $f \otimes g : X \times Y \to \mathbb{C}$ by $f \otimes g(x,y) = f(x)g(y)$. We point out that for x in

---

[3]Guido Fubini was born in Venice in 1879. He received his doctorate in 1900 from the University of Pisa, writing a thesis in differential geometry. He took up a position at the University of Catania in Sicily. Shortly after that he was offered a professorship at the University of Genoa and then in 1908 he went to Turin. He began to change the course of his research and gravitated to analysis. During the 1930s he saw the political situation in Italy deteriorating with the rise of Mussolini's anti-semitism. He decided the future was looking bleak for his family, and, in spite of his deteriorating health, in 1939 the Fubini family emigrated to New York. His contributions were in several different areas including integral equations, group theory, geometry, continuous groups, applied mathematics, and he was influential in the development of mathematics in Italy. He died in 1943 in New York.

X and y in Y, $(f \otimes g)_x = g$ and $(f \otimes g)^y = f$. Note that if X and Y are σ-compact, $f \in C_0(X)$, and $g \in C_0(Y)$, then $f \otimes g \in C_0(X \times Y)$.

4.7.3. Proposition. *If X and Y are σ-compact locally compact metric spaces and $\mathcal{A}$ is the linear span of all the functions $f \otimes g$, where $f \in C_0(X)$ and $g \in C_0(Y)$, then $\mathcal{A}$ is dense in $C_0(X \times Y)$.*

Proof. Since $C_c(X \times Y)$ is dense in $C_0(X \times Y)$ (1.4.5(b)), it suffices to show that if $F \in C_c(X \times Y)$, then $F \in \mathrm{cl}\,\mathcal{A}$. Fix such an F and an $\epsilon > 0$. There are pairs of sets (K, U) and (L, V) from X and Y, respectively, such that K and L are compact, U and V are open, $\mathrm{cl}\,U$ and $\mathrm{cl}\,V$ are compact, $K \subseteq U$, $L \subseteq V$, and F vanishes off $K \times L$. Since F is uniformly continuous (1.4.5(c)) there is a $\delta > 0$ such that $|F(x, y) - F(x_1, y_1)| < \epsilon$ when $d((x, y), (x_1, y_1)) < \delta$.

Let $x_1, \ldots, x_n \in X$ and $y_1, \ldots, y_m \in Y$ such that $K \subseteq \bigcup_{i=1}^{n} B(x_i; \delta/2) \subseteq U$ and $L \subseteq \bigcup_{j=1}^{m} B(y_j; \delta/2) \subseteq V$. By Proposition 1.4.7 there are non-negative continuous functions $\{\phi_1, \ldots, \phi_n\}$ and $\{\psi_1, \ldots, \psi_m\}$ on X and Y, respectively, such that

$$\sum_{i=1}^{n} \phi_i(x) = 1 \quad \text{if} \quad x \in K \quad \text{and} \quad \phi_i(x) = 0 \quad \text{if} \quad x \notin B(x_i; \delta/2)$$

$$\sum_{j=1}^{m} \psi_j(y) = 1 \quad \text{if} \quad y \in L \quad \text{and} \quad \psi_j(y) = 0 \quad \text{if} \quad y \notin B(y_j; \delta/2)$$

$$0 \le \sum_{i=1}^{n} \phi_i \le 1$$

$$0 \le \sum_{j=1}^{m} \psi_j \le 1$$

So for $1 \le i \le n$ and $1 \le j \le m$, $\phi_i \in C_c(X)$ and $\psi_j \in C_c(Y)$.

Consider the functions $\{\phi_i \otimes \psi_j : 1 \le i \le n, 1 \le j \le m\}$. It follows that for all i, j:

(a) $\phi_i \otimes \psi_j \in C_c(X \times Y)$;

(b) $\sum_{i,j} \phi_i \otimes \psi_j(x, y) = 1$ for all (x, y) in $K \times L$;

(c) $\phi_i \otimes \psi_j(x, y) = 0$ when $(x, y) \notin B(x_i; \delta/2) \times B(y_j; \delta/2)$;

(d) $0 \le \sum_{i,j} \phi_i \otimes \psi_j(x, y) \le 1$ for all (x, y) in $X \times Y$.

When $x \in B(x_i; \delta/2)$ and $y \in B(y_j; \delta/2)$, then $d((x, y), (x_i, y_j)) < \delta$; thus with $a_{ij} = F(x_i, y_j)$, we have $|F(x, y) - a_{ij}| < \epsilon$ for such x and y.

Put $\Phi = \sum_{i,j} a_{ij} \phi_i \otimes \psi_j$; so $\Phi \in \mathcal{A}$. If $\phi_i(x)\psi_j(y) \neq 0$, then $(x,y) \in B(x_i; \delta/2) \times B(y_j; \delta/2)$ and so $|F(x,y) - a_{ij}| < \epsilon$. Therefore for any (x,y) in $X \times Y$,

$$
\begin{aligned}
|F(x,y) - \Phi(x,y)| &= \left| \sum_{i,j} \phi_i(x)\psi_j(y)[F(x,y) - a_{ij}] \right| \\
&\leq \sum_{i,j} \phi_i(x)\psi_j(y)|F(x,y) - a_{ij}| \\
&< \epsilon \sum_{i,j} \phi_i(x)\psi_j(y) \\
&\leq \epsilon \qquad\qquad\qquad\qquad\qquad \blacksquare
\end{aligned}
$$

Now we begin the process of proving Theorem 4.7.1. The first lemma establishes the existence of the measure η.

4.7.4. Lemma. *There is a regular Borel measure η on $X \times Y$ such that for every continuous function F on $X \times Y$ we have that*

4.7.5
$$
\begin{aligned}
\int F(x,y)d\eta(x,y) &= \int \left[\int F(x,y)d\mu(x) \right] d\nu(y) \\
&= \int \left[\int F(x,y)d\nu(y) \right] d\mu(x)
\end{aligned}
$$

Proof. Define $L : C_0(X \times Y) \to \mathbb{F}$ by

$$
L(F) = \int \left[\int F(x,y)d\mu(x) \right] d\nu(y)
$$

for all F in $C_0(X \times Y)$. We have to check first that this makes sense; that is, we need to show that the function $y \mapsto \int F(x,y)d\mu(x)$ is integrable with respect to ν. We will do this by showing that for a fixed F in $C_0(X \times Y)$, $y \to \int F(x,y)d\mu(x)$ defines a continuous function on Y and, therefore, we can integrate it. In fact, if $y_n \to y_0$ in Y, then $F(x,y_n) \to F(x,y)$ for every x in X and $|F(x,y_n)|$ is dominated by the constant function $\|F\|$. Therefore the DCT implies $\int F(x,y_n)d\mu(x) \to \int F(x,y)d\mu(x)$ so that $y \mapsto \int F(x,y)d\mu(x)$ is continuous. Therefore the integral $\int \left[\int F(x,y)d\mu(x) \right] d\nu(y)$ used to define $L(F)$ makes sense.

It is easily checked that L is a positive linear functional on $C_0(X \times Y)$ and that $|L(F)| \leq \|\mu\|\|\nu\|\|F\|$. Therefore by Theorem 4.4.1 there is a finite positive Radon measure η on $X \times Y$ such that

$$
\int F(x,y)d\eta(x,y) = \int \left[\int F(x,y)d\mu(x) \right] d\nu(y)
$$

for all F in $C_0(X \times Y)$.

In a similar way there is a finite positive Radon measure τ on $X \times Y$ such that

$$\int F(x,y) d\tau(x,y) = \int \left[\int F(x,y) d\nu(y) \right] d\mu(x)$$

for all F. We want to show that $\eta = \tau$. But if $f \in C_0(X)$ and $g \in C_0(Y)$, we have

$$\int f \otimes g \, d\eta = \left[\int f \, d\mu \right] \left[\int g \, d\nu \right] = \int f \otimes g \, d\tau$$

Thus the integral with respect to η and τ of every continuous function that is in the linear span of such functions $f \otimes g$ is the same. By Proposition 4.7.3 these measures integrate every function in $C_0(X \times Y)$ in the same way; by Corollary 4.3.5, $\eta = \tau$ and thus (4.7.5) holds. $\blacksquare$

To complete the proof of Theorem 4.7.1 let's first extend the notation for functions introduced at the start of this section to sets. If $E \subseteq X \times Y$ and $x \in X$, let $E_x = \{ y \in Y : (x,y) \in E \}$. Similarly if $y \in Y$, let $E^y = \{ x \in X : (x,y) \in E \}$. This is actually just a particular example of the function notation since $(\chi_E)^y = \chi_{E^y}$ and $(\chi_E)_x = \chi_{E_x}$. Let $\mathcal{A}$ and $\mathcal{B}$ denote the μ-measurable and ν-measurable subsets of X and Y, respectively. Let $\mathcal{C}$ denote the σ-algebra of η-measurable subsets of $X \times Y$. Recall that these algebras contain all the sets of measure 0 and every one of the measurable sets can be written as the union of a Borel set and a set of measure zero (Exercise 2.3.4).

Fix the notation η. The next result is the first step in completing the proof. It furnishes a bit more information than required and so we label it a proposition rather than a lemma.

4.7.6. Proposition. *If $\mathcal{N}$ is the collection of all subsets E of $X \times Y$ such that for a.e. $[\mu]$ x in X and a.c.$[\nu]$ y in Y, $E^y \in \mathcal{A}$ and $E_x \in \mathcal{B}$, then $\mathcal{N}$ is a σ-algebra that contains the Borel sets.*

Proof. Clearly $\emptyset$ and $X \times Y$ belong to $\mathcal{N}$. If $E, F \in \mathcal{N}$, then $(E \backslash F)^y = E^y \backslash F^y$ and $(E \backslash F)_x = E_x \backslash F_x$, so $E \backslash F \in \mathcal{N}$. Similarly, $[\bigcup_{n=1}^{\infty} E_n]^y = \bigcup_{n=1}^{\infty} (E_n)^y$ and $[\bigcup_{n=1}^{\infty} E_n]_x = \bigcup_{n=1}^{\infty} (E_n)_x$, so that $\mathcal{N}$ is closed under countable unions. Therefore $\mathcal{N}$ is a σ-algebra. Finally, if K is a compact subset of $X \times Y$, then both K^y and K_x are compact, so that $\mathcal{N}$ contains all the compact sets. Hence $\mathcal{N}$ contains the Borel sets. $\blacksquare$

As a consequence of the preceding proposition, whenever E is a Borel set in $X \times Y$, we can form $\mu(E^y)$ and $\nu(E_x)$.

4.7.7. Lemma. *If E is either an open or a compact subset of $X \times Y$, then $y \mapsto \mu(E^y)$ is $[\nu]$-measurable, $x \mapsto \nu(E_x)$ is $[\mu]$-measurable, and*

$$\eta(E) = \int \mu(E^y) d\nu(y) = \int \nu(E_x) d\mu(x)$$

Proof. If E is an open subset of $X \times Y$ with compact closure, then there is a sequence of functions $\{F_n\}$ in $C_c(X \times Y)$ such that $0 \le F_n \le \chi_E$ and $F_n(x,y) \nearrow \chi_E(x,y)$ for all (x,y) in $X \times Y$. Hence $\eta(E) = \lim_n \int F_n d\eta$. For an arbitrary but fixed y in Y, $F_n^y(x) \nearrow \chi_{E^y}(x)$ for all x in X, so that $\mu(E^y) = \lim_n \int F_n^y d\mu$ by the MCT. If we set $g_n(y) = \int F_n^y d\mu$, then as before the DCT implies that $g_n \in C(Y)$. So $\int F_n d\eta = \int g_n d\nu \to \int \mu(E^y) d\nu(y)$; therefore $\eta(E) = \int \mu(E^y) d\nu(y)$. Similarly $\eta(E) = \int \nu(E_x) d\mu(x)$. If E is open but does not have compact closure, we can write $E = \bigcup_{n=1}^{\infty} E_n$ where each E_n is open with compact closure and $E_n \subseteq E_{n+1}$ for all $n \ge 1$. (Why?) A simple application of the MCT shows that the result holds for such a E. (Verify!)

The proof when E is a compact set is similar (Exercise 1). ■

4.7.8. Lemma. *If $E \subseteq X \times Y$ and $\eta(E) = 0$, then $E^y \in \mathcal{A}$ a.e. $[\nu]$, $E_x \in \mathcal{B}$ a.e. $[\mu]$, $\mu(E^y) = 0$ a.e. $[\nu]$, and $\nu(E_x) = 0$ a.e. $[\mu]$. Consequently, if $E \in \mathcal{C}$, then $E^y \in \mathcal{A}$ a.e. $[\nu]$ and $E_x \in \mathcal{B}$ a.e. $[\mu]$.*

Proof. Since $\eta(E) = 0$, there is a sequence of open sets $\{G_n\}$ in $X \times Y$ such that $E \subseteq G_{n+1} \subseteq G_n$ and $\eta(G_n) < \frac{1}{n}$ for all n; put $A = \bigcap_{n=1}^{\infty} G_n$ and note that A is a Borel set. For every y in Y, $\bigcap_{n=1}^{\infty} G_n^y = A^y$ and the sets $\{G_n^y\}$ are decreasing; hence $\mu(G_n^y) \searrow \mu(A^y)$. By the preceding lemma, $y \mapsto \mu(G_n^y)$ is $[\nu]$-measurable and so the MCT implies that $y \mapsto \mu(A^y)$ is $[\nu]$-measurable and $\int \mu(G_n^y) d\nu \searrow \int \mu(A^y) d\nu$. By the preceding lemma we also have that $\int \mu(G_n^y) d\nu(y) = \eta(G_n)$; also $\eta(G_n) \to 0$. Thus $\int \mu(A^y) d\nu = 0$ and there is a subset Z of Y with $\nu(Z) = 0$ and $\mu(A^y) = 0$ whenever $y \notin Z$. Since $E^y \subseteq A^y$, whenever $y \notin Z$ we have that $E^y \in \mathcal{A}$ and $\mu(E^y) = 0$. A similar argument shows that $E_x \in \mathcal{B}$ and $\nu(E_x) = 0$ a.e. $[\mu]$.

If $E \in \mathcal{C}$, then we can write $E = F \cup Z$, where F is a Borel subset of $X \times Y$ and $\eta(Z) = 0$ (Exercise 2.3.4). By Lemma 4.7.6, $E^y \in \mathcal{A}$ for all y. The first part of the proof shows that $Z^y \in \mathcal{A}$ a.e. $[\nu]$. Thus $E^y = F^y \cup Z^y \in \mathcal{A}$ a.e. $[\nu]$. Similarly, $E_x \in \mathcal{B}$ a.e. $[\mu]$. ■

Proof of Fubini's Theorem. Lemma 4.7.4 gives us the existence of the measure η. We will first establish part (b). Assume that $f = \chi_E$ for some $[\eta]$-measurable set E. By Lemma 4.7.7, (b) is true if E is either open or compact. In the general case let $\{K_n\}$ be a sequence of compact sets such that $K_n \subseteq K_{n+1} \subseteq E$ and $\eta(K_n) \to \eta(E)$. It follows that $\chi_{K_n} \to \chi_E$ a.e. $[\eta]$. Now if $A = \bigcup_{n=1}^{\infty} K_n$, then $\eta(E \setminus A) = 0$ and so Lemma 4.7.8 implies there is a set $Z \subseteq Y$ with $\nu(Z) = 0$ such that $\mu(A^y) = 0$ when $y \notin Z$. Therefore $\chi_{K_n}^y \nearrow \chi_E^y$ a.e. $[\mu]$ when $y \notin Z$. Using the MCT we have that $y \mapsto \mu(E^y)$ is $[\nu]$-measurable and $\eta(K_n) = \int \mu(K_n^y) d\nu(y) \to \int \mu(E^y) d\nu(y)$. But $\eta(K_n) \to \eta(E)$ so that $\eta(E) = \int \mu(E^y) d\nu(y)$ when $y \notin Z$. Similarly $\eta(E) = \int \nu(E_x) d\mu(x)$ a.e. $[\nu]$.

It follows easily that part (b) holds for positive simple η-measurable functions. Now assume that f is an arbitrary positive measurable function and use Proposition 2.4.17 to obtain a sequence $\{u_n\}$ in $L_s^1(\eta)_+$ such that $u_n \nearrow f$ a.e. $[\eta]$ and $\int u_n d\eta \to \int f d\eta$. Let E be a set in $X \times Y$ with $\eta(E) = 0$ and such that $u_n(x, y) \to f(x, y)$ whenever $(x, y) \notin E$. According to Lemma 4.7.8 there is a subset Z of Y such that $\nu(Z) = 0$ and $\mu(E^y) = 0$ when $x \notin Z$. Hence when $y \notin Z$ and $x \notin E^y$, $u_n^y(x) \nearrow f^y(x)$ and so $\int u_n^y d\mu \nearrow \int f^y d\mu$ whenever $y \notin Z$. This implies the function $y \to \int f^y d\mu$ is $[\nu]$-measurable and

$$\int f d\eta = \lim_n \int u_n d\eta$$

$$= \lim_n \int \left[\int u_n^y d\mu \right] d\nu$$

$$= \int \left[\int f^y d\mu \right] d\nu$$

The proof of the other equality in the theorem is similar. Since every function in $L^1(\eta)$ is the difference of two non-negative functions in $L^1(\eta)$, we have established part (b).

To prove (a) simply take $f = \chi_{A \times B}$ in part (b). ∎

Note that we actually proved that part (b) of Fubini's Theorem holds for non-negative measurable functions, not just the integrable ones. Let's record this.

4.7.9. Corollary. *With the notation of Theorem 4.7.1, if f is a non-negative η-measurable function, then f_x is ν-measurable for a.e. $[\mu]$ value of x in X and f^y is μ-measurable for a.e. $[\nu]$ value of y in Y. Moreover (4.7.2) is valid.*

We observe that Fubini's Theorem can be extended to all regular Borel measures on locally compact σ-compact metric spaces X and Y by using the Jordan decompositions. Namely, if μ and ν are signed measures on X and Y, respectively, and $\mu = \mu_+ - \mu_-, \nu = \nu_+ - \nu_-$ are the Jordan decompositions, then

$$\mu \times \nu \equiv (\mu_+ \times \nu_+) - (\mu_+ \times \nu_-) - (\mu_- \times \nu_+) + (\mu_- \times \nu_-)$$

Similarly if the measures are complex-valued and we decompose them into their real and imaginary parts, we can do the straightforward complex arithmetic to define their product.

There is a more general version of Fubini's Theorem than the one presented in (4.7.1) that does not involve metric spaces. We state it here but the reader is referred to [**18**] for a proof. Let $(X, \mathcal{A}, \mu)$ and $(Y, \mathcal{B}, \nu)$ be two σ-finite measure spaces. If $A \in \mathcal{A}$ and $B \in \mathcal{B}$, we call $A \times B \subseteq X \times Y$ a *measurable rectangle*.

4.7.10. Theorem (Fubini's Theorem). *If $(X, \mathcal{A}, \mu)$ and $(Y, \mathcal{B}, \nu)$ are two σ-finite measure spaces and $\mathcal{M}$ is the σ-algebra of subsets of $X \times Y$ generated by the measurable rectangles, then there is a unique measure $\mu \times \nu$ defined on $\mathcal{M}$ with the following properties.*

(a) $\mu \times \nu(A \times B) = \mu(A)\nu(B)$ *for every measurable rectangle $A \times B$.*

(b) *If $f \in L^1(\mu \times \nu)$, then $f_x \in L^1(\nu)$ for a.e. $[\mu]$ value of x in X and $f^y \in L^1(\mu)$ for a.e. $[\nu]$ value of y in Y. Moreover*

$$\int_{X \times Y} f d\mu \times \nu = \int_Y \left[\int_X f^y d\mu \right] d\nu(y) = \int_X \left[\int_Y f_x d\nu \right] d\mu(x)$$

Before we leave this section we use this abstract statement of Fubini's Theorem to derive a result for $L^p(\mu \times \nu)$.

4.7.11. Proposition. *If $(X, \mathcal{A}, \mu), (Y, \mathcal{B}, \nu)$ are measure spaces and $1 \le p \le \infty$, then the linear span of the functions*

$$\{f \otimes g : f \in L^p(\mu), g \in L^p(\nu)\}$$

is dense in $L^p(\mu \times \nu)$.

Proof. By Theorem 2.7.12 the simple functions that belong to $L^p(\mu \times \nu)$ are dense. So to prove this result when p is finite we need only show that each characteristic function χ_E, where $\mu \times \nu(E) < \infty$, can be approximated by a linear combination of functions of the form $f \otimes g$. Let $\epsilon > 0$. According to Exercise 6 we can find pairwise disjoint subsets of E, $\{A_k \times B_k : 1 \le k \le n\}$ with $A_k \in \mathcal{A}$, $B_k \in \mathcal{B}$, and such that $\mu \times \nu \left(E \backslash \bigcup_{k=1}^n A_k \times B_k \right) < \epsilon^p$. But note that this says that

$$\left\| \chi_E - \sum_{k=1}^n \chi_{A_k} \otimes \chi_{B_k} \right\|_p < \epsilon$$

The proof when $p = \infty$ is different and left to the reader. ∎

Exercises.

(1) Furnish the details for the proof of Lemma 4.7.7 when E is compact.

(2) For all m, n in $\mathbb{N}$ let $a_{mn} \in [0, \infty)$ and show that

$$\sum_{m=1}^{\infty} \sum_{n=1}^{\infty} a_{mn} = \sum_{n=1}^{\infty} \sum_{m=1}^{\infty} a_{mn}$$

(Use Theorem 4.7.1.)

(3) For m, n in $\mathbb{N}$ let $a_{nn} = 1, a_{n,n+1} = -1$, and $a_{mn} = 0$ otherwise. Show that $\sum_{m=1}^{\infty} \sum_{n=1}^{\infty} a_{mn} = 0$ and $\sum_{n=1}^{\infty} \sum_{m=1}^{\infty} a_{mn} = 1$.

(4) Let X and Y be compact metric spaces and let μ and ν be positive regular Borel measures. If $f \in L^1(\mu)$, $g \in L^1(\nu)$, and $\eta = \mu \times \nu$, show that $h : X \times Y \to \mathbb{R}$ defined by $h(x, y) = f(x)g(y)$ is a measurable function, $h \in L^1(\eta)$, and $\int h d\eta = \left[\int f d\mu\right] \left[\int g d\nu\right]$.

(5) Let λ be Lebesgue measure on $\mathbb{R}$. (a) If E is a Borel subset of $\mathbb{R}$, show that $\widetilde{E} = \{x - y : x, y \in E\}$ is a Borel subset of $\mathbb{R}^2$. (b) If $f : \mathbb{R} \to \mathbb{R}$ is a Borel function, show that $(x, y) \mapsto f(x - y)$ is a Borel function on $\mathbb{R}^2$.

(6) Adopt the notation in Theorem 4.7.10. (a) Show that $\mathcal{M}$, the σ-algebra generated by the measurable rectangles, consists of all countable unions of measurable rectangles. (b) Show that the countable union of measurable rectangles can be written as the countable union of measurable rectangles that are pairwise disjoint. (The bookkeeping for this strikes me as overly taxing. Doing this for finite unions might be easier and that's all we need for the next part.) (c) (Here we are assuming the validity of Theorem 4.7.10.) If $E \in \mathcal{M}$ with $\mu \times \nu(E) < \infty$ and $\epsilon > 0$, then there are pairwise disjoint measurable rectangles $A_1 \times B_1, \ldots, A_n \times B_n$ contained in E such that $\mu \times \nu(E \backslash \bigcup_{k=1}^{n} A_k \times B_k) < \epsilon$. (If $E = \bigcup_{j=1}^{\infty} C_j \times D_j$, then $\mu(\bigcup_{j=1}^{m} C_j \times D_j) \nearrow \mu(E)$. Choose m sufficiently large that $\mu(E \backslash [\bigcup_{j=1}^{m} C_j \times D_j]) < \epsilon$ and apply part (b).)

4.8. **Lebesgue measure on** $\mathbb{R}^d$

Consider d-dimensional Euclidean space, $\mathbb{R}^d$. Lebesgue measure on $\mathbb{R}^d$, denoted by λ, is the product measure defined by taking one-dimensional Lebesgue measure on each coordinate (§4.7). We will use Fubini's Theorem (4.7.1). More accurately, we will rely on Theorem 4.7.1 to write $\lambda = \lambda_1 \times \cdots \times \lambda_1$, where λ_1 is Lebesgue measure on the real line.

We recall a topic from Advanced Calculus that is assumed familiar to the reader. Let G be an open subset of $\mathbb{R}^d$ and suppose $\Phi : G \to \mathbb{R}^d$. So we can write $\Phi(x) = (\phi_1(x), \ldots, \phi_d(x))$, where each $\phi_i : G \to \mathbb{R}$. A function Φ is said to be of class C^1 if each coordinate function ϕ_i is of class C^1; that is, for $1 \le i \le d$, ϕ_i has continuous first partial derivative $\partial\phi_i/\partial x_j$ for $1 \le j \le d$. Denote the algebra of linear transformations from $\mathbb{R}^d$ into itself by $\mathcal{L}(\mathbb{R}^d)$, and recall that the derivative of Φ is the matrix-valued function $D\Phi : G \to \mathcal{L}(\mathbb{R}^d)$ where for each x in G, $D\Phi(x)$ is defined by

$$D\Phi(x) = \left[\frac{\partial\phi_i}{\partial x_j}(x) \right]_{ij}$$

That is, $D\Phi(x)$ is the linear transformation defined by the $d \times d$ matrix given above. The *Jacobian* of Φ is the determinant of $D\Phi(x)$: $\Delta(x) = \det D\Phi(x)$. This leads to the Change of Variables formula that if Φ is one-to-one with non-vanishing Jocobian, then for every continuous function $f : \Phi(G) \to \mathbb{R}$,

4.8.1
$$\int_{\Phi(G)} f \, d\lambda = \int_G f \circ \Phi \, |\det D\Phi(x)| \, d\lambda$$

This can be extended as follows.

4.8.2. Theorem (Change of Variables Formula). *If G is an open subset of $\mathbb{R}^d$ and $\Phi : G \to \mathbb{R}^d$ is a one-to-one C^1 function with $\det D\Phi(x) \ne 0$ for all x in G, then whenever $f : \Phi(G) \to \mathbb{R}$ is Lebesgue measurable, so is $f \circ \Phi : G \to \mathbb{R}$. If in addition $f \ge 0$ or if $f \in L^1(\Phi(G), \lambda)$, then (4.8.1) holds for f.*

Proof. Recall from Advanced Calculus that when Φ satisfies the hypothesis, then $\Phi(G)$ is an open subset of $\mathbb{R}^d$ and $\Phi : G \to \phi(G)$ is a homeomorphism (the Inverse Mapping Theorem). We use Corollary 4.3.5. Define two measures μ and ν on the Borel subsets of $\Phi(G)$ by $\mu(E) = \int_G \chi_E \circ \Phi(x) \, |\det D\Phi(x)| \, d\lambda(x) = \int_{\Phi^{-1}(E)} |\det D\Phi(x)| \, d\lambda$ and $\nu(E) = \lambda(E)$. It is straightforward to use the fact that Φ is a homeomorphism to prove that μ is a regular Borel measure. Formula (4.8.1) establishes that $\int f \, d\mu = \int f \, d\nu$ for every f in $C_c(G)$, hence $\mu = \nu$ and this proves the theorem. ∎

4.8.3. Corollary. *If T is an invertible linear transformation of $\mathbb{R}^d$ onto itself and $f : \mathbb{R}^d \to \mathbb{R}$ is a λ-measurable function, then $f \circ T$ is λ-measurable. If $f \in L^1(\lambda)$ or if $f \ge 0$, then $\int f \, d\lambda = |\det T| \int f \circ T \, d\lambda$. In particular, $\lambda(T(E)) = |\det T| \lambda(E)$ for every λ-measurable set E.*

Recall that a linear transformation T of $\mathbb{R}^d$ into itself is called unitary if $T^*T = I$, the identity linear transformation. The next corollary is immediate from the fact that when T is unitary, $\det T = \pm 1$.

4.8.4. Corollary. *If T is a unitary linear transformation on $\mathbb{R}^d$, then $\lambda(T(E)) = \lambda(E)$ for every λ-measurable subset E of $\mathbb{R}^d$.*

4.8.5. Corollary. *If $x \in \mathbb{R}^d$ and $r > 0$, then for every $t > 0$ in $\mathbb{R}$, $\lambda(B(x; tr)) = t^d \lambda(B(x; r))$.*

Proof. Use the linear transformation $T = tI$. ∎

Maps of $\mathbb{R}^d$ that will play an important role in the study of Lebesgue measure are the translations. If $a \in \mathbb{R}^d$, $\tau_a : \mathbb{R}^d \to \mathbb{R}^d$ is defined by

$$\tau_a(x) = x - a$$

Note that the Jacobian of τ_a is 1, so we get that Lebesgue measure is invariant under translation.

4.8.6. Corollary. *For every measurable set E, $\lambda(\tau_a(E)) = \lambda(E)$. Moreover if $f \in L^1(\lambda)$, then $f \circ \tau_a \in L^1(\lambda)$ and $\int f \circ \tau_a d\lambda = \int f d\lambda$.*

We mention that for any radius r, $\lambda(\text{ball } \mathbb{R}^d) = \pi^{\frac{d}{2}}/\Gamma(\frac{1}{2}d + 1)$, where Γ is the Gamma-function. Sometimes this is established in an Advanced Calculus course and sometimes it is not. We will not use this fact and only mention it for the curious.

For a discussion in $\mathbb{R}^2$ of the relationship between the condition that a function Φ have a non-vanishing Jacobian and the condition that Φ is injective, see [**27**].

Next is a result that will be of use later.

4.8.7. Proposition. *If $1 \leq p < \infty$ and $f \in L^p(\lambda)$, then $\lim_{x \to 0} \|\tau_{x+y} f - \tau_y f\|_p = 0$ for every y in $\mathbb{R}^d$.*

Proof. Note that $\tau_{x+y} = \tau_x \tau_y$. So if we prove the proposition for $y = 0$ and then replace the function f in the statement by $\tau_y f$, we have the conclusion.

Let $\phi \in C_c(\mathbb{R}^d)$ and put $K = \{x : \text{dist}(x, \text{spt } \phi) \leq 1\}$. By Proposition 1.4.5(c), ϕ is uniformly continuous. Thus if $\epsilon > 0$, there is a δ with $0 < \delta < 1$ such that $|\phi(x) - \phi(y)| < \epsilon$ when $|x - y| < \delta$. This implies $\|\tau_y \phi - \phi\|_\infty < \epsilon$ whenever $|y| < \delta$. Therefore for $|y| < \delta$, we have that $\|\tau_y \phi - \phi\|_p^p = \int_K |\phi(x - y) - \phi(x)|^p d\lambda \leq \epsilon^p \lambda(K)$. Thus the proposition follows when $f = \phi \in C_c(\mathbb{R}^d)$. But if $f \in L^p(\lambda)$ and $\epsilon > 0$, there is a function ϕ in $C_c(\mathbb{R}^d)$ with $\|f - \phi\|_p < \epsilon/3$ (4.4.2). Thus

$$\|\tau_y f - f\|_p \leq \|\tau_y f - \tau_y \phi\|_p + \|\tau_y \phi - \phi\|_p + \|\phi - f\|_p < 2\epsilon/3 + \|\tau_y \phi - \phi\|_p$$

whence the conclusion. ∎

Exercises.

(1) Show that if $\mathbb{R}^d$ is given any of the norms $\|x\|_1 = \sum_{k=1}^n |x_k|$, $\|x\|_\infty = \max\{|x_k| : 1 \leq k \leq n\}$, or $\|x\|_2 = \left[\sum_{k=1}^n |x_k|^2\right]^{\frac{1}{2}}$, then

a sequence in $\mathbb{R}^d$ converges with respect to one of these norms if and only if it converges with respect to each of the others.

(2) Call a set R in $\mathbb{R}^d$ a *strict rectangle* if $R = I_1 \times \cdots \times I_d$, where each I_j is an interval in $\mathbb{R}$. (This is to distinguish it from a measurable rectangle as defined in §4.7.) Show that if E is a Borel set in $\mathbb{R}^d$ and $\epsilon > 0$, then there are pairwise disjoint open strict rectangles $R_1, \ldots, R_n$ such that $\lambda\left(E \backslash \bigcup_{k=1}^n R_k\right) < \epsilon$.

(3) Show that the measure μ in the proof of Theorem 4.8.2 is a regular Borel measure on G.

(4) Show that for a fixed dimension d there is an absolute constant C depending only on d such that for every x in $\mathbb{R}^d$ and every radius r, $\lambda(B(x; r)) = C r^d$.

(5) If $x \in \mathbb{R}^d$ and $r > 0$, let $S = \partial B(x; r) = \{y \in \mathbb{R}^d : d(x, y) = r\}$. Show that $\lambda(S) = 0$.

4.9. Differentiation on $\mathbb{R}^d$

This section is based on the treatment in [**15**]. Here we will explore notions of the differentiation of functions $F : \mathbb{R}^d \to \mathbb{R}$ as well as look at signed measures on $\mathbb{R}^d$. This requires some preliminary work. In fact the reader will find this section more technical than most in this book. Why is that? When you begin looking for special information about specific objects in mathematics, the level of details and technicalities in the proofs increases; the more special and specific, the greater the level of detail. After all, we are not dealing with general concepts and proving results that apply to a whole class of objects. In this case we are looking at differentiating functions defined on sets in Euclidean space, and neither the functions nor the sets are what was treated in advanced calculus; moreover, we want specific information about this process.

Throughout the section λ denotes Lebesgue measure on $\mathbb{R}^d$.

4.9.1. Definition. Say that a function $f : \mathbb{R}^d \to \mathbb{C}$ is *locally integrable* if f is Lebesgue measurable and for every compact subset K of $\mathbb{R}^d$, $\int_K |f| d\lambda < \infty$. Let L^1_{loc} denote the space of all locally integrable functions.

It is easily verified that L^1_{loc} is a vector space. Observe that a measurable function f on $\mathbb{R}^d$ is locally integrable if and only if for every ball $B = B(x; r)$ with finite radius, $\int_B |f| d\lambda < \infty$. (Why?) For such a ball B and a function f in L^1_{loc} we define the *average* of f over B as

$$A_r f(x) = \frac{1}{\lambda(B(x; r))} \int_{B(x;r)} f(y) d\lambda(y)$$

Note that $|A_r f(x)| < \infty$ for any finite non-zero radius.

We might hope that as $r \to 0$, the average $A_r f(x)$ converges to $f(x)$. This is not the case as we can see with a simple example when $d = 1$: if $f = \chi_{[0,1]}$, then $A_r f(1) = (2r)^{-1} \int_{(1-r,1+r)} f \, d\lambda = (2r)^{-1} \lambda((1-r,1]) = \frac{1}{2}$ for all $r > 0$. (See Exercise 1.) However we will see below that for any function f in L^1_{loc}, $A_r f(x) \to f(x)$ a.e. $[\lambda]$ (4.9.6). Let's start with a basic result about the average.

4.9.2. Proposition. *If $f \in L^1_{\text{loc}}$, the map of $\mathbb{R}^d \times (0, \infty) \to \mathbb{C}$ defined by $(x, r) \mapsto A_r f(x)$ is continuous.*

Proof. Fix x_0 in $\mathbb{R}^d$ and $r_0 > 0$ and put $B_0 = B(x_0; r_0)$. Suppose $x_n \to x_0$ and $r_n \to r_0$, and put $B_n = B(x_n; r_n)$. Note that $\chi_{B_n}(y) \to \chi_{B_0}(y)$ whenever $y \notin \partial B_0$. By Exercise 4.8.5, $\chi_{B_n} \to \chi_{B_0}$ a.e. $[\lambda]$. It is also the case that $|\chi_{B_n}| \le \chi_{B(x_0; r_0+1)}$ for all sufficiently large n. Therefore the DCT implies that $\int_{B_n} f \, d\lambda \to \int_{B_0} f \, d\lambda$. But Exercise 4.8.4 implies that $\lambda(B_n) \to \lambda(B_0)$. Hence $A_{r_n} f(x_n) \to A_{r_0} f(x_0)$ and the proof is complete. ∎

Here we introduce a function associated with the average. Such quantities are playing an increasingly important role in mathematics as a means of making estimates, one of the primary weapons in developing analysis.

4.9.3. Definition. If $f \in L^1_{\text{loc}}$, define the *Hardy*[4]–*Littlewood*[5] *maximal function $Hf : \mathbb{R}^d \to \mathbb{R}_+$* by

$$Hf(x) = \sup_{r>0} A_r |f|(x) = \sup_{r>0} \frac{1}{\lambda(B(x; r))} \int_{B(x;r)} |f(y)| d\lambda(y)$$

---

[4]Godfrey Harold Hardy was born in Cranleigh, England in 1877. In 1901 he was elected a fellow of Trinity College, Cambridge and began a period where he wrote many papers, most of which he later termed unimportant. A watershed event in his life came in 1911 when he began his collaboration with Littlewood; it lasted 35 years and produced outstanding mathematics. In 1913 he received the first letter from Ramanujan and a year later succeeded in getting him to Cambridge. This collaboration also produced wonderful mathematics but ended prematurely with Ramanujan's untimely death. Because of political disagreements during World War I, Hardy became unhappy at Cambridge and secured a chair at Oxford in 1919. In 1931 he returned to Cambridge. The literature abounds with stories of his eccentricities. One such was that he continually tried to fool God and was convinced that He disliked and worked against him. For example, during a trip home from Denmark, where the sea was particularly violent, he sent back a postcard claiming that he had proved the Riemann hypothesis. He reasoned that God would not allow the boat to sink on the return journey and give him the same fame that Fermat had achieved with his "last theorem". He never married and died in Cambridge in 1947.

[5]John Edensor Littlewood was born in Rochester, England in 1885. When he was seven years old his family emigrated to South Africa, where his father was to be the headmaster of a school. At the age of 15 he returned to England for better schooling. He entered Trinity College in 1903. He lectured at the University of Manchester from 1907 to 1910 and then returned to Cambridge where he met Hardy. In addition to his collaboration with Hardy, Littlewood is also known for the extensive work he did on Fourier analysis with Paley as well as many other works. Throughout his life he struggled with depression and only seems to have received adequate treatment after he retired. He died in Cambridge in 1977.

The next lemma, needed in the proof of the main theorem about the maximal function, is typical of a collection of results called covering lemmas or theorems. Such results say that given a covering of a set by a collection of subsets with some property, we can extract a subcollection that may not be a covering but has some specific attributes including one that is close to being a covering.

4.9.4. Lemma. *If $\mathcal{D}$ is a collection of open balls in $\mathbb{R}^d$ that cover a measurable set E and $0 < c < \lambda(E)$, then there are pairwise disjoint balls $B_1, \ldots, B_n$ in $\mathcal{D}$ such that $\sum_{k=1}^{n} \lambda(B_k) > 3^{-d}c$.*

Proof. By the regularity of λ there is a compact subset K of E such that $\lambda(K) > c$. Let $D_1, \ldots, D_m \in \mathcal{D}$ such that $K \subseteq \bigcup_{i=1}^{m} D_i$. Let B_1 be the ball in $\{D_1, \ldots, D_m\}$ with the largest radius. Now suppose balls $B_1, \ldots, B_k$ are chosen from $\{D_1, \ldots, D_m\}$. If none of the remaining balls is disjoint from the chosen ones, stop this process; otherwise choose B_{k+1} to be the ball in $\{D_1, \ldots, D_m\}$ different from the ones already chosen and such that $B_{k+1} \cap B_j = \emptyset$ for $1 \leq j \leq k$ and B_{k+1} has the largest possible radius. Continue this process until we are forced to stop. The resulting balls $\{B_1, \ldots, B_n\}$ are pairwise disjoint with decreasing radii. Put $B_k = B(x_k; r_k)$.

Suppose one of the balls D_i is not among the chosen subfamily. So it must be that $D_i \cap B_k \neq \emptyset$ for at least one k, $1 \leq k \leq n$; let k be the first integer with $D_i \cap B_k \neq \emptyset$. Note that if ρ is the radius of D_i, then $\rho \leq r_k$. In fact suppose this is not the case. Since $D_i \cap B_j = \emptyset$ for $1 \leq j < k$, if $\rho > r_k$, then we would have chosen D_i rather than B_k at this step in the choice algorithm. Because $\rho \leq r_k$ and $D_i \cap B_k \neq \emptyset$, it follows that $D_i \subseteq B(x_k; 3r_k)$. Thus we arrive at the fact that $K \subseteq \bigcup_{k=1}^{n} B(x_k; 3r_k)$ and so $c < \lambda(K) \leq \sum_{k=1}^{n} \lambda(B(x_k; 3r_k)) = 3^d \sum_{k=1}^{n} \lambda(B_k)$, establishing the lemma. ∎

So the preceding lemma says that if we are presented with a cover of an open set G by open balls, though we may not be able to find a subcover by disjoint balls, in that cover we can find disjoint balls that fill up a fixed percentage of the volume of G. In fact that fixed fraction only depends on the dimension d of the containing Euclidean space and is the same for all open sets and all covers by open balls. The interested reader might want to consult [15] and [18] for a statement of the Vitali Covering Theorem, which has the same flavor.

4.9.5. Theorem (The Maximal Theorem). *If $f \in L^1_{\text{loc}}$ and $a > 0$, then*

$$\lambda(\{x : Hf(x) > a\}) \leq \frac{3^d}{a}\|f\|_1$$

Proof. Put $E_a = \{x : Hf(x) > a\}$ and assume $\lambda(E_a) > 0$. For each x in E_a, there is a radius r_x such that $A_{r_x}|f|(x) > a$; so $\lambda(B(x;r_x)) < a^{-1} \int_{B(x;r_x)} |f| d\lambda$. Let c be an arbitrary positive number with $c < \lambda(E_a)$. Since $\{B(x;r_x) : x \in E_a\}$ is an open cover of E_a, Lemma 4.9.4 implies there are points $x_1, \ldots, x_n$ in E_a such that if $B_k = B(x_k; r_{x_k})$, then $B_1, \ldots, B_n$ are pairwise disjoint and $\sum_{k=1}^{n} \lambda(B_k) > 3^{-d}c$. Therefore

$$c < 3^d \sum_{k=1}^{n} \lambda(B_k)$$

$$< \frac{3^d}{a} \sum_{k=1}^{n} \int_{B_k} |f| d\lambda$$

$$\leq \frac{3^d}{a} \|f\|_1$$

Since c was an arbitrary number less than $\lambda(E_a)$, we have finished the proof. ∎

A small item from the analyst's bag of tricks is to observe that when $F(r)$ is a positive function of r and we want to show that $\lim_{r \to 0} F(r) = 0$, it suffices to show that $\limsup_{r \to 0} F(r) = 0$. This allows us to use inequalities involving $F(r)$ even though we do not know that the limit exists. This rather abstract statement is amply illustrated in the next proof.

4.9.6. Theorem. *If* $f \in L^1_{\text{loc}}$, *then*

$$\lim_{r \to 0} A_r f(x) = f(x) \text{ a.e.} [\lambda]$$

Equivalently

$$\lim_{r \to 0} \frac{1}{\lambda(B(x;r))} \int_{B(x;r)} [f(y) - f(x)] d\lambda(y) = 0 \text{ a.e.} [\lambda]$$

Proof. To avoid confusion, denote the norm of an element x in $\mathbb{R}^d$ by $|x|$ rather than $\|x\|$; that is $|x| = \left[\sum_{j=1}^{d} |x_j|^2\right]^{\frac{1}{2}}$. We begin by showing that without loss of generality we may assume that $f \in L^1(\lambda)$. In fact, to establish the theorem we need only show that for any $n \geq 1$, $A_r f(x) \to f(x)$ a.e. on $\text{cl } B(0;n)$. But notice that for any x with $|x| \leq n$, the value of $\lim_{r \to 0} A_r f(x)$ depends only on the values of $f(y)$ for y in $B(0; n+1)$. Now $f \chi_{B(0;n+1)} \in L^1(\lambda)$ and if we prove the theorem with $f \chi_{B(0;n+1)}$ replacing f, we are done. Hence we can assume $f \in L^1(\lambda)$.

Fix an arbitrary $a > 0$ and put $E_a = \{x : \limsup_{r \to 0} |A_r f(x) - f(x)| > a\}$. The proof reduces to showing that $\lambda(E_a) = 0$. In fact, once this is done, we set $Z = \bigcup_{n=1}^{\infty} E_{\frac{1}{n}}$ so that $\lambda(Z) = 0$. It follows that when $x \in \mathbb{R}^d \backslash Z$, $\limsup_{r \to 0} |A_r f(x) - f(x)| = 0$ and so $\lim_{r \to 0} A_r f(x) = f(x)$.

Let $\epsilon > 0$. According to Proposition 4.4.2, there is a function g in $C_c(\mathbb{R}^d)$ such that $\|g - f\|_1 < \epsilon$. Since g is uniformly continuous, for every $\epsilon_1 > 0$ there is a δ such that $|g(x) - g(y)| < \epsilon_1$ whenever $|x - y| < \delta$. Hence for $0 < r < \delta$

$$|A_r g(x) - g(x)| = \frac{1}{\lambda(B(x;r))} \left| \int_{B(x;r)} [g(y) - g(x)] d\lambda(y) \right| < \epsilon_1$$

This implies that $\lim_{r \to 0} A_r g(x) = g(x)$ for every x in $\mathbb{R}^d$. Therefore

$$\limsup_{r \to 0} |A_r f(x) - f(x)|$$
$$\leq \limsup_{r \to 0} |A_r(f - g)(x)| + \limsup_{r \to 0} |A_r g(x) - g(x)|$$
$$+ \limsup_{r \to 0} |g(x) - f(x)|$$
$$\leq H(f - g)(x) + |f(x) - g(x)|$$

Put $F_a = \{x : |f(x) - g(x)| > a/2\}$, and note that $E_a \subseteq F_a \cup \{x : H(f - g)(x) > a/2\}$. Now $\epsilon > \|f - g\|_1 \geq \int_{F_a} |f - g| d\lambda \geq (a/2)\lambda(F_a)$. Also the Maximal Theorem implies that

$$\lambda(\{x : H(f - g)(x) > a/2\}) \leq \frac{3^d}{a/2}\|f - g\|_1 < \frac{2\epsilon 3^d}{a}$$

Hence

$$\lambda(E_a) \leq \lambda(F_a) + \lambda\left(\{x : H(f - g)(x) > a/2\}\right) < \frac{2\epsilon}{a} + \frac{2\epsilon 3^d}{a}$$

Since ϵ was arbitrary, we have that $\lambda(E_a) = 0$, completing the proof. ∎

Now we'll get a somewhat stronger version of the preceding proposition, though we need the first result to prove the second.

4.9.7. Definition. If $f \in L^1_{\text{loc}}$, define the *Lebesgue set* of f to be set

$$L_f = \left\{ x : \lim_{r \to 0} \frac{1}{\lambda(B(x;r))} \int_{(B(x;r)} |f(y) - f(x)| d\lambda(y) = 0 \right\}$$

4.9.8. Proposition. *If $f \in L^1_{\text{loc}}$, then $\lambda(\mathbb{R}^d \backslash L_f) = 0$.*

Proof. If $c \in \mathbb{R}$ and we apply Theorem 4.9.6 to the function $y \mapsto |f(y) - c|$, then we see that there is a set Z_c with $\lambda(Z_c) = 0$ such that for $x \notin Z_c$, $\lim_{r \to 0}[\lambda((B(x;r))]^{-1} \int_{(B(x;r)} |f(y) - c| d\lambda(y) = |f(x) - c|$. Let C be a countable dense subset of $\mathbb{C}$ and put $Z = \bigcup_{c \in C} Z_c$, so that $\lambda(Z) = 0$. If

$x \notin Z$ and $\epsilon > 0$, there is a c in C such that $|f(x) - c| < \epsilon/2$; thus for every y in X, $|f(y) - f(x)| < |f(y) - c| + \epsilon/2$. Hence

$$\int_{B(x;r)} |f(y) - f(x)| d\lambda(y) \leq \int_{B(x;r)} [|f(y) - c| + \epsilon/2] d\lambda(y)$$

and so

$$\limsup_{r \to 0} \frac{1}{\lambda((B(x;r))} \int_{B(x;r)} |f(y) - f(x)| d\lambda(y) \leq |f(x) - c| + \epsilon/2 < \epsilon$$

Since ϵ was arbitrary, we have that $x \in L_f$. That is, $\mathbb{R}^d \backslash Z \subseteq L_f$, so $\lambda(\mathbb{R}^d \backslash L_f) \leq \lambda(Z) = 0$. ∎

For a variety of reasons we want to "close down" on a point x in $\mathbb{R}^d$ using sets other than the balls $B(x;r)$. If $x \in \mathbb{R}^d$, say that a family of Borel sets $\{E_r : r > 0\}$ *shrinks nicely* to x if: (a) $E_r \subseteq B(x;r)$ for every $r > 0$; (b) there is a positive constant c that does not depend on r with $\lambda(E_r) > c\lambda(B(x;r))$. It should be emphasized that the sets E_r are not required to contain the point x. One example of such a collection of Borel sets is the balls $B(x;r)$ themselves. Another can be obtained as follows. Let E be any Borel set contained in $\{y \in \mathbb{R}^d : |y| < 1\}$ with $\lambda(E) > 0$ and put $E_r = x + rE = \{x + ry : y \in E\}$. (What is the constant c in this example?) A specific instance of this is obtained by letting the dimension $d = 1$ and $E = [0, 1)$; in this case the sets E_r become $[x, x+r)$.

4.9.9. Theorem (Lebesgue Differentiation Theorem). *If* $f \in L^1_{\text{loc}}$ *and* $x \in L_f$, *then whenever* $\{E_r\}$ *shrinks nicely to* x *we have*

$$\lim_{r \to 0} \frac{1}{\lambda(E_r)} \int_{E_r} |f(y) - f(x)| d\lambda(y) = 0$$

and

$$\lim_{r \to 0} \frac{1}{\lambda(E_r)} \int_{E_r} f(y) d\lambda(y) = f(x)$$

In particular, these equations hold a.e. $[\lambda]$.

Proof. We have already done most of the work for this proof. In fact if c is the constant such that $\lambda(E_r) > c\lambda(B(x;r))$, then

$$\frac{1}{\lambda(E_r)} \int_{E_r} |f(y) - f(x)| d\lambda(y) < \frac{1}{c\lambda(B(x;r))} \int_{B(x;r)} |f(y) - f(x)| d\lambda(y)$$

From the definition of L_f we get the first equality. The second follows from the first by taking the difference of both sides of the equation and rewriting this difference as an integral. The fact that these equalities hold a.e. $[\lambda]$ is a consequence of Proposition 4.9.8. ∎

Now we turn our attention to signed regular Borel measures on $\mathbb{R}^d$ and use what we have done to obtain a formula for the Radon–Nikodym derivative with respect to Lebesgue measure. So according to Theorem 4.1.6 if μ is a signed measure on $\mathbb{R}^d$, there is a signed measure ν with $\nu \perp \lambda$ and a signed measure $\eta \ll \lambda$ such that $\mu = \nu + \eta$. Both ν and η must also be regular Borel signed measures (Exercise 4.3.2). If f is the Radon–Nikodym derivative of η with respect to λ, then for every compact subset K of $\mathbb{R}^d$ we have $\int_K |f| d\lambda = |\eta|(K) < \infty$. Therefore $f \in L^1_{\text{loc}}$ and $\mu = \nu + f\lambda$. (When we have a measure η and a measurable function g, we use the notation $g\eta$ to mean the measure $E \mapsto \int_E g d\eta$, provided this is well defined for all measurable sets E.)

4.9.10. Proposition. *If μ is any signed regular Borel measure on $\mathbb{R}^d$ and $\mu = \nu + f\lambda$ is its Lebesgue decomposition with respect to λ with $\nu \perp \lambda$, then there is a set Z with $\lambda(Z) = 0$ such that when $x \notin Z$ and when $\{E_r\}$ shrinks nicely to x we have*

$$\lim_{r \to 0} \frac{\mu(E_r)}{\lambda(E_r)} = f(x)$$

Proof. By looking at the Jordan decomposition of μ, we quickly see that it suffices to assume that $\mu \geq 0$; hence $\nu \geq 0$ and $f \geq 0$. Since

$$\frac{\mu(E_r)}{\lambda(E_r)} = \frac{\nu(E_r)}{\lambda(E_r)} + \frac{1}{\lambda(E_r)} \int_{E_r} f d\lambda$$

the preceding theorem allows us to finish the proof by showing that

$$\lim_{r \to 0} \frac{\nu(E_r)}{\lambda(E_r)} = 0$$

If c is the constant in the definition of shrinking nicely to x, we have

$$\frac{\nu(E_r)}{\lambda(E_r)} \leq \frac{\nu(B(x;r))}{\lambda(E_r)} \leq \frac{1}{c} \frac{\nu(B(x;r))}{\lambda(B(x;r))}$$

Therefore it suffices to prove that

$$\lim_{r \to 0} \frac{\nu(B(x;r))}{\lambda(B(x;r))} = 0$$

Let A be a Borel set such that $\lambda = \lambda_A$ and $\nu = \nu_{\mathbb{R}^d \setminus A}$. For each $n \geq 1$, let

$$Z_n = \left\{ x \in A : \limsup_{r \to 0} \frac{\nu(B(x;r))}{\lambda(B(x;r))} > \frac{1}{n} \right\}$$

If we can show that $\lambda(Z_n) = 0$, then we let $Z = \bigcup_{n=1}^\infty Z_n$; this implies that $\lambda(Z) = 0$ and completes the proof.

To show that $\lambda(Z_n) = 0$, let $\epsilon > 0$. Since $\nu(A) = 0$, we can find an open set G containing A with $\nu(G) < \epsilon$. For each x in Z_n let B_x be an open ball centered at x such that $B_x \subseteq G$ and $\nu(B_x) > \frac{1}{n}\lambda(B_x)$; put

$H = \bigcup\{B_x : x \in Z_n\} \subseteq G$. Let $\{c_i\}$ be a sequence of positive real numbers such that $c_i < \lambda(H)$ and $c_i \to \lambda(H)$. For any fixed c_i, Lemma 4.9.4 implies there are points $x_1, \ldots, x_m$ in Z_n such that $B_{x_1}, \ldots, B_{x_m}$ are pairwise disjoint and

$$c_i < 3^d \sum_{j=1}^{m} \lambda(B_{x_j}) \leq n3^d \sum_{j=1}^{m} \nu(B_{x_j}) \leq n3^d \nu(G) < n3^d \epsilon$$

Letting $i \to \infty$ we have that $\lambda(H) \leq n3^d \epsilon$; since ϵ was arbitrary we have that $0 = \lambda(H) \geq \lambda(Z_n)$. $\blacksquare$

Exercises.

(1) Let $d = 1$, let $f = \chi_{[0,1]}$, and calculate $A_r f(x)$ for all $r > 0$ and all x in $\mathbb{R}$.

(2) Let

$$H^* f(x) = \sup\left\{\frac{1}{\lambda(B)}\int_B |f|d\lambda : x \in B \text{ and } B \text{ is a ball}\right\}$$

Show that this maximal function and the Hardy–Littlewood maximal function are comparable by proving that $Hf(x) \leq H^* f(x) \leq 2^d Hf(x)$ for all x in $\mathbb{R}^d$.

(3) If $f \in L^1_{\text{loc}}$ and f is continuous at x, show that $x \in L_f$, the Lebesgue set of f.

(4) For any Borel set define the *density* of E at x in $\mathbb{R}^d$ to be the quantity

$$D_E(x) = \lim_{r \to 0} \frac{\lambda(E \cap B(x;r))}{\lambda(B(x;r))}$$

whenever the limit exists. Show that $D_E(x) = 1$ a.e. $[\lambda]$ on E and $D_E(x) = 0$ a.e. $[\lambda]$ on the complement of E.

4.10. Absolutely continuous functions

In this section we will use what we did in the preceding one to examine closely what happens when the dimension is $d = 1$. In the process we will obtain the Fundamental Theorem of Calculus. Throughout this section λ denotes Lebesgue measure on the real line and J is the closed interval $[a, b]$.

We begin by once again discussing functions of bounded variation. Here we assume that α is a normalized function of bounded variation (§4.5): $\alpha(a) = 0$ and α is left-continuous at each point of (a, b). Recall that there is a unique regular Borel measure $\mu = \mu_\alpha$ on J such that $\int f d\alpha = \int f d\mu$ for every f in $C(J)$, $\alpha(t) = \mu([a, t))$ for $a \leq t < b$, and $\alpha(b) = \mu([a, b])$ (Proposition 4.5.2).

4.10.1. Theorem. *Let α be a normalized function of bounded variation on J.*

(a) *α is differentiable a.e. $[\lambda]$ and $\alpha' \in L^1(\lambda)$.*

(b) *$\mu_\alpha \perp \lambda$ if and only if $\alpha' = 0$ a.e. $[\lambda]$.*

(c) *$\mu_\alpha \ll \lambda$ if and only if $\mu_\alpha(E) = \int_E \alpha' d\lambda$ for every Borel set E.*

Proof. Let $\mu_\alpha = \mu = \nu + f\lambda$ be the Lebesgue decomposition of μ with respect to λ. For any t in $[a, b)$, let $E_h = [t, t+h)$ for $h > 0$. The reader can check that $\{E_h\}_{h>0}$ shrinks nicely to t and

$$\frac{\alpha(t + h) - \alpha(t)}{h} = \frac{\mu(E_h)}{\lambda(E_h)}$$

By Proposition 4.9.10 as $h \to 0+$ this limit exists and equals $f(t)$ a.e. $[\lambda]$. Similarly using the sets $F_h = [t - h, t)$ for $h > 0$

$$\frac{\alpha(t - h) - \alpha(t)}{-h} = \frac{-\mu(F_h)}{-\lambda(F_h)}$$

has a limit equal to $f(t)$ a.e. $[\lambda]$ as $h \to 0+$. Therefore α is differentiable a.e. $[\lambda]$ on J with $\alpha' = f$ a.e. $[\lambda]$, proving (a).

(b) $\mu_\alpha \perp \lambda$ if and only if $\mu_\alpha = \nu$; equivalently, $f = 0$ a.e. $[\lambda]$. Thus (b) is immediate from the preceding discussion.

(c) Again, $\mu_\alpha \ll \lambda$ if and only if $\nu = 0$; that is, if and only if $\mu_\alpha = f\lambda$. Since we know that $f = \alpha'$, (c) follows.   ∎

4.10.2. Definition. A function $f : [a, b] \to \mathbb{C}$ is *absolutely continuous* if for every $\epsilon > 0$ there is a $\delta > 0$ such that for any finite set of pairwise disjoint subintervals $\{(x_k, y_k) : 1 \le k \le n\}$ with $\sum_{k=1}^n (y_k - x_k) < \delta$, we have $\sum_{k=1}^n |f(y_k) - f(x_k)| < \epsilon$.

It is immediate that an absolutely continuous function is continuous; in fact such functions are uniformly continuous. So obtaining examples of functions that are not absolutely continuous is easy. However here are some examples of absolutely continuous functions as well as of a continuous function (and, therefore, uniformly continuous function) that is not absolutely continuous.

4.10.3. Example. (a) If f is a Lipschitz function, then f is absolutely continuous. In fact from the definition of a Lipschitz function there is a constant M with $|f(y) - f(x)| \le M|y - x|$. Thus $\sum_{k=1}^n |f(y_k) - f(x_k)| \le M \sum_{k=1}^n |y_k - x_k| \le M\delta$ when $\sum_{k=1}^n (y_k - x_k) < \delta$.

(b) From the preceding example and the Fundamental Theorem of Calculus we have that whenever f is a continuously differentiable function on $[a, b]$, then f is absolutely continuous.

(c) If $f(x) = x \sin x^{-1}$ when $x \neq 0$ and $f(0) = 0$, then f is not absolutely continuous on $[0, 1]$. This can be seen directly but it may be easier to use the fact that f is not a function of bounded variation (Exercise 1.1.3) and rely on the next result.

4.10.4. Proposition. *If f is absolutely continuous, then f is a function of bounded variation.*

Proof. Let $\delta > 0$ such that when $\sum_{j=1}^{n}(y_j - x_j) < \delta$, we have $\sum_{j=1}^{n} |f(y_j) - f(x_j)| < 1$. Let N be an integer such that $N > (b - a)/\delta$ and let $a = x_0 < \cdots < x_N = b$ be a partition with $x_k - x_{k-1} \leq (b - a)/N$ for $1 \leq k \leq N$. It is easy to see that $\mathrm{Var}(f, [x_{k-1}, x_k]) \leq 1$ for $1 \leq k \leq N$. Therefore $\mathrm{Var}(f, [a, b]) \leq \sum_{k=1}^{N} \mathrm{Var}(f, [x_{k-1}, x_k]) \leq N$. (See Exercise 1.) ∎

Since an absolutely continuous function α is continuous, let's point out that its normalization is simply obtained as $\tilde{\alpha} = \alpha - \alpha(a)$.

4.10.5. Proposition. *Let $J = [a, b]$ be any interval in the real line.*

(a) *The set of all absolutely continuous functions from J into the complex numbers is a vector space over $\mathbb{C}$.*

(b) *If $f : J \to \mathbb{C}$ is absolutely continuous, then so are $\mathrm{Re}\, f, \mathrm{Im}\, f : [a, b] \to \mathbb{R}$. If f is real-valued and absolutely continuous, then f can be written as the difference of two increasing absolutely continuous functions.*

Proof. The proof of (a) is trivial as is the first part of (b). So assume f is real-valued and consult the proof of Proposition 1.1.6, where it is shown that a function of bounded variation can be written as the difference of two increasing functions. It suffices to show that $g(x) = \mathrm{Var}(f, [a, x])$ defines an absolutely continuous function on J. Let $\epsilon > 0$ and choose $\delta > 0$ such that $\{(x_k, y_k) : 1 \leq k \leq n\}$ disjoint and $\sum_{k=1}^{n}(y_k - x_k) < \delta$ implies $\sum_{k=1}^{n} |f(y_k) - f(x_k)| < \epsilon/2$. Fix such a sequence of pairwise disjoint subintervals. For $1 \leq k \leq n$, let $x_k = a_{k0} < a_{k1} < \cdots < a_{km_k} = y_k$ be a partition of $[x_k, y_k]$ such that $\mathrm{Var}(f, [x_k, y_k]) < \sum_{j=1}^{m_k} |f(a_{kj}) - f(a_{k(j-1)})| + \epsilon/2n$. Therefore

$$\sum_{k=1}^{n} |g(y_k) - g(x_k)| = \sum_{k=1}^{n} \mathrm{Var}(f, [x_k, y_k])$$

$$< \sum_{k=1}^{n} \sum_{j=1}^{m_k} \left[|f(a_{kj}) - f(a_{k,(j-1)})| + \epsilon/2n \right]$$

$$< \epsilon/2 + \sum_{k=1}^{n} \sum_{j=1}^{m_k} \left[|f(a_{kj}) - f(a_{k,(j-1)})| \right]$$

$$< \epsilon$$

since $\sum_{k=1}^{n}\sum_{j=1}^{m_k}(a_{kj} - a_{k,(j-1)}) < \delta$.   ∎

4.10.6. Theorem (Fundamental Theorem of Calculus). *If α is a function of bounded variation on the interval $J = [a, b]$, then the following statements are equivalent.*

(a) α *is absolutely continuous.*

(b) $\mu_\alpha \ll \lambda$.

(c) $\alpha' \in L^1(\lambda)$ *and for every x in J, $\alpha(x) = \alpha(a) + \int_a^x \alpha'(t)dt$.*

Proof. Put $\mu = \mu_\alpha$.

(c) *implies* (a). Since $\alpha' \in L^1(\lambda)$, $E \mapsto \int_E \alpha' d\lambda$ defines a measure absolutely continuous with respect to λ. Therefore if $\epsilon > 0$ there is a $\delta > 0$ such that $\int_E |\alpha'| d\lambda < \epsilon$ when $\lambda(E) < \delta$ (Proposition 4.1.11). If $\{(x_k, y_k) : 1 \le k \le n\}$ are pairwise disjoint intervals in J with $\sum_{k=1}^{n}(y_k - x_k) < \delta$, then this says that $\lambda\left(\bigcup_{k=1}^{n}[x_k, y_k]\right) < \delta$. Hence $\sum_{k=1}^{n}|\alpha(y_k) - \alpha(x_k)| = \sum_{k=1}^{n}\left|\int_{[x_k, y_k]}\alpha' d\lambda\right| \le \int_{\bigcup_{k=1}^{n}[x_k, y_k]}|\alpha'| d\lambda < \epsilon$.

(b) *implies* (c). This is a restatement of part (c) of Theorem 4.10.1.

(a) *implies* (b). Without loss of generality we can assume that α is increasing so that μ is a positive measure. (Why?) Let $\epsilon > 0$, and let E be any set with $\lambda(E) = 0$. From (a), there is a $\delta > 0$ such that if $\{(x_k, y_k) : 1 \le k \le n\}$ satisfies $\sum_{k=1}^{n}(y_k - x_k) < \delta$, then $\sum_{k=1}^{n}[\alpha(y_k) - \alpha(x_k)] < \epsilon/2$. By the regularity of λ there is an open set G containing E with $\lambda(G) < \delta$. Now G is the union of a sequence of pairwise disjoint open intervals, so there are finitely many such open intervals $\{(x_k, y_k) : 1 \le k \le n\}$ with $\mu\left(G \backslash \bigcup_{k=1}^{n}(x_k, y_k)\right) < \epsilon/2$. Now $\sum_{k=1}^{n}(y_k - x_k) \le \lambda(G) < \delta$. Hence $\epsilon/2 > \sum_{k=1}^{n}[\alpha(y_k) - \alpha(x_k)] = \mu\left(\bigcup_{k=1}^{n}(x_k, y_k)\right)$, and, therefore,

$$\mu(E) \le \mu(G) = \mu\left(\bigcup_{k=1}^{n}(x_k, y_k)\right) + \mu\left(G \backslash \bigcup_{k=1}^{n}(x_k, y_k)\right) < \epsilon$$

Since ϵ was arbitrary, $\mu(E) = 0$ and so $\mu \ll \lambda$.   ∎

In light of the Fundamental Theorem of Calculus, the total opposite of absolute continuity occurs when we have a function of bounded variation whose derivative vanishes a.e. In this vein, see Exercise 4. We also call attention to the existence of a strictly increasing continuous function on the unit interval that maps the interval onto itself but has derivative that is 0 a.e. See Example 18.8 in [18]. For further information on absolutely continuous functions see [32] and [33].

Exercises.

(1) If $f : [a, b] \to \mathbb{R}$ and $a < c < b$, show that $\text{Var}(f, [a, b]) \leq \text{Var}(f, [a, c]) + \text{Var}(f, [c, b])$. What about equality?

(2) If α is a function of bounded variation on J and $\mu = \mu_\alpha$, show that $\mu = \mu_{ac} + \mu_d + \mu_s$, where $\mu_{ac} \ll \lambda$; $\mu_d = \sum_{n=1}^{\infty} w_n \delta_{t_n}$, where $\sum_{n=1}^{\infty} |w_n| < \infty$ and each δ_{t_n} is the unit point mass at the point t_n in J; $\mu_s \perp \lambda$ but $\mu_s(\{t\}) = 0$ for all t in J. Moreover these three measures are unique. What is the corresponding decomposition of the function α?

(3) Show that if α is an increasing function on J, then $\alpha(b) - \alpha(a) \geq \int_a^b \alpha' d\lambda$.

(4) If c is the Cantor function on the unit interval, show that $\mu_c \perp \lambda$.

(5) (Integration by Parts) If α and β are two functions of bounded variation on J that are left-continuous with at least one of them continuous, show that $\int \alpha d\mu_\beta + \int \beta d\mu_\alpha = \alpha(b)\beta(b) - \alpha(a)\beta(a)$. Hint: Let $E = \{(s, t) \in \mathbb{R}^2 : a \leq s \leq t < b\}$ and use Fubini's Theorem to calculate $\mu_\alpha \times \mu_\beta(E)$ in two different ways.

(6) If α and β are absolutely continuous functions on J, show that $\alpha\beta$ is absolutely continuous and $\int_{[a,b]} (\alpha\beta' + \alpha'\beta) d\lambda = \alpha(b)\beta(b) - \alpha(a)\beta(a)$.

(7) Is there a Mean Value Theorem for the derivative of an absolutely continuous function?

(8) If $\{\alpha_n\}$ is a sequence of normalized increasing functions on J such that $\alpha(t) = \sum_n \alpha_n(t) < \infty$ for every t in J, show that α is increasing and $\alpha'(t) = \sum_n \alpha_n'(t)$ a.e. $[\lambda]$. (Hint: Look at the measures μ_{α_n}.)

(9) Let α be a continuous increasing function on J with $c = \alpha(a)$, $d = \alpha(b)$. (a) Show that for every Borel set E contained in $[c, d]$, $\mu_\alpha(\alpha^{-1}(E)) = \lambda(E)$. (Hint: Show first that it works when E is an interval and go from there to any Borel set.) (b) If $f \in L^1([c, d], \lambda)$, show that $\int_J f \circ \alpha d\alpha = \int_{[c,d]} f d\lambda$. (c) What happens in (b) if we assume that α is absolutely continuous?

(10) Show that $f : J \to \mathbb{R}$ is a Lipschitz function if and only if f is absolutely continuous and f' is bounded.

(11) Is the maximum of two real-valued absolutely continuous functions absolutely continuous?

4.11. Convolution*

This section and the next are closely tied together. Everything done in these sections carries over to $\mathbb{R}^d$ with small adjustments but more complicated notation and extra constants. We restrict our attention to the one-dimensional case to give the reader a feel for the subject and leave it to him/her to seek more information in the literature.

Notation. We are going to modify Lebesgue measure a bit by multiplying it by a constant. Let

$$\omega = \frac{1}{\sqrt{2\pi}}\lambda$$

and use this everywhere we discuss Lebesgue measure in this section and the next. Why? Actually it has little to do with the material in this section but will affect certain results in the next; without this adjustment we would have an extra constant to keep track of as we compute various quantities. We will usually write $L^p(\mathbb{R})$ instead of $L^p(\omega)$. Of course as a set of functions this is the same as $L^p(\lambda)$, but the norm of each function is different: $\|f\|_{L^p(\mathbb{R})} = \|f\|_{L^p(\omega)} = (2\pi)^{-1/2p}\|f\|_{L^p(\lambda)}$. This difference in the norms will not surface in our discussion.

Naturally we begin with the definition of convolution, but first we have to establish that the definition makes sense.

4.11.1. Proposition. *If $f, g \in L^1(\mathbb{R})$, then*

$$h(x) = \int f(x-t)g(t)d\omega(t)$$

defines a function h in $L^1(\mathbb{R})$ with $\|h\|_1 \leq \|f\|_1\|g\|_1$.

Proof. We use Fubini's Theorem and the invariance of Lebesgue measure under translation to get

$$\int |h(x)|d\omega(x) = \int \left| \int f(x-t)g(t)d\omega(t) \right| d\omega(x)$$

$$\leq \int \int |f(x-t)||g(t)|d\omega(t)d\omega(x)$$

$$= \int \left[\int |f(x-t)|d\omega(x) \right] |g(t)|d\omega(t)$$

$$= \|f\|_1\|g\|_1 \qquad \blacksquare$$

4.11.2. Definition. If $f, g \in L^1(\mathbb{R})$, define the *convolution* of f and g to be the function

$$f * g(x) = \int f(x-t)g(t)d\omega(t) = \frac{1}{\sqrt{2\pi}}\int_{-\infty}^{\infty} f(x-t)g(t)dt$$

Here are the basic properties of convolution.

4.11.3. Proposition. *If $f, g, h \in L^1(\mathbb{R})$, the following hold:*

(a) $f * g \in L^1(\mathbb{R})$ *and* $\|f * g\|_1 \leq \|f\|_1 \|g\|_1$;

(b) $f * g = g * f$;

(c) $f * (g * h) = (f * g) * h$;

(d) $f * (g + h) = f * g + f * h$.

Proof. Part (a) is immediate from Proposition 4.11.1. For (b), we perform the change of variable $t \mapsto x - t$ and get $f * g(x) = \int f(x - t)g(t)d\omega(t) = \int f(t)g(x - t)d\omega(t) = g * f(x)$. To prove (c) we first apply Fubini's Theorem and then do a change of variables:

$$[f * (g * h)](x) = \int f(x - t)(g * h)(t)d\omega(t)$$

$$= \int f(x - t) \left(\int g(t - s)h(s)d\omega(s) \right) d\omega(t)$$

$$= \int \left(\int f(x - t)g(t - s)d\omega(t) \right) h(s)d\omega(s)$$

$$= \int \left(\int f(x - s - t)g(t)d\omega(t) \right) h(s)d\omega(s)$$

$$= \int (f * g)(x - s)h(s)d\omega(s)$$

$$= [(f * g) * h](x)$$

Part (d) is left as an exercise. ∎

So the preceding proposition says that the convolution is an associative multiplication on $L^1(\mathbb{R})$. This multiplication does not have an identity, but it almost does – in a literal sense as we now will see. Let $\phi \in L^1(\mathbb{R})$ such that $\int \phi d\omega = 1$. For $n \geq 1$, define the function ϕ_n by

4.11.4 $$\phi_n(x) = n\phi(nx)$$

Note that we still have $\int \phi_n d\omega = 1$. What's happening here as $n \to \infty$? Since $\phi \in L^1(\mathbb{R})$, for any $\epsilon > 0$ there is a compact set K such that $\left| \int_{\mathbb{R} \backslash K} \phi d\omega \right| < \epsilon$. That is, most of ϕ is concentrated on K. For large values of $n > 0$, $n^{-1}K$ gets very close to 0. Thus $\phi(nx)$ is mostly concentrated near the origin. When we multiply $\phi(nx)$ by n we normalize it. So the functions $\{\phi_n\}$ are, in some sense, approaching a unit point mass at the origin. (It might be instructive at this point to look at Exercise 9.)

4.11.5. Theorem. *If $\phi \in L^1(\mathbb{R})$ with $\int \phi d\omega = 1$, $\{\phi_n\}$ is defined as above, and $f \in L^1(\mathbb{R})$, then $f * \phi_n \to f$ in $L^1(\mathbb{R})$ as $n \to \infty$.*

Proof. We use the translation operators τ_a introduced in §4.8. For any x in $\mathbb{R}$, using the substitution $t \mapsto n^{-1}s$ we have that

$$f * \phi_n(x) - f(x) = \int [f(x-t) - f(x)]\phi_n(t)d\omega(t)$$

$$= \int [f(x - n^{-1}s) - f(x)]\phi(s)d\omega(s)$$

$$= \int [\tau_{n^{-1}s}f(x) - f(x)]\phi(s)d\omega(s)$$

Therefore

$$\|f * \phi_n - f\|_1 = \int \left| \int [\tau_{n^{-1}s}f(x) - f(x)]\phi(s)d\omega(s) \right| d\omega(x)$$

$$\leq \int \int |\tau_{n^{-1}s}f(x) - f(x)| \, |\phi(s)|d\omega(s)d\omega(x)$$

$$= \int \left[\int |\tau_{n^{-1}s}f(x) - f(x)|d\omega(x) \right] |\phi(s)|d\omega(s)$$

$$= \int \|\tau_{n^{-1}s}f - f\|_1 |\phi(s)|d\omega(s)$$

According to Proposition 4.8.7, for each s we have that $\lim_n \|\tau_{n^{-1}s}f - f\|_1 = 0$. But $\|\tau_{n^{-1}s}f - f\|_1 |\phi(s)| \leq 2\|f\|_1|\phi(s)|$, a function in $L^1(\mathbb{R})$. Therefore the DCT implies the conclusion of the theorem. ∎

A sequence $\{\phi_n\}$ in $L^1(\mathbb{R})$ with $\int \phi_n d\omega = 1$ for all n and satisfying the conclusion of the preceding theorem is called an *approximate identity* for obvious reasons. The next result on differentiating under the integral sign could have easily been presented earlier in this book.

4.11.6. Theorem. *Let $(X, \mathcal{A}, \mu)$ be a σ-finite measure space and let $[a, b]$ be a bounded interval in $\mathbb{R}$. Let $f : X \times [a, b] \to \mathbb{C}$ be a function such that for $a \leq t \leq b$, $x \mapsto f(x, t)$ is μ-integrable, and define the function $F : [a, b] \to \mathbb{C}$ by*

$$F(t) = \int_X f(x, t)d\mu(x)$$

(a) *If for each x in X, $t \mapsto f(x, t)$ is continuous and there is a function g in $L^1(\mu)$ with $|f(x, t)| \leq g(x)$ for all x and t, then F is a continuous function on the interval $[a, b]$.*

(b) *If for each x in X the function $t \mapsto f(x, t)$ is differentiable and there exists a function g in $L^1(\mu)$ with $|\partial f/\partial t(x, t)| \leq g(x)$ on X, then F is differentiable and*

$$F'(t) = \int_X \frac{\partial f}{\partial t}(x, t)d\mu(x)$$

Proof. The hypothesis gives the clue that the proof will use the DCT. The proof of (a) is immediate: if $t_n \to t$, then $F(t_n) = \int f(x, t_n) d\mu(x) \to F(t)$. The proof of (b) needs a little more effort. Again let $t_n \to t$. So

$$\frac{F(t_n) - F(t)}{t_n - t} = \int_X \left[\frac{f(x, t_n) - f(x, t)}{t_n - t} \right] d\mu(x)$$

By the Mean Value Theorem for derivatives we have that

$$\left| \frac{f(x, t_n) - f(x, t)}{t_n - t} \right| \leq \sup \left\{ \left| \frac{\partial f(x, s)}{\partial s} \right| : a \leq s \leq b \right\} \leq g(x)$$

so that the DCT once again applies to yield the conclusion. ∎

4.11.7. Corollary. *If $f \in L^1(\mathbb{R})$ and ϕ is a continuously differentiable function on $\mathbb{R}$ that has compact support, then $f * \phi$ is continuously differentiable and*

$$(f * \phi)' = f * \phi'$$

Proof. This is immediate from the preceding theorem. ∎

Now the idea is to combine Corollary 4.11.7 with Theorem 4.11.5 to prove some approximation results about infinitely differentiable functions. For that we need an example of an infinitely differentiable functions with some specific properties.

4.11.8. Example. Define a function ϕ on $\mathbb{R}$ by

$$\phi(x) = \begin{cases} ce^{(1-|x|^2)^{-1}} & \text{when } |x| < 1 \\ 0 & \text{when } |x| \geq 1 \end{cases}$$

where c is a positive constant chosen so that $\int \phi \, d\omega = 1$. It is left to the reader to verify that ϕ is infinitely differentiable on $\mathbb{R}$. (This is similar to the proof of Exercise 1.1.1.) We also see that ϕ is a positive function, $\phi(-x) = \phi(x)$, and ϕ has compact support.

For any open set G in $\mathbb{R}$ let $C_c^\infty(G)$ denote the infinitely differentiable functions on G that have compact support inside G.

4.11.9. Proposition. *If G is an open subset of $\mathbb{R}$, then $C_c^\infty(G)$ is uniformly dense in $C_0(G)$.*

Proof. Since $C_c(G)$ is dense in $C_0(G)$, it suffices to prove that $C_c^\infty(G)$ is dense in $C_c(G)$. Fix f in $C_c(G)$ and $\epsilon > 0$. Let ϕ be as in Example 4.11.8 and consider the functions ϕ_n as in Theorem 4.11.5. Fix an arbitrary x in

$\mathbb{R}$ and let $E_\epsilon = \{y \in \mathbb{R} : |f(x) - f(y)| < \epsilon\}$ and $F_\epsilon = \mathbb{R} \backslash E_\epsilon$. We have

$$|f(x) - f * \phi_n(x)| \leq \int_{E_\epsilon} |f(x) - f(y)| \phi_n(x - y) d\omega(y)$$

$$+ \int_{F_\epsilon} |f(x) - f(y)| \phi_n(x - y) d\omega(y)$$

$$\leq \epsilon \int_{E_\epsilon} \phi_n(x - y) d\omega(y)$$

$$+ \int_{F_\epsilon} |f(x) - f(y)| \phi_n(x - y) d\omega(y)$$

$$\leq \epsilon + \int_{F_\epsilon} |f(x) - f(y)| \phi_n(x - y) d\omega(y)$$

Since f is uniformly continuous, there is a $\delta > 0$ such that $|f(t) - f(y)| < \epsilon$ when $|t - y| < \delta$. Let $n > \delta^{-1}$. Note that for y in F_ϵ it must be that $|x - y| \geq \delta > n^{-1}$ and so $\phi_n(x - y) = 0$. (Let's underline the fact that δ, and hence n, is independent of x.) Therefore the surviving integral in the above displayed inequality in fact does not survive for such values of n and it vanishes. Thus we have that $|f(x) - f * \phi_n(x)| \leq \epsilon$ when $n > \delta^{-1}$ and x is arbitrary.

So we have that $\|f - f * \phi_n\|_\infty \to 0$. But according to Corollary 4.11.7 $f * \phi_n$ is differentiable and $(f * \phi_n)' = f * \phi_n'$. Continuing we see that each $f * \phi_n$ is infinitely differentiable and so for all large n, $f * \phi_n \in C_c^\infty(G)$. ∎

4.11.10. Corollary. *If G is any open set in $\mathbb{R}$, μ is a positive regular Borel measure on G, and $1 \leq p < \infty$, then $C_c^\infty(G)$ is dense in $L^p(\mu)$.*

Proof. Since Proposition 4.4.2 says that $C_c(\mathbb{R})$ is dense in $L^p(\mu)$, the preceding proposition gives the result. ∎

Exercises.

(1) For $a > 0$, compute $\chi_{[-1,1]} * \chi_{[-a,a]}$.

(2) Prove part (d) of Proposition 4.11.3.

(3) If $f \in L_{\text{loc}}^1$ and g is a bounded measurable function with compact support, show that $f * g$ is a function in $L^1(\mathbb{R})$.

(4) In Corollary 4.11.7, how far can you weaken the conditions on ϕ and still have the same conclusion?

(5) Show that the function ϕ in Example 4.11.8 is infinitely differentiable.

(6) If $f \in L_{\text{loc}}^1$ and ϕ is as in Example 4.11.8, show that for $n \geq 1$, $f_n(x) = f * \phi_n$ is infinitely differentiable. What can you say as $n \to \infty$?

(7) Look at the proof of Proposition 4.11.9 and show that if $f \in C_c(\mathbb{R})$, then $f * \phi_n \to f$ uniformly on $\mathbb{R}$.

(8) If $1 \le p < \infty$, $p^{-1} + q^{-1} = 1$, $f \in L^p(\mathbb{R})$, and $g \in L^q(\mathbb{R})$, show that $f * g(x)$ is defined for all x and that $f * g$ is a bounded uniformly continuous function on $\mathbb{R}$ with $\|f * g\|_\infty \le \|f\|_p \|g\|_q$. If $1 < p < \infty$, show that $f * g \in C_0(\mathbb{R})$.

(9) Recall that $M(\mathbb{R})$ denotes the space of finite regular Borel measures on $\mathbb{R}$. (a) If $\mu, \nu \in M(\mathbb{R})$, show that

$$L(f) = \int \int f(x+y) d\mu(x) d\nu(y)$$

defines a bounded linear functional $L : C_0(\mathbb{R}) \to \mathbb{C}$ with $\|L\| \le \|\mu\| \|\nu\|$. As such it is given by a measure, which we denote by $\mu * \nu$, called the *convolution* of μ and ν. (b) If $f, g \in \mathbb{R}$, $\mu = f\lambda$, and $\nu = g\nu$, show that $\mu * \nu = (f * g)\lambda$. Show that this definition of multiplication on $M(\mathbb{R})$ given by convolution is abelian and associative. (d) If δ_0 is the unit point mass at 0, show that $\delta_0 * \mu = \mu$ for every μ in $M(\mathbb{R})$.

4.12. The Fourier transform*

Here we look at the Fourier transform on $\mathbb{R}$. The treatment here is largely based on what appears in [**30**]. We maintain the definition of the weighted Lebesgue measure ω from the last section.

4.12.1. Definition. For f in $L^1(\mathbb{R})$, define

$$\widehat{f}(t) = \int f(x) e^{-ixt} d\omega(x) = \frac{1}{\sqrt{2\pi}} \int_{-\infty}^{\infty} f(x) e^{-ixt} dx$$

for all t in $\mathbb{R}$. The function $\widehat{f}$ is called the *Fourier*[6] *transform* of f.

---

[6] Jean Baptiste Joseph Fourier was born in 1768 in Auxerre, France, which lies in Burgundy about half-way between Paris and Dijon. In 1780 he was enrolled in the École Royale Militaire of Auxerre, where he soon exhibited talent for mathematics. Later he decided to study for the priesthood at the Benedictine abbey of St. Benoit-sur-Loire, though he was never ordained. He seems to have become consumed with mathematics and was determined to have a major impact on the subject. In 1789 he went to Paris where he presented a paper, but returned to the abbey the next year as a mathematics instructor. Shortly, however, he became involved in the political tumult of the time, spurred on by his idealism. In 1793 he joined the local Revolutionary Committee, though the prevalence of the Terror soon disillusioned him. He defended a faction of the revolution in Orléans, which led to his arrest in 1794. He seemed headed to the guillotine but, fortunately, Robespierre preceded him, the Terror abated, and he was released. In 1795 he was one of the first to study at the École Normale in Paris where he came under the mentorship of Lagrange. He began to teach at the College de France, continued his research, and succeeded Lagrange in the chair in analysis at the École Polytechnique in 1797. The next year he accompanied Napoleon to Egypt, where he witnessed that debacle. Fourier returned to France in 1801 and reclaimed his chair. However Napoleon appointed him Prefect of the Department of Isère and he transferred to

Caveat. When you consult the literature you will sometimes see a different definition of the Fourier transform. It is often defined as

$$t \mapsto \int_{-\infty}^{\infty} f(x) \exp(-2\pi i x t) dx$$

No matter the choice of definition, the results are the same except for the presence or absence of certain constants. In fact it is for such reasons that Lebesgue measure has been weighted as it was in the preceding section. Each definition has advantages and drawbacks. I don't have a dog in this fight; I see advantages and drawbacks with each definition.

4.12.2. Theorem. *Let* $f \in L^1(\mathbb{R})$.

(a) $\widehat{f} \in C_0(\mathbb{R})$ *and* $\|\widehat{f}\|_\infty \leq \|f\|_1$.

(b) *If* g *is also a function in* $L^1(\mathbb{R})$, *then* $\widehat{f * g} = \widehat{f}\, \widehat{g}$.

Proof. Since $|e^{-ixt}| = 1$ for all values of x and t, we have that

$$|\widehat{f}(t)| = \left| \int f(x) e^{-ixt} d\omega(x) \right| \leq \int |f(x)| d\omega(x) = \|f\|_1$$

To show that $\widehat{f}$ is continuous, let $t_n \to t$ in $\mathbb{R}$. So

$$|\widehat{f}(t_n) - \widehat{f}(t)| \leq \int |f(x)||e^{-ixt_n} - e^{-ixt}| d\omega(x)$$

Now the integrand is bounded by $2|f|$, an integrable function. Hence the DCT implies $\widehat{f}(t_n) \to \widehat{f}(t)$ and $\widehat{f}$ is continuous. To see that $\widehat{f}(t) \to 0$ as $t \to \pm\infty$, use the fact that $e^{i\pi} = -1$ to get that

$$\widehat{f}(t) = -\int f(x) e^{-it(x+\pi/t)} d\omega(x)$$

$$= -\int f\left(x - \frac{\pi}{t}\right) e^{-ixt} d\omega(x)$$

$$= \frac{1}{2} \int \left[f(x) - f\left(x - \frac{\pi}{t}\right) \right] e^{-ixt} d\omega(x)$$

Thus $|\widehat{f}(t)| \leq \frac{1}{2}\|f - \tau_{\pi/t}f\|_1$, which, by Proposition 4.8.7, converges to 0 as $t \to \pm\infty$. This proves (a).

Grenoble. During this time he did his famous work on heat and in 1807 submitted his memoir *On the Propagation of Heat in Solid Bodies*. This was met with skepticism by the establishment. He defended himself in this controversy over his revolutionary idea of expanding functions in trigonometric series. Eventually the mathematical world accepted his approach and he received a prize for the work. Nevertheless he had difficulty getting his memoir published. In 1817 he was elected to the Académie des Sciences, becoming its secretary in 1822, the same year that the Académie published his work *Théorie analytique de la chaleur*. Fourier's work commands the attention of the mathematical world to the present day, and it was part of the motivation in Lebesgue's discovery of measure theory. Joseph Fourier died in 1830 in Paris.

To prove (b) we use Fubini's Theorem and the invariance of Lebesgue measure under translation. So

$$\widehat{(f * g)}(t) = \int f * g(x)e^{-ixt}d\omega(x)$$

$$= \int \left[\int f(x-y)g(y)d\omega(y) \right] e^{-ixt}d\omega(x)$$

$$= \int \left[\int f(x-y)e^{-i(x-y)t}d\omega(x) \right] g(y)e^{-iyt}d\omega(y)$$

$$= \widehat{f}(t)\,\widehat{g}(t) \qquad \blacksquare$$

The preceding result that $\widehat{f} \in C_0(\mathbb{R})$ is called the *Riemann–Lebesgue Lemma*. The reader can look ahead to Theorem 5.3.9 and Theorem 5.3.10 where another Riemann–Lebesgue Lemma is presented as well as another Fourier transform. The common setting that treats both these cases is obtained by discussing the group algebra of a locally compact abelian group. As an introduction to this, see §VII.9 in [**8**] where there are additional references.

4.12.3. Definition. A *Schwartz*[7] *function* or a *rapidly decreasing function* is an infinitely differentiable function ϕ on $\mathbb{R}$ such that for all $m, n \geq 0$

$$\sup\{|x^m \phi^{(n)}(x)| : x \in \mathbb{R}\} < \infty$$

The set of all Schwartz functions is denoted by $\mathcal{S}$.

4.12.4. Example. Note that any infinitely differentiable function with compact support belongs to $\mathcal{S}$. So the function defined in Example 4.11.8 is in $\mathcal{S}$ as are the functions $\phi(x) = x^m \exp(-ax^2)$, where $a > 0$.

---

[7]Laurent Schwartz was born in 1915 in Paris. After early schooling in Paris, he entered graduate school at Strasbourg and received his doctorate in 1943. In addition to mathematics, politics was central for him. His early admiration for the Soviet Union turned to disillusionment during Stalin's show trials. He spent 1944–45 lecturing at the university in Grenoble and then moved to Nancy as a professor. He returned to Paris in 1953 as a professor with positions at various universities until his retirement in 1983. His early work inventing the theory of distributions in the late 1940s was pivotal in mathematics and led to his receiving the Fields Medal in 1950 at the International Congress in Cambridge, Mass. Unfortunately the US, because of his involvement in communist political activities, decided not to allow him to enter the country to receive the medal. Distributions generalize the notion of a function and are omnipresent in differential equations. His activities in both mathematics and politics continued throughout his life. He was a leader in opposition to the French repression in Algeria and the US war in Vietnam. He also did significant work in stochastic differential calculus. He had 16 PhD students, including several subsequent leaders in French mathematics; he has 2156 mathematical descendants. (I had the pleasure of taking my first class in functional analysis from him.) In addition he was an avid butterfly collector with over 20,000 specimens. We might end with a quote: "To discover something in mathematics is to overcome an inhibition and a tradition. You cannot move forward if you are not subversive." He died in 2002 in Paris.

The proof of the following basic properties of Schwartz functions is routine.

4.12.5. Proposition. (a) *If $\phi \in S$, then for any $n \geq 1$, there is a constant C such that $|\phi(x)| \leq C|x|^{-n}$.*

(b) *Each derivative of a Schwartz function is also a Schwartz function.*

(c) *S is an algebra when addition and multiplication are defined pointwise.*

(d) *$S \subseteq L^1(\mathbb{R})$ for $1 \leq p \leq \infty$.*

(e) *Each Schwartz function is uniformly continuous.*

We want to invest the symbol x with an ambiguous character. We continue to use it to designate a real variable, but we also want to use it to designate the function on $\mathbb{R}$ whose value at t is t:

$$x(t) = t$$

The context should usually remove the ambiguity, and the convenience far outweighs the small possibility of confusion.

The symmetry of the first two parts of the next theorem is a precursor.

4.12.6. Theorem. *If $\phi \in S$ and $n \geq 1$, then $\widehat{\phi}$ is infinitely differentiable and the following hold:*

(a) $\widehat{(\phi^{(n)})} = i^n x^n \widehat{\phi}$;

(b) $\left[\widehat{\phi}\right]^{(n)} = (-i)^n \widehat{x^n \phi}$;

(c) *if $\phi \in S$, then $\widehat{\phi} \in S$.*

Proof. (a) Consider the case where $n = 1$. Using integration by parts and the fact that $\phi(x)e^{-ixt} \to 0$ as $x \to \pm\infty$, we have that

$$\widehat{(\phi')}(t) = \frac{1}{\sqrt{2\pi}} \int_{-\infty}^{\infty} \phi'(x)e^{-ixt}dx = \frac{it}{\sqrt{2\pi}} \int_{-\infty}^{\infty} \phi(x)e^{-ixt}dx = it\widehat{\phi}(t)$$

Now use induction to complete the proof.

(b) Assume $n = 1$. Differentiating under the integral sign gives

$$\left[\widehat{\phi}\right]'(t) = \frac{d}{dt}\frac{1}{\sqrt{2\pi}} \int_{-\infty}^{\infty} \phi(x)e^{-ixt}dx$$

$$= -i\frac{1}{\sqrt{2\pi}} \int_{-\infty}^{\infty} \phi(x)xe^{-ixt}dx$$

$$= -i\widehat{x\phi}(t)$$

Also $\left[\widehat{\phi}\right]'' = (d/dt)\left[\widehat{\phi}\right]' = -i\left[\widehat{x\phi}\right]' = (-i)^2\widehat{x^2\phi}$. The induction argument proceeds in the same way.

(c) Let $\phi \in \mathcal{S}$ and put $\psi = x^n\phi$; clearly $\psi \in \mathcal{S}$. Using the previous parts of the theorem we have that $\widehat{\psi} = i^n[\widehat{\phi}]^{(n)}$. So $x^m[\widehat{\phi}]^{(n)} = x^m(-i)^n\widehat{\psi} = (-i)^{m+n}[\widehat{\psi}]^{(m)}$. But $\psi \in L^1(\mathbb{R})$ (4.12.5(c)) and so $x^m[\widehat{\phi}]^{(n)} \in L^\infty(\mathbb{R})$ (4.12.2(a)). Thus $\widehat{\phi} \in \mathcal{S}$ whenever $\phi \in \mathcal{S}$. $\blacksquare$

We are almost at the point where we can prove one of the main results about the Fourier transform, but first we need a lemma.

4.12.7. Lemma. *The function*

$$\theta(x) = e^{-\frac{1}{2}x^2}$$

belongs to $\mathcal{S}$, $\widehat{\theta}(0) = 1$, *and* $\widehat{\theta} = \theta$.

Proof. The fact that $\theta \in \mathcal{S}$ is clear. The proof that

$$\widehat{\theta}(0) = \frac{1}{\sqrt{2\pi}}\int_{-\infty}^{\infty} e^{-\frac{1}{2}x^2}\,dx = 1$$

is sketched in Exercise 8. Also see

    http://en.wikipedia.org/wiki/Gaussian_integral#Computation

To show that $\widehat{\theta} = \theta$, the reader can verify that $\theta' = -x\theta$. By (4.12.6(b)), $[\widehat{\theta}]' = -i\widehat{(x\theta)}$. But $x\theta = -\theta'$, so, by 4.12.6(a), $[\widehat{\theta}]' = -i\widehat{(-\theta')} = i\widehat{\theta'} = i(ix\widehat{\theta}) = -x\widehat{\theta}$. That is both θ and $\widehat{\theta}$ satisfy the differential equation $y' + xy = 0$. Hence they differ by a constant. Now $\theta(0) = 1$ and we also have that $\widehat{\theta}(0) = 1$; since the two functions agree at 0, they agree everywhere. $\blacksquare$

4.12.8. Theorem (Inversion Theorem). (a) *If* $\phi \in \mathcal{S}$, *then*

4.12.9 $$\phi(x) = \frac{1}{\sqrt{2\pi}}\int_{-\infty}^{\infty} \widehat{\phi}(t)e^{ixt}\,dt$$

(b) *If* $\mathcal{F} : \mathcal{S} \to \mathcal{S}$ *is the Fourier transform,* $\mathcal{F}(\phi) = \widehat{\phi}$, *then* $\mathcal{F}$ *is a surjective linear mapping of* $\mathcal{S}$ *onto itself satisfying* $\mathcal{F}^4 = \mathcal{F}$.

(c) *If* $f \in L^1(\mathbb{R})$ *such that we also have that* $\widehat{f} \in L^1(\mathbb{R})$, *then*

4.12.10 $$f(x) = \frac{1}{\sqrt{2\pi}}\int_{-\infty}^{\infty} \widehat{f}(t)e^{ixt}\,dt$$

a.e. $[\lambda]$.

Before starting the proof, some comments are in order. First, (4.12.9) is called the *Inversion Formula*. Second, the Inversion Formula says that

$\phi(-x) = \widehat{\widehat{\phi}}(x)$, so that the use of the word "inversion" is justified. Third, note that Equation 4.12.10 says that

$$f(x) = \widehat{\widehat{\widetilde{f}}}(x)$$

a.e., where for any function g we put $\widetilde{g}(x) = g(-x)$. Since f agrees a.e. with a Fourier transform, it follows that when both f and $\widehat{f}$ belong to $L^1(\mathbb{R})$, then f agrees almost everywhere with a function in $C_0(\mathbb{R})$ (4.12.2(a)).

Proof. A straightforward application of Fubini's Theorem shows that for all f and g in $L^1(\mathbb{R})$ we have

4.12.11 $$\int \widehat{f} g \, d\omega = \int f \widehat{g} \, d\omega$$

(Verify!) We will use this several times below.

(a) If $f(x) = \phi(x/n)$ for some ϕ in $\mathcal{S}$ and $n \geq 1$, then

$$\widehat{f}(t) = \frac{1}{\sqrt{2\pi}} \int_{-\infty}^{\infty} \phi(n^{-1}t) e^{-ixt} dx$$

$$= \frac{1}{\sqrt{2\pi}} \int_{-\infty}^{\infty} \phi(y) e^{-inyt} n \, dy$$

$$= n\widehat{\phi}(nt)$$

In (4.12.11) take $g = \psi \in \mathcal{S}$ and f as above and we get

$$\int \phi(n^{-1}s)\widehat{\psi}(s) d\omega(s) = n \int \psi(t)\widehat{\phi}(nt) d\omega(t) = \int \psi(n^{-1}r)\widehat{\phi}(r) d\omega(r)$$

Now we want to let $n \to \infty$; so $\phi(n^{-1}s) \to \phi(0)$ and $\psi(n^{-1}r) \to \psi(0)$ for all s, r. Since $\widehat{\psi}, \widehat{\phi} \in \mathcal{S} \subseteq L^1(\mathbb{R})$, $|\phi(n^{-1}s)\widehat{\psi}(s)| \leq |\widehat{\psi}(s)|$, and $|\psi(n^{-1}r)\widehat{\phi}(r)| \leq |\widehat{\phi}(r)|$ for all s, r, the DCT implies that $\psi(0) \int \widehat{\phi} \, d\omega = \phi(0) \int \widehat{\psi} \, d\omega$ for all ϕ, ψ in $\mathcal{S}$. If we take $\psi = \theta$ as in Lemma 4.12.7, we get

$$\phi(0) = \int \widehat{\phi} \, d\omega$$

for all ϕ in $\mathcal{S}$. For any x in $\mathbb{R}$ let $\phi_x(y) = \phi(y+x)$; so $\phi(x) = \phi_x(0) = \int \widehat{\phi_x} \, d\omega$. On the other hand changing the variable $s = y + x$ below we have

$$\widehat{\phi_x}(t) = \frac{1}{\sqrt{2\pi}} \int_{-\infty}^{\infty} \phi_x(y) e^{-iyt} dy = \frac{1}{\sqrt{2\pi}} \int_{-\infty}^{\infty} \phi(y+x) e^{-iyt} dy$$

$$= \frac{1}{\sqrt{2\pi}} \int_{-\infty}^{\infty} \phi(s) e^{-i(s-x)t} ds = \frac{1}{\sqrt{2\pi}} \int_{-\infty}^{\infty} \phi(s) e^{-ist} e^{ixt} ds$$

$$= e^{ixt}\widehat{\phi}(t)$$

Hence $\phi(x) = \int \widehat{\phi}(t) e^{ixt} d\omega(t)$, as was to be shown.

(b) We know (4.12.6) that $\mathcal{F}(\mathcal{S}) \subseteq \mathcal{S}$. By (4.12.9), $\mathcal{F}$ is injective. For ϕ in $\mathcal{S}$ let $\tilde{\phi}$ denote the function in $\mathcal{S}$ defined by $\tilde{\phi}(x) = \phi(-x)$. By (4.12.9)

$$\widehat{\phi}(x) = \int \widehat{\widehat{\phi}}(t) e^{ixt} d\omega(t)$$

Also $\widehat{\phi}(x) = \int \phi(t) e^{-ixt} d\omega(t) = \int \tilde{\phi}(t) e^{ixt} d\omega(t)$. Thus if $\phi \in \mathcal{S}$,

$$0 = \int \left[\widehat{\widehat{\phi}}(t) - \tilde{\phi}(t) \right] e^{ixt} d\omega(t) = \int \left[\widehat{\widehat{\phi}}(t) - \phi(t) \right] e^{-ixt} d\omega(t)$$

But since $\mathcal{F}$ is injective, this implies $0 = \widehat{\widehat{\widehat{\phi}}} - \phi$ or $\mathcal{F}^2 \phi = \tilde{\phi}$. Therefore $\mathcal{F}$ is surjective on $\mathcal{S}$ and $\mathcal{F}^4$ is the identity.

(c) Since $\widehat{f} \in L^1(\mathbb{R})$, $f_0(x) = \int \widehat{f}(t) e^{ixt} d\omega(t)$ is defined. If ϕ is an arbitrary Schwartz function, (4.12.11) implies that $\int f \widehat{\phi} d\omega = \int \widehat{f} \phi d\omega$. Applying (4.12.9) to ϕ, this implies that

$$\int f \widehat{\phi} d\omega = \int \widehat{f}(x) \phi(x) d\omega(x)$$

$$= \int \widehat{f}(x) \left[\int \widehat{\phi}(s) e^{isx} d\omega(s) \right] d\omega(x)$$

$$= \int \widehat{\phi}(s) \left[\int \widehat{f}(x) e^{isx} d\omega(x) \right] d\omega(s)$$

$$= \int \widehat{\phi}(s) f_0(s) d\omega(s)$$

By part (b), this implies that

$$\int f \phi d\omega = \int f_0 \phi d\omega$$

for every ϕ in $\mathcal{S}$.

Here we have a bit of technical difficulty. We do not know that $f_0 \in L^1(\mathbb{R})$. We do know that f_0 is a bounded function since $f_0 = \widehat{\widehat{f}}(-x)$ and we can invoke (4.12.2(a)). Therefore for any bounded open set G, $f_0 \chi_G \in L^1(\mathbb{R})$. By Proposition 4.11.9 we have that $\int_G [f - f_0] \phi d\omega = 0$ for every ϕ in $C_c(G)$, hence $f - f_0 = 0$ a.e. on G; since G was arbitrary, f and f_0 agree almost everywhere, proving (c). ■

In part (c) of the Inversion Theorem we assumed that both f and its Fourier transform belong to $L^1(\mathbb{R})$. This is clearly not satisfied by every function in $L^1(\mathbb{R})$, since, as we pointed out, it implies that f agrees a.e. with a function in $C_0(\mathbb{R})$. Here is a specific example.

4.12.12. Example. If $f(x) = 0$ for $x < 0$ and $f(x) = e^{-x}$ when $x \geq 0$, then $f \in L^1(\mathbb{R})$ but $\widehat{f} \notin L^1(\mathbb{R})$.

Recall that $C_c^\infty(\mathbb{R})$ is dense in $L^p(\mathbb{R})$ (4.11.10); hence $\mathcal{S}$ is dense in $L^p(\mathbb{R})$ for all finite p.

4.12.13. Theorem (Plancherel's[8] Theorem). *The Fourier transform $\phi \mapsto \widehat{\phi}$ extends to a surjective isometry of $L^2(\mathbb{R})$ onto itself.*

Proof. The Inversion Theorem implies that for any ϕ, ψ in $\mathcal{S}$ we have that

$$\langle \phi, \psi \rangle = \int \phi \bar{\psi} d\omega = \int \left[\int \widehat{\phi}(t) e^{ixt} d\omega(t) \right] \overline{\psi(x)} \, d\omega(x)$$

$$= \int \left[\int \overline{\psi(x)} \, \overline{e^{-ixt}} d\omega(x) \right] \widehat{\phi}(t) \, d\omega(t)$$

$$= \int \widehat{\phi} \, \overline{\widehat{\psi}} \, d\omega$$

$$= \langle \widehat{\phi}, \widehat{\psi} \rangle$$

If we let $\psi = \phi$ in this equation, we see that $U\phi = \widehat{\phi}$ defines an isometry when $\mathcal{S}$ is considered as a linear manifold contained in $L^2(\mathbb{R})$. Since $\mathcal{S}$ is dense in $L^2(\mathbb{R})$ and the Fourier transform is surjective when applied to $\mathcal{S}$, we have the proof of the theorem.   ∎

It should be emphasized that the extension U of the Fourier transform from $\mathcal{S}$ to $L^2(\mathbb{R})$ in Plancherel's Theorem is no longer the Fourier transform. Indeed there are functions in $L^2(\mathbb{R})$ for which the Fourier transform is not defined. See Exercise 1. However for f in $L^1(\mathbb{R}) \cap L^2(\mathbb{R})$, it is the case that $Uf = \widehat{f}$. (Exercise 5.)

Exercises.

(1) (a) Give an example of a function f in $L^2(\mathbb{R})$ such that $\widehat{f}$ is not defined; that is, such that $x \mapsto f(x)e^{-ixt}$ is not integrable for almost every t in $\mathbb{R}$. (b) If $1 < p \le \infty$, show that there is a function f in $L^p(\mathbb{R})$ such that $\widehat{f}$ is not defined.

(2) Prove Proposition 4.12.5.

(3) Verify the information in Example 4.12.12.

---

[8]Michel Plancherel was born in 1885 in Bussy, Switzerland. At the age of 7 his family moved to nearby Fribourg, where he attended school. He received his doctorate in 1907 from the university in Fribourg with a thesis on quadratic forms, part of which was written while he was in the Swiss army. He then went to Göttingen (1907–1909) and Paris (1909–1910) to continue his research. He returned to Switzerland for a position at Geneva in 1910 and then at Fribourg the next year. In 1920 he was appointed to a professorship at the ETH Zürich, where he remained until he retired in 1955. He met his wife while in school and they had 9 children. His research extended to include analysis and mathematical physics, and he enjoyed an international reputation. He had 30 doctoral students. He died in 1967 in Zurich as a result of injuries received in an automobile accident.

(4) If $1 < p < \infty$ and $(X, \mathcal{A}, \mu)$ is any σ-finite measure space, show that $L^1(\mu) \cap L^p(\mu)$ is dense in $L^p(\mu)$. What happens when $p = \infty$?

(5) If $U : L^2(\mathbb{R}) \to L^2(\mathbb{R})$ is the isometry obtained in Plancherel's Theorem, show that $Uf = \hat{f}$ when $f \in L^1(\mathbb{R}) \cap L^2(\mathbb{R})$.

(6) Show that for any f in $L^1(\mathbb{R})$ we have that $\widehat{\widehat{f}} = \widetilde{\widetilde{f}}$.

(7) Find a necessary and sufficient condition on f in $L^1(\mathbb{R})$ such that $\hat{f}$ is real-valued.

(8) Let θ be the function defined in Lemma 4.12.7. Show that $\theta(0) = 1$ as follows. Observe that

$$2\pi\theta(0)^2 = \left[\int e^{-\frac{1}{2}x^2}\,dx\right]\left[\int e^{-\frac{1}{2}y^2}\,dy\right]$$
$$= \int_{\mathbb{R}^2} e^{-\frac{x^2+y^2}{2}}\,d\lambda_2(x, y)$$

Now evaluate this integral using polar coordinates.

Chapter 5

Linear Transformations

In this chapter we return to a discussion of normed spaces and Hilbert spaces and their continuous linear transformations. After discussing the rudiments of linear transformations on normed and Banach spaces, most of our effort here will be directed toward Hilbert spaces – a continuation of Chapter 3, though these results will incorporate many things from measure theory. In Chapter 9 we will address some of these topics in the Banach space setting.

5.1. Basics

5.1.1. Definition. If $\mathcal{X}$ and $\mathcal{Y}$ are normed spaces over $\mathbb{F}$, a *linear transformation* or *operator* $T : \mathcal{X} \to \mathcal{Y}$ is a function satisfying $T(a_1 x_1 + a_2 x_2) = a_1 T x_1 + a_2 T x_2$ for all a_1, a_2 in $\mathbb{F}$ and all x_1, x_2 in $\mathcal{X}$. T is *bounded* if there is a constant M with $\|Tx\| \leq M\|x\|$ for all x in $\mathcal{X}$. We denote the set of all bounded operators from $\mathcal{X}$ into $\mathcal{Y}$ by $\mathcal{B}(\mathcal{X}, \mathcal{Y})$ and we let $\mathcal{B}(\mathcal{X}) = \mathcal{B}(\mathcal{X}, \mathcal{X})$.

5.1.2. Example. Let $\mathcal{X}$ and $\mathcal{Y}$ be normed spaces.

(a) Every bounded linear functional L on $\mathcal{X}$ is a bounded linear transformation from $\mathcal{X}$ into $\mathbb{F}$.

(b) If A is an $n \times m$ matrix over $\mathbb{F}$ and $T : \mathbb{F}^m \to \mathbb{F}^n$ is defined by $T(x) = Ax$, then T is a bounded linear transformation.

(c) Let $(X, \mathcal{A}, \mu)$ be a measure space and let $1 \leq p \leq \infty$. If $\phi \in L^\infty(\mu)$ and $M_\phi : L^p(\mu) \to L^p(\mu)$ is defined by $M_\phi f = \phi f$, then $\|M_\phi f\| \leq \|\phi\|_\infty \|f\|_p$ and $M_\phi \in \mathcal{B}(L^p(\mu))$.

(d) Let X be a metric space and let $g \in C_b(X)$. If $Tf = fg$ for every f in $C_b(X)$, then $T \in \mathcal{B}(C_b(X))$ and $\|Tf\| \leq \|g\|\|f\|$. Since $T1 = g$, it follows that $\|T\| = \|g\|$.

(e) If $g \in L^1(\mathbb{R})$ and $T : L^1(\mathbb{R}) \to C_0(\mathbb{R})$ is defined by $Tf = f * g$, then Proposition 4.11.1 implies T is a bounded operator and $\|T\| \leq \|g\|_1$. (See Exercise 1.)

For the rest of this section, $\mathcal{X}$ and $\mathcal{Y}$ will denote normed spaces unless it is stipulated otherwise. The next result and its proof are similar to Proposition 1.5.3 on linear functionals.

5.1.3. Proposition. *If $T : \mathcal{X} \to \mathcal{Y}$ is a linear transformation, then the following statements are equivalent.*

(a) *T is bounded.*

(b) *T is continuous.*

(c) *T is continuous at 0.*

(d) *T is continuous at some point of $\mathcal{X}$.*

5.1.4. Definition. If $T \in \mathcal{B}(\mathcal{X}, \mathcal{Y})$,

$$\|T\| = \sup\{\|Tx\| : x \in \text{ball } \mathcal{X}\}$$

is called the *norm* of T.

Of course this just extends the definition of the norm of a bounded linear functional. Again, the proof of the next result is similar to that of the corresponding result for a linear functional (1.5.4).

5.1.5. Proposition. *If $T \in \mathcal{B}(\mathcal{X}, \mathcal{Y})$, then*

$$\|T\| = \sup\{\|Tx\|/\|x\| : x \in \mathcal{X} \text{ and } x \neq 0\}$$
$$= \sup\{\|Tx\| : x \in \mathcal{X} \text{ and } \|x\| = 1\}$$

Also it is easy to see that $\|Tx\| \leq \|T\|\|x\|$ for all vectors x.

5.1.6. Proposition. (a) *If addition and scalar multiplication are defined pointwise for operators in $\mathcal{B}(\mathcal{X}, \mathcal{Y})$, then $\mathcal{B}(\mathcal{X}, \mathcal{Y})$ is a normed space.*

(b) *If $\mathcal{Y}$ is a Banach space, then $\mathcal{B}(\mathcal{X}, \mathcal{Y})$ is a Banach space.*

(c) *If $T \in \mathcal{B}(\mathcal{X}, \mathcal{Y})$, and $S \in \mathcal{B}(\mathcal{Y}, \mathcal{Z})$ for some normed space $\mathcal{Z}$, and if $ST : \mathcal{X} \to \mathcal{Z}$ is defined as composition, then $ST \in \mathcal{B}(\mathcal{X}, \mathcal{Z})$ and $\|ST\| \leq \|S\|\|T\|$.*

Proof. The proofs of (a) and (b) are similar to that of Proposition 1.5.5. For x in $\mathcal{X}$ we note that $\|STx\| \leq \|S\|\|Tx\| \leq \|S\|\|T\|\|x\|$, and so it follows that $\|ST\| \leq \|S\|\|T\|$. ∎

5.1.7. Example. Let $1 \leq p \leq \infty$ and let $\{a_n\} \in \ell^\infty$. If $T : \ell^p \to \ell^p$ is defined by $T\{x_n\} = \{a_n x_n\}$, then $T \in \mathcal{B}(\ell^p)$ and $\|T\| = \sup_n |a_n|$.

The next result gives some specifics about Example 5.1.2(c). We also observe that if we use counting measure on $\mathbb{N}$ for the measure μ in the next proposition, then this establishes the statement in the preceding example.

5.1.8. Proposition. *Let $(X, \mathcal{A}, \mu)$ be a σ-finite measure space and let $1 \leq p \leq \infty$. If $\phi \in L^\infty(\mu)$ and $M_\phi : L^p(\mu) \to L^p(\mu)$ is defined by $M_\phi f = \phi f$, then $M_\phi \in \mathcal{B}(L^p(\mu))$ and $\|M_\phi\| = \|\phi\|_\infty$.*

Proof. It is routine to show that M_ϕ is linear and $\|M_\phi f\| \leq \|\phi\|_\infty \|f\|_p$. If $\epsilon > 0$, then the fact that μ is σ-finite implies there is an E in $\mathcal{A}$ such that $0 < \mu(E) < \infty$ and $|\phi(x)| > \|\phi\|_\infty - \epsilon$ for all x in E. (Details?) If $1 \leq p < \infty$, let $f = [\mu(E)]^{-\frac{1}{p}} \chi_E$. So $f \in \text{ball } L^p(\mu)$ and $\|M_\phi\|^p \geq [\mu(E)]^{-1} \int_E |\phi|^p d\mu > (\|\phi\|_\infty - \epsilon)^p$. Since ϵ was arbitrary, this completes the proof when $p < \infty$. For $p = \infty$, we take the same definition of E and see that $\|M_\phi\| \geq \|M_\phi \chi_E\|_\infty \geq \|\phi\|_\infty - \epsilon$. ∎

We might note that in the proof of the preceding proposition when $p = \infty$, we did not need to take the set E to have finite measure. So the proposition is valid for $p = \infty$ in arbitrary measure spaces. When the measure space is not σ-finite, however, this result is not true for finite p. See Exercise 2.

5.1.9. Proposition. *Let $(X, \mathcal{A}, \mu)$ be a σ-finite measure space, let $1 \leq p < \infty$, and let $k : X \times X \to \mathbb{F}$ be a $\mu \times \mu$-measurable function such that there are constants c_1 and c_2 with*

$$\int |k(x, y)| d\mu(y) \leq c_1 \quad a.e. \ [\mu]$$

$$\int |k(x, y)| d\mu(x) \leq c_2 \quad a.e. \ [\mu]$$

If Kf is defined for every f in $L^p(\mu)$ by

$$(Kf)(x) = \int k(x, y) f(y) d\mu(y)$$

then $K \in \mathcal{B}(L^p(\mu))$ and $\|K\| \leq c_1^{\frac{1}{q}} c_2^{\frac{1}{p}}$.

Proof. It must be shown that $Kf \in L^p(\mu)$, but this will surface as we proceed. Let $\frac{1}{p} + \frac{1}{q} = 1$. For $f \in L^p(\mu)$ and $x \in X$,

$$|Kf(x)| \leq \int |k(x, y)| |f(y)| d\mu(y)$$

$$= \int |k(x, y)|^{\frac{1}{q}} [|k(x, y)|^{\frac{1}{p}} |f(y)|] d\mu(y)$$

$$\leq \left[\int |k(x, y)| d\mu(y) \right]^{\frac{1}{q}} \left[\int |k(x, y)| |f(y)|^p d\mu(y) \right]^{\frac{1}{p}}$$

$$\leq c_1^{\frac{1}{q}} \left[\int |k(x,y)||f(y)|^p d\mu(y) \right]^{\frac{1}{p}}$$

Hence, using Fubini's Theorem,

$$\int |Kf(x)|^p d\mu(x) \leq c_1^{\frac{p}{q}} \int \left[\int |k(x,y)||f(y)|^p d\mu(y) \right] d\mu(x)$$

$$= c_1^{\frac{p}{q}} \int |f(y)|^p \left[\int |k(x,y)| d\mu(x) \right] d\mu(y)$$

$$\leq c_1^{\frac{p}{q}} c_2 \|f\|_p^p$$

This shows two things. First, the formula that defines $Kf(x)$ is finite-valued a.e. $[\mu]$ and belongs to $L^p(\mu)$. Second it establishes that $\|K\|_p \leq c_1^{\frac{1}{q}} c_2^{\frac{1}{p}}$ as desired. ∎

An operator K as defined in the preceding proposition is called an *integral operator* and the function k is its *kernel*. The reader might find it curious and perhaps instructive to compare the preceding proposition with Proposition 4.11.1 where the kernel is $k(x,y) = f(x-y)$.

5.1.10. Example. For (x,y) in $[0,1] \times [0,1]$, let $k(x,y)$ be the characteristic function of the upper half of the unit square: $\{(x,y) \in [0,1] \times [0,1] : y < x\}$. The integral operator with this kernel is called the *Volterra*[1] *operator*. So

$$Vf(x) = \int_0^x f(y)dy$$

which is just the indefinite integral.

Exercises.

(1) This exercise requires that you have gone through §4.11. Show that the operator T defined in Example 5.1.2(e) satisfies $\|T\| = \|g\|_1$. (Hint: Use an approximate identity.)

(2) Let $X = [0,1]$ and define μ on the Borel sets by letting $\mu(E) = \lambda(E)$ when $0 \notin E$ (λ is Lebesgue measure), and $\mu(E) = \infty$ when $0 \in E$. In other words, $\mu = \lambda + \infty \delta_0$; we have an infinite atom at 0. If $\phi = \chi_{\{0\}}$ and $1 \leq p < \infty$, show that $M_\phi = 0$ while $\|\phi\|_\infty = 1$.

(3) If M_ϕ is the operator defined on $L^p(\mu)$ as in Example 5.1.2(c), show that $M_\phi^2 = M_\phi$ if and only if ϕ is a characteristic function.

(4) If X is a metric space, $f \in C_b(X)$, and $M_f(g) = fg$ for every g in $C_b(X)$, show that if X is connected and $f \in C_b(X)$ such that $M_f^2 = M_f$, then either $f = 0$ or $f = 1$.

---

[1]See Example 10.5.14 for a biographical note.

(5) Show that Proposition 5.1.8 is valid when $p = \infty$ even if the measure space is not assumed to be σ-finite.

(6) Let $(X, \mathcal{A}, \mu)$ be a σ-finite measure space and assume $1 \leq p < \infty$. Let k_1 and k_2 be two kernels satisfying the hypothesis of Proposition 5.1.9 and let K_1 and K_2 be the corresponding integral operators. Define k on $X \times X$ by

$$k(x, y) = \int k_1(x, z)k_2(z, y)d\mu(z)$$

Show that k also satisfies the hypothesis of (5.1.9) and if K is the integral operator with this kernel, then $K = K_1 K_2$. What does this remind you of? Is there more than an analogy happening here?

(7) In Proposition 5.1.9, assume that the measure space is finite and that $k \in L^p(\mu \times \mu)$, for some p, $1 \leq p \leq \infty$. Show that k satisfies the hypothesis of Proposition 5.1.9.

5.2. Orthonormal basis

Throughout this section $\mathcal{H}$ denotes a Hilbert space over the scalars $\mathbb{F}$. We won't see many linear transformations in the current section, but they will dominate our attention after we get a basic notion of a Hilbert space established.

5.2.1. Definition. If $\mathcal{H}$ is a Hilbert space, an *orthonormal* subset is a subset $\mathcal{E}$ of $\mathcal{H}$ such that: (a) $\|e\| = 1$ for all e in $\mathcal{E}$; (b) $e_1 \perp e_2$ whenever $e_1, e_2 \in \mathcal{E}$ and $e_1 \neq e_2$. A *basis* is a maximal orthonormal subset.

We know from linear algebra and Zorn's Lemma that every vector space has a Hamel basis; that is, a maximal linearly independent subset. Note that a basis in a Hilbert space, as defined above, is not a Hamel basis. For this reason we will always use the term Hamel basis to refer to a maximal linearly independent set and reserve "basis" for the concept introduced in the preceding definition.

If $\mathcal{B}$ is a Hamel basis for the vector space $\mathcal{V}$, then every vector v in $\mathcal{V}$ can be written as a finite linear combination of elements of $\mathcal{B}$. Since $\mathcal{H}$ is a vector space, it has a Hamel basis. But we want more. We seek to establish an analogous state of affairs in a Hilbert space, but one that has additional structure. A Hilbert space has an inner product so we focus on a basis that reflects this. Also the inner product defines a norm and therefore we have a concept of convergence; so unlike the case with a Hamel basis, we can allow convergent infinite sums.

5.2.2. Proposition. *If $\mathcal{E}$ is an orthonormal subset of a Hilbert space $\mathcal{H}$, then $\mathcal{E}$ is contained in a basis. Hence every Hilbert space has a basis.*

The proof of this proposition is a straightforward application of Zorn's Lemma.

5.2.3. Proposition. *In a separable Hilbert space, every orthonormal set is countable.*

Proof. Suppose $\mathcal{E}$ is an orthonormal set in the Hilbert space $\mathcal{E}$. If e_1 and e_2 are distinct elements of $\mathcal{E}$, then the Pythagorean Theorem implies that $\|e_1 - e_2\|^2 = 2$. Thus $\{B(e; \sqrt{2}) : e \in \mathcal{E}\}$ is a collection of pairwise disjoint open sets in $\mathcal{H}$. If $\mathcal{H}$ is separable, this collection must be countable (Exercise 1.2.5). ∎

We now make the following agreement.

Agreement: All Hilbert spaces considered from now on in this book are assumed to be separable.

This agreement is arrived at in the interest of simplifying the exposition and facilitating the digestion of the reader. Virtually none of what is done in this section is untrue in the non-separable case. If we were to treat the non-separable case, it would necessitate discussing cardinal numbers and making greater use of nets. The interested reader may look at [**8**] where the material for non-separable Hilbert spaces is presented. Note that we will still consider Banach spaces that are not separable.

5.2.4. Example. (a) If $\mathcal{H} = \mathbb{F}^d$ and for $1 \leq j \leq d$ we let e_j be the vector with a 1 in the j-th spot and zeros elsewhere, then $\{e_1, \ldots, e_d\}$ is a basis for $\mathbb{F}^d$.

(b) If λ is Lebesgue measure on $[0, 2\pi]$ and $\mathcal{H} = L^2_{\mathbb{C}}([0, 2\pi], (2\pi)^{-1}\lambda)$, then for n in $\mathbb{Z}$ define $e_n(t) = \exp(int)$. $\mathcal{E} = \{e_n : n \in \mathbb{Z}\}$ is an orthonormal set. Later in Theorem 5.3.6 we will show that in fact $\mathcal{E}$ is a basis for $\mathcal{H}$.

(c) Recall the definition of the Hilbert space ℓ^2 (3.1.7(c)). For each $n \geq 1$ define e_n by $e_n(n) = 1$ and $e_n(k) = 0$ when $k \neq n$. It follows that $\{e_n : n \geq 1\}$ is a basis for ℓ^2.

(d) If λ is Lebesgue measure on $[0, 1]$, $\{I_1, I_2, \ldots\}$ is a collection of pairwise disjoint non-degenerate intervals in the closed unit interval, and $f_n = [\lambda(I_n)]^{-1}\chi_{I_n}$, then $\{f_1, f_2, \ldots\}$ is an orthonormal sequence. It is not a basis, however. In fact this is clear since any linear combination of these functions must be constant on each of the intervals I_n.

Recall (Exercise 1.3.2) that for a subset $\mathcal{S}$ of a normed space, $\bigvee \mathcal{S}$ is the closed linear span of $\mathcal{S}$.

5.2.5. Proposition. *If $\{e_n\}$ is an orthonormal sequence and $x \in \mathcal{H}$, then $\sum_{n=1}^{\infty}\langle x, e_n\rangle e_n$ converges in $\mathcal{H}$. In fact if $\mathcal{M} = \bigvee\{e_n : n \geq 1\}$ and P is the* *projection of $\mathcal{H}$ onto $\mathcal{M}$, then*

$$Px = \sum_{n=1}^{\infty} \langle x, e_n \rangle e_n$$

In order to prove this proposition we need a lemma that has historical significance.

5.2.6. Lemma (Bessel's[2] Inequality). *If $\{e_n : n \in \mathbb{N}\}$ is an orthonormal sequence and $x \in \mathcal{H}$, then*

$$\sum_{n=1}^{\infty} |\langle x, e_n \rangle|^2 \leq \|x\|^2$$

Proof. If $x_n = x - \sum_{k=1}^{n} \langle x, e_k \rangle e_k$, then $x_n \perp e_k$ for $1 \leq k \leq n$. (Details?) By the Pythagorean Theorem

$$\|x\|^2 = \|x_n\|^2 + \left\| \sum_{k=1}^{n} |\langle x, e_k \rangle e_k \right\|^2$$

$$= \|x_n\|^2 + \sum_{k=1}^{n} |\langle x, e_k \rangle|^2$$

$$\geq \sum_{k=1}^{n} |\langle x, e_k \rangle|^2$$

Since n was arbitrary, the result follows. ∎

Proof of Proposition 5.2.5. Since $\sum_{n=1}^{\infty} |\langle x, e_n \rangle|^2 \leq \|x\|^2 < \infty$, if $\epsilon > 0$, there is an N such that $\sum_{n=N+1}^{\infty} |\langle x, e_n \rangle|^2 < \epsilon^2$. Put $F_0 = \{e_k : 1 \leq k \leq N\}$. For any $n \geq 1$, let $x_n = \sum_{k=1}^{n} \langle x, e_k \rangle e_k$. If $m > n > N$,

$$\|x_m - x_n\|^2 = \sum_{k=n+1}^{m} |\langle x, e_k \rangle|^2 < \epsilon^2$$

So $\{x_n\}$ is a Cauchy sequence and must converge to $\sum_{n=1}^{\infty} \langle x, e_n \rangle e_n$.

---

[2]Wilhelm Bessel was born in 1784 in Minden, Westphalia (now Germany). His early schooling was unremarkable, and at the age of 14 he became an apprentice to an import-export firm in Bremen. With time he took an interest in the position of the firm's ships at sea, he studied navigation, and was led to the study of astronomy, the subject to which he made his major contributions. He used published data to calculate the orbit of Halley's comet and submitted his work to the astronomer Olbers; this led to a publication that was considered at the doctoral level. Eventually Bessel left the firm and pursued astronomy full time. His work gained recognition and he was offered the post of director of the new Prussian observatory at Königsberg (now Kaliningrad, Russia) in 1809. He launched a systematic attack on the motion of stars and introduced additional mathematics to the study. Here he defined the famous Bessel functions and, presumably, was led to the present inequality. He had a happy marriage and produced two sons and three daughters. Unfortunately sickness entered this family. Both sons died prematurely and his health began to deteriorate in 1840. He died of cancer in 1846 in Königsberg.

Now observe that if $x \in \mathcal{H}$ and $k \geq 1$, then $\langle x - \sum_{n=1}^{\infty} \langle x, e_n \rangle e_n, e_k \rangle = \langle x, e_k \rangle - \sum_{n=1}^{\infty} \langle \langle x, e_n \rangle e_n, e_k \rangle = \langle x, e_k \rangle - \langle x, e_k \rangle = 0$. Hence Px has the desired form by the uniqueness of the projection. $\blacksquare$

Though the next result is important, its proof is easy.

5.2.7. Theorem. *If $\{e_n\}$ is an orthonormal sequence in $\mathcal{H}$, then the following statements are equivalent.*

(a) *$\{e_n\}$ is a basis.*

(b) *If $x \in \mathcal{H}$ and $x \perp e_n$ for all $n \geq 1$, then $x = 0$.*

(c) *$\bigvee_{n=1}^{\infty} \{e_n\} = \mathcal{H}$.*

(d) *If $x \in \mathcal{H}$, then $x = \sum_{n=1}^{\infty} \langle x, e_n \rangle e_n$.*

(e) *If $x, y \in \mathcal{H}$, then $\langle x, y \rangle = \sum_{n=1}^{\infty} \langle x, e_n \rangle \langle e_n, y \rangle$.*

(f) *If $x \in \mathcal{H}$, then $\|x\|^2 = \sum_{n=1}^{\infty} |\langle x, e_n \rangle|^2$ (Parseval's[3] identity).*

Proof. (b) *is equivalent to* (c). See Corollary 3.2.11.

(a) *implies* (b). If $x \perp \{e_n\}$ and $x \neq 0$, then $\{e_n\} \cup \{x/\|x\|\}$ is a properly larger orthonormal set, violating the maximality of $\{e_n\}$.

(b) *implies* (d). By Proposition 5.2.5, $Px = \sum_{n=1}^{\infty} \langle x, e_n \rangle e_n$ is the orthogonal projection of x onto $\bigvee \{e_n\} = \mathcal{H}$, so $Px = x$.

(d) *implies* (e). Exercise.

(e) *implies* (f). $\|x\|^2 = \langle x, x \rangle$, so the result is immediate from (e).

(f) *implies* (a). If $\{e_n\}$ is not a basis there is a vector $x \perp \{e_n\}$ with $\|x\| = 1$. But by (f), $1 = \sum_{n=1}^{\infty} |\langle x, e_n \rangle|^2 = 0$, a contradiction. $\blacksquare$

Part (d) in the preceding theorem makes the connection with our usual idea of a basis. Every vector in the Hilbert space is a linear combination of basis elements, though here that linear combination is infinite.

There are many approaches to defining what is meant by a basis in a Banach space, none as simple and successful as what happens in a Hilbert space. In fact it is fair to say that much of the early work on Banach spaces was motivated by a desire to carry over to this setting as much as possible of what is known in Hilbert space. Now the subject has gone in its own

---

[3] Marc-Antoine Parseval des Chênes was born to a well-to-do family in 1755 in Rosi-res-aux-Salines in the Lorraine region of France. It seems relatively little is known of him. He married in 1795, though the marriage shortly ended in divorce. He was a royalist and was imprisoned in 1792, though he did not remain there. Later he began to write poetry against Napoleon and when he realized his arrest had been ordered, he successfully escaped from France. He had only five mathematical publications and this identity was discovered for certain Fourier series, not in the generality we have here. He died in 1836 in Paris.

direction. Almost any book on Banach spaces will discuss the existence of a basis of some type, but the reader might start with [**11**].

Exercises.

(1) Verify the statements in Example 5.2.4.

(2) Let λ be Lebesgue measure on the open unit disk $\mathbb{D} = \{z \in \mathbb{C} : |z| < 1\}$ and show that $\{1, z, z^2, \dots\}$ are pairwise orthogonal vectors in $L^2_{\mathbb{C}}(\mathbb{D}, \lambda)$. Is $\{z^n / \|z^n\| : n \geq 0\}$ a basis for $L^2_{\mathbb{C}}(\mathbb{D}, \lambda)$? (Hint: What is the relation of the function $z \mapsto \bar{z}$ to this orthonormal set?)

(3) Show that if λ is Lebesgue measure on $\mathbb{R}^d$, then $L^2(\lambda)$ is separable.

(4) If X is any compact metric space and μ is a Radon measure on $L^2(\mu)$, then $L^2(\mu)$ is separable.

(5) Show that if ball $\mathcal{H}$ is compact, then $\mathcal{H}$ is finite-dimensional.

5.3. Isomorphic Hilbert spaces

Here we introduce the natural concept for identifying Hilbert spaces, the bijections between Hilbert spaces that preserve their structure as Hilbert spaces.

5.3.1. Definition. An *isomorphism* between two Hilbert spaces $\mathcal{H}$ and $\mathcal{K}$ is a linear surjection $U : \mathcal{H} \to \mathcal{K}$ such that

$$\langle Ux, Uy \rangle = \langle x, y \rangle$$

for all x, y in $\mathcal{H}$. Two such spaces are said to be *isomorphic*. In symbols we write $\mathcal{H} \cong \mathcal{K}$. An isomorphism is also called a *unitary*.

By taking $y = x$ in the above equation, we get that $\|Ux\|^2 = \|x\|^2$ for every x in $\mathcal{H}$. Therefore an isomorphism is a bijection and an isometry. It is easy to see that if U is an isomorphism, then so is U^{-1}. From this the reader can easily check that the relation of being isomorphic is an equivalence relation among Hilbert spaces. Of course this relation is also an equivalence relation between all inner product spaces (including the incomplete ones), but since an isomorphism is an isometry, a Hilbert space cannot be equivalent to an incomplete inner product space.

5.3.2. Proposition. *If $\mathcal{H}$ and $\mathcal{K}$ are Hilbert spaces, a linear map $V : \mathcal{H} \to \mathcal{K}$ is an isometry if and only if $\langle Vx, Vy \rangle = \langle x, y \rangle$ for all x, y in $\mathcal{H}$.*

Proof. If V satisfies this equation, it is easy to see that it is an isometry. Now assume that V is an isometry. So for x, y in $\mathcal{H}$ and a in $\mathbb{F}$, $\|V(x + ay)\|^2 = \|x + ay\|^2$. Using the polar identity we get

$$\|x\|^2 + 2\mathrm{Re}\,\bar{a}\langle x, y \rangle + |a|^2\|y\|^2 = \|Vx\|^2 + 2\mathrm{Re}\,\bar{a}\langle Vx, Vy \rangle + |a|^2\|Vy\|^2$$

Since V is an isometry, after cancellations this equation becomes

$$\operatorname{Re} \bar{a}\langle x, y\rangle = \operatorname{Re} \bar{a}\langle Vx, Vy\rangle$$

If $\mathbb{F} = \mathbb{R}$, take $a = 1$. If $\mathbb{F} = \mathbb{C}$, first take $a = 1$ and then take $a = i$ to see that $\langle x, y\rangle$ and $\langle Vx, Vy\rangle$ have the same real and imaginary parts. ∎

The keen reader may have noticed in this last proposition that we did not assume the isometry was surjective. In finite-dimensional spaces surjectivity is automatic for a linear isometry since it is injective. In infinite-dimensional spaces this is not the case.

5.3.3. Example. If $S : \ell^2 \to \ell^2$ is defined by

$$S(a_0, a_1, \dots) = (0, a_0, a_1, \dots)$$

then S is an isometry that is not surjective. This particular isometry has been highly studied in operator theory and is called the *unilateral shift*. Define $U : \ell^2(\mathbb{Z}) \to \ell^2(\mathbb{Z})$ by

$$U(\dots, a_{-2}, a_{-1}, \widehat{a_0}, a_1, \dots) = (\dots, a_{-2}, \widehat{a_{-1}}, a_0, a_1, \dots)$$

where $\widehat{}$ denotes the entry in the 0-th place. This operator is called the *bilateral shift* and is a surjective isometry. That is, it is a unitary.

5.3.4. Proposition. *All separable Hilbert spaces are isomorphic.*

Proof. let $\mathcal{H}$ and $\mathcal{K}$ be separable Hilbert spaces with orthonormal bases $\{e_n\}$ and $\{f_n\}$, respectively. Define $U : \mathcal{H} \to \mathcal{K}$ by $U(\sum_{n=1}^{\infty} a_n e_n) = \sum_{n=1}^{\infty} a_n f_n$. We leave it to the reader to verify that U is an isomorphism. ∎

If $\mathcal{H}$ is a separable Hilbert space, then $\mathcal{H}$ is isomorphic to ℓ^2. So rather than study separable Hilbert spaces, couldn't we just study the space ℓ^2? This is true if all we were interested in is the geometry of a Hilbert space. However the real task is to study the operators on a Hilbert space. Even though all separable Hilbert spaces are isomorphic, for a given Hilbert space there are operators that are natural when the space is represented in a certain way while other operators are not so natural in that representation. For example, both ℓ^2 and $L^2[0,1]$ are separable. If $\phi \in L^\infty[0,1]$, then the operator M_ϕ is naturally defined on $L^2[0,1]$, but what does this operator look like on ℓ^2? Of course the answer to that question depends on which isomorphism we use to identify the two Hilbert spaces. The right choice might shed light on the structure of M_ϕ, but the wrong choice obscures everything and cloaks the operator in undecipherable camouflage.

We consider the Hilbert space $L_{\mathbb{C}}^2[0, 2\pi]$ to illustrate isomorphisms. For this we need a result whose proof will be postponed.

5.3.5. Theorem. *The set of all polynomials in the complex variable z and its conjugate, $\bar{z}$, is uniformly dense in $C_{\mathbb{C}}(\partial \mathbb{D})$.*

Later we'll see a proof of this result as a consequence of the Stone–Weierstrass Theorem (8.5.3). Another proof can be manufactured using complex analysis (see [**7**], p. 263). For the moment, however, we will assume the result without a proof. We ask the reader to trust the author that the proof of the Stone–Weierstrass Theorem does not depend on what we derive as a consequence of assuming Theorem 5.3.5. (Trust, but verify!)

If $z \in \partial \mathbb{D}$, then $\bar{z} = z^{-1}$, so a polynomial in z and $\bar{z}$ becomes a function of the form

$$p(z, \bar{z}) = \sum_{k=-n}^{n} a_k z^k$$

for some $n \geq 0$. Putting $z = e^{i\theta}$, this becomes $\sum_{k=-n}^{n} a_k e^{-ik\theta}$. If we write $e^{i\theta} = \cos\theta + i\sin\theta$ so that $e^{ik\theta} = \cos k\theta + i\sin k\theta$, we readily understand that such a polynomial in z and $\bar{z}$ on $\partial \mathbb{D}$ is a *trigonometric polynomial*.

5.3.6. Theorem. *If for each n in $\mathbb{Z}$ we define $e_n(t) = \exp(int)$, then $\{e_n : n \in \mathbb{Z}\}$ is a basis for $\mathcal{H} = L^2_{\mathbb{C}}([0, 2\pi], (2\pi)^{-1}\lambda)$.*

Proof. We already saw in Example 5.2.4(b) that the vectors $\mathcal{E} = \{e_n : n \in \mathbb{Z}\}$ form an orthonormal family. Let $\mathcal{T}$ denote the linear span of $\mathcal{E}$; so $\mathcal{T}$ is precisely the trigonometric polynomials. Note that this is a subalgebra of $C_{\mathbb{C}}[0, 2\pi]$ and $f(0) = f(2\pi)$ for every f in $\mathcal{T}$. If $\mathcal{C} = \{g \in C_{\mathbb{C}}[0, 2\pi] : g(0) = g(2\pi)\}$ and $g \in \mathcal{C}$, define $f : \partial \mathbb{D} \to \mathbb{C}$ by $f(e^{it}) = g(t)$. The fact that f is continuous follows from the periodicity of g. (Verify!) By Theorem 5.3.5 there is a sequence $\{p_n(z, \bar{z})\}$ that converges uniformly on $\partial \mathbb{D}$ to $f(z)$. It follows that $p_n(e^{it}, e^{-it}) \in \mathcal{T}$ and this sequence converges to $g(t)$ uniformly on $[0, 2\pi]$. That is, $\mathcal{T}$ is uniformly dense in $\mathcal{C}$. But $\mathcal{C}$ is dense in $L^2_{\mathbb{C}}[0, 2\pi]$ (Exercise 5), which implies that $\mathcal{T}$ is dense in $L^2_{\mathbb{C}}[0, 2\pi]$. ∎

Continue the notation of the preceding theorem. If $f \in \mathcal{H}$, define $\widehat{f} : \mathbb{Z} \to \mathbb{C}$ by

5.3.7 $$\widehat{f}(n) = \langle f, e_n \rangle = \frac{1}{2\pi} \int_0^{2\pi} f(t) e^{-int} dt$$

The function $\widehat{f} : \mathbb{Z} \to \mathbb{C}$ is called the *Fourier*[4] *transform* of f and $\widehat{f}(n)$ is called the n-th *Fourier coefficient* of f. The interested reader can consult §4.12 for the corresponding notion of the Fourier transform of a function in $L^1(\mathbb{R})$. There is a relationship between this Fourier transform and the one in §4.12, which the reader can explore in most books on harmonic analysis.

---

[4]See the beginning of §4.12 for a biographical note on Fourier.

By Theorem 5.2.7(d),

5.3.8
$$f = \sum_{n=-\infty}^{\infty} \widehat{f}(n)e_n$$

where the convergence is in the L^2-norm. This is called the *Fourier series* of f. In fact the expansion of the vector x that appears in Theorem 5.2.7(d) for an arbitrary Hilbert space is often called the Fourier series of x.

Parseval's identity applied to this situation says that $\sum_{n=-\infty}^{\infty}|\widehat{f}(n)|^2 = \|f\|^2 < \infty$. This proves a classical result.

5.3.9. Theorem (The Riemann–Lebesgue Lemma). *If $f \in L^2[0, 2\pi]$, then*

$$\int_0^{2\pi} f(t)e^{-int}dt \to 0$$

as $n \to \pm\infty$.

If $f \in L^2[0, 2\pi]$, then its Fourier series converges to f in the L^2 norm. As we know, this does not mean that the series converges a.e. However Lusin conjectured that this does indeed happen. This was proved in [**5**]. [**19**] extended this to the case for f in $L^p[0, 2\pi]$ if $1 < p \le \infty$. Long before all this activity Kolmogorov gave an example of an L^1 function whose Fourier series does not converge a.e.

The linear transformation $U : L^2[0, 2\pi] \to \ell^2(\mathbb{Z})$ given by $Uf = \widehat{f}$ is also called the *Fourier transform*. From Parseval's Identity, we know that U is an isometry. On the other hand if $\{a_n : n \in \mathbb{Z}\}$ and all but a finite number of the a_n vanish, then it is easy to see that $f = \sum_{n=-\infty}^{\infty} a_n e_n \in L^2[0, 2\pi]$, f is a trigonometric polynomial, and $Uf = \{a_n\}$. Thus U has dense range. Since U is an isometry, its range must be closed and therefore U is surjective. This proves the following.

5.3.10. Theorem. *The Fourier transform is an isomorphism from $L^2_{\mathbb{C}}[0, 2\pi]$ onto $\ell^2(\mathbb{Z})$.*

The important thing about this last result is not that the two Hilbert spaces are isomorphic. In fact they are both separable so we already knew that. The content of this theorem is that it is the Fourier transform that implements an isomorphism.

The reader who has covered §4.12 might have seen a similarity between the last theorem and Plancherel's Theorem (4.12.13). In fact this last result is often referred to by that name. As we mentioned before, there is a relation between the Fourier transform here and that in §4.12 and these two isomorphisms between Hilbert spaces are consequences of that more general theory that covers both – the theory of harmonic analysis on locally compact topological groups.

Exercises.

(1) Define $V : L^2(0, \infty) \to L^2(0, \infty)$ by $Vf(t) = f(t - 1)$ if $t > 1$ and $Vf(t) = 0$ if $t \leq 1$. Show that V is an isometry that is not an isomorphism.

(2) If $V : L^2(\mathbb{R}) \to L^2(\mathbb{R})$ is defined by $Vf(t) = f(t - 1)$ for all t, show that V is an isomorphism.

(3) Let $\mathcal{H}$ denote the space of all absolutely continuous functions f on the unit interval that satisfy $f(0) = 0$ and $f' \in L^2[0, 1]$. (a) If for f, g in $\mathcal{H}$ we define $\langle f, g \rangle = \int_0^1 f'(t)g'(t)dt$, show that $\mathcal{H}$ is a Hilbert space. (b) If $U : \mathcal{H} \to L^2[0, 2\pi]$ is defined by $Uf = f'$, show that U is an isomorphism. (c) Find a formula for U^{-1}.

(4) If $(X, \mathcal{A}, \mu)$ is a σ-finite measure space and $u \in L^\infty(\mu)$, show that the operator M_u on $L^2(\mu)$ is an isometry if and only if $|u| = 1$ a.e. $[\mu]$. When this is the case, show that M_u must be surjective. Did you use the fact that the measure space is σ-finite?

(5) Show that $\mathcal{C} = \{g \in C_{\mathbb{C}}[0, 2\pi] : g(0) = g(2\pi)\}$ is dense in $L^2_{\mathbb{C}}[0, 2\pi]$.

(6) Show that $\{1/\sqrt{2\pi}, (1/\sqrt{\pi}) \cos nt, (1/\sqrt{\pi}) \sin nt : 1 \leq n < \infty\}$ is a basis for $L^2_{\mathbb{R}}[-\pi, \pi]$.

(7) Let $\mathcal{A}$ be a σ-algebra of subsets of X and let μ and ν be two measures defined on $\mathcal{A}$ such that $\nu \ll \mu$. If $\phi = d\mu/d\nu$, show that $U : L^2(\mu) \to L^2(\nu)$ defined by $Uf = f\sqrt{\phi}$ is an isometry. Show that U is surjective if and only if $\mu \ll \nu$.

5.4. The adjoint

At this point we are going to slightly modify some of the notation we have been using. We no longer will customarily designate the elements of a Hilbert space by $x, y, \ldots$ but rather $f, g, h, \ldots$. This reflects the fact that most of the Hilbert spaces we will encounter, like $L^2(\mu)$, will be spaces of functions.

5.4.1. Definition. If $\mathcal{H}$ and $\mathcal{K}$ are Hilbert spaces, a function $u : \mathcal{H} \times \mathcal{K} \to \mathbb{F}$ is a *sesquilinear form* if whenever we have a, b in $\mathbb{F}$, h, g in $\mathcal{H}$, and k, f in $\mathcal{K}$, it follows that:

(a) $u(ah + bg, k) = au(h, k) + bu(g, k)$;

(b) $u(h, ak + bf) = \bar{a}u(h, k) + \bar{b}u(h, f)$.

A sesquilinear form is *bounded* if for some constant M we have that $|u(h, k)| \leq M\|h\|\|k\|$ whenever $h \in \mathcal{H}$ and $k \in \mathcal{K}$.

The prefix "sesqui" means "one and a half" and it is used here to distinguish it from a bilinear form which satisfies these conditions but where the complex conjugates are removed.

Here is a way to get examples of bounded sesquilinear forms. If $A \in \mathcal{B}(\mathcal{H}, \mathcal{K})$, let $u(h, k) = \langle Ah, k \rangle$. Here we are using the inner product in $\mathcal{K}$. This form is bounded if we take for the constant $M = \|A\|$. Similarly if $B \in \mathcal{B}(\mathcal{K}, \mathcal{H})$, $(h, k) \mapsto \langle h, Bk \rangle$ defines a bounded sesquilinear form on $\mathcal{H} \times \mathcal{K}$. What we show now is that this is the only way to obtain a bounded sesquilinear form. The proof involves applying the Riesz Representation Theorem.

5.4.2. Theorem. *If $u : \mathcal{H} \times \mathcal{K} \to \mathbb{F}$ is a bounded sesquilinear form with bound M, then there are unique operators A in $\mathcal{B}(\mathcal{H}, \mathcal{K})$ and B in $\mathcal{B}(\mathcal{K}, \mathcal{H})$ such that*

5.4.3 $u(h, k) = \langle Ah, k \rangle = \langle h, Bk \rangle$

for all h in $\mathcal{H}$ and all k in $\mathcal{K}$ and both $\|A\|$ and $\|B\|$ are dominated by M.

Proof. We only show the existence of A; the proof of the existence of B is similar. For each h in $\mathcal{H}$ define $L_h : \mathcal{K} \to \mathbb{F}$ by $L_h(k) = \overline{u(h, k)}$. The reader can check that L_h is a linear functional on $\mathcal{K}$ and $|L_h(k)| \leq M\|h\|\|k\|$. Therefore L_h is a bounded linear functional with $\|L_h\| \leq M\|h\|$. By the Riesz Representation Theorem there is a unique f in $\mathcal{K}$ such that $\langle k, f \rangle = L_h(k) = \overline{u(h, k)}$ and $\|f\| \leq \|L_h\| \leq M\|h\|$. Thus $u(h, k) = \overline{\langle k, f \rangle} = \langle f, k \rangle$. Since the vector f is unique, we can define a function $A : \mathcal{H} \to \mathcal{K}$ by $A(h) = f$; so

$$u(h, k) = \langle Ah, k \rangle$$

We must show that A is linear. If $h, g \in \mathcal{H}$, then we have $\langle A(h + g), k \rangle = u(h + g, k) = u(h, k) + u(g, k) = \langle Ah, k \rangle + \langle Ag, k \rangle = \langle Ah + Ag, k \rangle$. This says that $L_{h+g} = L_h + L_g$, so by the uniqueness part of the Riesz Representation Theorem we have that $A(h + g) = Ah + Ag$. Similarly $A(ah) = aAh$ for a in $\mathbb{F}$ and h in $\mathcal{H}$ and A is a linear transformation. The above inequality also shows that $\|A\| \leq M$ and so $A \in \mathcal{B}(\mathcal{H}, \mathcal{K})$.

Finally we must verify the uniqueness of A. That is, if A_1 is an operator in $\mathcal{B}(\mathcal{H}, \mathcal{K})$ such that $u(h, k) = \langle A_1 h, k \rangle$ for all h and k, we want to show that $A = A_1$. But this condition implies $\langle Ah, k \rangle = \langle A_1 h, k \rangle$ for all h and k. So for each h, $Ah = A_1 h$ and this says that $A = A_1$.   ∎

Now assume we are given an operator $A : \mathcal{H} \to \mathcal{K}$ and define $u(h, k) = \langle Ah, k \rangle$. This defines a bounded sesquilinear form and so there exists an operator B in $\mathcal{B}(\mathcal{K}, \mathcal{H})$ with $\langle Ah, k \rangle = \langle h, Bk \rangle$ for all h in $\mathcal{H}$ and all k in $\mathcal{K}$. This ensures that the following definition makes sense.

5.4.4. Definition. If $A \in \mathcal{B}(\mathcal{H}, \mathcal{K})$, then the *adjoint* of A is the unique operator $A^* : \mathcal{K} \to \mathcal{H}$ such that

$$\langle Ah, k \rangle = \langle h, A^* k \rangle$$

for all h in $\mathcal{H}$ and all k in $\mathcal{K}$.

5.4.5. Example. (a) If μ is a σ-finite measure, $\phi \in L^\infty(\mu)$, and M_ϕ is defined on $L^2(\mu)$ as in Example 5.1.2(c), then $M_\phi^* = M_{\bar\phi}$. In fact if $f \in L^2(\mu)$ and $h = M_\phi^* f$, then for every g in $L^2(\mu)$ we have that $\langle h, g \rangle = \langle f, M_\phi g \rangle = \int f \overline{\phi g} d\mu = \int [\bar\phi f] \bar{g} d\mu$. Since g was arbitrary, $\bar\phi f = M_\phi^* f$.

(b) If $\mathcal{H} = \mathbb{F}^d$ and $T \in \mathcal{B}(\mathbb{F}^d)$ is defined by the matrix A, then T^* is defined by the conjugate transpose of A.

(c) If K is the integral operator defined on $L^2(\mu)$ with kernel $k(x, y)$ as in Proposition 5.1.9, then K^* is the integral operator with kernel $(x, y) \mapsto \overline{k(y, x)}$.

Recall the unilateral shift on ℓ^2 (5.3.3).

5.4.6. Proposition. *If S is the unilateral shift on ℓ^2, then its adjoint is given by $S^*(a_0, a_1, \dots) = (a_1, a_2, \dots)$.*

Proof. For $(a_n), (b_n)$ in ℓ^2,

$$
\begin{aligned}
\langle S^*(a_n), (b_n) \rangle &= \langle (a_n), S(b_n) \rangle \\
&= \langle (a_0, a_1, \dots), (0, b_0, b_1, \dots) \rangle \\
&= a_1 \bar{b}_0 + a_2 \bar{b}_1 + \cdots \\
&= \langle (a_1, a_2, \dots), (b_0, b_1, \dots) \rangle
\end{aligned}
$$

This holds for all (b_n) and so the proposition is proven. ∎

For obvious reasons the adjoint of the unilateral shift is often called the *backward shift*. Here is a variation on the unilateral shift. Let $\{\alpha_n\} \in \ell^\infty$, where $\alpha_n \neq 0$ for all $n \geq 1$; define T on ℓ^2 by $T(x_1, x_2, \dots) = (0, \alpha_1 x_1, \alpha_2 x_2, \dots)$. The operator T is called the *unilateral weighted shift* or just the *weighted shift* with weight sequence $\{\alpha_n\}$.

5.4.7. Example. If T is the weighted shift on ℓ^2 with weight sequence $\{\alpha_n\}$, then T^* is given by $T^*(x_1, x_2, \dots) = (\overline{\alpha_1} x_2, \overline{\alpha_2} x_3, \dots)$.

5.4.8. Proposition. *If $A, B \in \mathcal{B}(\mathcal{H}, \mathcal{K})$, $C \in \mathcal{B}(\mathcal{K}, \mathcal{L})$, and $a \in \mathbb{F}$, then the following hold.*

(a) $(aA + B)^* = \bar{a}A^* + B^*$.

(b) $(CA)^* = A^*C^*$.

(c) $A^{**} \equiv (A^*)^* = A$.

(d) *If A is invertible, then $A^* : \mathcal{K} \to \mathcal{H}$ is invertible and $(A^*)^{-1} = (A^{-1})^*$.*

The proof of the preceding proposition is left as an exercise as is the proof of the next one. Here, however, before stating the next proposition we establish the following.

Agreement: We use 1 to denote the identity operator rather than the customary notation I.

This is rather standard practice and is done mostly for convenience. Similarly if $a \in \mathbb{F}$, a will denote the operator $h \mapsto ah$ or aI.

5.4.9. Proposition. *If $T \in \mathcal{B}(\mathcal{H}, \mathcal{K})$, then T is an isomorphism if and only if T is invertible and $T^* = T^{-1}$; T is an isometry if and only if $T^*T = 1$.*

We saw in Example 5.3.3 that the unilateral shift is a non-surjective isometry.

The reader must thoroughly digest this next result because it is very important, even though its proof is easy.

5.4.10. Theorem. *If $A \in \mathcal{B}(\mathcal{H})$, then*

$$\ker A = (\operatorname{ran} A^*)^{\perp} \quad and \quad \ker A^* = (\operatorname{ran} A)^{\perp}$$

Proof. First note that the second equality follows from the first and the fact that $A^{**} = A$; that is, assuming the first equality we have that $\ker A^* = (\operatorname{ran} A^{**})^{\perp} = (\operatorname{ran} A)^{\perp}$. If $h \in \ker A$ and $g \in \mathcal{H}$, then $\langle h, A^*g \rangle = \langle Ah, g \rangle = 0$ so $\ker A \subseteq (\operatorname{ran} A^*)^{\perp}$. On the other hand, if $h \perp \operatorname{ran} A^*$ and $g \in \mathcal{H}$, then $\langle Ah, g \rangle = \langle h, A^*g \rangle = 0$. Since g was arbitrary, $Ah = 0$ and we have that $(\operatorname{ran} A^*)^{\perp} \subseteq \ker A$. ∎

5.4.11. Definition. An operator A in $\mathcal{B}(\mathcal{H})$ is called *self-adjoint* or *hermitian*[5] if $A = A^*$. The operator is *unitary* if $A^*A = 1 = AA^*$. The operator is called *normal* if $A^*A = AA^*$.

The term unitary was introduced above at the start of §5.3, where it is used as a synonym for isomorphism. It is a routine exercise to show that an

---

[5]Charles Hermite was born in 1822 in Dieuze, France, which is west of Paris near the German border. In 1840 he went to school at Collège Louis-le-Grand in Paris, 15 years after Galois studied there. His tendency was to read original papers of mathematicians rather than work to pass the exams. Nevertheless, with a somewhat average performance on the entrance exam, he was admitted to the École Polytechnique. Unfortunately he had a birth defect that resulted in a malformed foot and because of this he was told he had to leave. (From today's perspective, this is truly amazing; but such things happened.) An appeal led to a reversal of the decision, but strict conditions were imposed on him and he decided to leave. On his own he pursued his studies, all the while doing research. In 1847 he passed the exams to receive the baccalauréat. A year later he was appointed to the faculty at École Polytechnique, the same school that had made life difficult for him. He worked on number theory and algebra, orthogonal polynomials, and elliptic functions, with several important contributions. The Hermite polynomials are named after him, and he was the first to prove that the number e is transcendental – that is, it is not the zero of any polynomial with rational coefficients. He had 9 doctoral students, including Poincaré and Stieltjes. He died in 1901 in Paris.

operator in $\mathcal{B}(\mathcal{H})$ is an isomorphism if and only if it satisfies the definition of unitary just given. Generally an isomorphism operates between two different spaces, but we'll often refer to an isomorphism as a unitary. It is clear that a hermitian operator and a unitary are both normal. Also a unitary is invertible and $A^{-1} = A^*$ (5.4.9).

The use of the term self-adjoint is self-explanatory. The reason these operators are also called hermitian is not clear to the author and would be worth a historical study. I regard the use of the term "normal" in this context as rather unfortunate, though a reasonable alternative doesn't pop readily to mind. In fact we will see these operators are anything but the normally occurring ones. Nevertheless the tradition is long established, so it is pointless to fight it.

In fact we can make an analogy of sorts. Taking the adjoint of an operator is like taking the conjugate of a complex number. So hermitian operators are the analogs of real numbers, and unitaries are the analogs of numbers having absolute value 1 (5.4.9). It turns out that normal operators are the true analogs of complex numbers. With this in mind, consider the following.

5.4.12. Example. (a) The orthogonal projection P of $\mathcal{H}$ onto a subspace $\mathcal{M}$ is a hermitian operator.

(b) Let M_ϕ be the multiplication operator defined on $L^2(\mu)$ as in Example 5.1.2. M_ϕ is always normal; M_ϕ is hermitian if and only if ϕ is real-valued a.e. $[\mu]$; M_ϕ is unitary if and only if $|\phi| = 1$ a.e. $[\mu]$.

The next result is the first we have seen that requires the underlying scalar field to be the complex numbers.

5.4.13. Proposition. *If $\mathcal{H}$ is a $\mathbb{C}$-Hilbert space and $A \in \mathcal{B}(\mathcal{H})$, then A is hermitian if and only if $\langle Ah, h \rangle \in \mathbb{R}$ for all h in $\mathcal{H}$.*

Proof. If $A^* = A$, the fact $\langle Ah, h \rangle \subset \mathbb{R}$ is easy. Now assume $\langle Ah, h \rangle \in \mathbb{R}$ for every h in $\mathcal{H}$. If $a \in \mathbb{C}$ and $h, g \in \mathcal{H}$, then $\langle A(h + ag), h + ag \rangle = \langle Ah, h \rangle + \bar{a}\langle Ah, g \rangle + a\langle Ag, h \rangle + |a|^2\langle Ag, g \rangle \in \mathbb{R}$. So this expression equals its complex conjugate. Since the first and last terms on the right hand side of this equation are also real, if we equate the number with its conjugate and do some cancellation we get

$$a\langle Ag, h \rangle + \bar{a}\langle Ah, g \rangle = \bar{a}\langle h, Ag \rangle + a\langle g, Ah \rangle$$
$$= \bar{a}\langle A^*h, g \rangle + a\langle A^*g, h \rangle$$

If we first let $a = 1$ and then let $a = i$, we get

$$\langle Ag, h \rangle + \langle Ah, g \rangle = \langle A^*h, g \rangle + \langle A^*g, h \rangle$$
$$i\langle Ag, h \rangle - i\langle Ah, g \rangle = -i\langle A^*h, g \rangle + i\langle A^*g, h \rangle$$

After some arithmetic, we get $\langle Ag, h \rangle = \langle A^*g, h \rangle$ for all h, g in $\mathcal{H}$. Hence A is hermitian. ∎

We return to the environment of an arbitrary field.

5.4.14. Proposition. *If A is a hermitian operator, then*

$$\|A\| = \sup\{|\langle Ah, h \rangle| : \|h\| = 1\}$$

Proof. Let M denote the supremum in the statement. Since $|\langle Ah, h \rangle| \leq \|A\|$ whenever $h \in$ ball $\mathcal{H}$, we have $M \leq \|A\|$.

If $\|h\| = \|g\| = 1$, then

$$\langle A(h \pm g), h \pm g \rangle = \langle Ah, h \rangle \pm \langle Ah, g \rangle \pm \langle Ag, h \rangle \pm \langle Ag, g \rangle$$
$$= \langle Ah, h \rangle \pm \langle Ah, g \rangle \pm \langle g, A^*h \rangle \pm \langle Ag, g \rangle$$
$$= \langle Ah, h \rangle \pm \langle Ah, g \rangle \pm \overline{\langle A^*h, g \rangle} \pm \langle Ag, g \rangle$$

Since $A = A^*$, when we subtract one of these equations from the other and do some simplifying, we get $4\mathrm{Re}\,\langle Ah, g \rangle = \langle A(h+g), h+g \rangle - \langle A(h-g), h-g \rangle$. Since $\langle Af, f \rangle| \leq M\|f\|^2$ for all f in $\mathcal{H}$, using the parallelogram law shows that this last equation yields

$$4\mathrm{Re}\,\langle Ah, g \rangle \leq M(\|h + g\|^2 + \|h - g\|^2)$$
$$= 2M(\|h\|^2 + \|g\|^2)$$
$$= 4M$$

since h, g are unit vectors. If $\langle Ah, g \rangle = e^{i\theta}|\langle Ah, g \rangle|$, then substituting $e^{-i\theta}h$ for h in the above inequality gives $|\langle Ah, g \rangle| \leq M$ whenever $\|h\| = \|g\| = 1$. If we take the supremum over all such g, we get $\|Ah\| \leq M$; now take the supremum over all such h and the proof is complete. ∎

5.4.15. Corollary. *If A is hermitian and $\langle Ah, h \rangle = 0$ for all h, then $A = 0$.*

The preceding proposition and corollary are decidedly false if the operator A is not hermitian. For example consider the 2×2 matrix $A = \begin{bmatrix} 0 & 1 \\ -1 & 0 \end{bmatrix}$ as an operator on $\mathbb{R}^2$. It is easy to check that $\langle Ah, h \rangle = 0$ for every h. If, however, the underlying scalar field is $\mathbb{C}$, then we have a bonus as the next result shows. The proof is left to the reader.

5.4.16. Proposition. *If $\mathcal{H}$ is a $\mathbb{C}$-Hilbert space and $A \in \mathcal{B}(\mathcal{H})$ such that $\langle Ah, h \rangle = 0$ for all h in $\mathcal{H}$, then $A = 0$.*

Proof. If $h, g \in \mathcal{H}$ and $a \in \mathbb{C}$, then we have that $0 = \langle A(ah + g), ah + g \rangle = |a|^2\langle Ah, h \rangle + a\langle Ah, g \rangle + \bar{a}\langle Ag, h \rangle + \langle Ag, g \rangle = a\langle Ah, g \rangle + \bar{a}\langle Ag, h \rangle$. First take $a = 1$, then $a = i$, and we get $\langle Ah, g \rangle = -\langle Ag, h \rangle$ and $\langle Ah, g \rangle = \langle Ag, h \rangle$ so that $\langle Ah, g \rangle = 0$ for all h, g. ∎

When $\mathcal{H}$ is a $\mathbb{C}$-Hilbert space, for any operator A in $\mathcal{B}(\mathcal{H})$ we observe that $B = (A + A^*)/2$ and $C = (A - A^*)/2i$ are hermitian and $A = B + iC$. The operators B and C are called the *real part* and the *imaginary part* of A.

5.4.17. Proposition. *If $A \in \mathcal{B}(\mathcal{H})$, then the following statements are equivalent.*

(a) *A is normal.*

(b) *$\|Ah\| = \|A^*h\|$ for every h in $\mathcal{H}$.*

   If $\mathcal{H}$ is a $\mathbb{C}$-Hilbert space, then these two statements are equivalent to

(c) *The real and imaginary parts of A commute.*

Proof. Note that $\|Ah\|^2 - \|A^*h\|^2 = \langle (A^*A - AA^*)h, h \rangle$ and $A^*A - AA^*$ is hermitian. Thus the equivalence of (a) and (b) is immediate from Corollary 5.4.15. The proof that (c) and (a) are equivalent is Exercise 13. ∎

Exercises.

(1) Verify the statements in Example 5.4.5(b) and (c).

(2) Verify the statement in Example 5.4.7.

(3) Prove Propositions 5.4.8 and 5.4.9.

(4) Give a necessary and sufficient condition that an integral operator be hermitian.

(5) Show that the unilateral shift is not normal.

(6) Let $T \in \mathcal{B}(\mathcal{H})$ for a complex Hilbert space $\mathcal{H}$. Show that $A = \frac{1}{2}(T + T^*)$ and $B = \frac{1}{2i}(T - T^*)$ are hermitian operators and $T = A + iB$.

(7) If S is the unilateral shift, compute $S^n S^{*n}$ and $S^{*n}S^n$ for all $n \geq 1$.

(8) Verify Example 5.4.7.

(9) If $A \in \mathcal{B}(\mathcal{H})$, show that A is injective if and only if A^* has dense range.

(10) Show that a $d \times d$ matrix whose only non-zero entries are on the main diagonal is a normal operator.

(11) Verify the statements in Example 5.4.12.

(12) Show that the Volterra operator (5.1.10) is not normal.

(13) In Proposition 5.4.17 show that (a) and (c) are equivalent.

(14) When A is a normal operator, show that $\ker A = \ker A^*$. Is the converse true?

(15) Show that A is unitary if and only if it is a normal isometry.

(16) Let $\phi \in L^\infty(\mu)$ and let M_ϕ be the multiplication operator on $L^2(\mu)$. Show that ker $M_\phi = (0)$ if and only if $\phi = 0$ a.e. $[\mu]$. Give a necessary and sufficient condition that ran M_ϕ is closed.

5.5. The direct sum of Hilbert spaces

The thrust of this section is to define what we mean by $\mathcal{H} \oplus \mathcal{K}$, the direct sum of Hilbert spaces $\mathcal{H}$ and $\mathcal{K}$, and explore the concept of an invariant subspace and a reducing subspace for an operator on a Hilbert space. In a sense the purpose of this section, besides introducing these important concepts, is to provide two ways of manufacturing new operators from old ones.

5.5.1. Definition. If $\mathcal{H}$ and $\mathcal{K}$ are two Hilbert spaces, let $\mathcal{H} \oplus \mathcal{K} = \{h \oplus k : h \in \mathcal{H}, k \in \mathcal{K}\}$ and define addition and scalar multiplication on $\mathcal{H} \oplus \mathcal{K}$ coordinatewise:

$$h_1 \oplus k_1 + h_2 \oplus k_2 = (h_1 + h_2) \oplus (k_1 + k_2)$$
$$a(h \oplus k) = (ah) \oplus (ak)$$

Define an inner product on this space by

$$\langle h_1 \oplus k_1, h_2 \oplus k_2 \rangle = \langle h_1, k_1 \rangle + \langle h_2, k_2 \rangle$$

The reader is expected to verify that with these definitions, $\mathcal{H} \oplus \mathcal{K}$ is a vector space and we have a true inner product on it. We also note that with this definition of an inner product, $\|h \oplus k\| = (\|h\|^2 + \|k\|^2)^{\frac{1}{2}}$. It is left as an exercise for the reader to show that $\mathcal{H} \oplus \mathcal{K}$ is a Hilbert space; that is, it is complete.

Now we make sense of the direct sum of a sequence of Hilbert spaces.

5.5.2. Proposition. *Let $\{\mathcal{H}_n\}$ be a sequence of separable Hilbert spaces and let $\mathcal{H} = \{(h_n)_{n=1}^\infty : h_n \in \mathcal{H}_n$ for all $n \geq 1$ and $\sum_{n=1}^\infty \|h_n\|^2 < \infty\}$; define addition and scalar multiplication on $\mathcal{H}$ coordinatewise. If for $h = (h_n), g = (g_n)$ in $\mathcal{H}$ we define*

5.5.3 $$\langle h, g \rangle = \sum_{n=1}^\infty \langle h_n, g_n \rangle,$$

then (5.5.3) defines an inner product on $\mathcal{H}$ and, with respect to this inner product, $\mathcal{H}$ is a separable Hilbert space.

Proof. Note that by the CBS Inequality,

$$\sum_{n=1}^\infty |\langle h_n^\infty, g_n \rangle| \leq \sum_{n=1}^\infty \|h_n\| \|g_n\| \leq \left[\sum_{n=1}^\infty \|h_n\|^2\right]^{\frac{1}{2}} \left[\sum_{n=1}^\infty \|g_n\|^2\right]^{\frac{1}{2}} < \infty$$

so that the series used to define (5.5.3) converges absolutely. It is now rather straightforward to verify that this is, indeed, an inner product on $\mathcal{H}$ and with this inner product $\mathcal{H}$ is complete. (The proof of completeness is similar to the proof in (1.3.5) that ℓ^1 is complete.)

To show that $\mathcal{H}$ is separable, for each $n \geq 1$ let $\{f_{nk} : k \geq 1\}$ be a dense sequence in $\mathcal{H}_n$ and let $\mathcal{L}$ denote all those elements $\{h_n\}$ in $\mathcal{H}$ such that there is an integer N such that $h_n = 0$ for $n \geq N+1$ and $h_n \in \{f_{nk} : k \geq 1\}$ for $1 \leq n \leq N$. It follows that $\mathcal{L}$ is countable and dense in $\mathcal{H}$. ∎

The direct sum defined in the preceding proposition is denoted by

$$\mathcal{H} = \mathcal{H}_1 \oplus \mathcal{H}_2 \oplus \cdots = \bigoplus_{n=1}^{\infty} \mathcal{H}_n$$

We will sometimes denote the elements of such a direct sum as $h_1 \oplus h_2 \oplus \cdots$ or $\bigoplus_n h_n$ rather than (h_n).

5.5.4. Proposition. *Let $\{\mathcal{H}_n\}$ be a sequence of Hilbert spaces and put $\mathcal{H} = \bigoplus_{n=1}^{\infty} \mathcal{H}_n$. If for each $n \geq 1$, $A_n \in \mathcal{B}(\mathcal{H}_n)$ and $\sup_n \|A_n\| < \infty$, then $A : \mathcal{H} \to \mathcal{H}$ defined by $A(\bigoplus_{n=1}^{\infty} h_n) = \bigoplus_{n=1}^{\infty} A_n h_n$ is a bounded operator on $\mathcal{H}$ with $\|A\| = \sup_n \|A_n\|$.*

The proof of this result is straightforward and follows the reasoning used to verify the statements in Example 5.1.7. In this case we denote the operator defined in the preceding proposition by $A = \bigoplus_{n=1}^{\infty} A_n$.

Suppose $T \in \mathcal{B}(\mathcal{H} \oplus \mathcal{K})$ and let P, Q denote the orthogonal projections of $\mathcal{H} \oplus \mathcal{K}$ onto $\mathcal{H}$ and $\mathcal{K}$, respectively: $P(h \oplus k) = h$, $Q(h \oplus k) = k$. Define operators A in $\mathcal{B}(\mathcal{H})$ and D in $\mathcal{B}(\mathcal{K})$ by $Ah = PT(h \oplus 0)$ and $Dk = QT(0 \oplus k)$. Similarly define $B : \mathcal{K} \to \mathcal{H}$ and $C : \mathcal{H} \to \mathcal{K}$ by $Bk = PT(0 \oplus k)$ and $Ch = QT(h \oplus 0)$. The operator T can then be represented by the matrix

5.5.5
$$T = \begin{bmatrix} A & B \\ C & D \end{bmatrix}$$

The converse is also true: if A, B, C, D are operators on the appropriate spaces, then the matrix in (5.5.5) defines an operator T in $\mathcal{B}(\mathcal{H} \oplus \mathcal{K})$. Similarly if $T \in \mathcal{B}(\mathcal{H})$ and $\mathcal{M} \leq \mathcal{H}$, we can write $\mathcal{H} = \mathcal{M} \oplus \mathcal{M}^{\perp}$ and then T can be written as the matrix in (5.5.5).

A *projection* is an operator P that is *idempotent* (that is, it satisfies $P^2 = P$) and such that $\ker P = (\operatorname{ran} P)^{\perp}$. If $\mathcal{M} \leq \mathcal{H}$, then the orthogonal projection of $\mathcal{H}$ onto $\mathcal{M}$ as in Theorem 3.2.8 is an example of a projection as just defined. Conversely, if P is a projection as defined above, then P is precisely the orthogonal projection of $\mathcal{H}$ onto $\mathcal{M} = \operatorname{ran} P$. It is not difficult to manufacture an example of an idempotent that is not a projection (see Exercise 7).

5.5.6. Definition. If $A \in \mathcal{B}(\mathcal{H})$ and $\mathcal{M} \leq \mathcal{H}$, say that $\mathcal{M}$ is an *invariant subspace* for A if $Ah \in \mathcal{M}$ whenever $h \in \mathcal{M}$. Say that $\mathcal{M}$ is a *reducing subspace* for A if both $\mathcal{M}$ and $\mathcal{M}^\perp$ are invariant.

5.5.7. Example. (a) If $\phi \in L^\infty(\mu)$ and E is a μ-measurable set, then $\mathcal{M} = \{f \in L^2(\mu) : f(x) = 0 \text{ a.e. } [\mu] \text{ off } E\}$ is a reducing subspace for the multiplication operator M_ϕ. If $Pf = \chi_E f = M_{\chi_E} f$ for all f in $L^2(\mu)$, then P is the projection of $L^2(\mu)$ onto $\mathcal{M}$.

(b) If V is the Volterra operator on $L^2[0,1]$ (5.1.10), $a \in [0,1]$, and $\mathcal{M} = \{f \in L^2[0,1] : f(t) = 0 \text{ for } 0 \leq t \leq a\}$, then $\mathcal{M}$ is an invariant subspace for V, but it is not reducing.

(c) Let A be the $d \times d$ matrix with zeros everywhere except on the first subdiagonal, where there are only ones; and consider A as acting on $\mathbb{F}^d$. If $\{e_1, \ldots, e_d\}$ are the standard basis elements, then for $2 \leq k \leq d$, $\mathcal{M}_k = \bigvee \{e_k, \ldots, e_d\}$ is invariant but not reducing.

5.5.8. Proposition. *If $T \in \mathcal{B}(\mathcal{H})$, P is a projection in $\mathcal{B}(\mathcal{H})$, and $\mathcal{M} = \operatorname{ran} P$, then the following statements are equivalent.*

(a) $\mathcal{M}$ *is an invariant subspace for T.*

(b) $PTP = TP$.

(c) *In the representation of T as in (5.5.5) with respect to $\mathcal{H} = \mathcal{M} \oplus \mathcal{M}^\perp$, $C = 0$.*

Proof. (a) *implies* (b). If $h \in \mathcal{H}$, then $Ph \in \mathcal{M}$; hence $TPh \in \mathcal{M}$. But this implies that $PTPh = TPh$.

(b) *implies* (c). If we represent P as a 2×2 matrix as in (5.5.5), then

$$P = \begin{bmatrix} 1 & 0 \\ 0 & 0 \end{bmatrix}$$

Hence

$$PTP = \begin{bmatrix} A & 0 \\ 0 & 0 \end{bmatrix} = AP = \begin{bmatrix} A & 0 \\ C & 0 \end{bmatrix}$$

Therefore $C = 0$.

(c) *implies* (a). Since $C = 0$, for each h in $\mathcal{M}$ we have

$$Th = \begin{bmatrix} A & B \\ 0 & D \end{bmatrix} \begin{bmatrix} h \\ 0 \end{bmatrix} = \begin{bmatrix} Ah \\ 0 \end{bmatrix} \in \mathcal{M} \qquad \blacksquare$$

5.5.9. Proposition. *If $T \in \mathcal{B}(\mathcal{H})$, P is a projection in $\mathcal{B}(\mathcal{H})$, and $\mathcal{M} = \operatorname{ran} P$, then the following statements are equivalent.*

(a) $\mathcal{M}$ *reduces T.*

(b) $PT = TP$.

(c) *In the representation of T as in (5.5.5) with respect to $\mathcal{H} = \mathcal{M} \oplus \mathcal{M}^{\perp}$,
$C = 0 = B$.*

(d) *$\mathcal{M}$ is an invariant subspace for both T and T^*.*

Proof. (a) *implies* (b). Recall that the projection onto $\mathcal{M}^{\perp}$ is $1 - P$. Thus part (b) of the preceding proposition implies $PTP = TP$ as well a $(1 - P)T(1 - P) = T(1 - P)$. If you do the multiplication and simplify, we get part (b).

(b) *implies* (c). This proof is similar to the corresponding part of the preceding proof.

(c) *implies* (d). Since (c) holds, Exercise 6 implies

$$T^* = \begin{bmatrix} A^* & 0 \\ 0 & D^* \end{bmatrix}$$

and it follows that $\mathcal{M}$ is invariant for T^*.

(d) *implies* (a). If $h \in \mathcal{M}^{\perp}$ and $f \in \mathcal{M}$, then $\langle Th, f \rangle = \langle h, T^* f \rangle = 0$ since $\mathcal{M}$ is invariant for T^*. Thus $\mathcal{M}^{\perp}$ is an invariant subspace for T. ∎

As stated at the commencement of this section, we have found two ways to get new operators from old ones. One is to take the direct sum of a bounded sequence of operators and the other is to restrict a given operator to an invariant subspace.

Exercises.

(1) Give the details of the proof of Proposition 5.5.2.

(2) Give the details of the proof of Proposition 5.5.4.

(3) Let $(X_n, \mathcal{A}_n, \mu_n)$ be a measure space for each $n \geq 1$. Let X denote the disjoint union of the spaces $\{X_n\}$. (The notion of the disjoint union has been encountered before in Exercise 1.2.7. In this case, if you want to be precise, you must embed each X_n in a common space Y and write $X = \bigcup_{n=1}^{\infty}(X_n \times \{n\})$ in $Y \times \mathbb{N}$. But the way we'll think of X is as $X = \bigcup_{n=1}^{\infty} X_n$, where $X_n \cap X_m = \emptyset$ if $n \neq m$.) Let $\mathcal{A} = \{E \subseteq X : E \cap X_n \in \mathcal{A}_n \text{ for all } n\}$. (a) Show that $\mathcal{A}$ is a σ-algebra of subsets of X. For each E in $\mathcal{A}$, define $\mu(E) = \sum_{n=1}^{\infty} \mu_n(E \cap X_n)$. (b) Show that $(X, \mathcal{A}, \mu)$ is a measure space and that this measure space is σ-finite if and only if each μ_n is a σ-finite measure. (c) Show that $V : L^2(\mu) \to \bigoplus_{n=1}^{\infty} L^2(\mu_n)$ defined by $Vf = f|X_1 \oplus f|X_2 \oplus \cdots$ is an isomorphism. Describe $V^* = V^{-1}$. (d) If for each n, $\phi_n \in L^{\infty}(\mu_n)$ and we define ϕ on X by $\phi(x) = \phi_n(x)$ when $x \in X_n$, show that $\phi \in L^{\infty}(\mu)$ if and only if $\sup_n \|\phi_n\|_{\infty} < \infty$, in which case

$\|\phi\|_\infty = \sup_n \|\phi_n\|_\infty$. (e) With the notation established in (c) and (d), show that $VM_\phi V^* = \bigoplus_{n=1}^\infty M_{\phi_n}$.

(4) Let X be a set, $\mathcal{A}$ a σ-algebra of subsets, and μ_1, μ_2 two σ-finite measures defined on $\mathcal{A}$. Put $\mu = \mu_1 + \mu_2$ and define the map $V : L^2(\mu) \to L^2(\mu_1) \oplus L^2(\mu_2)$ by $Vf = f_1 \oplus f_2$, where f_k is the equivalence class in $L^2(\mu_k)$ corresponding to f. Show that V is an isomorphism if and only if $\mu_1 \perp \mu_2$.

(5) In Proposition 5.5.4, show that A is hermitian (normal) if and only if each A_n is hermitian (normal).

(6) If T is given by the matrix (5.5.5), show that

$$ T^* = \begin{bmatrix} A^* & C^* \\ B^* & D^* \end{bmatrix} $$

(7) Let $\mathcal{H} = \mathbb{R}^2$, $\mathcal{M} = \{(x,0) : x \in \mathbb{R}\}$, and let $\mathcal{N} = \{(x, x\tan\theta) : x \in \mathbb{R}\}$, where $0 < \theta < \frac{\pi}{2}$. (a) Show that $\mathcal{H} = \mathcal{M} + \mathcal{N}$ and $\mathcal{M} \cap \mathcal{N} = (0)$. (b) By (a), every vector h in $\mathcal{H}$ has a unique decomposition $h = f + g$, where $f \in \mathcal{M}$ and $g \in \mathcal{N}$. Define $E_\theta h = f$ and find a formula for the operator E_θ. (c) Show that E_θ is a bounded operator with $E_\theta^2 = E_\theta$. (That is, E_θ is an idempotent.) (d) Show that $\|E_\theta\| = (\sin\theta)^{-1}$.

(8) Let P and Q be projections. (a) Show that $PQ = QP$ if and only if $P + Q - PQ$ is a projection. In this situation, show that $\operatorname{ran}(P + Q - PQ) = \operatorname{ran} P + \operatorname{ran} Q$ and $\ker(P + Q - PQ) = \ker P \cap \ker Q$. (b) Show that the following statements are equivalent. (i) $P - Q$ is a projection. (ii) $\operatorname{ran} Q \le \operatorname{ran} P$. (iii) $PQ = Q$. (iv) $QP = Q$. In this situation show that $\operatorname{ran}(P - Q) = \operatorname{ran} P \cap (\operatorname{ran} Q)^\perp$ and $\ker(P - Q) = \operatorname{ran} Q + \ker P$.

(9) Verify the statements in Example 5.5.7.

(10) Give an example of two projections that do not commute.

(11) Let $T \in \mathcal{B}(\mathcal{H})$, let $\mathcal{M} = \mathcal{H} \oplus (0)$, and let $\mathcal{N} = \{(h, Th) : h \in \mathcal{H}\}$, the *graph* of T. Verify that both $\mathcal{M}$ and $\mathcal{N}$ are closed subspaces of $\mathcal{H} \oplus \mathcal{H}$. (a) Show that $\mathcal{M} \cap \mathcal{N} = (0)$ if and only if $\ker T = (0)$. (b) Show that $\mathcal{M} + \mathcal{N}$ is dense in $\mathcal{H} \oplus \mathcal{H}$ if and only if $\operatorname{ran} T$ is dense in $\mathcal{H}$. (c) Show that $\mathcal{M} + \mathcal{N} = \mathcal{H} \oplus \mathcal{H}$ if and only if T is surjective.

(12) Find two closed subspaces $\mathcal{M}$ and $\mathcal{N}$ of an infinite-dimensional Hilbert space $\mathcal{H}$ such that $\mathcal{M} \cap \mathcal{N} = (0)$, $\mathcal{M} + \mathcal{N}$ is dense, but $\mathcal{M} + \mathcal{N} \ne \mathcal{H}$. (Use Exercise 11.)

(13) Find an invariant subspace of the bilateral shift (5.3.3) that is not reducing.

5.6. Compact linear transformations

The theory of operators on a finite-dimensional space is well understood, even though there remain unanswered questions. There are difficult problems in matrix theory, but that is actually a different subject. In infinite-dimensional spaces it is fair to say that more is unknown about operators than is known. This applies not only to operators on Banach spaces, where the structure of the spaces themselves presents mysteries without ever worrying about the operators on them; it also applies to operators on a Hilbert space as well, in spite of the fact that we completely understand the linear structure of the spaces. However there is one class of operators where we know quite a lot.

5.6.1. Definition. For Banach spaces $\mathcal{X}$ and $\mathcal{Y}$ an operator $T : \mathcal{X} \to \mathcal{Y}$ is called *compact* if the closure of $T(\text{ball}\,\mathcal{X})$ is a compact subset of $\mathcal{Y}$. Denote the set of all compact operators from $\mathcal{X}$ into $\mathcal{Y}$ by $\mathcal{B}_0(\mathcal{X}, \mathcal{Y})$ and let $\mathcal{B}_0(\mathcal{X}) = \mathcal{B}_0(\mathcal{X}, \mathcal{X})$.

Initially in this section we'll focus on the compact operators on a Banach space, but after some preliminary work we'll focus on Hilbert space. Later in this book in Chapter 9 we'll return to the general theory. Some examples of compact operators are those having *finite rank*; that is, those operators T such that $\text{ran}\,T$ is finite-dimensional. This follows because a bounded subset of a finite-dimensional Banach space has compact closure. Other examples will be given later in this section.

5.6.2. Proposition. *Let $\mathcal{X}$ and $\mathcal{Y}$ be Banach spaces.*

(a) $\mathcal{B}_0(\mathcal{X}, \mathcal{Y})$ *is a closed linear subspace of $\mathcal{B}(\mathcal{X}, \mathcal{Y})$.*

(b) *If $A \in \mathcal{B}(\mathcal{X})$, $B \in \mathcal{B}(\mathcal{Y})$, and $T \in \mathcal{B}_0(\mathcal{X}, \mathcal{Y})$, then TA and $BT \in \mathcal{B}_0(\mathcal{X}, \mathcal{Y})$.*

Proof. (a) That $\mathcal{B}_0(\mathcal{X}, \mathcal{Y})$ is contained in the space of bounded operators is clear since a subset of $\mathcal{Y}$ that has compact closure must be bounded. We only show that $\mathcal{B}_0(\mathcal{X}, \mathcal{Y})$ is closed and leave the proof that it is a vector space to the reader. Let $\{T_n\}$ be a sequence in $\mathcal{B}_0(\mathcal{X}, \mathcal{Y})$ that converges to T in $\mathcal{B}(\mathcal{X}, \mathcal{Y})$, and let $\{x_k\}$ be a sequence in ball $\mathcal{X}$; we want to show that $\{Tx_k\}$ has a convergent subsequence. Needless to say a direct proof of this involves considerable bookkeeping, but we can let Tykhonov's Theorem do most of that for us. Let $Z_n = \text{cl}\,[T_n(\text{ball}\,\mathcal{X})] \subseteq \mathcal{Y}$; so Z_n is compact. Therefore $Z = \prod_{n=1}^{\infty} Z_n$ is a compact metric space. For each $k \geq 1$ let $\zeta_k = ([T_n x_k]_n) \in Z$; thus $\{\zeta_k\}$ is a sequence in Z. So there is a subsequence $\{\zeta_{k_j}\}$ and a point $\zeta = (y_n)$ in Z such that $\zeta_{k_j} \to \zeta$. Note that each y_n in the definition of ζ is a point in $\mathcal{Y}$ and the convergence statement says that for each $n \geq 1$, $\|T_n x_{k_j} - y_n\| \to 0$ as $k_j \to \infty$.

We first show that $\{y_n\}$ is a Cauchy sequence in $\mathcal{Y}$ and therefore converges. In fact $\|y_n - y_m\| \leq \|y_n - T_n x_{k_j}\| + \|T_n x_{k_j} - T_m x_{k_j}\| + \|T_m x_{k_j} - y_m\| \leq \|y_n - T_n x_{k_j}\| + \|T_n - T_m\| + \|T_m x_{k_j} - y_m\|$. Clearly this can be made as small as desired for all large n, m so that $\{y_n\}$ is a Cauchy sequence; let $y = \lim_n y_n$.

Now we show that this point y in $\mathcal{Y}$ is the sought after point such that $T x_{k_j} \to y$, thus showing that $\{T x_k\}$ has a convergent subsequence and hence that T is a compact operator. In fact we have that $\|T x_{k_j} - y\| \leq \|T x_{k_j} - T_n x_{k_j}\| + \|T_n x_{k_j} - y_n\| + \|y_n - y\| \leq \|T - T_n\| + \|T_n x_{k_j} - y_n\| + \|y_n - y\|$. If $\epsilon > 0$, fix an n such that $\|T - T_n\| < \epsilon/3$ and $\|y_n - y\| < \epsilon/3$. For this n choose a k_0 such that $\|T_n x_{k_j} - y_n\| < \epsilon/3$ whenever $k_j \geq k_0$. It follows that $\|T x_{k_j} - y\| < \epsilon$ for all $k_j \geq k_0$.

(b) Without loss of generality we can assume that $\|A\|, \|B\| \leq 1$. Now if $\{x_k\}$ is a sequence in ball $\mathcal{X}$, $\{A x_k\} \subseteq$ ball $\mathcal{X}$. Therefore there is a subsequence $\{x_{k_j}\}$ and points y and z in $\mathcal{Y}$ such that $T x_{k_j} \to y$ and $T A x_{k_j} \to z$; hence $B T x_{k_j} \to B y$ and both $T A$ and $B T$ are compact. ∎

We now focus on compact operators on a Hilbert space.

5.6.3. Lemma. *Let* $\mathcal{H}$ *and* $\mathcal{K}$ *be Hilbert spaces and let* $T \in \mathcal{B}_0(\mathcal{H}, \mathcal{K})$. *If* $\{e_1, e_2, \dots\}$ *is a basis for* $\mathrm{cl}\,[\mathrm{ran}\,T]$ *and* P_n *is the projection of* $\mathcal{K}$ *onto the span of the first* n *of these basis vectors, then* $\|P_n T - T\| \to 0$.

Proof. Let $\mathcal{M} = \mathrm{cl}\,[\mathrm{ran}\,T]$. For each f in $\mathcal{M}$, $P_n f \to f$. Therefore $\|P_n T h - T h\| \to 0$ for every h in $\mathcal{H}$. If $\epsilon > 0$, let $h_1, \dots, h_m \in$ ball $\mathcal{H}$ such that $T(\mathrm{ball}\,\mathcal{H}) \subseteq \bigcup_{j=1}^m B(T h_j; \epsilon/3)$. There is an integer N such that $\|T h_j - P_n T h_j\| < \epsilon/3$ for $1 \leq j \leq m$ and $n \geq N$. Let $h \in$ ball $\mathcal{H}$, and fix a j with $\|T h - T h_j\| < \epsilon/3$. Therefore for $n \geq N$,

$$\|T h - P_n T h\| \leq \|T h - T h_j\| + \|T h_j - P_n T h_j\| + \|P_n T h_j - P_n T h\|$$
$$< \frac{2\epsilon}{3} + \|P_n (T h_j - T h)\|$$
$$< \epsilon$$

Since h was arbitrary, $\|T - P_n T\| < \epsilon$ when $n \geq N$. ∎

5.6.4. Theorem. *If* $\mathcal{H}$ *and* $\mathcal{K}$ *are Hilbert spaces and* $T \in \mathcal{B}(\mathcal{H}, \mathcal{K})$, *the following statements are equivalent.*

(a) *T is compact.*

(b) *T^* is compact.*

(c) *There is a sequence of finite-rank operators that converges to T.*

Proof. (c) *implies* (a). This is immediate from Proposition 5.6.2(a) and the fact that a finite rank operator is compact.

(a) *implies* (c). If $\{P_n\}$ is as in Lemma 5.6.3 and $T_n = P_nT$, then T_n has finite rank and $\|T - T_n\| \to 0$.

(c) *implies* (b). In fact if $\{T_n\}$ is a sequence of finite rank operators such that $\|T_n - T\| \to 0$, then each T_n^* has finite rank (Why?) and $\|T_n^* - T^*\| = \|T_n - T\| \to 0$.

(b) *implies* (a). Since T^* is compact and we know that (a) and (c) are equivalent for T^*, there is a sequence of finite rank operators $\{S_n\}$ in $\mathcal{B}(\mathcal{K}, \mathcal{H})$ such that $\|T^* - S_n\| \to 0$. Thus $\|T - S_n^*\| \to 0$ and each S_n^* has finite rank. Hence T is compact. ∎

Is the preceding theorem true for compact operators on a Banach space? The equivalence of (a) and (b) for Banach spaces will be shown to be true once we establish the notion of the adjoint of a linear transformation between Banach spaces. (See Theorem 9.2.1.) Since each operator of finite rank is compact, Proposition 5.6.2 shows that (c) implies (a) in the Banach space setting as well. However there are Banach spaces in which every compact operator is not the limit of finite rank operators. See [14] and the subsequent literature on the "approximation problem."

The next result has more importance than just furnishing a collection of examples.

5.6.5. Proposition. *Let $\{e_n : n \geq 1\}$ be a basis for $\mathcal{H}$ and let $\{a_n\}$ be a sequence of scalars. For each n define $Te_n = a_ne_n$. T extends to a bounded operator on $\mathcal{H}$ if and only if $\sup_n |a_n| < \infty$, in which case $\|T\| = \sup_n |a_n|$. This operator is compact if and only if $a_n \to 0$.*

Proof. If $T \in \mathcal{B}(\mathcal{H})$, then $\infty > \|T\| \geq \sup_n \|Te_n\| = \sup_n |a_n|$. Conversely if $M = \sup_n |a_n| < \infty$ and $h \in \mathcal{H}$, then

$$\left\| \sum_{n=1}^{\infty} a_n \langle h, e_n \rangle \right\|^2 = \sum_{n=1}^{\infty} |a_n|^2 |\langle h, e_n \rangle|^2 \leq M^2 \|h\|^2,$$

so that $Th = \sum_{n=1}^{\infty} a_n \langle h, e_n \rangle$ is a bounded operator with $\|T\| \leq M$ and $Te_n = a_ne_n$ for all $n \geq 1$. Also $\|T\| \geq \|Te_n\| = |a_n|$, so that $\|T\| \geq M$.

Let P_n be the projection of $\mathcal{H}$ onto $\bigvee\{e_1, \ldots, e_n\}$. We have that $(T - TP_n)e_k = a_ke_k$ when $k > n$ and $(T - TP_n)e_k = 0$ when $k \leq n$. By the first part of the proof, $\|T - TP_n\| = \sup\{|a_k| : k > n\}$. So if $T \in \mathcal{B}_0(\mathcal{H})$, Lemma 5.6.3 implies $\|T - TP_n\| \to 0$; hence $a_n \to 0$. Conversely, if $a_n \to 0$, then $\|T - TP_n\| \to 0$ and so T is the limit of finite rank operators and is thus compact. ∎

An operator T defined as in the preceding proposition is said to be *diagonalizable*. That is, T is diagonalizable when there is a basis $\{e_n\}$ and a sequence of scalars $\{a_n\}$ such that $Te_n = a_ne_n$ for all $n \geq 1$.

The proof of this next lemma is straightforward. See Exercise 9.

5.6.6. Lemma. *If $(X, \mathcal{A}, \mu)$ is a measure space, $\{e_n\}$ is a basis for $L^2(\mu)$, and $\phi_{mn}(x, y) = \overline{e_n(x)}e_m(y)$ for all x, y in X and $m, n \geq 1$, then $\{\phi_{mn} : m, n \geq 1\}$ is an orthonormal set for $L^2(\mu \times \mu)$.*

5.6.7. Proposition. *If $(X, \mathcal{A}, \mu)$ is a measure space and $k \in L^2(\mu \times \mu)$, then the integral operator*

$$Kf(x) = \int k(x, y)f(y)d\mu(y)$$

is compact and $\|K\| \leq \|k\|$.

Proof. First we show that K defines a bounded operator. If $f \in L^\infty(\mu)$, then

$$\|Kf\|^2 = \int \left| \int k(x, y)f(y)d\mu(y) \right|^2 d\mu(x)$$

$$\leq \int \left(\int |k(x, y)|^2 d\mu(y) \right) \left(\int |f(y)|^2 d\mu(y) \right) d\mu(x)$$

$$= \|k\|^2 \|f\|^2$$

So $\|K\| \leq \|k\|$.

It is easily verified that if $\{e_n\}$ and ϕ_{mn} are as in the preceding lemma, then $\langle Ke_m, e_n \rangle = \langle k, \phi_{mn} \rangle$. Hence

5.6.8
$$\|k\|^2 \geq \sum_{m,n=1}^\infty |\langle k, \phi_{mn} \rangle|^2 = \sum_{m,n=1}^\infty |\langle Ke_m, e_n \rangle|^2$$

Denote the orthogonal projection of $L^2(\mu)$ onto $\bigvee\{e_k : 1 \leq k \leq n\}$ by P_n and put $K_n = KP_n + P_nK - P_nKP_n$. Note that K_n has finite rank; we will show that K is compact by showing that $\|K - K_n\| \to 0$.

If $f \in L^2(\mu)$ with $\|f\| \leq 1$ and $f = \sum_{k=1}^\infty a_k e_k$, then

$$\|Kf - K_nf\|^2 = \sum_{k=1}^\infty |\langle Kf - K_nf, e_k \rangle|^2$$

$$= \sum_{k=1}^\infty \left| \sum_{m=1}^\infty a_m \langle (K - K_n)e_m, e_k \rangle \right|^2$$

$$\leq \sum_{k=1}^\infty \left[\sum_{m=1}^\infty |a_m|^2 \right] \left[\sum_{m=1}^\infty |\langle (K - K_n)e_m, e_k \rangle|^2 \right]$$

$$\leq \|f\|^2 \left[\sum_{k=1}^\infty \sum_{m=1}^\infty |\langle (K - K_n)e_m, e_k \rangle|^2 \right]$$

Now

$$\langle (K - K_n)e_m, e_k \rangle = \langle Ke_m, e_k \rangle - \langle KP_n e_k, e_m \rangle - \langle P_n K e_k, e_m \rangle$$
$$+ \langle P_n K P_n e_k, e_m \rangle$$
$$= \langle (1 - P_n)K(1 - P_n)e_k, e_m \rangle$$

Hence

$$\|Kf - K_n f\|^2 \leq \|f\|^2 \sum_{k=n+1}^{\infty} \sum_{m=n+1}^{\infty} |\langle Ke_m, e_k \rangle|^2$$
$$= \sum_{k=n+1}^{\infty} \sum_{m=n+1}^{\infty} |\langle k, \psi_{mk} \rangle|^2 .$$

Since f was arbitrary, $\|K - K_n\|^2$ is dominated by this last sum, which converges to 0 as $n \to \infty$ by (5.6.8). ∎

Observe that the kernel used to define the Volterra operator (5.1.10) satisfies the hypothesis of the preceding proposition and as a consequence we have that the Volterra operator on $L^2[0,1]$ is compact.

An important concept for all operators, but especially for the compact ones, is the following. The concept is valid in the context of a Banach space and is valid in linear algebra, where the reader likely first encountered it.

5.6.9. Definition. If $\mathcal{X}$ is a Banach space and $T \in \mathcal{B}(\mathcal{X})$, a scalar λ is called an *eigenvalue* for T provided there is a non-zero vector x with $Tx = \lambda x$. Such a vector x is called an *eigenvector* and the set of all such vectors, $\ker(T - \lambda)$, is called the *eigenspace* of T corresponding to λ. Let $\sigma_p(T)$ denote the set of eigenvalues of T. (Later we will see the reason this particular notation was chosen to denote the set of eigenvalues.)

5.6.10. Example. (a) For the diagonalizable operator T defined in Proposition 5.6.5, each a_n is an eigenvalue and e_n is a corresponding eigenvector. If $a \in \sigma_p(T)$, then $a = a_n$ for some $n \geq 1$. If $J_a = \{n : a_n = a\}$, then $\ker(T - a) = \bigvee\{e_n : n \in J_a\}$.

(b) The Volterra operator does not have any eigenvalues.

(c) If $(X, \mathcal{A}, \mu)$ is a σ-finite measure space, $1 \leq p \leq \infty$, $\phi \in L^\infty(\mu)$, and M_ϕ is the multiplication operator on $L^p(\mu)$, then λ is an eigenvalue of M_ϕ if and only if $E = \{x : \phi(x) = \lambda\}$ has $\mu(E) > 0$. In this case $\ker(M_\phi - \lambda) = \{f \in L^p(\mu) : f = f\chi_E\}$.

(d) Let $h \in L^2_{\mathbb{C}}[0, 2\pi]$ and define $K : L^2_{\mathbb{C}}[0, 2\pi] \to L^2_{\mathbb{C}}[0, 2\pi]$ by

$$Kf(x) = \frac{1}{2\pi} \int_{-\pi}^{\pi} h(x - y)f(y)dy.$$

If $\lambda_n = (2\pi)^{-1/2} \int_{-\pi}^{\pi} h(x) \exp(-inx) dx = \widehat{h}(n)$, the n-th Fourier coefficient of h, then $Ke_n = \lambda_n e_n$, where $e_n(x) = (2\pi)^{-1/2} \exp(-inx)$. (To see this, extend the functions in $L^2_{\mathbb{C}}[0, 2\pi]$ to all of $\mathbb{R}$ by periodicity and perform a change of variables in the formula for Ke_n.)

5.6.11. Proposition. *If $T \in \mathcal{B}_0(\mathcal{H})$ and λ is a non-zero eigenvalue, then* $\ker(T - \lambda)$ *is finite-dimensional.*

Proof. If the eigenspace were infinite-dimensional, then there would be an infinite orthonormal sequence $\{e_n\}$ in $\ker(T-\lambda)$; so $Te_n = \lambda e_n$ for all $n \geq 1$. But $T(\mathrm{ball}\,\mathcal{H})$ has compact closure, so there is a subsequence of $\{Te_n\}$ that converges. But when $n \neq m$, $\|Te_n - Te_m\| = |\lambda|\|e_n - e_m\| = \sqrt{2}|\lambda| > 0$. So $\{Te_n\}$ cannot have a convergent subsequence, a contradiction. ■

5.6.12. Proposition. *If $T \in \mathcal{B}_0(\mathcal{H})$, $\lambda \neq 0$, and $\inf\{\|(T - \lambda)h\| : \|h\| = 1\} = 0$, then λ is an eigenvalue of T.*

Proof. Let $\{h_n\}$ be a sequence of unit vectors in $\mathcal{H}$ such that $\|(T-\lambda)h_n\| \to 0$. Since T is compact, by passing to a subsequence we may assume that there is a vector f in $\mathcal{H}$ such that $\|Th_n - f\| \to 0$. Thus $h_n = \lambda^{-1}[(\lambda - T)h_n + Th_n] \to \lambda^{-1}f$. This tells us two things. First we have that $Th_n \to \lambda^{-1}Tf$, and, since $Th_n \to f$, this implies $Tf = \lambda f$. Also $1 = \lim_n \|h_n\| = \|\lambda^{-1}f\| = |\lambda|^{-1}\|f\|$ and we have that $f \neq 0$. Therefore λ is an eigenvalue of T with non-zero eigenvector f. ■

The preceding proposition is not a practical way to show the existence of a non-zero eigenvalue, but we will find it useful. Here is a lemma that applies in Banach spaces as well as Hilbert space.

5.6.13. Lemma. *If $\mathcal{X}$ and $\mathcal{Y}$ are Banach spaces and $T \in \mathcal{B}(\mathcal{X}, \mathcal{Y})$ such that $\inf\{\|Tx\| : \|x\| = 1\} > 0$, then $\mathrm{ran}\,T$ is closed in $\mathcal{Y}$.*

Proof. If $c = \inf\{\|Tx\| : \|x\| = 1\} > 0$, then for any $x \neq 0$ in $\mathcal{X}$ we have that $c \leq \|T(\|x\|^{-1}x)\| = \|x\|^{-1}\|Tx\|$; thus $\|Tx\| \geq c\|x\|$ for all x in $\mathcal{X}$. Hence if $Tx_n \to y$ in $\mathcal{Y}$, then $\|x_n - x_m\| \leq c^{-1}\|Tx_n - Tx_m\|$ and it follows that $\{x_n\}$ is a Cauchy sequence in $\mathcal{X}$. If $x \in \mathcal{X}$ such that $x_n \to x$, then $y = Tx$ and so $\mathrm{ran}\,T$ is closed. ■

5.6.14. Corollary. *If $T \in \mathcal{B}_0(\mathcal{H})$, $\lambda \neq 0$, $\lambda \notin \sigma_p(T)$, and $\bar{\lambda} \notin \sigma_p(T^*)$, then $T - \lambda$ is an invertible operator.*

Proof. Since $\lambda \notin \sigma_p(T)$, Proposition 5.6.12 implies that $\inf\{\|(T - \lambda)h\| : \|h\| = 1\} > 0$; by the preceding lemma, $\mathrm{ran}\,(T - \lambda)$ is closed. But by Theorem 5.4.10, $\mathrm{ran}\,(T - \lambda)^{\perp} = \ker(T^* - \bar{\lambda})$, and this must be (0) by hypothesis; therefore $T - \lambda$ is surjective. Since it is also injective, we have that $T - \lambda$

has an (algebraic) inverse A; we must show that this inverse A is a bounded operator. This follows by the IMT, but here is a direct, elementary proof.

If $c = \inf\{\|(T-\lambda)h\| : \|h\| = 1\}$, then we know from the proof of (5.6.13) that $\|(T - \lambda)h\| \geq c\|h\|$ for all h; equivalently, $\|Af\| \leq c^{-1}\|f\|$ for all f in $\mathcal{H}$. This says that A is bounded with $\|A\| \leq c^{-1}$.   ■

It will be proved in a later chapter that when T is compact, $\lambda \neq 0$, and $\lambda \notin \sigma_p(T)$, then $\bar{\lambda}$ is not an eigenvalue for T^*. So the last part of the hypothesis of the preceding corollary becomes superfluous.

Exercises.

(1) Show that if $T \in \mathcal{B}(\mathcal{H}, \mathcal{K})$ and T has finite rank, then $T^* : \mathcal{K} \to \mathcal{H}$ has finite rank and $\dim(\operatorname{ran} T) = \dim(\operatorname{ran} T^*)$.

(2) Show that an idempotent operator is compact if and only if it has finite rank.

(3) If $\phi \in L^\infty[0,1]$ and $\phi \neq 0$, show that M_ϕ is not a compact operator on $L^2[0,1]$.

(4) If T is a compact operator and $\mathcal{M}$ is an invariant subspace, show that $T|\mathcal{M}$ is a compact operator.

(5) (a) If $T \in \mathcal{B}_0(\mathcal{H})$ with $\dim \mathcal{H} = \infty$ and $\{e_n\}$ is an orthonormal sequence in $\mathcal{H}$, show that $\|Te_n\| \to 0$. (b) Give an example of a non-compact operator on a Hilbert space $\mathcal{H}$ and an orthonormal sequence $\{e_n\}$ in $\mathcal{H}$ such that $\{Te_n\}$ does not converge to 0. (c) Show that if $T \in \mathcal{B}(\mathcal{H})$ is such that $Te_n \to 0$ for every orthonormal sequence $\{e_n\}$, then T is compact.

(6) Verify the statements in Example 5.6.10.

(7) Show that if λ is an eigenvalue of T, then $|\lambda| \leq \|T\|$.

(8) Let $T_n \in \mathcal{B}(\mathcal{H}_n)$ for all n in $\mathbb{N}$ such that $\sup_n \|T_n\| < \infty$, so that $T = \bigoplus_{n=1}^\infty T_n$ defines a bounded operator on $\mathcal{H} = \bigoplus_{n=1}^\infty \mathcal{H}_n$. (a) Show that T is compact if and only if each T_n is compact and $\|T_n\| \to 0$. (b) Show that λ is an eigenvalue of T if and only if λ is an eigenvalue of some T_n. In this case what is $\dim \ker(T - \lambda)$? (c) If T is compact and $\lambda \in \sigma_p(T_n)$ for all n, what can you say about λ?

(9) In Lemma 5.6.6, show that $\{\phi_{mn}\}$ is a basis for $L^2(\mu \times \mu)$.

5.7. The Spectral Theorem

The main result of this section is the following.

5.7.1. Theorem (The Spectral Theorem). *If T is a compact hermitian operator on $\mathcal{H}$, then T has a countable number of distinct non-zero eigenvalues, $\{\lambda_1, \lambda_2, \dots\}$. If P_n is the projection onto $\ker(T - \lambda_n)$, then P_n has finite rank, $P_n P_m = P_m P_n$ for all n, m, this product being 0 when $n \neq m$, and*

$$\textbf{5.7.2} \qquad\qquad\qquad T = \sum_{n=1}^{\infty} \lambda_n P_n$$

where convergence of this series is in the norm of $\mathcal{B}(\mathcal{H})$.

The virtue of this theorem is that it gives a concise representation of a compact hermitian operator. Note that the ranges of the finite rank projections $P_1, P_2, \dots$ need not span $\mathcal{H}$. Indeed, T may have a kernel and, in fact, $\ker T$ may be infinite-dimensional. It is also useful to recall the definition of a diagonalizable operator (5.6.5) and note that this theorem says that every hermitian compact operator is diagonalizable.

Actually there are more general versions of the above theorem, all of which are called the Spectral Theorem. To begin, we can replace the word hermitian in the above statement with the word normal, and the exact same statement is true. However in this case it is required that the underlying Hilbert space is over the complex numbers. (See Theorem 5.8.2 in the next section.) More generally there is a Spectral Theorem for non-compact normal operators, where the eigenvalues need not exist but are replaced by a compact subset of $\mathbb{C}$ and the sum is replaced by an integral with respect to a projection-valued measure on the Borel subsets of this compact set. Needless to say, this takes considerable preparation. This is carried out in [**8**], though a somewhat modified form, avoiding the projection-valued measures, appears in §11.4.

The proof of Theorem 5.7.1 needs several preliminary results. Before initiating this process we will examine a few consequences.

5.7.3. Corollary. *With the notation of the Spectral Theorem:*

(a) $\ker T = [\bigvee \{P_n \mathcal{H} : n \in \mathbb{N}\}]^{\perp} = (\operatorname{ran} T)^{\perp}$;

(b) $\|T\| = \sup_n |\lambda_n|$ *and* $\lambda_n \to 0$.

Proof. (a) Because the projections P_n are pairwise orthogonal, if $h \in \mathcal{H}$, then $\|Th\|^2 = \sum_{n=1}^{\infty} |\lambda_n|^2 \|P_n h\|^2$. Hence $Th = 0$ if and only if $P_n h = 0$ for all n. That is, $h \in \ker T$ if and only if $h \in \bigcap_{n=1}^{\infty} \ker P_n$; equivalently, if and only if $h \perp P_n \mathcal{H}$ for all n. That establishes the first equality in (a). The second follows from Theorem 5.4.10 and the fact that T is hermitian.

(b) From Exercise 5.6.7 we have that $\|T\| \geq \sup_n |\lambda_n| = M$. Also from the equation established in the proof of (a), $\|Th\|^2 \leq M \sum_{n=1}^{\infty} \|P_n h\|^2 \leq M\|h\|^2$ by the Pythagorean Theorem, establishing the equation for the norm of $\|T\|$. To complete the proof of part (b), for each $n \geq 1$ let e_n be a unit vector in ran P_n; the theorem says that $P_n P_m = 0$ when $n \neq m$, so that $e_n \perp e_m$ when $n \neq m$. By Exercise 5.6.5, $\|Te_n\| \to 0$; but $\|Te_n\| = |\lambda_n|$. ∎

5.7.4. Corollary. *If T is a compact hermitian operator, then there is a sequence of non-zero real scalars $\{\mu_k\}$ and an orthonormal basis $\{e_k\}$ for $(\ker T)^{\perp}$ such that*

$$Th = \sum_{k=1}^{\infty} \mu_k \langle h, e_k \rangle e_k$$

for all h in $\mathcal{H}$.

Proof. We use the notation in the Spectral Theorem. Each $P_n \mathcal{H}$ is finite-dimensional, so pick an orthonormal basis for $P_n \mathcal{H}$ and let $\{e_k\}$ be a numbering of the union of these bases for all $n \geq 1$. Let $\mu_k = \lambda_n$ when $e_k \in P_n \mathcal{H}$. The formula for Th given in this corollary is precisely the same as the formula in (5.7.2). By the preceding corollary, $\{e_k\}$ is a basis for $(\ker T)^{\perp}$. ∎

Now we start the path to a proof of the Spectral Theorem. In some of the steps we take, we will prove results that apply to more than hermitian compact operators.

5.7.5. Proposition. *If T is a normal operator and $\lambda \in \mathbb{F}$, then $\ker(T - \lambda) = \ker(T^* - \bar{\lambda})$ and $\ker(T - \lambda)$ is a reducing subspace for T.*

Proof. The fact that $\ker(T - \lambda) = \ker(T^* - \bar{\lambda})$ follows from Proposition 5.4.17(b) and the fact that $T - \lambda$ is normal. If $h \in \ker(T - \lambda)$, then $Th = \lambda h \in \ker(T - \lambda)$; also $T^* h = \bar{\lambda} h \in \ker(T - \lambda)$. Therefore $\ker(T - \lambda)$ reduces T. ∎

5.7.6. Proposition. *If T is a normal operator and λ, μ are distinct eigenvalues, then $\ker(T - \lambda) \perp \ker(T - \mu)$.*

Proof. If $Th = \lambda h$ and $Tg = \mu g$, then the fact that $T^* g = \bar{\mu} g$ (5.7.5) implies that $\lambda \langle h, g \rangle = \langle Th, g \rangle = \langle h, T^* g \rangle = \mu \langle h, g \rangle$. Since $\lambda \neq \mu$, the only way this can happen is if $\langle h, g \rangle = 0$. ∎

5.7.7. Proposition. *If T is hermitian, then $\sigma_p(T) \subseteq \mathbb{R}$.*

Proof. If $h \in \ker(T - \lambda)$ and $h \neq 0$, then $\lambda h = Th = T^* h = \bar{\lambda} h$ by (5.7.5). Hence $(\lambda - \bar{\lambda}) h = 0$. Since $h \neq 0$, $\lambda - \bar{\lambda} = 0$. ∎

So far our steps toward the proof have been straightforward. Now we take a substantial one by showing that a hermitian compact operator has an eigenvalue that is as large as possible.

5.7.8. Proposition. *If T is a hermitian compact operator, then $\pm\|T\|$ is an eigenvalue.*

Proof. Without loss of generality we may assume that $T \neq 0$. By (5.4.14) there is a sequence of unit vectors $\{h_n\}$ in $\mathcal{H}$ such that $|\langle Th_n, h_n \rangle| \to \|T\|$; by passing to a subsequence we may assume that $\langle Th_n, h_n \rangle \to \lambda$, where $|\lambda| = \|T\|$. We will show that λ is an eigenvalue by using Proposition 5.6.12. In fact, since $|\lambda| = \|T\| \in \mathbb{R}$, we have that $\|(T - \lambda)h_n\|^2 = \|Th_n\|^2 - 2\lambda\langle Th_n, h_n \rangle + \lambda^2 \leq 2\lambda^2 - 2\lambda\langle Th_n, h_n \rangle \to 0$. ∎

Proof of the Spectral Theorem. We generate the eigenvalues by using an inductive process. By the preceding proposition, we have an eigenvalue λ_1 with $|\lambda_1| = \|T\|$. Put $\mathcal{P}_1 = \ker(T - \lambda_1)$ and let P_1 be the projection of $\mathcal{H}$ onto $\mathcal{P}_1$. According to Proposition 5.7.5, $\mathcal{P}_1$ reduces T; let $\mathcal{H}_2 = \mathcal{P}_1^{\perp}$ and put $T_2 = T|\mathcal{H}_2$. So T_2 is a compact hermitian operator (Verify). Again Proposition 5.7.8 implies there is an eigenvalue λ_2 for T_2 with $|\lambda_2| = \|T_2\|$. Note that $\lambda_2 \neq \lambda_1$ since $\mathcal{P}_1 \perp \mathcal{H}_2$. Put $\mathcal{P}_2 = \ker(T_2 - \lambda_2) = \ker(T - \lambda_2)$ and let P_2 be the projection of $\mathcal{H}$ onto $\mathcal{P}_2$. Since $\|T_2\| \leq \|T\|$, $|\lambda_2| \leq |\lambda_1|$. Put $\mathcal{H}_3 = [\mathcal{P}_1 \oplus \mathcal{P}_2]^{\perp}$; so $\mathcal{H}_3$ reduces T and we put $T_3 = T|\mathcal{H}_3$. Using induction (supply the details) we obtain a sequence of eigenvalues $\{\lambda_n\}$ of T such that the following hold:

(i) $|\lambda_1| \geq |\lambda_2| \geq \cdots$;
(ii) $\lambda_n \neq \lambda_m$ for $n \neq m$;
(iii) if $\mathcal{P}_n = \ker(T - \lambda_n)$, then $|\lambda_{n+1}| = \|T|[\mathcal{P}_1 \oplus \cdots \oplus \mathcal{P}_n]^{\perp}\|$.

By (i) there is a scalar $\alpha \geq 0$ such that $|\lambda_n| \to \alpha$. If e_n is a unit vector in $\mathcal{P}_n$, then the fact that T is compact implies there is a subsequence $\{e_{n_k}\}$ and a vector h in $\mathcal{H}$ such that $Te_{n_k} \to h$. Since the e_{n_k} are eigenvectors, $\|Te_{n_k} - Te_{n_j}\|^2 = \lambda_{n_k}^2 + \lambda_{n_j}^2 \geq 2\alpha^2$. But $\{Te_{n_k}\}$ is convergent, so it must be that $\alpha = 0$.

If $1 \leq k \leq n$ and $h \in \mathcal{P}_k$, then $(T - \sum_{k=1}^{n} \lambda_k P_k) h = 0$. That is $\mathcal{P}_1 \oplus \cdots \oplus \mathcal{P}_n \subseteq \ker\left(T - \sum_{k=1}^{n} \lambda_k P_k\right)$. Moreover $\mathcal{P}_1 \oplus \cdots \oplus \mathcal{P}_n$ reduces $(T - \sum_{k=1}^{n} \lambda_k P_k)$; thus

$$\left\|\left(T - \sum_{k=1}^{n} \lambda_k P_k\right)\right\| = \left\|\left(T - \sum_{k=1}^{n} \lambda_k P_k\right) | [\mathcal{P}_1 \oplus \cdots \oplus \mathcal{P}_n]^{\perp}\right\|$$

But if $h \in [\mathcal{P}_1 \oplus \cdots \oplus \mathcal{P}_n]^\perp$, then $(T - \sum_{k=1}^n \lambda_k P_k) h = Th$. Therefore

$$\left\| \left(T - \sum_{k=1}^n \lambda_k P_k \right) \right\| = \left\| T \mid [\mathcal{P}_1 \oplus \cdots \oplus \mathcal{P}_n]^\perp \right\| = |\lambda_{n+1}| \to 0$$

This says that the series $\sum_{n=1}^\infty \lambda_n P_n$ converges to T in the norm of $\mathcal{B}(\mathcal{H})$. ∎

Exercises.

(1) Determine the decomposition (5.7.2) for a diagonalizable (5.6.5) compact operator.

(2) Can a compact hermitian operator have an infinite-dimensional kernel? If so, give an example; if not, prove it.

(3) Adopt the notation of Proposition 5.6.7 and assume $k(x, y) = \overline{k(y, x)}$. Show that K is hermitian. If $\{\mu_n\}$ are the eigenvalues of K each repeated $\dim \ker(K - \mu_n)$ times, show that $\sum_{n=1}^\infty |\mu_n|^2 < \infty$.

(4) With the notation of Corollary 5.7.4, show that $h \in \operatorname{ran} T$ if and only if $h \perp \ker T$ and $\sum_{n=1}^\infty \mu_n^{-2} |\langle h, e_n \rangle|^2 < \infty$.

(5) With the notation of Corollary 5.7.4, if $\lambda \neq 0$ and $\lambda \neq \mu_n$ for any $n \geq 1$, show that $T - \lambda$ is bijective and find a formula for $(\lambda - T)^{-1} h$.

5.8. Some applications of the Spectral Theorem*

Here we give some applications of the Spectral Theorem, including its extension to compact normal operators acting on a complex Hilbert space.

5.8.1. Theorem. *If $A \in \mathcal{B}(\mathcal{H})$, T is a compact hermitian operator, and we adopt the notation in the Spectral Theorem, then $AT = TA$ if and only if for each $n \geq 1$, $\operatorname{ran} P_n$ is a reducing subspace for A.*

Proof. Assume $AT = TA$. If $h \in \operatorname{ran} P_n = \ker(T - \lambda_n)$, then $TAh = ATh = \lambda_n Ah$; so $\operatorname{ran} P_n$ is invariant for A. On the other hand $A^*T = (TA)^* = (AT)^* = TA^*$. So the same argument shows that $\operatorname{ran} P_n$ is also invariant for A^*; that is, $\operatorname{ran} P_n$ reduces A.

For the converse assume each $\operatorname{ran} P_n$ reduces A; but $\bigoplus_{n=1}^\infty \operatorname{ran} P_n$ reduces A and hence so does the orthogonal complement of this space, which is precisely $\ker T$. Put $\mathcal{H}_0 = \ker T$ and $\lambda_0 = 0$. Thus for any vector h in $\mathcal{H}$, $TAh = TA \sum_{n=0}^\infty P_n h = \sum_{n=0}^\infty TAP_n h = \sum_{n=0}^\infty \lambda_n AP_n h = A \sum_{n=0}^\infty \lambda_n P_n h = ATh$. ∎

It is good to keep in mind the fact that when each $\operatorname{ran} P_n$ reduces an operator A, so does $\ker T$, as was seen in the preceding proof. It may surprise the reader that we can now extend the Spectral Theorem to compact normal operators.

5.8.2. Theorem (The Spectral Theorem for compact normal operators). *If T is a compact normal operator on a complex Hilbert space $\mathcal{H}$, then T has a countable number of distinct non-zero eigenvalues, $\{\lambda_1, \lambda_2, \dots\}$ in $\mathbb{C}$, where $\lambda_n \to 0$. If P_n is the projection onto $\ker(T - \lambda_n)$, then P_n has finite rank, $P_n P_m = P_m P_n$ for all n, m, this product being 0 when $n \neq m$. If $\lambda_0 = 0$ and P_0 is the projection onto $\ker T = \left[\bigvee_{n \geq 1} P_n \mathcal{H} \right]^{\perp}$, then*

5.8.3
$$T = \sum_{n=0}^{\infty} \lambda_n P_n$$

where convergence of this series is in the norm of $\mathcal{B}(\mathcal{H})$.

Proof. By Exercise 5.4.6 we can write $T = A + iB$, where A and B are hermitian. Since T is compact, so are A and B. By the same exercise, $AB = BA$ since T is normal. Applying the Spectral Theorem to A we get $A = \sum_{n=1}^{\infty} \alpha_n P_n$; also let $\mathcal{A}_0 = \ker A$ and put $\mathcal{A}_n = \ker(A - \alpha_n) = \operatorname{ran} P_n$ for $n \geq 1$. By Theorem 5.8.1, each $\mathcal{A}_n$ reduces B. If $B_n = B|\mathcal{A}_n$ for $n \geq 0$, then each B_n is a hermitian compact operator and we can apply the Spectral Theorem to it. (The reader is urged to justify each of the statements made here and for the rest of the proof, as details are omitted.) Thus for $n \geq 0$, $B_n = \sum_{k=1}^{m_n} \beta_{nk} Q_{nk}$, where the β_{nk} are distinct eigenvalues of B_n and Q_{nk} are pairwise orthogonal projections having finite rank when $n \geq 1$. Here we have to depart somewhat from the notation in the Spectral Theorem for hermitian compact operators, since it could be that some of the operators B_n have a non-trivial kernel. So allow the possibility that some of the eigenvalues β_{nk} are 0. We also note that for $n \geq 1$, m_n is finite and $\sum_{k=1}^{m_n} Q_{nk} = P_n$. For $n = 0$ we put $B_0 = \sum_{k=0}^{m_0} \beta_{0k} Q_{0k}$, where $\beta_{00} = 0$ and it may be that $m_0 = \infty$. Putting $\alpha_0 = 0$, we have

$$\sum_{n=0}^{\infty} \sum_{k=1}^{m_n} (\alpha_n + i\beta_{nk}) Q_{nk} = A + iB = T$$

It is left to the reader to verify the remaining parts of the theorem. ∎

It is necessary for the Hilbert space in the preceding theorem to be complex. To begin with, in the proof we cannot make sense of the decomposition $T = A + iB$ unless we have a complex Hilbert space. Moreover, consider the operator acting on $\mathbb{R}^2$ defined by the matrix

5.8.4
$$T = \begin{bmatrix} 0 & 1 \\ -1 & 0 \end{bmatrix}$$

The reader can check that it is normal, but T has no eigenvalues.

The Spectral Theorem allows us to define a functional calculus for compact normal operators on a complex Hilbert space. If T is such an operator and (5.8.3) holds, with the eigenvalues λ_n in $\mathbb{C}$, let P_0 be the projection on $\ker T = [\bigvee_n \operatorname{ran} P_n]^\perp$. For every bounded function $\phi : \mathbb{C} \to \mathbb{C}$, we define the operator $\phi(T)$ by

5.8.5
$$\phi(T) = \phi(0)P_0 + \sum_{n=1}^{\infty} \phi(\lambda_n)P_n$$

5.8.6. Theorem (Functional Calculus for Compact Normal Operators). *If T is a compact normal operator on a complex Hilbert space, then the map $\phi \mapsto \phi(T)$ as defined in (5.8.5) has the following properties.*

(a) *$\phi \mapsto \phi(T)$ is a multiplicative linear map of $\ell^\infty(\mathbb{C})$ into $\mathcal{B}(\mathcal{H})$. If $\phi(z) = 1$ on $\sigma_p(T) \cup \{0\}$, then $\phi(T) = 1$; if $\phi(z) = z$ on $\sigma_p(T) \cup \{0\}$, then $\phi(T) = T$.*

(b) *For every ϕ in $\ell^\infty(\mathbb{C})$, $\|\phi(T)\| = \sup\{|\phi(z)| : z \in \sigma_p(T) \cup \{0\}\}$.*

(c) *If for each ϕ in $\ell^\infty(\mathbb{C})$ we define $\phi^*(z) = \overline{\phi(z)}$, then $\phi(T)^* = \phi^*(T)$.*

(d) *If $A \in \mathcal{B}(\mathcal{H})$ and $AT = TA$, then $A\phi(T) = \phi(T)A$ for every ϕ in $\ell^\infty(\mathbb{C})$.*

Proof. For convenience, let $\lambda_0 = 0$.

(a) If $\phi, \psi \in \ell^\infty(\mathbb{C})$, then, since $P_n P_m = 0$ when $n \neq m$,

$$\phi(T)\psi(T) = \left(\sum_{n=0}^{\infty} \phi(\lambda_n)P_n \right) \left(\sum_{n=0}^{\infty} \psi(\lambda_n)P_n \right) = \sum_{n=0}^{\infty} \phi(\lambda_n)\psi(\lambda_n)P_n$$

This last operator is precisely equal to $(\phi\psi)(T)$. The rest of the proof of (a) is left as an exercise.

(b) For any h in $\mathcal{H}$, $\|\phi(T)h\|^2 = \sum_{n=0}^{\infty} |\phi(\lambda_n)|^2 \|P_n h\|^2$, since the vectors $P_n h$ are pairwise orthogonal. But this implies that if $M = \sup\{|\phi(\lambda_n)| : n \geq 0\}$, then $\|\phi(T)h\|^2 \leq M^2 \sum_{n=0}^{\infty} \|P_n h\|^2 = M^2 \|h\|^2$; so $\|\phi(T)\| \leq M$. On the other hand if h is a unit vector in $\operatorname{ran} P_n$, then $\|\phi(T)\| \geq \|\phi(T)h\| = |\phi(\lambda_n)|$; hence $\|\phi(T)\| \geq M$.

(c) For $h, g \in \mathcal{H}$,

$$\langle \phi(T)^* h, g \rangle = \langle h, \phi(T)g \rangle$$
$$= \sum_{n=0}^{\infty} \overline{\phi(\lambda_n)} \langle h, P_n g \rangle$$
$$= \sum_{n=0}^{\infty} \overline{\phi(\lambda_n)} \langle P_n h, g \rangle$$
$$= \left\langle \sum_{n=0}^{\infty} \overline{\phi(\lambda_n)} P_n h, g \right\rangle$$
$$= \langle \phi^*(T)h, g \rangle$$

Since this holds for all h and g, we have the desired equality.

(d) Just as in the proof of Theorem 5.8.1 we know that if $AT = TA$, then $AP_n = P_n A$ for all $n \geq 0$. Therefore the proof of (d) is a simple matter of distributing A over the sum $\sum_{n=0}^{\infty} \phi(\lambda_n) P_n$. ∎

In §11.4 this functional calculus is extended to arbitrary normal operators. When presented with a result such as the functional calculus for compact normal operators that associates with T a collection of additional operators, it is good to know if there is some internal characterization of the collection of operators so obtained. Here is that characterization.

5.8.7. Theorem. *If T is a compact normal operator on a complex Hilbert space $\mathcal{H}$ and $B \in \mathcal{B}(\mathcal{H})$, then there is a function ϕ in $\ell^{\infty}(\mathbb{C})$ such that $B = \phi(T)$ if and only if $BA = AB$ for every operator A such that $AT = TA$.*

Proof. If $B = \phi(T)$ for some function ϕ, then Theorem 5.8.6(d) implies $BA = AB$ whenever $AT = TA$. Now assume $BA = AB$ whenever $AT = TA$. Since $P_n T = TP_n$ for all $n \geq 0$, ran P_n reduces B; let $B_n = B|\text{ran } P_n$. Let $A_n \in \mathcal{B}(\text{ran } P_n)$ and define an operator A on all of $\mathcal{H}$ by setting $A|\text{ran } P_n = A_n$, $A|\text{ran } P_m = 0$ when $m \neq n$, and extending by linearity. Since $T|\text{ran } P_n$ is a multiple of the identity, it easily follows that $AT = TA$. Therefore $BT = TB$ and this implies that $B_n A_n = A_n B_n$ for every A_n in $\mathcal{B}(\text{ran } P_n)$. It follows (Exercise 1) that there is a scalar β_n such that $B_n = \beta_n P_n$; hence $B = \bigoplus_{n=0}^{\infty} \beta_n P_n$. If we define $\phi(\lambda_n) = \beta_n$ for $n \geq 0$ and $\phi(z) = 0$ when $z \notin \sigma_p(T) \cup \{0\}$, then $B = \phi(T)$. ∎

Which functions ϕ give that $\phi(T)$ is compact? The perspicacious reader may have suspected something close to the answer; in any case, the proof of the next result is left as an exercise.

5.8.8. Proposition. *Let T be a compact normal operator on a complex Hilbert space and let $\phi \in \ell^{\infty}(\mathbb{C})$.*

(a) *If $\dim \ker T = \infty$, then $\phi(T)$ is compact if and only if $\phi(0) = 0$ and $\phi(\lambda_n) \to 0$.*

(b) *If $\dim \ker T < \infty$, then $\phi(T)$ is compact if and only if $\phi(\lambda_n) \to 0$.*

5.8.9. Definition. An operator T on $\mathcal{H}$ is *positive* if $\langle Th, h \rangle \geq 0$ for all h in $\mathcal{H}$. In notation we write $T \geq 0$.

The reader may encounter the term *positive definite* in the literature. This means that T is positive in the sense just defined as well as invertible. We won't use this term.

By Proposition 5.4.13, on a complex Hilbert space every positive operator is hermitian. By again examining our friend in (5.8.4) we see that when the Hilbert space is not complex, a positive operator need not be hermitian. We might also note that the notion of positivity will be more fully explored in a more general context later in §11.3.

5.8.10. Proposition. *If T is a compact normal operator on a complex Hilbert space, then T is positive if and only if all its non-zero eigenvalues are positive.*

Proof. Let $T = \sum_{n=0}^{\infty} \lambda_n P_n$ as in the Spectral Theorem. If $T \geq 0$ and $h \in \operatorname{ran} P_n$, then $0 \leq \langle Th, h \rangle = \lambda_n \|h\|^2$, which implies $\lambda_n \geq 0$. Now assume that $\lambda_n \geq 0$ for all $n \geq 0$. We have $\langle Th, h \rangle = \langle \sum_{n=0}^{\infty} \lambda_n P_n h, \sum_{n=0}^{\infty} P_n h \rangle = \sum_{n=0}^{\infty} \lambda_n \|P_n\|^2 \geq 0$. ∎

Now we put the functional calculus to work. We might point out that when the compact operator T is hermitian, we do not need to assume we have a complex Hilbert space as long as we only use real-valued functions in $\ell^{\infty}(\mathbb{R})$.

5.8.11. Theorem. *If T is a hermitian compact operator, then there are positive hermitian compact operators A and B such that $T = A - B$.*

Proof. Define $\phi_{\pm} : \mathbb{R} \to [0, \infty)$ by $\phi_{\pm}(x) = \pm x$ when $\pm x \geq 0$ and $\phi_{\pm}(x) = 0$ otherwise. Put $A = \phi_+(T)$ and $B = \phi_-(T)$. Since $\phi_+(x) - \phi_-(x) = x$ for all x, Theorem 5.8.6 implies that $T = A - B$. Proposition 5.8.8 implies that both A and B are compact. The non-zero eigenvalues of both operators are all positive, so both are positive operators (5.8.10). ∎

5.8.12. Theorem. *If T is a positive hermitian compact operator, then there is a unique positive hermitian operator A such that $T = A^2$.*

Proof. Let $\phi(x) = \sqrt{x}$ for $x \geq 0$ and $\phi(x) = 0$ elsewhere; put $A = \phi(T)$ and check that it has the properties desired. The proof of uniqueness is not difficult (Exercise 11). ∎

Exercises.

(1) If $B \in \mathcal{B}(\mathcal{H})$ and $BT = TB$ for every T in $\mathcal{B}_0(\mathcal{H})$, show that T is a scalar multiple of the identity.

(2) In the functional calculus for compact normal operators, what is the kernel of the map $\phi \mapsto \phi(T)$?

(3) Let T be a compact normal operator on the complex Hilbert space $\mathcal{H}$ and let $X = \sigma_p(T) \cup \{0\}$. Show that $\phi \mapsto \phi(T)$ is a Banach space isometry between $\ell^{\infty}(X)$ and $\{B \in \mathcal{B}(\mathcal{H}) : BA = AB \text{ for every } A \text{ such that } AT = TA\}$.

(4) If T is a compact normal operator on a complex Hilbert space such that $\dim \ker(T - \lambda) \leq 1$ for all λ in $\mathbb{C}$, show that when $A \in \mathcal{B}(\mathcal{H})$ and $AT = TA$, then there is a function ϕ in $\ell^\infty(\mathbb{C})$ such that $A = \phi(T)$.

(5) Prove a converse to the preceding exercise by showing that if $\{\phi(T) : \phi \in \ell^\infty(\mathbb{C})\} = \{A \in \mathcal{B}(\mathcal{H}) : AT = TA\}$, then $\dim \ker(T - \lambda) \leq 1$ for all λ in $\mathbb{C}$.

(6) (a) If T is a compact normal operator on a complex Hilbert space such that $\dim \ker(T - \lambda) \leq 1$ for all λ in $\mathbb{C}$, show that there is a vector h in $\mathcal{H}$ such that $\mathcal{H} = \mathrm{cl}\,\{p(T)h : p \text{ is a polynomial}\}$. Such a vector h is called a *cyclic vector* for T. (b) Find all the cyclic vectors for such an operator T.

(7) Prove Proposition 5.8.8.

(8) Discuss the uniqueness of the operators A and B in Theorem 5.8.11.

(9) If T is a compact normal operator on a complex Hilbert space and $n \in \mathbb{N}$, show that there is a compact normal operator A such that $T = A^n$. Discuss the uniqueness of such an operator A. Is there an operator B such that $T = e^B$?

(10) If T is a compact normal operator on a complex Hilbert space, show that there is a compact positive operator A and a unitary operator U such that $T = UA = AU$. Discuss the uniqueness of A and U.

(11) Prove the uniqueness part of Theorem 5.8.12.

(12) Let T be a compact normal operator on a complex Hilbert space. (a) Show that the functional calculus $\phi \mapsto \phi(T)$ satisfies the following continuity property: if $\phi, \phi_1, \phi_2, \ldots \in \ell^\infty(\mathbb{C})$ such that $\phi_n(\lambda) \to \phi(\lambda)$ for every λ in $\mathbb{C}$ and $\sup_n \|\phi_n\|_\infty < \infty$, then $\langle \phi_n(T)h, g \rangle \to \langle \phi(T)h, g \rangle$ for all h, g in $\mathcal{H}$. (b) Show that the functional calculus is unique in the following sense. If $\tau : \ell^\infty(\mathbb{C}) \to \mathcal{B}(\mathcal{H})$ is a multiplicative linear map satisfying: (i) $\tau(1) = 1$; (ii) $\tau(\psi) = T$, where $\psi(z) = z$ for all z in $\mathbb{C}$; (iii) $\|\tau(\phi)\| = \sup\{|\phi(\lambda)| : \lambda \in \sigma_p(T) \cup \{0\}\}$ for all ϕ in $\ell^\infty(\mathbb{C})$; (iv) when $\phi, \phi_1, \phi_2, \ldots \in \ell^\infty(\mathbb{C})$ such that $\sup_n \|\phi_n\|_\infty < \infty$ and $\phi_n(\lambda) \to \phi(\lambda)$ for every λ in $\mathbb{C}$, then $\langle \tau(\phi_n)h, g \rangle \to \langle \tau(\phi)h, g \rangle$ for all h, g in $\mathcal{H}$; then $\tau(\phi) = \phi(T)$ for every ϕ in $\ell^\infty(\mathbb{C})$.

5.9. Unitary equivalence*

In mathematics we seek characterizations of equivalence classes. In matrix theory, for example, canonical Jordan forms characterize when two matrices are similar. If you will, we associate to each square matrix a series of pairs consisting of a scalar and an integer indicating the size of the associate

Jordan block. We then get that a necessary and sufficient condition for two matrices to be similar is that their two sequences of pairs are the same after a permutation. In this section we introduce one of the basic forms of equivalence for operators on a Hilbert space. We will then see how the Spectral Theorem allows us to attach to each compact normal operator a certain function and such functions parametrize the equivalence classes of such operators.

In this section, all Hilbert spaces are over $\mathbb{C}$. The next definition is easily seen to be an equivalence relation between operators on separable Hilbert spaces.

5.9.1. Definition. If $T \in \mathcal{B}(\mathcal{H})$ and $S \in \mathcal{B}(\mathcal{K})$, say that T and S are *unitarily equivalent* if there is an isomorphism (unitary) $U : \mathcal{H} \to \mathcal{K}$ such that $UTU^* = S$. In symbols we write $T \cong S$.

If T is a compact normal operator, define its *multiplicity function* $m_T : \mathbb{C} \to \mathbb{N} \cup \{0, \infty\}$ by $m_T(\lambda) = \dim(T - \lambda)$ for all λ in $\mathbb{C}$.

5.9.2. Proposition. *If T is a compact normal operator, then $\{\lambda \in \mathbb{C} : m_T(\lambda) \neq 0\}$ is either a finite set, a sequence of non-zero points in $\mathbb{C}$ that converges to 0, or such a sequence together with its limit point $\{0\}$ and whenever $\lambda \neq 0$, $m_T(\lambda) \in \mathbb{N} \cup \{0\}$. Moreover, if any function $m : \mathbb{C} \to \mathbb{N} \cup \{0, \infty\}$ has such properties, then there is a compact normal operator T such that $m_T = m$.*

Proof. The first part of this proposition follows from the material presented in the preceding sections. Now suppose we have any function $m : \mathbb{C} \to \mathbb{N} \cup \{0, \infty\}$ having the properties in the statement of the proposition, and let $\{\lambda : m(\lambda) \neq 0\} = \{0 = \lambda_0, \lambda_1, \lambda_2, \dots\}$. (We are assuming here that $m(0) \neq 0$; the proof when this is not the case is easily obtained from the proof of this case.) For each $n \geq 0$, let $\mathcal{H}_n$ be a Hilbert space with $\dim \mathcal{H}_n = m(\lambda_n)$. Put $\mathcal{H} = \bigoplus_{n=0}^{\infty} \mathcal{H}_n$ and define T on $\mathcal{H}$ by $T(h_0 \oplus h_1 \oplus h_2 \oplus \cdots) = 0 \oplus \lambda_1 h_1 \oplus \lambda_2 h_2 \oplus \cdots$. If P_n is the projection of $\mathcal{H}$ onto $\mathcal{H}_n$, it is easy to check that $T = \sum_{n=1}^{\infty} \lambda_n P_n$. Because $\lambda_n \to 0$, it follows that this series converges to T in the operator norm; since the finite partial sums are finite rank operators, it follows that T is the limit of a sequence of finite rank operators and is, therefore, compact. The fact that it is normal is easily checked. Since $\ker(T - \lambda_n) = \mathcal{H}_n$, $m_T = m$. ∎

5.9.3. Theorem. *If T and S are compact normal operators, then $T \cong S$ if and only if $m_T = m_S$.*

Proof. If U is a unitary such that $UTU^* = S$, it follows that $U[\ker(T-\lambda)] = \ker(S - \lambda)$ for every λ in $\mathbb{C}$; therefore $m_T = m_S$. For the converse, let $T = \sum_{n=0}^{\infty} \lambda_n P_n$ be the spectral decomposition of T, where $\lambda_0 = 0$. If we assume that $m_T = m_S$, then $\lambda_0, \lambda_1, \ldots$ are the only complex numbers where m_S takes a non-zero value. Therefore S has a spectral decomposition $S = \sum_{n=0}^{\infty} \lambda_n Q_n$ and $\dim \operatorname{ran} P_n = \dim \operatorname{ran} Q_n$ for each n since $m_T(\lambda_n) = m_S(\lambda_n)$ for all n. Thus for each $n \geq 0$ there is an isomorphism $U_n : \operatorname{ran} P_n \to \operatorname{ran} Q_n$. Define $U : \mathcal{H} \to \mathcal{K}$ by $Uh = \sum_{n=0}^{\infty} U_n(P_n h)$. It is routine that U is an isomorphism and $UTU^* = S$. ∎

In §11.6.15 we will see a characterization of the unitary equivalence classes of all normal operators, of which the preceding is a special case. The latter characterization involves measures as well as a multiplicity function. (When applied to compact normal operators the measures are purely atomic with atoms at the eigenvalues of the compact operator.)

Exercises.

(1) Show that unitary equivalence is an equivalence relation on $\mathcal{B}(\mathcal{H})$.

(2) Show that the weighted shift with weight sequence $\{\alpha_n\}$ in Example 5.4.7 is unitarily equivalent to the one with weight sequence $\{|\alpha_n|\}$.

(3) Let $U : \mathcal{H} \to \mathcal{H}$ be an isomorphism and define $\rho : \mathcal{B}(\mathcal{H}) \to \mathcal{B}(\mathcal{K})$ by $\rho(T) = UTU^*$. Prove the following. (a) ρ is an isomorphism of the algebra $\mathcal{B}(\mathcal{H})$, $\rho(T)^* = \rho(T^*)$, and ρ is an isometry. (b) $\rho(T)$ is compact if and only if T is compact. (c) For T and A in $\mathcal{B}(\mathcal{H})$, $\rho(T)\rho(A) = \rho(A)\rho(T)$ if and only if $AT = TA$. (d) If $\mathcal{M} \leq \mathcal{H}$, then $\mathcal{M}$ is an invariant subspace for T if and only if $U\mathcal{M}$ is an invariant subspace for $\rho(T)$. (e) If $\mathcal{M} \leq \mathcal{H}$, then $\mathcal{M}$ is a reducing subspace for T if and only if $U\mathcal{M}$ is a reducing subspace for $\rho(T)$.

(4) Show that two projections P and Q are unitarily equivalent if and only if $\dim \operatorname{ran} P = \dim \operatorname{ran} Q$ and $\dim \ker P = \dim \ker Q$.

(5) Say that an operator T on $\mathcal{H}$ is *irreducible* if the only reducing subspaces for T are $\mathcal{H}$ and (0). (a) Prove that the Volterra operator and the unilateral shift are irreducible operators. (b) Show that for a measure μ and ϕ in $L^\infty(\mu)$, M_ϕ is never irreducible. (c) Suppose that $T = \bigoplus_{n=1}^{\infty} T_n$ and $S = \bigoplus_{n=1}^{\infty} S_n$, where each T_n and S_n is irreducible. Show that $T \cong S$ if and only if there is a bijection $\tau : \mathbb{N} \to \mathbb{N}$ such that $T_n \cong S_{\tau(n)}$ for each $n \geq 1$.

(6) Define T on $L^2[0,1]$ by $Tf(x) = xf(x)$ for all x. Show that $T \cong T^2$.

Chapter 6

Banach Spaces

In §1.3 we saw the definition and elementary properties of normed and Banach spaces. Later in §1.5 we initiated the study of bounded linear functionals. In this chapter we will go more deeply into these subjects and establish the basics of the theory of Banach spaces.

6.1. Finite-dimensional spaces

In any part of functional analysis it is a good idea to see what any concept means in a finite-dimensional space. In a sense this is the same with the study of any new mathematics – how does it relate to what we already know. Even more important, however, from what we learn about finite-dimensional normed spaces we will derive additional information about all normed spaces.

Before examining finite-dimensional spaces, we introduce a new concept that will be used in §6.3 and examined more extensively in the next chapter. At present we will only use one aspect of this material, but efficiency dictates introducing the concept now.

6.1.1. Definition. If $\mathcal{X}$ is a vector space over $\mathbb{F}$, a *seminorm* is a function $p : \mathcal{X} \to [0, \infty)$ that satisfies: (a) $p(x + y) \leq p(x) + p(y)$ for all x, y in $\mathcal{X}$; (b) $p(\alpha x) = |\alpha| p(x)$ for all x in $\mathcal{X}$ and all scalars α.

Note that as a consequence of (b), $p(0) = 0$. However it may be that there are vectors $x \neq 0$ such that $p(x) = 0$; hence the prefix "semi".

6.1.2. Example. (a) If X is a locally compact space, $x \in X$, and $p : C(X) \to [0, \infty)$ is defined by $p(f) = |f(x)|$, then p is a seminorm. More generally, if μ is a finite positive measure, then $p(f) = \left| \int f d\mu \right|$ is a seminorm.

(b) If X is as in (a), K is a compact subset of X, and we define $p(f) = \sup\{|f(x)| : x \in K\}$ for every f in $C(X)$, then p is a seminorm on $C(X)$.

(c) If $\mathcal{X}$ is any normed space and L is a bounded linear functional on $\mathcal{X}$, then $p_L(x) = |L(x)|$ defines a seminorm on $\mathcal{X}$.

Also note that any finite linear combination of seminorms is a seminorm as long as we only use positive coefficients. In §7.1 we'll look at seminorms more closely. What we will actually study is a vector space with a whole collection of seminorms; for example, the spaces in (6.1.2) with all the seminorms defined by those processes. One thing we will want to do is compare seminorms, but we also want to compare two norms on a vector space. The next lemma is the needed tool.

6.1.3. Lemma. *If p and q are two seminorms of a vector space $\mathcal{X}$, then the following statements are equivalent.*

(a) $p \le q$.

(b) $\{x : q(x) < 1\} \subseteq \{x : p(x) < 1\}$.

(c) $\{x : q(x) \le 1\} \subseteq \{x : p(x) \le 1\}$.

(d) $\{x : q(x) < 1\} \subseteq \{x : p(x) \le 1\}$.

Proof. We'll only prove the equivalence of (a) and (b) and leave the remainder as an exercise. The fact that (a) implies (b) is clear. Assume (b) holds and let $x \in \mathcal{X}$ with $\alpha = q(x)$. If $\epsilon > 0$, then $q((\alpha + \epsilon)^{-1}x) < 1$, so that $p((\alpha + \epsilon)^{-1}x) < 1$ and hence that $p(x) < \alpha + \epsilon = q(x) + \epsilon$. Since ϵ was arbitrary, part (a) follows. ∎

To discuss finite-dimensional spaces, or any vector spaces for that matter, it sometimes helps to be able to compare norms. We therefore start with the following. If $\mathcal{X}$ is a vector space over $\mathbb{F}$ and $\|\cdot\|_1$ and $\|\cdot\|_2$ are two norms on $\mathcal{X}$, we say they are *equivalent norms* if they define the same convergent sequences. That is, $\|x_n - x\|_1 \to 0$ if and only if $\|x_n - x\|_2 \to 0$. The fact that the convergent sequences are the same for the two norms means that the two metrics define the same collection of open sets. That is, a set is open relative to the first norm if and only if it is open relative to the second norm. The reader might want to examine Exercise 1 and compare the concepts of equivalent norms and equivalent metrics, especially as it relates to the next proposition.

6.1.4. Proposition. *Two norms $\|\cdot\|_1$ and $\|\cdot\|_2$ on a vector space are equivalent if and only if there are constants c and C such that*

$$c\|x\|_1 \le \|x\|_2 \le C\|x\|_1$$

for all x in $\mathcal{X}$.

Proof. If there are constants c, C with $c\|x\|_1 \leq \|x\|_2 \leq C\|x\|_1$ for all x in $\mathcal{X}$, then $c\|x_n - x\|_1 \leq \|x_n - x\|_2 \leq C\|x_n - x\|_1$ and it is easy to see that $\|x_n - x\|_1 \to 0$ if and only if $\|x_n - x\|_2 \to 0$. Now assume the two norms are equivalent. Hence, as discussed before the statement of this proposition, $B = \{x : \|x\|_1 < 1\}$ is an open set in the topology defined by $\| \cdot \|_2$ and contains 0; thus there is a $c > 0$ with $\{x : \|x\|_2 < c\} \subseteq B$. By the preceding lemma, $\|x\|_1 \leq c^{-1}\|x\|_2$ for all x, giving us one of the inequalities. The existence of C is obtained in a similar way. ∎

6.1.5. Theorem. *On a finite-dimensional space, any two norms are equivalent.*

Proof. Let $\mathcal{X}$ be the finite-dimensional space and let $\| \cdot \|$ be any norm on $\mathcal{X}$. Fix a basis $\{e_1, \ldots, e_d\}$ for $\mathcal{X}$ and define a new norm on $\mathcal{X}$ by $\|\sum_{k=1}^{d} x_k e_k\|_\infty = \max_k |x_k|$. (Verify that this is a norm.) We'll prove that the arbitrary norm is equivalent to this norm $\| \cdot \|_\infty$ and this proves the theorem. First note that if $C = \sum_{k=1}^{d} \|e_k\|$ and $x = \sum_{k=1}^{d} x_k e_k$, then $\|x\| \leq \sum_{k=1}^{d} |x_k|\|e_k\| \leq C\|x\|_\infty$.

Now let $\mathcal{T}$ be the topology defined by the given norm and let $\mathcal{T}_\infty$ be the topology defined by the norm $\| \cdot \|_\infty$. What we have just proved is that the identity map $i : (\mathcal{X}, \mathcal{T}_\infty) \to (\mathcal{X}, \mathcal{T})$ is continuous so that $\mathcal{T} \subseteq \mathcal{T}_\infty$. If $B = \{x : \|x\|_\infty \leq 1\}$, then B is $\mathcal{T}_\infty$-compact (Why?) and hence it is $\mathcal{T}$-compact and the two topologies agree on B since the subsets of B that are closed in either topology are the same. Thus if $A = \{x : \|x\|_\infty < 1\}$, then A is open in $(B, \mathcal{T})$ and there is a set U in $\mathcal{T}$ with $U \cap B = A$. Since $0 \in U$, there is an $r > 0$ such that $\{x : |x\| < r\} \subseteq U$. That is,

6.1.6 $\qquad \|x\| < r$ and $\|x\|_\infty \leq 1$ implies $\|x\|_\infty < 1$.

Claim. $\|x\| < r$ implies $\|x\|_\infty < 1$.

In fact suppose $x = \sum_{k=1}^{d} x_k e_k$ satisfies $\|x\| < r$ and put $\alpha = \|x\|_\infty$; so $\|\alpha^{-1}x\|_\infty = 1$. If $\alpha \geq 1$, then $\|\alpha^{-1}x\| < \alpha^{-1}r \leq r$ and so we have that $\|\alpha^{-1}x\|_\infty < 1$ by (6.1.6), a contradiction. Thus $\alpha < 1$ and we have established the claim.

If $x \in \mathcal{X}$, $\beta = \|x\|$, and $\epsilon > 0$, then $\|r(\beta+\epsilon)^{-1}x\| < r$, so the claim implies that $\|r(\beta + \epsilon)^{-1}x\|_\infty < 1$. It follows that $\|x\|_\infty \leq r^{-1}\|x\|$ for all x in $\mathcal{X}$ and so by Proposition 6.1.4 we have that the two norms are equivalent. ∎

6.1.7. Corollary. *A finite-dimensional subspace of any normed space is closed.*

Proof. Let $\mathcal{M}$ be a finite-dimensional subspace of the normed space $\mathcal{X}$. If we define the norm $\| \cdot \|_\infty$ on $\mathcal{M}$ as in the preceding proof, we see that with respect to this norm $\mathcal{M}$ is a complete metric space. Since the norm $\| \cdot \|_\infty$

is equivalent to the given norm, $\mathcal{M}$ is complete in its original norm and therefore must be closed. ∎

6.1.8. Corollary. *If $\mathcal{X}$ is a finite-dimensional normed space and $\mathcal{Y}$ is any normed space, then every linear transformation $T : \mathcal{X} \to \mathcal{Y}$ is continuous.*

Proof. Again let $\{e_1, \ldots, e_d\}$ be a basis for $\mathcal{X}$ and define the norm $\| \cdot \|_\infty$ as in the proof of Theorem 6.1.5. By that result we need only prove that there is a constant M such that $\|Tx\| \leq M\|x\|_\infty$. If $T : \mathcal{X} \to \mathcal{Y}$ is a linear transformation, $\left\| T \left(\sum_{k=1}^{d} x_k e_k \right) \right\| \leq \sum_{k=1}^{d} |x_k| \|Te_k\| \leq M\|x\|_\infty$, where $M = \sum_{k=1}^{d} \|Te_k\|$. ∎

Let's point out that the last corollary requires the domain of the linear transformation to be finite-dimensional. If instead we only require the range space $\mathcal{Y}$ to be finite-dimensional, then there are discontinuous linear transformations from an infinite-dimensional normed space into $\mathcal{Y}$. For example see Exercise 1.5.3.

Exercises.

(1) If X is a set, say that two metrics d_1 and d_2 are equivalent if they define the same convergent sequences; equivalently, they both define the same collection of open sets. (a) Show that if there are constants c and C such that $cd_1(x, y) \leq d_2(x, y) \leq Cd_1(x, y)$ for all x, y in X, then d_1 and d_2 are equivalent metrics. (b) Show that the converse of (a) is false by giving a counterexample. (c) Show that two equivalent metrics on a set define the same continuous functions but not necessarily the same uniformly continuous functions. (d) Show that if the two metrics satisfy the condition stated in part (a), then they do define the same uniformly continuous functions. Is the converse of this true?

(2) Show that if $\mathcal{X}$ is a normed space and ball $\mathcal{X}$ is compact, then $\mathcal{X}$ is finite-dimensional. That is, the only locally compact normed spaces are the finite-dimensional ones.

(3) If $\mathcal{X}$ is a Banach space and $\mathcal{M} \leq \mathcal{X}$ such that $\dim \mathcal{X}/\mathcal{M} < \infty$, then there is a closed subspace $\mathcal{N}$ of $\mathcal{X}$ such that $\mathcal{M} \cap \mathcal{N} = (0)$ and $\mathcal{M} + \mathcal{N} = \mathcal{X}$.

6.2. Sums and quotients of normed spaces

Think of this section as showing how to manufacture new normed spaces from old ones. If we compare this to what happened with Hilbert spaces, there is a great contrast. We never considered the quotient of a Hilbert space $\mathcal{H}$ by a closed subspace $\mathcal{M}$. Instead we just had the orthogonal complement of $\mathcal{M}$, $\mathcal{M}^{\perp}$. But for normed spaces we have no concept of orthogonality. We did, however, explore the concept of the direct sum of Hilbert spaces. Unlike with Hilbert spaces, there is no standard way to define the direct sum of normed spaces. With a sequence of Hilbert spaces $\{\mathcal{H}_n : n \geq 1\}$ we define the direct sum as the vector space $\mathcal{H}$ of all sequences $\{h_n\}$, where $h_n \in \mathcal{H}_n$ for each $n \geq 1$ and $\sum_{n=1}^{\infty} \|h_n\|^2 < \infty$; then we defined the inner product on $\mathcal{H} = \bigoplus_{n=1}^{\infty} \mathcal{H}_n$ in such a way that the norm becomes $\|\{h_n\}\|^2 = \sum_{n=1}^{\infty} \|h_n\|^2$. For normed spaces there are many ways to define the direct sum of a collection of normed spaces. Some of these are explored in the exercises at the end of this section.

For the concept of the quotient of normed spaces, there is a canonical way to proceed and we explore this in some detail. (Attention! My experience is that many students have difficulty with this.) If $\mathcal{X}$ is a normed space and $\mathcal{M}$ is a linear manifold in $\mathcal{X}$, we form the quotient space $\mathcal{X}/\mathcal{M}$ as we do in linear algebra: $\mathcal{X}/\mathcal{M} = \{x + \mathcal{M} : x \in \mathcal{X}\}$. Just as in linear algebra, this is a vector space. We also define the *quotient map* $Q : \mathcal{X} \to \mathcal{X}/\mathcal{M}$ as $Q(x) = x + \mathcal{M}$; it is easy to see that Q is a linear transformation. Understand that certain algebraic facts hold; for example, if $B \subseteq \mathcal{X}$, $Q^{-1}(Q(B)) = \bigcup_{y \in \mathcal{M}}(y + B)$. We attempt to define a norm on $\mathcal{X}/\mathcal{M}$ by

6.2.1 $$\|x + \mathcal{M}\| = \inf\{\|x + y\| : y \in \mathcal{M}\} = \operatorname{dist}(x, \mathcal{M})$$

It is easy to see that this has the properties of a norm on the quotient space except for one: if $\mathcal{M}$ fails to be closed it is not a norm since if $x \in \operatorname{cl}\mathcal{M}$ but $x \notin \mathcal{M}$, then $\|x + \mathcal{M}\| = 0$ but $x + \mathcal{M} \neq 0$ in the quotient space. For this reason we always assume that $\mathcal{M}$ is closed when we take a quotient space.

6.2.2. Theorem. *If $\mathcal{M} \leq \mathcal{X}$, then (6.2.1) defines a norm on $\mathcal{X}/\mathcal{M}$ and Q is a linear map. The following are valid.*

(a) *$\|Q(x)\| \leq \|x\|$ for all x in $\mathcal{X}$ and so Q is continuous.*

(b) *A subset W of $\mathcal{X}/\mathcal{M}$ is open in the topology defined by its norm if and only if $Q^{-1}(W)$ is open in $\mathcal{X}$.*

(c) *If G is open in $\mathcal{X}$, then $Q(G)$ is open in the quotient space.*

(d) *If $\mathcal{X}$ is a Banach space, then the quotient space is also a Banach space.*

Proof. The initial statement in this theorem is easily verified.

(a) In the definition of the norm on $\mathcal{X}/\mathcal{M}$, 0 is one possible choice when we form the infimum; so clearly (a) holds.

(b) Let $W \subseteq \mathcal{X}/\mathcal{M}$. If W is open, then $Q^{-1}(W)$ is open in $\mathcal{X}$ since Q is continuous. To prove the reverse inclusion we first establish the following.

Claim. If $r > 0$ and $B_r = \{x \in \mathcal{X} : \|x\| < r\}$, then $Q(B_r) = \{x + \mathcal{M} : \|x + \mathcal{M}\| < r\}$.

If $\|x\| < r$, then $\|Q(x)\| = \|x + \mathcal{M}\| \leq \|x\| < r$, so that half is easy. For the other half, if $\|x + \mathcal{M}\| < r$, then from the definition of the norm there is a y in $\mathcal{M}$ with $\|x + y\| < r$; that is, $x + y \in B_r$. Thus $x + \mathcal{M} = Q(x + y) \in Q(B_r)$, and the claim is established.

Now assume that $Q^{-1}(W)$ is open in $\mathcal{X}$. If $x_0 + \mathcal{M} \in W$, then $x_0 \in Q^{-1}(W)$ and so there is an $r > 0$ such that $x_0 + B_r \subseteq Q^{-1}(W)$. Hence, by the claim, $W \supseteq Q(x_0 + B_r) = \{x + \mathcal{M} : \|x - x_0 + \mathcal{M}\| < r\}$ and so W is open.

(c) If G is open in $\mathcal{X}$, then $Q^{-1}(Q(G)) = \{x + y : x \in G, y \in \mathcal{M}\} = \bigcup_{y \in \mathcal{M}} y + G$. Since each $y + G$ is open, so is $Q^{-1}(Q(G))$. By (b), $Q(G)$ is open in $\mathcal{X}/\mathcal{M}$.

(d) Let $\{x_n + \mathcal{M}\}$ be a Cauchy sequence in $\mathcal{X}/\mathcal{M}$. We will show there is a subsequence x_{n_k} and a sequence $\{y_k\}$ in $\mathcal{M}$ such that $\{x_{n_k} + y_k\}$ is a Cauchy sequence in $\mathcal{X}$. Since $\mathcal{X}$ is complete, it will follow that there is an x in $\mathcal{X}$ such that $\|x_{n_k} + y_k - x\| \to 0$. But this implies $x_{n_k} + \mathcal{M} = Q(x_{n_k} + y_k) \to Q(x) = x + \mathcal{M}$. But in any metric space, if a Cauchy sequence has a convergent subsequence, then the original sequence must converge (Exercise 2); so it will follow that $x_n + \mathcal{M} \to x + \mathcal{M}$ and $\mathcal{X}/\mathcal{M}$ is a Banach space.

To carry this out, choose a subsequence $\{x_{n_k} + \mathcal{M}\}$ such that

$$\|(x_{n_k} + \mathcal{M}) - (x_{n_{k-1}} + \mathcal{M})\| = \|x_{n_k} - x_{n_{k-1}} + \mathcal{M}\| < 2^{-k}$$

Let $y_1 = 0$. From the definition of the norm on $\mathcal{X}/\mathcal{M}$, there is a y_2 in $\mathcal{M}$ with $\|x_{n_1} - x_{n_2} + y_2\| \leq \|x_{n_1} - x_{n_2} + \mathcal{M}\| + 2^{-1} < 2 \cdot 2^{-1}$. Similarly there is a y_3 in $\mathcal{M}$ with $\|(x_{n_2} + y_2) - (x_{n_3} + y_3)\| \leq \|x_{n_3} - x_{n_2} + \mathcal{M}\| + 2^{-2} < 2 \cdot 2^{-2}$. Continuing we get a sequence $\{y_k\}$ in $\mathcal{M}$ with

$$\|(x_{n_k} + y_k) - (x_{n_{k-1}} + y_{k-1})\| < 2 \cdot 2^{-k}$$

and it follows that $\{x_{n_k} + y_k\}$ is a Cauchy sequence.   ∎

Let's point out that part (b) of the preceding theorem says that the topology defined on the quotient space $\mathcal{X}/\mathcal{M}$ is exactly the quotient topology as defined in general topology.

6.2.3. Corollary. *If $\mathcal{X}$ is a normed space, $\mathcal{M} \leq \mathcal{X}$, and $\mathcal{N}$ is a finite-dimensional subspace of $\mathcal{X}$, then $\mathcal{M} + \mathcal{N}$ is closed.*

Proof. Consider $Q : \mathcal{X} \to \mathcal{X}/\mathcal{M}$. Because $Q(\mathcal{N})$ is finite-dimensional it is closed (6.1.7). Hence $\mathcal{M} + \mathcal{N} = Q^{-1}(Q(\mathcal{N}))$ is closed.   ∎

Again taking our lead from linear algebra, we use the quotient to look at linear transformations.

6.2.4. Proposition. *If* $\mathcal{X}$ *and* $\mathcal{Y}$ *are normed spaces and* $T : \mathcal{X} \to \mathcal{Y}$ *a bounded linear transformation, then there is a unique bounded linear transformation* $\widetilde{T} : \mathcal{X}/\ker T \to \mathcal{Y}$ *such that* $\widetilde{T}Q = T$. *Moreover* $\|\widetilde{T}\| = \|T\|$.

Proof. Define $\widetilde{T}(x + \ker T) = Tx$; just as in linear algebra, $\widetilde{T}$ is a well-defined linear transformation – it is assumed that the reader will verify the details. By definition, $\widetilde{T}Q = T$. Also for any y in $\ker T$, $\|\widetilde{T}(x + \ker T)\| = \|Tx\| = \|T(x+y)\| \leq \|T\|\|x+y\|$. Taking the infimum over all such y gives that $\|\widetilde{T}(x + \ker T)\| \leq \|T\|\|x + \ker T\|$ and so $\|\widetilde{T}\| \leq \|T\|$. On the other hand, if $\epsilon > 0$ and $x \in \text{ball}\,\mathcal{X}$ with $\|Tx\| > \|T\| - \epsilon$, then $\|x + \ker T\| \leq 1$ and $\|\widetilde{T}\| \geq \|\widetilde{T}(x + \ker T)\| = \|Tx\| > \|T\| - \epsilon$. Since ϵ was arbitrary we get that $\|\widetilde{T}\| \geq \|T\|$. If T_1 is a second linear transformation from $\mathcal{X}/\ker T$ into $\mathcal{Y}$ such that $T_1 Q = T$, then $(\widetilde{T} - T_1)Q = 0$ and by what we just saw with the norms, $\|\widetilde{T} - T_1\| = 0$ so that $T_1 = \widetilde{T}$. Thus $\widetilde{T}$ is unique. ∎

Exercises.

(1) Let $\{\mathcal{X}_n\}$ be a sequence of normed spaces. (a) If $1 \leq p < \infty$, show that

$$\mathcal{X}_p = \bigoplus_{n=1}^{\infty} \mathcal{X}_n = \left\{ x \in \prod_{n=1}^{\infty} \mathcal{X}_n : \|x\| \equiv \left[\sum_{n=1}^{\infty} \|x(n)\|^p \right]^{\frac{1}{p}} < \infty \right\}$$

is a normed space that is a Banach space if and only if each $\mathcal{X}_n$ is a Banach space. Sometimes we write $\mathcal{X} = \bigoplus_p \mathcal{X}_n$. (b) Show that

$$\mathcal{X}_\infty = \bigoplus_{n=1}^{\infty} \mathcal{X}_n = \left\{ x \in \prod_{n=1}^{\infty} \mathcal{X}_n : \|x\| \equiv \sup_n \|x(n)\| < \infty \right\}$$

is a normed space that is a Banach space if and only if each $\mathcal{X}_n$ is a Banach space. Sometimes we write $\mathcal{X} = \bigoplus_\infty \mathcal{X}_n$. (c) Show that

$$\mathcal{X}_0 = \bigoplus_{n=1}^{\infty} \mathcal{X}_n = \left\{ x \in \bigoplus_\infty \mathcal{X}_n : \|x(n)\| \to 0 \right\}$$

is a closed subspace of $\bigoplus_\infty \mathcal{X}_n$. Sometimes we write $\mathcal{X} = \bigoplus_0 \mathcal{X}_n$.

(2) Show that a metric space is complete if and only if every Cauchy sequence has a convergent subsequence.

(3) Let X be a compact space and F a closed subset. If $\mathcal{M} - \{f \in C(X) : f(x) = 0$ for all x in $F\}$, show that $C(X)/\mathcal{M}$ is isometrically isomorphic to $C(F)$.

(4) If $\mathcal{X}$ is a normed space and $\mathcal{M} \leq \mathcal{X}$ such that both $\mathcal{M}$ and $\mathcal{X}/\mathcal{M}$ are Banach spaces, is $\mathcal{X}$ a Banach space?

(5) Let $\mathcal{X}$ be a normed space and let $\mathcal{M}$ be a closed subspace. (a) Show that if $\mathcal{X}$ is separable, so is $\mathcal{X}/\mathcal{M}$. (b) If both $\mathcal{M}$ and $\mathcal{X}/\mathcal{M}$ are separable, show that $\mathcal{X}$ is separable. (c) Give an example such that $\mathcal{X}/\mathcal{M}$ is separable, but $\mathcal{X}$ is not.

(6) Consider a Hilbert space $\mathcal{H}$ as a Banach space and show that if $\mathcal{M} \leq \mathcal{H}$, then $\mathcal{H}/\mathcal{M}$ and $\mathcal{M}^{\perp}$ are isometrically isomorphic.

6.3. The Hahn–Banach Theorem

In §1.5 we introduced the concept of a bounded linear functional on a normed space $\mathcal{X}$ and showed that $\mathcal{X}^*$, the space of all bounded linear functionals on $\mathcal{X}$, is a Banach space with its norm

$$\|x^*\| = \sup\{|x^*(x)| : x \in \text{ball }\mathcal{X}\}$$

It perhaps will seem to the reader as tardy to now point out that we do not know that every infinite-dimensional normed space has a non-zero bounded linear functional. In this section we'll establish this and much more when we prove one of the pillars of functional analysis, the Hahn–Banach Theorem. But first we revisit the concept of a linear functional and show that there is a geometric condition equivalent to the boundedness of a linear functional.

For a vector space $\mathcal{X}$ a *hyperplane* is a linear manifold $\mathcal{M}$ in $\mathcal{X}$ such that $\dim(\mathcal{X}/\mathcal{M}) = 1$. If $f : \mathcal{X} \to \mathbb{F}$ is a linear functional, and $f \neq 0$, then $\mathcal{M} = \ker f$ is a hyperplane. In fact f induces a linear isomorphism of $\mathcal{X}/\ker f$ onto $\mathbb{F}$. Conversely, if $\mathcal{M}$ is a hyperplane in $\mathcal{X}$, then there is an isomorphism $\tau : \mathcal{X}/\mathcal{M} \to \mathbb{F}$ so that $f(x) = \tau(x + \mathcal{M})$ defines a linear functional with $\ker f = \mathcal{M}$. Also note that if f and g are two linear functionals with $\ker f = \ker g$, then there is a non-zero scalar α with $f = \alpha g$. In fact let $x_0 \in \mathcal{X}$ that does not belong to $\ker f = \ker g$. By multiplying x_0 by a suitable scalar we may assume that $g(x_0) = 1$. If $\alpha = f(x_0)$, then $\alpha \neq 0$ and for any x in $\mathcal{X}$ we have that $g(g(x)x_0 - x) = 0$; so $0 = f(g(x)x_0 - x) = \alpha g(x) - f(x)$. Let's gather this together.

6.3.1. Proposition. *A linear manifold in a vector space $\mathcal{X}$ is a hyperplane if and only if it is the kernel of a non-zero linear functional. Two linear functionals have the same kernel if and only if each is a non-zero multiple of the other.*

Now let's put a norm on $\mathcal{X}$ and see what happens.

6.3.2. Proposition. *A hyperplane in a normed space is either closed or it is dense. If f is a linear functional on $\mathcal{X}$, then f is bounded if and only if $\ker f$ is closed.*

Proof. Suppose $\mathcal{M}$ is a hyperplane that is not closed. Thus $\operatorname{cl}\mathcal{M}$ is a linear space that properly contains $\mathcal{M}$. Since $\dim \mathcal{X}/\mathcal{M} = 1$, it has to be that $\operatorname{cl}\mathcal{M} = \mathcal{X}$.

Now suppose $f : \mathcal{X} \to \mathbb{F}$ is a non-zero linear functional and $\mathcal{M}$ is its kernel. If f is bounded, then $\mathcal{M} = f^{-1}(0)$ must be closed. Conversely, assume that $\mathcal{M}$ is closed and let $Q : \mathcal{X} \to \mathcal{X}/\mathcal{M}$ be the quotient map. Since $\dim \mathcal{X}/\mathcal{M} = 1$, let $T : \mathcal{X}/\mathcal{M} \to \mathbb{F}$ be an isomorphism. By (6.1.8), T is continuous. Therefore $g = TQ : \mathcal{X} \to \mathbb{F}$ is a bounded linear functional with the same kernel, $\mathcal{M}$, as f. Therefore there is a scalar α with $f = \alpha g$ and f is continuous. ∎

A concept close to that of a seminorm is the following, which we will need in order to give the most general form of the Hahn–Banach Theorem.

6.3.3. Definition. If $\mathcal{X}$ is a vector space over $\mathbb{F}$ a *sublinear functional* is a function $q : \mathcal{X} \to \mathbb{R}$ such that for all x, y in $\mathcal{X}$ we have that $q(x + y) \le q(x) + q(y)$ and $q(\alpha x) = \alpha q(x)$ whenever α is a non-negative scalar.

It is easy to see that a seminorm is sublinear, but coming up with additional sublinear functionals is a bit more difficult. In §7.3 we'll see a general way to find such examples. Going for the most general form of the Hahn–Banach Theorem seems contrary to the philosophy, enunciated in the Preface of this book, of streamlining the material for the novice. However this is needed to obtain the geometric consequences of this result, which are both useful and exciting.

6.3.4. Theorem (Hahn–Banach Theorem). *Let $\mathcal{X}$ be a vector space over $\mathbb{R}$ and suppose q is a sublinear functional on $\mathcal{X}$. If $\mathcal{M}$ is a linear manifold in $\mathcal{X}$ and $f : \mathcal{M} \to \mathbb{R}$ is a linear functional satisfying $f(x) \le q(x)$ for all x in $\mathcal{M}$, then there is a linear functional $F : \mathcal{X} \to \mathbb{R}$ such that $F(x) = f(x)$ for all x in $\mathcal{M}$ and $F(x) \le q(x)$ for all x in $\mathcal{X}$.*

We will refer to the Hahn–Banach Theorem as the HBT. Let's emphasize a few points about the HBT. First the important point in the statement is not that the linear functional f can be extended to a linear functional F defined on all of $\mathcal{X}$, but that this extension continues to be dominated by the sublinear functional. If all we wanted was the extended functional, this could be done by just using a Hamel basis as follows. Let $\mathcal{B}_0$ be a basis for $\mathcal{M}$ and let $\mathcal{B}$ be a basis for $\mathcal{X}$ that contains $\mathcal{B}_0$. If $\mathcal{C} = \mathcal{B} \backslash \mathcal{B}_0$, then $F\left(\sum_{b \in \mathcal{B}_0} \alpha_b b + \sum_{c \in \mathcal{C}} \alpha_c c\right) = f\left(\sum_{b \in \mathcal{B}_0} \alpha_b b\right)$ is such an extension. Second, the theorem is not phrased for a normed space, only a vector space with a sublinear functional. In fact as stated there is no reference to any topology whatsoever, just to a vector space on which there is a sublinear functional. We will give the statement for normed spaces in one of the corollaries below, but we will, in fact, use this level of generality later in this book. Third, the theorem in the context of a Hilbert space is trivial to prove. That is, a continuous linear functional f defined on a subspace $\mathcal{M}$ of a Hilbert space $\mathcal{H}$ can be extended to a continuous linear functional F defined on all of $\mathcal{H}$. In fact, such an f can be extended to the closure of $\mathcal{M}$ since it is uniformly continuous and then we can use the Riesz Representation Theorem (see Exercise 3.3.1). Alternatively we could use an argument involving an orthonormal basis similar to the argument just given for arbitrary vector spaces.

I think the reader can be forgiven if (s)he is perplexed by my calling the HBT one of the pillars of functional analysis. It really seems unimposing, doesn't it? So we'll precede the proof by giving several corollaries. The reader can also look at some of the exercises at the end of this section, particularly Exercises 3, 5, 6, and 7; also examine §7.3 where several geometric consequences of the HBT are given. We'll see many other applications of this theorem during the course of this book so that such an accolade will readily be seen to be warranted.

As stated the HBT makes no sense for complex vector spaces since the inequality $f(x) \leq q(x)$ does not make sufficient sense. We will present in the first corollary, however, a version of the HBT for complex and real vector spaces where we replace the sublinear functional by a seminorm and use the inequality $|f(x)| \leq p(x)$. But before doing that we need a lemma. Remember that a vector space over $\mathbb{C}$ is also a vector space over $\mathbb{R}$.

6.3.5. Lemma. *Let $\mathcal{X}$ be a vector space over $\mathbb{C}$.*

(a) *If $L : \mathcal{X} \to \mathbb{R}$ is an $\mathbb{R}$-linear functional, then $\widetilde{L}(x) \equiv L(x) - iL(ix)$ is a $\mathbb{C}$-linear functional and $L = \operatorname{Re} \widetilde{L}$.*

(b) *If $S : \mathcal{X} \to \mathbb{C}$ is a $\mathbb{C}$-linear functional, $L = \operatorname{Re} S$, and $\widetilde{L}$ is defined as in part (a), then $\widetilde{L} = S$.*

(c) *With the notation of (a) if p is a seminorm on $\mathcal{X}$, then $|L(x)| \leq p(x)$ for all x in $\mathcal{X}$ if and only if $|\widetilde{L}(x)| \leq p(x)$ for all x in $\mathcal{X}$. If $\mathcal{X}$ is a normed space, then $\|L\| = \|\widetilde{L}\|$.*

Proof. The proofs of (a) and (b) are algebraic and left as exercises. To prove (c), assume that $|\widetilde{L}(x)| \leq p(x)$ for all x. Then $|L(x)| = |\operatorname{Re} \widetilde{L}(x)| \leq |\widetilde{L}(x)| \leq p(x)$. Now assume the converse: $|L(x)| \leq p(x)$ for all x. Choose θ such that $\widetilde{L}(x) = e^{i\theta}|\widetilde{L}(x)|$. Hence $|\widetilde{L}(x)| = \widetilde{L}(e^{-i\theta}x) = \operatorname{Re} \widetilde{L}(e^{-i\theta}x) = L(e^{-i\theta}x) \leq p(e^{-i\theta}x) = p(x)$. The proof of the statement involving norms is almost identical. ∎

6.3.6. Corollary. *Let $\mathcal{X}$ be a vector space over $\mathbb{F}$ and suppose p is a seminorm on $\mathcal{X}$. If $\mathcal{M}$ is a linear manifold in $\mathcal{X}$ and $f : \mathcal{M} \to \mathbb{F}$ is a linear* *functional satisfying $|f(x)| \leq p(x)$ for all x in $\mathcal{M}$, then there is a linear functional $F : \mathcal{X} \to \mathbb{F}$ such that $F(x) = f(x)$ for all x in $\mathcal{M}$ and $|F(x)| \leq p(x)$ for all x in $\mathcal{X}$.*

Proof. Suppose $\mathbb{F} = \mathbb{R}$. A seminorm is a sublinear functional and we have that $f(x) \leq |f(x)| \leq p(x)$ on $\mathcal{M}$. So by the HBT we get the extension F of f to all of $\mathcal{X}$ with $F(x) \leq p(x)$ for all x in $\mathcal{X}$. But $-F(x) = F(-x) \leq p(-x) = p(x)$, and we have that $|F(x)| \leq p(x)$ for all x.

   If $\mathbb{F} = \mathbb{C}$, let $f_1(x) = \operatorname{Re} f(x)$; note that $|f_1(x)| \leq |f(x)| \leq p(x)$. So the real case implies there is a linear functional $F_1 : \mathcal{X} \to \mathbb{R}$ that extends f_1 and satisfies $|F_1(x)| \leq p(x)$ for all x in $\mathcal{X}$. Put $F(x) = F_1(x) - iF_1(ix)$. According to the preceding lemma, $|F(x)| \leq p(x)$ for all x. That lemma also implies that F is an extension of f. ∎

6.3.7. Corollary. *If $\mathcal{X}$ is a normed space, $\mathcal{M}$ is a linear manifold in $\mathcal{X}$, and $f : \mathcal{M} \to \mathbb{F}$ is a bounded linear functional, then there is an F in $\mathcal{X}^*$ that is an extension of f and satisfies $\|F\| = \|f\|$.*

Proof. Use Corollary 6.3.6 with $p(x) = \|f\|\|x\|$. ∎

6.3.8. Corollary. *If $\mathcal{X}$ is a normed space, $\{x_1, \ldots, x_d\}$ are linearly independent vectors in $\mathcal{X}$, and $\alpha_1, \ldots, \alpha_d \in \mathbb{F}$, then there is an f in $\mathcal{X}^*$ with $f(x_k) = \alpha_k$ for $1 \leq k \leq d$.*

Proof. Let $\mathcal{M}$ be the linear span of $\{x_1, \ldots, x_d\}$ and define $g : \mathcal{M} \to \mathbb{F}$ by $g\left(\sum_{k=1}^{n} \beta_k x_k\right) = \sum_{k=1}^{n} \beta_k \alpha_k$. Since $\mathcal{M}$ is finite-dimensional, g is continuous (6.1.8). Now use the preceding corollary. ∎

6.3.9. Corollary. *If $\mathcal{X}$ is a normed space and $x \in \mathcal{X}$, then*

$$\|x\| = \sup\{|f(x)| : f \in \text{ball } \mathcal{X}^*\}$$

Moreover, this supremum is attained.

Proof. Fix x in $\mathcal{X}$ and put $s = \sup\{|f(x)| : f \in \text{ball } \mathcal{X}^*\}$. It is clear that $s \leq \|x\|$. If $\mathcal{M}$ is the one-dimensional space spanned by the vector x, define $g : \mathcal{M} \to \mathbb{F}$ by $g(\alpha x) = \alpha\|x\|$. So $g \in \mathcal{M}^*$ with $\|g\| = 1$. By Corollary 6.3.7 there is an f in $\mathcal{X}^*$ that extends g and satisfies $\|f\| = \|g\| = 1$. We have that $s \geq f(x) = g(x) = \|x\|$. ∎

   This emphatically says that there are non-zero bounded linear functionals on $\mathcal{X}$. In fact there are sufficiently many to determine the norm of each element of $\mathcal{X}$.

6.3.10. Corollary. *If $\mathcal{X}$ is a normed space, $\mathcal{M} \leq \mathcal{X}$, $x_0 \in \mathcal{X}\backslash\mathcal{M}$, and $d = \operatorname{dist}(x_0, \mathcal{M})$, then there is an f in $\mathcal{X}^*$ with $f(x_0) = 1$, $f(x) = 0$ for all x in $\mathcal{M}$, and $\|f\| = d^{-1}$.*

Proof. Note that using the norm on $\mathcal{X}/\mathcal{M}$, $\|x_0 + \mathcal{M}\| = d$. Therefore there is a g in ball $(\mathcal{X}/\mathcal{M})^*$ with $g(x_0 + \mathcal{M}) = d$. If Q is the quotient map of $\mathcal{X}$ onto $\mathcal{X}/\mathcal{M}$, we can define $f = d^{-1}g \circ Q : \mathcal{X} \to \mathbb{F}$. It follows that $f(x_0) = 1$, $f(x) = 0$ for all x in $\mathcal{M}$, and $\|f\| \le d^{-1}$. But there is a sequence $\{x_n\}$ in $\mathcal{X}$ with $\|x_n + \mathcal{M}\| \le 1$ for all n such that $g(x_n + \mathcal{M}) \to \|g\| = 1$. It follows that $\|x_n\| \le \|x_n + \mathcal{M}\| \le 1$ and $f(x_n) = d^{-1}g(x_n + \mathcal{M}) \to d^{-1}$; hence $\|f\| = d^{-1}$. ∎

Now it's time to start the proof of the HBT.

Proof of the Hahn–Banach Theorem. This is done by transfinite induction. The first step is to show we can extend the functional to a space that is larger by one dimension and then we will apply Zorn's Lemma. Like many induction proofs, the most difficult part of the proof is the first step.

Claim. If $\dim \mathcal{X}/\mathcal{M} = 1$, then F can be found.

Let $x_0 \in \mathcal{X} \backslash \mathcal{M}$; we have to figure out how to choose the value of $F(x_0)$. To do this suppose for the moment that F has been defined and let's see what this imposes on $\alpha_0 = F(x_0)$. Here for $t > 0$ and any y in $\mathcal{M}$, we have $t\alpha_0 + f(y) = F(tx_0 + y) \le q(tx_0 + y)$. Hence $\alpha_0 \le -t^{-1}f(y) + t^{-1}q(tx_0 + y) = -f(t^{-1}y) + q(x_0 + t^{-1}y)$. Since $t^{-1}y$ is an arbitrary vector in $\mathcal{M}$, we see that for any y_1 in $\mathcal{M}$

$$\alpha_0 \le q(x_0 + y_1) - f(y_1)$$

Observe that if α_0 is a real number satisfying this inequality, then by reversing the above argument we get that when $t \ge 0$, $t\alpha_0 + f(y_1) \le q(tx_0 + y_1)$ for all y_1 in $\mathcal{M}$ and $t \ge 0$. That is, the two inequalities are equivalent.

Now let $t > 0$ and let $y \in \mathcal{M}$. Once again if F exists we obtain that $-t\alpha_0 + f(y) = F(-tx_0 + y) \le q(-tx_0 + y)$. As in the last paragraph, this implies that

$$\alpha_0 \ge f(y_2) - q(-x_0 + y_2)$$

for every y_2 in $\mathcal{M}$. Conversely, this inequality is sufficient to have that $-t\alpha_0 + f(y_2) \le q(-tx_0 + y_2)$ for all $t \ge 0$ and all y_2 in $\mathcal{M}$.

Combining these inequalities we see that we must show that we can choose α_0 such that

6.3.11 $$f(y_2) - q(-x_0 + y_2) \le \alpha_0 \le -f(y_1) + q(x_0 + y_1)$$

for all y_1, y_2 in $\mathcal{M}$. But

$$f(y_1) + f(y_2) = f(y_1 + y_2)$$
$$\le q(y_1 + y_2)$$
$$= q((y_1 + x_0) + (-x_0 + y_2))$$
$$\le q(y_1 + x_0) + q(-x_0 + y_2)$$

That is
$$f(y_2) - q(-x_0 + y_2) \le -f(y_1) + q(x_0 + y_1)$$
for all y_1, y_2 in $\mathcal{M}$. Since y_1 and y_2 are independent, we can choose α_0 satisfying (6.3.11). If we put $F(tx_0 + y) = t\alpha_0 + f(y)$, we have established the claim.

To prove the theorem let $\mathcal{L}$ be the collection of all pairs $(\mathcal{N}, g)$, where $\mathcal{N}$ is a linear manifold containing $\mathcal{M}$ and g is a linear functional on $\mathcal{N}$ that extends f and satisfies $g(x) \le q(x)$ for all x in $\mathcal{N}$. Order $\mathcal{L}$ by declaring that $(\mathcal{N}_1, g_1) \le (\mathcal{N}_2, g_2)$ if $\mathcal{N}_1 \subseteq \mathcal{N}_2$ and g_2 is an extension of g_1. Let $\mathcal{L}_1$ be a chain in $\mathcal{L}$ and put $\mathcal{Y} = \bigcup\{\mathcal{N} : (\mathcal{N}, g) \in \mathcal{L}_1\}$. It is easy to see that $\mathcal{Y}$ is a linear manifold. Now define $G : \mathcal{Y} \to \mathbb{F}$ by setting $G(x) = g(x)$ when $x \in \mathcal{N}$ and $(\mathcal{N}, g) \in \mathcal{L}_1$. It also follows from the fact that $\mathcal{L}_1$ is a chain that G is well defined, linear, and $(\mathcal{Y}, G) \in \mathcal{L}$. So $(\mathcal{Y}, F)$ is an upper bound for $\mathcal{L}_1$. Therefore Zorn's Lemma implies that there is a maximal element $(\mathcal{Z}, F)$ in $\mathcal{L}$. But the claim implies that since $(\mathcal{Z}, F)$ is maximal, it must be that $\mathcal{Z} = \mathcal{X}$. ∎

The next result also follows from the HBT but it is sufficiently important we won't call it a corollary.

6.3.12. Theorem. *If $\mathcal{X}$ is a normed space and $\mathcal{M}$ is a linear manifold in $\mathcal{X}$, then*
$$\mathrm{cl}\, \mathcal{M} = \bigcap\{\ker f : f \in \mathcal{X}^* \text{ and } \mathcal{M} \subseteq \ker f\}$$

Proof. Let $\mathcal{N} = \bigcap\{\ker f : f \in \mathcal{X}^* \text{ and } \mathcal{M} \subseteq \ker f\}$; clearly $\mathrm{cl}\, \mathcal{M} \subseteq \mathcal{N}$. If $x_0 \notin \mathrm{cl}\, \mathcal{M}$, then Corollary 6.3.10 implies there is an f in $\mathcal{X}^*$ with $\mathrm{cl}\, \mathcal{M} \subseteq \ker f$ and $f(x_0) = 1$. Thus $x_0 \notin \mathcal{N}$ and we have the other inclusion. ∎

6.3.13. Corollary. *If $\mathcal{X}$ is a normed space and $\mathcal{S}$ is a subset of $\mathcal{X}$, then the linear span of $\mathcal{S}$ is dense in $\mathcal{X}$ if and only if the only bounded linear functional on $\mathcal{X}$ that vanishes on $\mathcal{S}$ is the zero functional.*

This corollary is proved by letting $\mathcal{M}$ be the linear span of $\mathcal{S}$ and applying the preceding theorem. Note the similarity between this last corollary and the fact that in a Hilbert space a set $\mathcal{S}$ has dense linear span if and only if $\mathcal{S}^\perp = (0)$.

Exercises.

(1) If $\mathcal{X}$ is a vector space over $\mathbb{C}$ and L_1 and L_2 are two linear functionals defined on $\mathcal{X}$, show that $L_1 = L_2$ if and only if $\mathrm{Re}\, L_1(x) = \mathrm{Re}\, L_2(x)$ for all x in $\mathcal{X}$.

(2) Let $\mathcal{X}$ be a vector space over $\mathbb{F}$ and suppose p is a seminorm on $\mathcal{X}$. If $\mathcal{M} = p^{-1}(0)$, show that $\|x + \mathcal{M}\| = \inf\{p(x + y) : y \in \mathcal{M}\}$ defines a norm on $\mathcal{X}/\mathcal{M}$.

(3) If μ is a regular Borel signed measure on $[0,1]$ and $\int x^n d\mu = 0$ for all $n \geq 0$, show that $\mu = 0$.

(4) Let $\mathcal{P}$ be the linear manifold of all polynomials $p(z)$ in the complex variable z. Consider $\mathcal{P}$ as contained in $C(\partial\mathbb{D})$, the space of continuous functions on the unit circle $\partial\mathbb{D} = \{z \in \mathbb{C} : |z| = 1\}$. Show that there is a non-zero complex measure μ such that $\mu \neq 0$ but $\int p(z)d\mu = 0$ for every $p(z)$ in $\mathcal{P}$.

(5) If $n \geq 1$, show that there is a signed measure μ on $[0,1]$ such that $\int p d\mu = p'(0)$ for every polynomial $p(x)$ of degree at most n. Can you find such a measure μ such that $\int p d\mu = p'(0)$ for every polynomial?

(6) Use the HBT to give another proof that if $1 \leq p < \infty$, X is a compact metric space, and μ is a positive Radon measure on X, then $C(X)$ is dense in $L^p(\mu)$. See Theorem 2.4.26 and Exercise 2.7.4. (Hint: Use Corollary 6.3.13.)

(7) Use the HBT to give another proof that for $1 \leq p < \infty$, the simple measurable functions belonging to $L^p(\mu)$ are dense in $L^p(\mu)$ (2.7.12), though when $p = 1$ we must assume the underlying measure space is σ-finite. (Hint: Use Corollary 6.3.13.)

6.4. Banach limits*

This section gives an application of the HBT, but is not needed for the remainder of the book.

Let $c = \{x : \mathbb{N} \to \mathbb{F} : \lim_n x(n) \text{ exists}\}$, the space of all convergent sequences of scalars; so $c \leq \ell^\infty$. On c there is a special linear functional, the limit: $L(x) = \lim_n x(n)$ for all x in c. This linear functional has several special properties. For example it is positive: if $x \geq 0$ (that is, $x(n) \geq 0$ for all n), then $L(x) \geq 0$. It is also shift invariant: if for $x \in c$ we define $x'(n) = x(n+1)$, then $L(x) = L(x')$. Finally it has norm 1 (Why?). We will show, using the HBT, that there is a linear functional on ℓ^∞ that extends these properties of the limit.

6.4.1. Theorem. *There is a linear functional $L : \ell^\infty \to \mathbb{F}$ having the following properties.*

(a) $\|L\| = 1$.

(b) *If $x \in c$, $L(x) = \lim_n x(n)$.*

(c) *If $x \in \ell^\infty$ and $x(n) \geq 0$ for all n, then $L(x) \geq 0$.*

(d) *If $x \in \ell^\infty$ and $x'(n) = x(n+1)$ for all $n \geq 1$, then $L(x) = L(x')$.*

Proof. First assume that $\mathbb{F} = \mathbb{R}$ and put $\mathcal{M} = \{x - x' : x \in \ell^\infty\}$, where x' is as in the statement of the theorem. It is easy to check that $(x + \alpha y)' = x' + \alpha y'$ for any x, y in ℓ^∞ and any scalar α. Thus $\mathcal{M}$ is a linear manifold in ℓ^∞. Denote by 1 the constantly one function in ℓ^∞.

6.4.2. Claim. $\operatorname{dist}(1, \mathcal{M}) = 1$.

Since $0 \in \mathcal{M}$, we have that $\operatorname{dist}(1, \mathcal{M}) \le 1$. Fix an x in ℓ^∞. If for some n, $(x - x')(n) \le 0$, then $\|1 - (x - x')\| \ge |1 - (x - x')(n)| \ge 1$. If we have an x in ℓ^∞ where this does not happen, then $x(n+1) \le x(n)$ for all n; that is, $\{x(n)\}$ is a bounded decreasing sequence and so $\alpha = \lim_n x(n)$ exists. Thus $\lim_n (x - x')(n) = 0$ and so $\|1 - (x - x')\| \ge 1$. Hence we see that $\|1 - (x - x')\| \ge 1$ for all x and this proves the claim.

We now apply Corollary 6.3.10 to get a bounded linear functional L on ℓ^∞ such that $L(y) = 0$ whenever $y \in \mathcal{M}$, $L(1) = 1$, and $\|L\| = [\operatorname{dist}(1, \mathcal{M})]^{-1} = 1$. That is, L satisfies (a) and (d) in the statement of the theorem.

6.4.3. Claim. $c_0 \le \ker L$.

For any x in ℓ^∞, let $x^{(1)} = x'$ and for all $n \ge 1$ let $x^{(n+1)} = (x^{(n)})'$. Since $x^{(n+1)} - x = [x^{(n+1)} - x^{(n)}] + \cdots + [x^{(1)} - x]$, we have that $L(x) = L(x^{(n)})$ for all $n \ge 1$. Now if $x \in c_0$ and $\epsilon > 0$, there is an $n_0 \ge 1$ such that $|x(n)| < \epsilon$ for all $n \ge n_0$. Thus $\|x^{(n)}\| < \epsilon$ for $n \ge n_0$, and so $|L(x)| = |L(x^{(n)}| < \|L\|\epsilon = \epsilon$ for $n \ge n_0$. Since ϵ was arbitrary we have proved Claim 6.4.3.

From here we see that if $x \in c$ and $\alpha = \lim_n x(n)$, then $x - \alpha 1 \in c_0$ and so $L(x) = \alpha$; that is, (b) is valid.

To show (c), suppose $x \ge 0$ in ℓ^∞ and $L(x) < 0$. Replacing x by $x/\|x\|$ we see that without loss of generality we may assume that $0 \le x(n) \le 1$ for all n. But then $\|1 - x\| \le 1$ and $L(1 - x) = 1 - L(x) > 1$, a contradiction. Thus (c) holds and the proof is complete in the case where $\mathbb{F} = \mathbb{R}$.

Now assume that $\mathbb{F} = \mathbb{C}$ and let $L_1 : \ell^\infty_{\mathbb{R}} \to \mathbb{R}$ be the functional obtained in the first part of the proof. If $x \in \ell^\infty$, write $x = x_1 + ix_2$, where $x_1, x_2 \in \ell^\infty_{\mathbb{R}}$. Define $L(x) = L_1(x_1) + iL_1(x_2)$. It is immediate that (b), (c), and (d) are satisfied. It remains to verify (a).

Note that for any x in ℓ^∞ there is a sequence $\{x_n\}$ in ℓ^∞ such that $\|x_n - x\| \to 0$ and each $(x_n(1), x_n(2), \dots)$ has only a finite number of distinct terms. (This can be seen directly by careful work as called for in Exercise 1. We can also use Theorem 2.7.12 and the observation that $\ell^\infty = L^\infty(\mu)$, where μ is counting measure on $\mathbb{N}$.) Thus if we show that $|L(x)| \le 1$ when $\|x\| \le 1$ and x is a bounded sequence with only a finite number of distinct terms, then we have proven (a). Now if x is such a sequence, we can write $x = \sum_{k=1}^n \alpha_k \chi_{E_k}$, where $E_1, \dots, E_n$ are pairwise disjoint subsets of $\mathbb{N}$ and

$|\alpha_k| \leq 1$ for $1 \leq k \leq n$. Thus $L(x) = \sum_{k=1}^{n} \alpha_k L_1(\chi_{E_k})$, and $L_1(\chi_{E_k}) \geq 0$ for each k. But $\sum_{k=1}^{n} L_1(\chi_{E_k}) = L_1(\chi_E) \leq 1$ where $E = \bigcup_{k=1}^{n} E_k$. But the fact that each α_k belongs to $\mathrm{cl}\,\mathbb{D} = \{z \in \mathbb{C} : |z| \leq 1\}$ and this is a convex set implies that $\sum_{k=1}^{n} \alpha_k L_1(\chi_{E_k}) \in \mathrm{cl}\,\mathbb{D}$, so that $|L(x)| \leq 1$. $\blacksquare$

A linear functional $L : \ell^\infty \to \mathbb{F}$ satisfying the conditions in the preceding theorem is called a *Banach limit*.

Exercises.

(1) Show directly that for any x in ℓ^∞ there is a sequence $\{x_n\}$ in ℓ^∞ such that $\|x_n - x\| \to 0$ and each $(x_n(1), x_n(2), \dots)$ has only a finite number of distinct terms.

(2) If L is a Banach limit on ℓ^∞, show that it cannot satisfy $L(xy) = L(x)L(y)$ for every x and y in ℓ^∞.

(3) Is the linear functional having the properties listed in Theorem 6.4.1 unique?

6.5. The Open Mapping and Closed Graph Theorems

Here we meet another of the pillars of functional analysis.

6.5.1. Theorem (Open Mapping Theorem). *If $\mathcal{X}$ an $\mathcal{Y}$ are Banach spaces and $A : \mathcal{X} \to \mathcal{Y}$ is a surjective bounded linear transformation, then A is an open mapping; that is, $A(G)$ is open in $\mathcal{Y}$ whenever G is open in $\mathcal{X}$.*

Proof. Let $B(r) = B(0;r) = \{x : \|x\| < r\}$. We begin by showing that it suffices to show that for any $r > 0$,

6.5.2 $$0 \in \mathrm{int}\, A(B(r))$$

In fact assume this has been done and let G be an open subset of $\mathcal{X}$. For each x in G let $r_x > 0$ such that $B(x; r_x) \subseteq G$. Since $0 \in \mathrm{int}\, A(B(r_x))$, $A(x) \in \mathrm{int}\, A(B(x; r_x))$. Thus there is an $s_x > 0$ with $U_x = \{y \in \mathcal{Y} : \|y - A(x)\| < s_x\} \subseteq A(B(x; r_x))$. But $A(G) = \bigcup_{x \in G} A(B(x; r_x)) \supseteq \bigcup_{x \in G} U_x$; on the other hand $A(x) \in U_x$ for each x in G so that we get $A(G) = \bigcup_{x \in G} U_x$, which is open.

We begin the proof of (6.5.2) by establishing something a bit easier.

Claim. $0 \in \mathrm{int}\,\mathrm{cl}\,[A(B(r))]$.

In fact note that because A is surjective,

$$\mathcal{Y} = \bigcup_{k=1}^{\infty} \mathrm{cl}\,[A(B(kr/2))] = \bigcup_{k=1}^{\infty} k\,\mathrm{cl}\,[A(B(r/2))].$$

By the Baire Category Theorem (A.1.3), there is an integer $k \geq 1$ such that $k \operatorname{cl}[A(B(r/2))]$ has non-empty interior. Multiplying by k^{-1} we get that $V = \operatorname{int} \operatorname{cl}[A(B(r/2)] \neq \emptyset$. If $y_0 \in V$, there is an $s > 0$ such that $\{y \in \mathcal{Y} : \|y - y_0\| < s\} \subseteq V \subseteq \operatorname{cl}[A(B(r/2))]$. Let $\{x_n\}$ be a sequence in $B(r/2)$ such that $A(x_n) \to y_0$. For a fixed y with $\|y\| < s$, there is also a sequence $\{z_n\}$ in $B(r/2)$ such that $A(z_n) \to y_0 + y$. Therefore $A(z_n - x_n) \to y$. Since $\|z_n - x_n\| < r$, this shows that $\{y : \|y\| < s\} \subseteq \operatorname{cl}[A(B(r/2))]$, establishing the claim.

In light of the claim we will have proven (6.5.2) if we can show that

$$\operatorname{cl}[A(B(r/2))] \subseteq A(B(r))$$

Fix a y_1 in $\operatorname{cl}[A(B(r/2))]$; by the claim $0 \in \operatorname{int}[\operatorname{cl} A(B(2^{-2}r))]$ and so $[y_1 - \operatorname{cl} A(B(2^{-2}r))] \cap A(B(r/2)) \neq \emptyset$. Hence there is an x_1 in $B(r/2)$ such that $A(x_1) \in y_1 - \operatorname{cl} A(B(2^{-2}r))$, and so $A(x_1) = y_1 - y_2$, where $y_2 \in \operatorname{cl} A(B(2^{-2}r))$. By the same reasoning $[y_2 - \operatorname{cl} A(B(2^{-3}r))] \cap A(B(r/2^{-2})) \neq \emptyset$. Continue. We get a sequence $\{x_n\}$ in $\mathcal{X}$ and a sequence $\{y_n\}$ in $\mathcal{Y}$ such that

6.5.3 $$\begin{cases} \text{(a)} & x_n \in B(2^{-n}r) \\ \text{(b)} & y_n \in \operatorname{cl} A(B(2^{-n}r)) \\ \text{(c)} & y_{n+1} = y_n - A(x_n) \end{cases}$$

Note that $\sum_{n=1}^{\infty} \|x_n\| < \sum_{n=1}^{\infty} 2^{-n}r = r < \infty$, so that $x = \sum_{n=1}^{\infty} x_n$ converges in $\mathcal{X}$ and $\|x\| < r$. Also $\sum_{k=1}^{n} A(x_k) = \sum_{k=1}^{n}(y_k - y_{k+1}) = y_1 - y_{n+1}$. But (6.5.3(b)) implies $\|y_n\| \leq \|A\| 2^{-n}r$ so that $y_n \to 0$. Therefore $y_1 = \sum_{n=1}^{\infty} A(x_n) = A(x) \in A(B(r))$ and we are done. ∎

We will refer to the Open Mapping Theorem as the OMT. A look at [17] will show a proof of the Tietze Extension Theorem that uses an approach similar to the preceding proof. The next result is a simple corollary of the Open Mapping Theorem but is used enough and important enough to earn its own moniker.

6.5.4. Theorem (Inverse Mapping Theorem). *If $\mathcal{X}$ and $\mathcal{Y}$ are Banach spaces and $A : \mathcal{X} \to \mathcal{Y}$ is a bijective bounded linear transformation, then A^{-1} is bounded.*

Proof. To prove that A^{-1} is continuous, we need only show that A is an open mapping. That is what the OMT says. ∎

We refer to this result as the IMT. For the OMT and IMT we need the spaces to be complete. For example, let $C_0^{(1)}[0,1]$ denote the continuously differentiable functions on the unit interval that vanish at $x = 0$. Give both $C_0^{(1)}[0,1]$ and $C[0,1]$ the supremum norm and define $T : C[0,1] \to C_0^{(1)}[0,1]$ by $T(f)(x) = \int_0^x f(t)dt$. (Yes, this is like the Volterra operator, but the domain and range are different.) The operator is bounded with $\|Tf\| \leq \|f\|$ for all f in $C[0,1]$. It is also surjective by the Fundamental Theorem of Calculus and clearly it is injective. However, T^{-1} is not bounded. (See Exercise 2.)

In forming the category of normed spaces, there are two possibilities for the definition of an isomorphism. We have seen several times already the idea of an isometric linear isomorphism. We'll continue to use that term but when we say that T is an *isomorphism* (no modifier before "isomorphism") between normed spaces $\mathcal{X}$ and $\mathcal{Y}$, we mean $T : \mathcal{X} \to \mathcal{Y}$ is a bounded bijection with a bounded inverse. So according to the IMT, T is an isomorphism if and only if T is a bounded bijection.

Here we look for a converse to Lemma 5.6.13 in the context of Hilbert space.

6.5.5. Proposition. *If $\mathcal{H}$ and $\mathcal{K}$ are Hilbert spaces and $T \in \mathcal{B}(\mathcal{H}, \mathcal{K})$, then T has closed range if and only if*

$$\inf\{\|Th\| : h \in (\ker T)^{\perp} \text{ and } \|h\| = 1\} > 0$$

Proof. Define $\widetilde{T} : (\ker T)^{\perp} \to \operatorname{ran} T$ to be the restriction of T to $(\ker T)^{\perp}$. By Lemma 5.6.13, if the stated condition in the proposition holds, then $\operatorname{ran} T = \operatorname{ran} \widetilde{T}$ is closed. Conversely, assume that $\operatorname{ran} T = \operatorname{ran} \widetilde{T}$ is closed. Clearly $\widetilde{T}$ is injective, so, by the IMT, $\widetilde{T}$ is invertible. Thus for $h \in (\ker T)^{\perp}$, $\|h\| = \|\widetilde{T}^{-1}Th\| \leq \|\widetilde{T}^{-1}\|\|Th\|$, and we have the stated condition. $\blacksquare$

The preceding proposition does not carry over to Banach spaces and a strict converse to Lemma 5.6.13 is not true. In fact the condition in (5.6.13) that $\inf\{\|Tx\| : \|x\| = 1\} > 0$ also implies that $\ker T = (0)$. What happens in the Hilbert space context is that we can get a closed subspace that is "complementary" to $\ker T$, namely $(\ker T)^{\perp}$. This is not always possible in the Banach space setting. More on this subject will be seen in the next section. Also see Exercise 5.

6.5.6. Example. (a) Let T be the diagonalizable operator defined by the sequence $\{a_n\}$ as in Proposition 5.6.5. T has closed range if and only if $\inf\{|a_n| : a_n \neq 0\} > 0$.

(b) If M_ϕ is the multiplication operator defined on $L^2(\mu)$ by a function ϕ in $L^\infty(\mu)$, then $\operatorname{ran} M_\phi$ is closed if and only if there is a $\delta > 0$ such that $|\phi(x)| \geq \delta$ for all $x \notin \{y : \phi(y) = 0\}$.

The next result gives a famous sufficient condition for a linear transformation to be bounded.

6.5.7. Theorem (Closed Graph Theorem). *If $\mathcal{X}$ and $\mathcal{Y}$ are Banach spaces and $A : \mathcal{X} \to \mathcal{Y}$ is a linear transformation such that the graph of A:*

$$\text{graph } A = \{x \oplus Ax : x \in \mathcal{X}\}$$

is closed in $\mathcal{X} \oplus \mathcal{Y}$, then A is bounded.

Proof. Let $\mathcal{G}$ denote the graph of A; since it is closed in $\mathcal{X} \oplus \mathcal{Y}$, it is a Banach space. Define $P : \mathcal{G} \to \mathcal{X}$ to be the projection onto the first coordinate: $P(x \oplus Ax) = x$. It is routine to check that P is continuous and bijective. By the IMT, $P^{-1} : \mathcal{X} \to \mathcal{G}$ is bounded. Note that $P^{-1}(x) = x \oplus Ax$. Therefore A is the composition of P^{-1} and the projection of $\mathcal{X} \oplus \mathcal{Y}$ onto the second coordinate space, and as such it must be continuous. ∎

We need to examine the delicacy of the hypothesis of the CGT. If we return to the example discussed after the IMT, we see that graph T^{-1} is closed since it is essentially the same as graph T but T^{-1} is not bounded. Thus we need the domain of the linear transformation under discussion to be complete. We also need the range to be complete as the following example, due to Alp Eden, shows. Let $\mathcal{X}$ be a separable Banach space and let $\{e_i : i \in I\}$ be a Hamel basis for $\mathcal{X}$; by multiplying each e_i by $\|e_i\|^{-1}$, we can assume that each basis vector has norm 1. An easy application of the Baire Category Theorem (A.1.3) shows that I must be uncountable. Define a new norm on $\mathcal{X}$ by $\|x\|_1 = \sum_i |\alpha_i|$, where $x = \sum_i \alpha_i e_i$ is the expansion of x with respect to the basis; this makes sense since $\alpha_i = 0$ except for at most a finite number of choices of i in I. Since $\|e_i\| = 1$ for all i, we have that $\|x\| \leq \|x\|_1$. Let $\mathcal{Y} = \mathcal{X}$ with the norm $\| \cdot \|_1$ and define $T : \mathcal{X} \to \mathcal{Y}$ by $Tx = x$. Since $\|x\| \leq \|x\|_1$, $T^{-1} : \mathcal{Y} \to \mathcal{X}$ is a contraction, so that graph T^{-1} is closed; therefore graph T is closed. But T cannot be bounded since then it would follow that T is a homeomorphism and that $\mathcal{Y}$ is separable. Moreover $\|e_i - e_j\|_1 - 2$ for $i \neq j$, so $\{B(e_i; 1) : i \in I\}$ is an uncountable number of pairwise disjoint open balls in $(\mathcal{Y}, \| \cdot \|_1)$, a contradiction to the separability of $\mathcal{Y}$.

The next result simplifies the task of showing that a graph is closed. The proof is left as an exercise.

6.5.8. Proposition. *If $\mathcal{X}$ and $\mathcal{Y}$ are normed spaces and A is a linear transformation from $\mathcal{X}$ into $\mathcal{Y}$, then* graph A *is closed if and only if whenever $x_n \to 0$ in $\mathcal{X}$ and $Ax_n \to y$ in $\mathcal{Y}$ we have that $y = 0$.*

Note that if we wanted to show that A is continuous, we have to show that $Ax_n \to 0$ whenever $x_n \to 0$. The virtue of the preceding proposition and the CGT is that we can also assume that $\{Ax_n\}$ is convergent.

Exercises.

(1) If $\mathcal{X}$ and $\mathcal{Y}$ are Banach spaces and $A \in \mathcal{B}(\mathcal{X}, \mathcal{Y})$, show that ran A is closed when it is second category.

(2) If $A : C^{(1)}[0,1] \to C[0,1]$ is defined by $Af = f'$ and both spaces have the supremum norm, show that A is unbounded.

(3) Verify the statements in Example 6.5.6. Is (6.5.6)(b) true for $L^p(\mu)$?

(4) Prove Proposition 6.5.8.

(5) If $\mathcal{X}$ and $\mathcal{Y}$ are Banach spaces and $A \in \mathcal{B}(\mathcal{X}, \mathcal{Y})$, show that

$$\inf\{\|Ax\| : x \in \mathcal{X}, \|x\| = 1\} > 0$$

if and only if A is injective and has closed range.

(6) If X is a compact space, let $\mathcal{X}$ be a Banach subspace of $C(X)$ and let E be a closed subset of X. Show that if for every g in $C(E)$ there is an f in $\mathcal{X}$ with $g = f|E$, the restriction of f to E, then there is a constant M such that for every g in $C(E)$, the extension f in $\mathcal{X}$ can be chosen with $\|f\| \le M\|g\|$.

(7) Let $(X, \mathcal{A}, \mu)$ be a σ-finite measure space and let ϕ be an $\mathcal{A}$-measurable function on X. If $1 \le p \le \infty$ and for each f in $L^p(\mu)$ we have that $\phi f \in L^p(\mu)$, show that $\phi \in L^\infty(\mu)$. (Hint: Use the CGT.)

(8) Let $(X, \mathcal{A}, \mu)$ be a σ-finite measure space and suppose that $k : X \times X \to \mathbb{F}$ is an $\mathcal{A} \times \mathcal{A}$-measurable function such that for every f in $L^p(\mu)$ and almost every x in X, $k(x, \cdot)f(\cdot) \in L^1(\mu)$ and $Kf(x) = \int k(x, y)f(y)d\mu(y)$ defines an element of $L^p(\mu)$. Show that $K : L^p(\mu) \to L^p(\mu)$ is a bounded operator.

(9) In this section we proved the OMT and showed that it implies the IMT and the CGT. (a) Assume the CGT is true and show that it implies the OMT and the IMT. (b) Assume the IMT and show that it implies the OMT and CGT. Thus all three results are equivalent.

6.6. Complemented subspaces*

Say that a linear manifold $\mathcal{M}$ in a vector space $\mathcal{X}$ is *algebraically complemented* if there is another linear manifold $\mathcal{N}$ such that $\mathcal{M} \cap \mathcal{N} = (0)$ and $\mathcal{M} + \mathcal{N} = \mathcal{X}$. In this case every vector in $\mathcal{X}$ has a unique decomposition as the sum of a vector in $\mathcal{M}$ and one from $\mathcal{N}$. If $\mathcal{M}$ is given and $\mathcal{B}_0$ is a Hamel basis for $\mathcal{M}$, let $\mathcal{B}$ be a basis for $\mathcal{X}$ that contains $\mathcal{B}_0$. If $\mathcal{N}$ is the linear span of $\mathcal{B} \setminus \mathcal{B}_0$, then it is easy to see that $\mathcal{M}$ and $\mathcal{N}$ are algebraically complementary. So every linear manifold in a vector space has an algebraic complement.

If $\mathcal{M}$ and $\mathcal{N}$ are algebraically complementary in $\mathcal{X}$, define $E : \mathcal{X} \to \mathcal{X}$ by $E(x + y) = x$ whenever $x \in \mathcal{M}$ and $y \in \mathcal{N}$. Because the spaces are complementary, E is well defined on all of $\mathcal{X}$. It is routine to show that E is a linear transformation. (Do it.) Moreover $E^2(x + y) = E(x + y) = x$, so that E is an idempotent. It follows that $\operatorname{ran} E = \mathcal{M}$ and $\ker E = \mathcal{N}$. That is, E is a type of projection of $\mathcal{X}$ onto $\mathcal{M}$ "along" $\mathcal{N}$.

What happens in the context of normed spaces? First, since we are in a normed space setting, let's only consider closed subspaces of the normed space. If $\mathcal{M}, \mathcal{N} \leq \mathcal{X}$ and they are complementary, consider $\mathcal{M} \oplus_1 \mathcal{N}$. Recall (Exercise 6.2.1) that $\mathcal{M} \oplus_1 \mathcal{N} = \{x \oplus y : x \in \mathcal{M}, y \in \mathcal{N}\}$ and we give the space the norm $\|x \oplus y\|_1 = \|x\| + \|y\|$. Let $A : \mathcal{M} \oplus_1 \mathcal{N} \to \mathcal{X}$ be defined by $A(x \oplus y) = x + y$. It is easy to check that A is a linear bijection and $\|A(x \oplus y)\| = \|x + y\| \leq \|x \oplus y\|_1$; so A is bounded. Say that $\mathcal{M}$ and $\mathcal{N}$ are *topologically complemented* if this map is a homeomorphism; equivalently, if $\|\|x + y\|\| \equiv \|x\| + \|y\|$ is an equivalent norm on $\mathcal{X}$. When the normed spaces are complete, the IMT has something to say here. In fact since A is a bounded bijection, it has a continuous inverse by the IMT. Thus we have the following.

6.6.1. Theorem. *If $\mathcal{X}$ is a Banach space and $\mathcal{M}$ and $\mathcal{N}$ are two closed subspaces of $\mathcal{X}$ that are algebraically complementary, then they are topologically complementary.*

This permits us to speak of *complementary subspaces* of a Banach space without a modifier.

In the Hilbert space setting we know that every subspace $\mathcal{M}$ of a Hilbert space $\mathcal{H}$ has a complement, namely $\mathcal{M}^\perp$, though there may be others. In a Banach space there may be subspaces that do not have a complement. In fact [**28**] shows that if $\mathcal{X}$ is a Banach space such that every closed subspace of $\mathcal{X}$ has a closed complement, then $\mathcal{X}$ is isomorphic to a Hilbert space.

The next result is left to the reader to prove.

6.6.2. Theorem. *Let $\mathcal{X}$ be a Banach space.*

(a) *If $\mathcal{M}$ and $\mathcal{N}$ are closed complementary subspaces of $\mathcal{X}$ and E is defined on $\mathcal{X}$ by $E(x + y) = x$ whenever $x \in \mathcal{M}$ and $y \in \mathcal{N}$, then E is a bounded linear transformation such that $E^2 = E$, $\operatorname{ran} E = \mathcal{M}$, and $\ker E = \mathcal{N}$.*

(b) *If $E \in \mathcal{B}(\mathcal{X})$ such that $E^2 = E$, then $\mathcal{M} = \operatorname{ran} E$ and $\mathcal{N} = \ker E$ are complementary subspaces of $\mathcal{X}$.*

Exercises.

(1) If $\mathcal{X}$ is a Banach space and E is a linear transformation on $\mathcal{X}$ such that $E^2 = E$ and both ran E and ker E are closed, show that E is bounded.

(2) If $\mathcal{X}$ is a Banach space and $\mathcal{M} \leq \mathcal{X}$ and if $\mathcal{M}$ has a complement, then every complement is isomorphic as a Banach space to $\mathcal{X}/\mathcal{M}$.

(3) Give the details of the proof of Theorem 6.6.2.

6.7. The Principle of Uniform Boundedness

This is the third pillar of functional analysis.

6.7.1. Theorem (Principle of Uniform Boundedness). *Let $\mathcal{X}$ be a Banach space and let $\mathcal{Y}$ be a normed space. If $\mathcal{A} \subseteq \mathcal{B}(\mathcal{X}, \mathcal{Y})$ such that $\sup\{\|Ax\| : A \in \mathcal{A}\} < \infty$ for every x in X, then*

$$\sup\{\|A\| : A \in \mathcal{A}\} < \infty$$

Proof. For each natural number n let

$$\mathcal{B}_n = \{x \in \mathcal{X} : \|Ax\| \leq n \text{ for all } A \text{ in } \mathcal{A}\}$$

According to the hypothesis, $\mathcal{X} = \bigcup_{n=1}^{\infty} \mathcal{B}_n$. It is easy to see that each $\mathcal{B}_n$ is closed so the Baire Category Theorem (A.1.3) implies there is an integer n with int $\mathcal{B}_n \neq \emptyset$. Thus there is an x_0 in $\mathcal{B}_n$ and an $r > 0$ with $B(x_0; r/2) \subseteq \mathcal{B}_n$.

Let $M = \sup\{\|Ax_0\| : A \in \mathcal{A}\}$. Let $x \in \text{ball } \mathcal{X}$ and let A be an arbitrary linear transformation in $\mathcal{A}$. Now $x_0 + (r/2)x \in \mathcal{B}_n$ and so $\|A((r/2)x + x_0)\| \leq n$. Therefore $\|Ax\| = 2r^{-1}\|A((r/2)x + x_0) - Ax_0\| \leq 2r^{-1}(n + M) \equiv C$. Thus $\|A\| \leq C$ for every A in $\mathcal{A}$, as desired. ∎

PUB will refer to Theorem 6.7.1 as well as any other result in this section with the same flavor. So the PUB says that a set of operators is uniformly bounded if and only if it is pointwise bounded, an unusual occurrence in analysis.

6.7.2. Corollary. *If $\mathcal{X}$ is a normed space and S is a subset of $\mathcal{X}$, then S is bounded if and only if for every f in $\mathcal{X}^*$, $\sup\{|f(x)| : x \in S\} < \infty$.*

Proof. Consider $\mathcal{X}$ as a subset of $\mathcal{B}(\mathcal{X}^*, \mathbb{F})$ as follows. For each x in $\mathcal{X}$ let $\widehat{x} : \mathcal{X}^* \to \mathbb{F}$ be defined as $\widehat{x}(f) = f(x)$ for each f in $\mathcal{X}^*$. Even though $\mathcal{X}$ may not be complete, $\mathcal{X}^*$ is a Banach space (1.5.6). Therefore we can now apply the PUB to conclude that $\sup\{\|\widehat{x}\| : x \in S\} < \infty$. But according to Corollary 6.3.9, $\|\widehat{x}\| = \|x\|$ for every x, so that S must be bounded. ∎

Clearly this is a useful and powerful tool and helps to explain our emphasis in the earlier parts of this book in determining the dual spaces of various Banach spaces such as $C(X)$ and $L^p(\mu)$.

6.7.3. Corollary. *If $\mathcal{X}$ is a Banach space and $\mathcal{F} \subseteq \mathcal{X}^*$, then $\sup\{\|f\| : f \in \mathcal{F}\} < \infty$ if and only if for each x in $\mathcal{X}$ we have $\sup\{|f(x)| : f \in \mathcal{F}\} < \infty$.*

Proof. Consider $\mathcal{X}^*$ as $\mathcal{B}(\mathcal{X}, \mathbb{F})$ and apply the PUB. ∎

6.7.4. Corollary. *If $\mathcal{X}$ is a Banach space and $\mathcal{Y}$ is a normed space, then a subset $\mathcal{A}$ of $\mathcal{B}(\mathcal{X}, \mathcal{Y})$ is uniformly bounded if and only if for each x in $\mathcal{X}$ and each f in $\mathcal{X}^*$,*

$$\sup\{|f(Ax)| : A \in \mathcal{A}\} < \infty$$

Proof. Combine Corollary 6.7.2 with Theorem 6.7.1. ∎

The next result could easily be classified as a corollary of the PUB, but it has a somewhat different flavor.

6.7.5. Theorem (Banach–Steinhaus[1] Theorem). *If $\mathcal{X}$ and $\mathcal{Y}$ are Banach spaces and $\{A_n\}$ is a sequence in $\mathcal{B}(\mathcal{X}, \mathcal{Y})$ with the property that for every x in $\mathcal{X}$, $\{A_n x\}$ is a convergent sequence in $\mathcal{Y}$; then there is an A in $\mathcal{B}(\mathcal{X}, \mathcal{Y})$ such that $\|A_n x - Ax\| \to 0$ for every x in $\mathcal{X}$ and $\sup_n \|A_n\| < \infty$.*

Proof. Define $A : \mathcal{X} \to \mathcal{Y}$ by $Ax = \lim_n A_n x$ for all x in $\mathcal{X}$; this is well defined by the hypothesis. It is easily checked that A is a linear transformation. The PUB implies that there is a constant M such that $\|A_n\| \le M$ for all n. Thus $\|Ax\| \le M\|x\|$ for all x, so $A \in \mathcal{B}(\mathcal{X}, \mathcal{Y})$. ∎

The next result would be challenging to prove without the aid of the PUB.

6.7.6. Proposition. *If X is a compact space and $\{f_n\}$ is a sequence in $C(X)$, then $\int f_n d\mu \to 0$ for every regular Borel signed measure μ if and only if the functions $\{f_n\}$ are uniformly bounded and $f_n(x) \to 0$ for every x in X.*

---

[1]Hugo Steinhaus was born in 1887 in Jaslo, Galicia in the Polish part of the Austrian Empire. The majority of his mathematical education was in Göttingen where he worked under the direction of Hilbert and received his degree in 1911. After service in the Polish army in World War I, he lived in Krakow. He liked taking long walks and relates the following story. "During one such walk I overheard the words 'Lebesgue measure'. I approached the park bench and introduced myself to the two young apprentices of mathematics. They told me they had another companion by the name of Witold Wilkosz, whom they extravagantly praised. The youngsters were Stefan Banach and Otto Nikodym. From then on we would meet on a regular basis, and ... we decided to establish a mathematical society." Later Steinhaus moved to Lvov and became a Professor at the university. He and Banach had many collaborations. After World War II he moved to the University in Wrocław. He wrote several excellent monographs, contributed to the development of Polish mathematics, and made many significant contributions to functional analysis, measure theory, and probability. Steinhaus died in 1972 in Wrocław, Poland.

Proof. Recall that $M(X)$ denotes the space of all regular Borel signed measures on X. The Riesz Representation Theorem says that $C(X)^*$ is isometrically isomorphic to $M(X)$ when $M(X)$ has its total variation norm. Therefore the statement that $\int f_n d\mu \to 0$ for every μ in $M(X)$ is the statement that $L(f_n) \to 0$ for every L in $C(X)^*$. By Corollary 6.7.3, $\sup_n \|f_n\| < \infty$. By taking $\mu = \delta_x$, the unit point mass at x, we get that $f_n(x) \to 0$ for every x in X. The converse is an immediate consequence of the DCT. ∎

Exercises.

(1) If $1 < p < \infty$ and $\{x_n\} \subseteq \ell^p$, show that $\sum_{j=1}^{\infty} x_n(j)y(j) \to 0$ for every y in ℓ^q with $\frac{1}{p} + \frac{1}{q} = 1$ if and only if $\sup_n \|x_n\|_p < \infty$ and $x_n(j) \to 0$ for all $j \geq 1$.

(2) If $\{x_n\} \subseteq \ell^1$, show that $\sum_{j=1}^{\infty} x_n(j)y(j) \to 0$ for every y in c_0 if and only if $\sup_n \|x_n\|_1 < \infty$ and $x_n(j) \to 0$ for all $j \geq 1$.

(3) If $(X, \mathcal{A}, \mu)$ is a measure space, $1 < p < \infty$, and $\{f_n\} \subseteq L^p(\mu)$, show that $\int f_n g d\mu \to 0$ for every g in $L^q(\mu)$ with $\frac{1}{p} + \frac{1}{q} = 1$ if and only if $\sup_n \|f_n\|_p < \infty$ and $\int_E f_n d\mu \to 0$ for every E in $\mathcal{A}$ with $\mu(E) < \infty$.

(4) If $(X, \mathcal{A}, \mu)$ is a σ-finite measure space and $\{f_n\} \subseteq L^1(\mu)$, show that $\int f_n g d\mu \to 0$ for every g in $L^\infty(\mu)$ if and only if $\sup_n \|f_n\|_1 < \infty$ and $\int_E f_n d\mu \to 0$ for every E in $\mathcal{A}$.

(5) Let $\mathcal{H}$ be a Hilbert space with orthonormal basis $\{e_k\}$. If $\{h_n\} \subseteq \mathcal{H}$, show that $\langle h_n, h \rangle \to 0$ for every h in $\mathcal{H}$ if and only if $\sup_n \|h_n\| < \infty$ and for every $k \geq 1$, $\langle h_n, e_k \rangle \to 0$ as $n \to \infty$.

(6) If (X, d) is a metric space and $\mathcal{X}$ is a normed space, recall that a function $f : X \to \mathcal{X}$ is *Lipschitz* if there is a constant M such that $\|f(x) - f(y)\| \leq M d(x, y)$ for all x, y in X. Show that $f : X \to \mathcal{X}$ is Lipschitz if and only if for every L in $\mathcal{X}^*$, $L \circ f : X \to \mathbb{F}$ is Lipschitz.

Locally Convex Spaces

This short chapter just scratches the surface of this subject. Its role is only to provide a framework for the examination of the weak and weak* topologies in the next chapter.

7.1. Basics of locally convex spaces

Here we will explore some basic results we have seen before in a more general setting.

7.1.1. Definition. A *locally convex space* or LCS is a vector space $\mathcal{X}$ over $\mathbb{F}$ together with a collection of seminorms $\mathcal{P}$ satisfying $\bigcap\{p^{-1}(0) : p \in \mathcal{P}\} = (0)$.

The use of the term "locally" here is typical of mathematics. It is usually used only in a topological space and refers to a phenomenon that occurs in arbitrarily small open neighborhoods of each point. We saw this when we defined a locally compact space as one such that inside each neighborhood of a point x there was another open set containing x that had compact closure. Here the same thing applies to "locally convex", though the full justification for this statement must wait until we define the topology on an LCS and establish Theorem 7.1.6 below. There are many examples of LCSs; for example, if we look at each of the examples in (6.1.2) and let $\mathcal{P}$ be all the seminorms defined in the manner indicated there, the resulting space is an LCS. In addition each normed space is an LCS where $\mathcal{P}$ consists of the norm alone.

There are examples of non-locally convex vector spaces that have a reasonable topology (that is, the topology coordinates well with the linear structure). Examples are the spaces $L^p(\mu)$ when $0 < p < 1$. We refer to the book [9] and any book on topological vector spaces.

The theory of LCSs is well developed with a rich literature. One of the main reasons we are looking at this generalization of normed spaces here is to explore the weak and weak* topologies on a Banach space and its dual in the next chapter. By examining these weak topologies we will see the force of functional analysis.

7.1.2. Example. (a) Every normed space is an LCS.

(b) Let X be a topological space, $\mathcal{X} = C(X)$, and for every compact subset K of X, let $p_K(f) = \sup\{|f(x)| : x \in K\}$. $C(X)$ with the family of seminorms $\{p_K : K$ is a compact subset of $K\}$ is an LCS.

(c) Let $\mathcal{S}$ denote the Schwartz space (4.12.3) and for all $m, n \geq 0$ let $p_{m,n}(\phi) = \sup\{|x^m \phi^{(n)}(x)| : x \in \mathbb{R}\} < \infty$ for each ϕ in $\mathcal{S}$. $\mathcal{S}$ with the seminorms $\{p_{m,n} : m, n \geq 0\}$ is an LCS.

There are many basic facts we have to verify, and these will be left to the reader who can do so and thereby fix the ideas in his/her head. To start, every LCS is a topological space. Here we define $\mathcal{T}$ to be the topology on $\mathcal{X}$ with a subbasis consisting of the sets $\{x : p(x - x_0) < r\}$, where p varies over the collection of seminorms $\mathcal{P}$, x_0 varies over the points in $\mathcal{X}$, and $r > 0$. That is, if $\mathcal{S}$ is the collection of all such sets $\{x : p(x - x_0) < r\}$ and $\mathcal{B}$ is the collection of all the finite intersections of sets from $\mathcal{S}$, then every set in $\mathcal{T}$ is the union of some collection of sets from $\mathcal{B}$. Equivalently, a set U belongs to $\mathcal{T}$ if and only if for each x_0 in U there are $p_1, \ldots, p_n$ in $\mathcal{P}$ and $r_1, \ldots, r_n > 0$ with $\bigcap_{k=1}^n \{x : p_k(x - x_0) < r_j\} \subseteq U$. A net $\{x_i\}$ in $\mathcal{X}$ converges to x if and only if $p(x_i - x) \to 0$ for every p in $\mathcal{P}$. (See the Appendix for information on nets. Starting here we will be using nets more frequently, so the reader should be sure (s)he feels comfortable with the concept. (S)He can start by writing out the details proving what was just stated.) $\mathcal{T}$ is a Hausdorff topology. In fact, if $x \neq y$ then there is a p in $\mathcal{P}$ with $r = p(x - y) \neq 0$; thus $U = \{z : p(z - x) < r/2\}$ and $V = \{z : p(z - y) < r/2\}$ are open sets and if $z \in U \cap V$, then the triangle inequality shows that $p(x - y) = p(x - z + z - y) \leq p(x - z) + p(z - y) < r$, a contradiction. Thus $U \cap V = \emptyset$ and $\mathcal{T}$ is Hausdorff. Whenever we have an LCS, we will always use the topology just defined. Here are some basic properties of this topology, the proofs of which are routine.

7.1.3. Proposition. *Let $\mathcal{X}$ be an LCS.*

(a) *The map $\mathcal{X} \times \mathcal{X} \to \mathcal{X}$ defined by $(x, y) \mapsto x + y$ is continuous.*

(b) *The map* $\mathbb{F} \times \mathcal{X} \to \mathcal{X}$ *defined by* $(a, x) \mapsto ax$ *is continuous.*

If $y \in \mathcal{X}$ and $a \in \mathbb{F}$ with $a \neq 0$, then the preceding proposition says $x \mapsto ax + y$ is continuous. But it also has a continuous inverse, so it is a homeomorphism. In particular, when U is an open set in $\mathcal{X}$, $aU + y$ is open. Put this in your memory banks.

If $\mathcal{X}$ is an LCS, we may encounter seminorms that do not belong to the original collection $\mathcal{P}$ that define the topology on $\mathcal{X}$ and we want to decide if such seminorms are continuous. The next result is useful for this.

7.1.4. Proposition. *If $\mathcal{X}$ is an LCS and p is a seminorm on $\mathcal{X}$, then the following statements are equivalent.*

(a) *p is continuous.*

(b) *$\{x : p(x) < 1\}$ is open.*

(c) *$0 \in \operatorname{int} \{x : p(x) < 1\}$.*

(d) *p is continuous at 0.*

(e) *There is a continuous seminorm q on $\mathcal{X}$ such that $p(x) \leq q(x)$ for all x.*

Proof. Clearly (a) implies (b) and this implies (c). Assume (c). If $\epsilon > 0$, then $\{x : p(x) < \epsilon\} = \epsilon\{x : p(x) < 1\}$ and so (c) implies there is an open set V in $\mathcal{X}$ that contains 0 such that $V \subseteq \{x : p(x) < \epsilon\}$. This says that (d) holds. Now assume (d). So there is an open set U in $\mathcal{X}$ such that $0 \in U \subseteq \{x : p(x) < 1\}$. From the definition of the topology there are seminorms $p_1, \ldots, p_n$ in $\mathcal{P}$ and positive $\epsilon_1, \ldots, \epsilon_n$ such that $\bigcap_{j=1}^n \{x : p_j(x) < \epsilon_j\} \subseteq U$. If $q(x) = \sum_{j=1}^n \epsilon_j p_j(x)$, then we have that $\{x : q(x) < 1\} \subseteq \{x : p(x) < 1\}$. Clearly q is a continuous seminorm and by Lemma 6.1.3, $p \leq q$. Now assume (e). To show that p is continuous, let $x \in \mathcal{X}$ and suppose $\{x_i\}$ is a net in $\mathcal{X}$ that converges to x. So $|p(x) - p(x_i)| \leq p(x - x_i) \leq q(x - x_i)$, and this converges to 0 since q is continuous. ∎

One virtue of this last proposition is that when we have an LCS it is often convenient to enlarge the family $\mathcal{P}$ by adding all the seminorms that are finite linear combinations of the original family. Alternatively, we can assume that $\mathcal{P}$ includes all continuous seminorms. That may strike you as a bit bizarre, but where is the harm?

Recall the definition of a convex set (3.2.5) as well as the definition of the convex hull of a set A, $\operatorname{co}(A)$, and the closed convex hull, $\overline{\operatorname{co}}(A)$. We will need the following intuitive result about convex sets in an LCS.

7.1.5. Proposition. *Let $\mathcal{X}$ be an LCS and let A be a convex subset of $\mathcal{X}$. If $a \in \operatorname{int} A$ and $b \in \operatorname{cl} A$, then $[a, b) \equiv \{tb + (1 - t)a : 0 \leq t < 1\} \subseteq \operatorname{int} A$.*

Proof. Fix t with $0 < t < 1$ and put $c = tb + (1-1)a$. Since a is an interior point there is an open set V in $\mathcal{X}$ with $0 \in V$ and $a + V \subseteq A$ (see Exercise 2). Hence for any x in A

$$
\begin{aligned}
A \supseteq\ & tx + (1-t)(a+V) \\
& = t(x-b) + tb + (1-t)(a+V) \\
& = [t(x-b) + (1-t)V] + c
\end{aligned}
$$

Thus the proof will be finished if we show that x can be chosen in A such that $0 \in t(x-b) + (1-t)V$, since this set is open for any choice of x. Now $t^{-1}(1-t)V$ is an open neighborhood of 0, and so $b - t^{-1}(1-t)V$ is an open neighborhood of b; since $b \in \operatorname{cl} A$, there is an x in A such that $x \in b - t^{-1}(1-t)V$. For this choice of x, $tx \in tb - (1-t)V$ and so $0 \in t(x-b) + (1-t)V$. ∎

We are now in a position to characterize the sets of the form $V = \{x \in \mathcal{X} : p(x) < 1\}$ when p is a continuous seminorm. This will be important when we use the HBT to establish a variety of geometric consequences. Of course one property that V has is that it is an open neighborhood of 0 and another is that it is *balanced*: when $x \in V$ and α is a scalar with $|\alpha| \leq 1$, then $\alpha x \in V$. These conditions turn out to be sufficient.

7.1.6. Proposition. *If $\mathcal{X}$ is an LCS and V is a non-empty open, convex, balanced subset of $\mathcal{X}$, then there is a unique continuous seminorm p on $\mathcal{X}$ such that*

$$
V = \{x \in \mathcal{X} : p(x) < 1\}
$$

Proof. Define $p(x)$ by

$$
p(x) = \inf\{t > 0 : x \in tV\}
$$

This makes sense because the fact that V is open implies $\mathcal{X} = \bigcup_{t>0} tV$. It is easily seen that $p(0) = 0$. Now suppose α is any scalar, and let's show that $p(\alpha x) = |\alpha| p(x)$; it suffices to assume that $\alpha \neq 0$. Since V is balanced, note that $\alpha^{-1}V = |\alpha|^{-1}V$, and so

$$
\begin{aligned}
p(\alpha x) &= \inf\{t > 0 : \alpha x \in tV\} \\
&= \inf\{t > 0 : x \in t\left(\alpha^{-1}V\right)\} \\
&= \inf\{t > 0 : x \in t\left(|\alpha|^{-1}V\right)\} \\
&= |\alpha| \inf\left\{\frac{t}{|\alpha|} > 0 : x \in \frac{t}{|\alpha|}V\right\} \\
&= |\alpha| p(x)
\end{aligned}
$$

Now let $x, y \in \mathcal{X}$ and put $\alpha = p(x), \beta = p(y)$. By the convexity of V, when $a, b \in V$ we have that

$$\alpha a + \beta b = (\alpha + \beta) \left(\frac{\alpha}{\alpha + \beta} a + \frac{\beta}{\alpha + \beta} b \right) \in (\alpha + \beta) V$$

Now for any positive δ we have $\alpha + \delta > p(x)$ so that $x \in (\alpha + \delta)V$; similarly $y \in (\beta + \delta)V$. Thus $x + y \in (\alpha + \delta)V + (\beta + \delta)V = (\alpha + \beta + 2\delta)V$. (Why?) Thus $x + y \in (\alpha + \beta + 2\delta)V$, and so $p(x + y) \leq \alpha + \beta + 2\delta = p(x) + p(y) + 2\delta$. Since δ was arbitrary, this completes the proof that p is a seminorm.

It remains to show that $V = \{x : p(x) < 1\}$, which, by Proposition 7.1.4, automatically implies that p is continuous. Suppose $\alpha = p(x) < 1$; if $\alpha < \beta < 1$, then $x \in \beta V \subseteq V$. Conversely, if $x \in V$, then $p(x) \leq 1$ by definition. But because V is open at x there is an $\epsilon > 0$ such that for $0 < t < \epsilon$, $y = x + tx \in V$, so that $p(y) \leq 1$. Thus $x = (1 + t)^{-1} y$ and so $p(x) = (1 + t)^{-1} p(y) \leq (1 + t)^{-1} < 1$.

The fact that p is unique follows from Lemma 6.1.3. ∎

The seminorm p obtained in the preceding result for the open convex set V is called the *gauge* of V. In light of this result we see that a vector space $\mathcal{X}$ with a topology such that vector addition and scalar multiplication are continuous is an LCS when about 0 there is a fundamental neighborhood system consisting of convex, balanced sets. This justifies the use of the word "local".

The proof of the next result is similar to the preceding proof and characterizes the set $\{x : q(x) < 1\}$ when q is a non-negative sublinear functional. Note that the difference between the open set G in the next proposition and the set V in the preceding one is that V is balanced while G is not so assumed. An examination of the proof of (7.1.6) shows that the only place where this was used is in showing that p is a seminorm rather than a semilinear functional.

7.1.7. Proposition. *If $\mathcal{X}$ is an LCS and G is an open convex subset of $\mathcal{X}$ that contains the origin, then*

$$q(x) = \inf\{t \geq 0 : x \in tG\}$$

is a non-negative sublinear functional such that $G = \{x : q(z) < 1\}$.

So to find an example of a sublinear functional that is not a seminorm we need only find an open convex set containing 0 that is not balanced. You can do this in $\mathbb{R}^2$.

Exercises.

(1) Prove Proposition 7.1.3.

(2) Let $\mathcal{X}$ be an LCS. This exercise asks the reader to prove several things stated in the section and a thing or two that was not previously mentioned. (a) Show that if $t \in \mathbb{F}$ and $t \neq 0$, then the map $x \mapsto tx$ is a homeomorphism of X. Also for any x_0 in $\mathcal{X}$, the map $x \mapsto x + x_0$ is a homeomorphism of $\mathcal{X}$. (b) If V is an open subset of $\mathcal{X}$ containing 0, then $\bigcup_{t>0} tV = \mathcal{X}$. (c) If G is an open subset of $\mathcal{X}$ that contains x_0, then there is an open set V containing 0 such that $x_0 + V \subseteq G$. (d) If K is a compact subset of $\mathcal{X}$ and V is an open neighborhood of 0, then there is an $\epsilon > 0$ such that $tK \subseteq V$ for $0 < t < \epsilon$.

(3) Show that if A is a convex set and $\alpha, \beta > 0$, then $\alpha A + \beta A = (\alpha + \beta)A$. Give an example of a non-convex set A where this is not true.

(4) If $\mathcal{X}$ is an LCS and A is a closed subset of $\mathcal{X}$, show that A is convex if and only if $\frac{1}{2}(x + y) \in A$ whenever $x, y \in A$.

(5) If $\mathcal{X}$ is an LCS and $A \subseteq \mathcal{X}$, show that
$$\operatorname{cl} A = \bigcap \{a + V : 0 \in V \text{ and } V \text{ is open}\}.$$

(6) If $\mathcal{X}$ is an LCS whose topology is defined by the family $\mathcal{P}$ of seminorms, show that a seminorm p on $\mathcal{X}$ is continuous if and only if there are $p_1, \ldots, p_n$ in $\mathcal{P}$ and positive scalars $t_1, \ldots, t_n$ such that
$$p(x) \leq \sum_{k=1}^{n} t_k p_k(x)$$
for all x in $\mathcal{X}$.

(7) Let X be a locally compact space and consider the space $C(X)$ of all (not just the bounded) continuous functions. As in Example 7.1.2, for each compact subset K of X define the seminorm $p_K(f) = \sup\{|f(x)| : x \in K\}$. Give $C(X)$ the topology defined by the family of seminorms $\{p_K : K \text{ is a compact subset of } X\}$. This is designated as (co) and referred to as the compact-open topology. Prove the following. (a) A net $\{f_i\}$ in $C(X)$ converges to f (co) if and only if $f_i(x) \to f(x)$ uniformly on compact subsets of X. (b) A linear functional $L : C(X) \to \mathbb{F}$ is (co)-continuous if and only if there a positive number M and a compact subset K such that $|L(f)| \leq M p_K(f)$ for every f in $C(X)$. (c) The subspace $C_0(X)$ is (co)-dense in $C(X)$. (d) L is a (co)-continuous linear functional on $C(X)$ if and only if there is an $\mathbb{F}$-valued measure μ on X that has compact support such that $L(f) = \int f d\mu$ for every f in $C(X)$. (Hint: If L is (co)-continuous, show that there is a compact subset K of X such that if we let $\widetilde{L} : C(K) \to \mathbb{F}$ be defined for each g

in $C(K)$ by $\widetilde{L}(g) = L(\widetilde{g})$, where $\widetilde{g}$ is any continuous extension of g to X, then $\widetilde{L}$ is a well-defined bounded linear functional on the Banach space $C(K)$.)

(8) If X is a locally compact space, for each ϕ in $C_0(X)$ define $p_\phi : C_b(X) \to [0,\infty)$ by $p_\phi(f) = \|f\phi\|$. (a) Show that p_ϕ is a seminorm. Let β be the topology defined on $C_b(X)$ by $\mathcal{P} = \{p_\phi : \phi \in C_0(X)\}$. (b) Show that $C_b(X)$ with the β-topology is a complete LCS. (That is, every Cauchy net converges.) β is called the *strict topology*.

(9) If $\mathcal{X}$ and $\mathcal{Y}$ are LCSs and $T : \mathcal{X} \to \mathcal{Y}$ is a linear transformation, show that T is continuous if and only if for every continuous seminorm p on $\mathcal{Y}$ we have that $p \circ T$ defines a continuous seminorm on $\mathcal{X}$.

(10) If $\mathcal{S}$ is the space of Schwartz functions on $\mathbb{R}$ and $\mathcal{S}$ has the topology defined by its seminorms as in Example 7.1.2(c), show that the Fourier transform $\mathcal{F} : \mathcal{S} \to \mathcal{S}$ is a homeomorphism (4.12.8).

7.2. Metrizable locally convex spaces*

If we are given an LCS, when is there a metric on the space that gives the same topology?

7.2.1. Proposition. *Let $p_1, p_2, \ldots$ be a sequence of seminorms on $\mathcal{X}$ such that $\bigcap_n p_n^{-1}(0) = (0)$. If for x, y in $\mathcal{X}$ we define*

$$d(x, y) = \sum_{n=1}^{\infty} 2^{-n} \frac{p_n(x-y)}{1 + p_n(x-y)}$$

then d is a metric on $\mathcal{X}$ and the topology on $\mathcal{X}$ defined by d is the same as the topology defined by the family of seminorms $\{p_1, p_2, \ldots\}$.

Proof. The proof that d is a metric is left as an exercise since it is similar to something proved in all courses on metric spaces. (In fact it is not difficult to see that when p is a seminorm, $p(x-y)[1+p(x-y)]^{-1}$ satisfies the triangle inequality.) Let $\mathcal{T}$ be the topology defined on $\mathcal{X}$ by this metric and let $\mathcal{U}$ be the topology defined by the seminorms $\{p_1, p_2, \ldots\}$. If $d(x_k, x) \to 0$, then the fact that for each $n \geq 1$, $2^{-n} p_n(x_k - x)[1 + p_n(x_k - x)]^{-1} \leq d(x_k, x)$ implies that $x_k \to x$ $(\mathcal{U})$. Conversely, assume $x_k \to x$ $(\mathcal{U})$; this implies that $p_n(x_k - x) \to 0$ for each $n \geq 1$. If $\epsilon > 0$, choose N such that $\sum_{n=N+1}^{\infty} 2^{-n} < \epsilon/2$. Now pick k_0 such that when $k \geq k_0$, $p_n(x_k - x)[1 + p_n(x_k - x)]^{-1} < \epsilon/2$ for $1 \leq n \leq N$. It follows that $d(x_k, x) < \epsilon$ when $k \geq k_0$. Thus $\mathcal{T} = \mathcal{U}$. ∎

7.2.2. Theorem. *If $\mathcal{X}$ is an LCS, then the following statements are equivalent.*

(a) *$\mathcal{X}$ is metrizable.*

(b) *There is a countable neighborhood base at 0. That is, there is sequence $\{U_n\}$ of neighborhoods of 0 such that if G is open and $0 \in G$, then there is a U_n contained in G.*

(c) *The topology of $\mathcal{X}$ is defined by a sequence of seminorms.*

Proof. Clearly (a) implies (b) and, by the preceding proposition, (c) implies (a). It remains to show that (b) implies (c). Let $\{U_n\}$ be as in (b). For each $n \geq 1$ there is a continuous seminorm p_n such that $\{x : p_n(x) < 1\} \subseteq U_n$. The claim is that the seminorms $\{p_n\}$ define the topology on $\mathcal{X}$. In fact suppose that $p_n(x_j) \to p_n(x)$ for each $n \geq 1$. If V is any neighborhood of x, then $V - x$ is a neighborhood of 0 and so there is an n with $\{y : p_n(y) < 1\} \subseteq U_n \subseteq V - x$. But there is a j_0 such that $p_n(x_j - x) < 1$ for all $j \geq j_0$. Thus, $x_j \in V$ for all $j \geq j_0$. ∎

Exercises.

(1) Say that a subset B of an LCS $\mathcal{X}$ is *bounded* if for every open neighborhood U of 0 there is an $\epsilon > 0$ such that $\epsilon B \subseteq U$. (a) Show that if U is a bounded, open, convex, and balanced set in $\mathcal{X}$, then there is a norm p such that $U = \{x : p(x) < 1\}$. (b) Show that an LCS $\mathcal{X}$ has a norm that defines its topology (that is, that $\mathcal{X}$ is normable) if and only if $\mathcal{X}$ has a bounded open set.

(2) Let $\mathcal{X}$ be an LCS and prove the following. (a) If B is a bounded set (preceding exercise) then so is $\operatorname{cl} B$. (b) All compact subsets of $\mathcal{X}$ are bounded and the finite union of bounded sets is bounded. (c) If $B \subseteq \mathcal{X}$, then B is bounded if and only if for every sequence $\{x_n\}$ in B and for every sequence $\{\alpha_n\}$ in c_0, $\alpha_n x_n \to 0$. (d) If $\mathcal{Y}$ is an LCS and $T : \mathcal{X} \to \mathcal{Y}$ is a continuous linear transformation, then $T(B)$ is bounded in $\mathcal{Y}$ whenever B is bounded in $\mathcal{X}$. (e) If $B \subseteq \mathcal{X}$, then B is bounded if and only if $\sup\{p(x) : x \in B\} < \infty$ for every continuous seminorm p on $\mathcal{X}$.

(3) Let X be a locally compact space and consider the strict topology β on $C_b(X)$. (See Exercise 7.1.8.) (a) Show that a subset of $C_b(X)$ is β-bounded if and only if it is norm bounded. (b) Show that the strict topology is metrizable if and only if X is compact.

7.3. Geometric consequences

In this section we explore various geometric results that are a consequence of the HBT, the correspondence between hyperplanes and linear functionals, and the characterization of open, convex, balanced neighborhoods of the origin as the sets $\{x : p(x) < 1\}$ for a continuous seminorm p. See (7.1.6) and the similar result (7.1.7) where the seminorm is replaced by the non-negative sublinear functional.

We will need certain facts in the setting of an LCS that were proved for linear functionals on normed spaces. In particular if $\mathcal{X}$ is an LCS, then a hyperplane $\mathcal{M}$ in $\mathcal{X}$ is either closed or dense and it is closed if and only if the corresponding linear functional is continuous (6.3.2). Also a linear functional on an LCS is continuous if and only if it is continuous at 0 (1.5.3). For an LCS whose topology is defined by the family $\mathcal{P}$ of seminorms, a linear functional $f : \mathcal{X} \to \mathbb{F}$ is continuous if and only if there are $p_1, \ldots, p_n$ in $\mathcal{P}$ and positive scalars $t_1, \ldots, t_n$ such that $|f(x)| \leq \sum_{k=1}^{n} t_k p_k(x)$ for all x in $\mathcal{X}$. The proof of this is similar to the proof that a linear functional is continuous on a normed space if and only if it is bounded. Another way to prove this is to show that f is continuous if and only if $x \mapsto |f(x)|$ is a continuous seminorm and then invoke Exercise 7.1.6.

7.3.1. Theorem. *If $\mathcal{X}$ is an LCS and G is an open convex subset of $\mathcal{X}$ such that $0 \notin G$, then there is a closed hyperplane $\mathcal{M}$ such that $\mathcal{M} \cap G = \emptyset$.*

Proof. We have to consider two cases depending on whether the underlying scalars are $\mathbb{R}$ or $\mathbb{C}$.

Case 1. $\mathbb{F} = \mathbb{R}$.

Pick any point x_0 in G and consider $H = x_0 - G$; it follows that H is an open convex set that contains 0. By Proposition 7.1.7 there is a non-negative sublinear functional q on $\mathcal{X}$ such that $H = \{x : q(x) < 1\}$; since $x_0 \notin H$, $q(x_0) \geq 1$. Let $\mathcal{Y} = \mathbb{R}x_0$ and define $f_0 : \mathcal{Y} \to \mathbb{R}$ by $f_0(\alpha x_0) = \alpha q(x_0)$. If $\alpha \geq 0$, then $f_0(\alpha x_0) = \alpha q(x_0) = q(\alpha x_0)$. On the other hand if $\alpha < 0$, then $f(\alpha x_0) = \alpha q(x_0) < 0 < q(\alpha x_0)$. By the HBT there is a linear functional f on $\mathcal{X}$ such that $f|\mathcal{Y} = f_0$ and $f(x) \leq q(x)$ for all x in $\mathcal{X}$. Put $\mathcal{M} = \ker f$; note that $\mathcal{M} \neq (0)$ so that $\mathcal{M}$ is a hyperplane.

The fact that H is open implies that q is continuous. If $f(x) < 0$, then $-f(x) = f(-x) \leq q(-x) \leq q(x) + q(-x)$. Hence $|f(x)| \leq q(x) + q(-x)$ for all x in $\mathcal{X}$ and $x \mapsto q(x) + q(-x)$ is continuous. It follows that f is continuous. (Details?) Thus $\mathcal{M}$ is a closed hyperplane. If $x \in G$, then $f(x_0) - f(x) = f(x_0 - x) \leq q(x_0 - x) < 1$, so $f(x) > f(x_0) - 1 = q(x_0) - 1 \geq 0$. Therefore $\mathcal{M} \cap G = \emptyset$.

Case 2. $\mathbb{F} = \mathbb{C}$.

Here $\mathcal{X}$ is also a real TVS and so by Case 1 there is a continuous linear functional $f : \mathcal{X} \to \mathbb{R}$ such that $G \cap \ker f = \emptyset$. Put $F(x) = f(x) - if(ix)$ (see Lemma 6.3.5). Then $F : \mathcal{X} \to \mathbb{C}$ is a continuous linear functional and $F = \operatorname{Re} f$. Clearly $F(x) = 0$ if and only if $f(x) = f(ix) = 0$, so $\mathcal{M} = \ker F = \ker f \cap \ker f(i\cdot)$. Hence $\mathcal{M} \cap G = \emptyset$ as required. ∎

We want to parlay this result and to do this we need some additional terminology and a concept that made its appearance in the last theorem. An *affine hyperplane* is a set $\mathcal{M}$ such that for every x_0 in $\mathcal{M}$, $\mathcal{M} - x_0$ is a hyperplane. (See Exercise 2.) An *affine manifold* is a set $\mathcal{Y}$ such that for every x_0 in $\mathcal{Y}$, $\mathcal{Y} - x_0$ is a linear manifold. An *affine subspace* is defined similarly.

7.3.2. Corollary. *If $\mathcal{X}$ is an LCS, G is an open convex subset, and $\mathcal{Y}$ is an affine subspace in $\mathcal{X}$ with $G \cap \mathcal{Y} = \emptyset$, then there is a closed affine hyperplane $\mathcal{M}$ such that $\mathcal{Y} \subseteq \mathcal{M}$ and $G \cap \mathcal{M} = \emptyset$.*

Proof. Fix a point x_0 in $\mathcal{Y}$ and consider $H = G - x_0$ and $\mathcal{Z} = \mathcal{Y} - x_0$. It follows that H is an open convex set, $\mathcal{Z}$ is a closed subspace, and $H \cap \mathcal{Z} = \emptyset$. If $Q : \mathcal{X} \to \mathcal{X}/\mathcal{Z}$ is the natural map, then $Q(H)$ is convex and, since $H \cap \mathcal{Z} = \emptyset$, $0 \notin Q(H)$. Since Q is an open map (proved for a normed space (6.2.2) and virtually the same proof works here), $Q(H)$ is open in the quotient space. By the preceding theorem there is a closed hyperplane $\mathcal{N}$ in $\mathcal{X}/\mathcal{Z}$ such that $Q(H) \cap \mathcal{N} = \emptyset$. Let $\mathcal{M}_0 = Q^{-1}(\mathcal{N})$. It is easy to check that $\mathcal{Y} - x_0 \subseteq \mathcal{M}_0$ and $\mathcal{M}_0 \cap (G - x_0) = \emptyset$. From here it follows that $\mathcal{M} = \mathcal{M}_0 + x_0$ has the desired properties. ∎

When it comes to discussing geometric properties in an LCS there is an advantage if $\mathcal{X}$ is a space over $\mathbb{R}$. Namely, when $f : \mathcal{X} \to \mathbb{R}$ is a continuous linear functional, then the hyperplane $\ker f$ disconnects the space, where $\mathcal{X} \backslash \ker f$ has the two components $\{x : f(x) < 0\}$ and $\{x : f(x) > 0\}$. See Exercise 3.

7.3.3. Definition. If $\mathcal{X}$ is a real LCS, a subset S of $\mathcal{X}$ is called an *open half-space* if there is a non-zero continuous linear functional $f : \mathcal{X} \to \mathbb{R}$ and a real number α such that $S = \{x : f(x) > \alpha\}$. Similarly S is a *closed half-space* if there is a continuous non-zero linear functional $f : \mathcal{X} \to \mathbb{R}$ and a real number α such that $S = \{x : f(x) \geq \alpha\}$.

Say that two subsets A and B of $\mathcal{X}$ are *strictly separated* if they are contained in disjoint open half-spaces. They are *separated* if they are contained in two closed half-spaces whose intersection is a closed affine hyperplane.

The reader is urged to consult Exercise 4.

The next result can easily lay claim to being the most important "separation" theorem. The one that is most often used, however, is Theorem 7.3.6 below.

7.3.4. Theorem. *If $\mathcal{X}$ is a real LCS and A and B are disjoint convex subsets with A open, then there is a continuous linear functional $f : \mathcal{X} \to \mathbb{R}$ and a real number α such that $A \subseteq \{x : f(x) < \alpha\}$ and $B \subseteq \{x : f(x) \geq \alpha\}$. If B is also open, then A and B are strictly separated.*

Proof. Put $G = A - B = \{a - b : a \in A, b \in B\} = \bigcup\{A - b : b \in B\}$, so that G is open. We leave it to the reader to show that G is convex and, since A and B are disjoint, $0 \notin G$. Therefore Theorem 7.3.1 implies that there is a closed hyperplane $\mathcal{M}$ such that $\mathcal{M} \cap G = \emptyset$. Let $f : \mathcal{X} \to \mathbb{R}$ be a continuous linear functional such that $\mathcal{M} = \ker f$. Since G is convex, $f(G)$ is an interval in the real line not containing 0. Thus $f(G)$ is contained in either $(0, \infty)$ or $(-\infty, 0)$; suppose $f(x) > 0$ for all x in G. Thus for a in A and b in B, $0 < f(a - b) = f(a) - f(b)$ or $f(a) < f(b)$. Therefore there is a real number α with $\sup\{f(a) : a \in A\} \le \alpha \le \inf\{f(b) : b \in B\}$. But since A is open, $f(A)$ must be an open interval – this last statement follows by the OMT but a direct, elementary proof is possible; see Exercise 5. Hence $A \subseteq \{x : f(x) < \alpha\}$ and $B \subseteq \{x : f(x) \ge \alpha\}$. If B is also open, $f(B)$ is also an open interval and so A and B are strictly separated. ∎

The next lemma has nothing to do (directly) with separation theorems, but is a basic tool in discussing compact subsets of an LCS.

7.3.5. Lemma. *If $\mathcal{X}$ is an LCS and K is a compact subset of $\mathcal{X}$, then for any open set V containing K there is an open neighborhood U of 0 such that $K \subseteq K + U = \{x + u : x \in K, u \in U\} \subseteq V$.*

Proof. Let $\mathcal{U}$ be the collection of all open sets containing 0 and order $\mathcal{U}$ by reverse inclusion. Suppose the lemma is untrue; thus for each U in $\mathcal{U}$ there is an x_U in K and a y_U in U such that $x_U + y_U \notin V$. We look at the nets (see (A.2.1(d)) in the Appendix) $\{x_U : U \in \mathcal{U}\}$ and $\{y_U : U \in \mathcal{U}\}$. By definition $y_U \to 0$; since K is compact, there is an x in K such that $x_U \to_{\text{cl}} x$. Thus $x_U + y_U \to_{\text{cl}} x$. Since $x \in V$, the definition of a clustering net implies there is a U in $\mathcal{U}$ such that $x_U + y_U \in V$, a contradiction. ∎

7.3.6. Theorem. *If $\mathcal{X}$ is a real LCS and A and B are two disjoint closed sets with one of them compact, then A and B are strictly separated.*

Proof. Assume that B is compact. Applying the preceding lemma to B and the open set $\mathcal{X} \backslash A$, there is an open neighborhood U_1 of the origin such that $(B + U_1) \cap A = \emptyset$. Since $\mathcal{X}$ is locally convex there is a continuous seminorm p such that $\{x : p(x) < 1\} \subseteq U_1$; put $U = \{x : p(x) < \frac{1}{2}\}$. It follows that $(A + U)$ and $(B + U)$ are disjoint open convex sets. The result now follows by applying Theorem 7.3.4. ∎

7.3.7. Corollary. *If $\mathcal{X}$ is a real LCS, A is a closed convex subset of $\mathcal{X}$, and $x \notin A$, then $\{x\}$ is strictly separated from A.*

Perhaps the perspicacious reader has noticed that this last corollary has a certain overlap with Corollary 6.3.10, though this earlier result has something missing from the present one and the earlier result has a metric property that is missing from the present one.

7.3.8. Corollary. *If $\mathcal{X}$ is a real LCS and A is a non-empty subset of $\mathcal{X}$, then $\overline{\mathrm{co}}(A)$ is the intersection of all closed half-spaces containing A.*

Proof. If $\mathcal{H}$ is the collection of all closed half-spaces containing A, then clearly $\overline{\mathrm{co}}(A) \subseteq \bigcap \{H : H \in \mathcal{H}\}$. If, however, $x_0 \notin \overline{\mathrm{co}}(A)$, then Corollary 7.3.7 implies there is a continuous linear functional f on $\mathcal{X}$ and a real number α such that $f(x_0) > \alpha$ and $f(x) < \alpha$ for all x in $\overline{\mathrm{co}}(A)$. Thus $H = \{x : f(x) \le \alpha\} \in \mathcal{H}$ and $x_0 \notin H$. ∎

The next corollary has an overlap with Theorem 6.3.12.

7.3.9. Corollary. *If $\mathcal{X}$ is a real LCS and A is a subset of $\mathcal{X}$, then the closed linear span of A is the intersection of all the closed hyperplanes containing A.*

Now it's time to see what kind of separation theorems we can get for complex spaces. The proof of the next result is easily accomplished by considering the complex space $\mathcal{X}$ as a real space, employing Theorem 7.3.6 to obtain the appropriate continuous linear functional into $\mathbb{R}$, and then finding the continuous linear functional $f : \mathcal{X} \to \mathbb{C}$ whose real part is the given real linear functional. (See (6.3.5).) The details are left to the reader.

7.3.10. Theorem. *If $\mathcal{X}$ is a complex LCS and A and B are two disjoint closed convex subsets with B compact, then there is a continuous linear functional $f : \mathcal{X} \to \mathbb{C}$, a real number α, and an $\epsilon > 0$ such that for a in A and b in B,*

$$\mathrm{Re}\, f(a) \le \alpha < \alpha + \epsilon \le \mathrm{Re}\, f(b)$$

7.3.11. Corollary. *If $\mathcal{X}$ is an LCS and $\mathcal{M}$ is a linear manifold in $\mathcal{X}$, then $\mathcal{M}$ is dense if and only if the only continuous linear functional on $\mathcal{X}$ that vanishes on $\mathcal{M}$ is the zero functional.*

7.3.12. Corollary. *If $\mathcal{X}$ is an LCS and $\mathcal{Y} \le \mathcal{X}$, then for any x_0 in $\mathcal{X} \backslash \mathcal{Y}$ there is a continuous linear functional f on $\mathcal{X}$ such that $f(x) = 0$ for every x in $\mathcal{Y}$ and $f(x_0) = 1$.*

Exercises.

(1) Let p be a sublinear function on $\mathcal{X}$, $G = \{x : p(x) < 1\}$, and let q be the non-negative sublinear function obtained for G as in Proposition 7.1.7. Show that $q(x) = \max\{p(x), 0\}$.

(2) Show that $\mathcal{M}$ is an affine hyperplane if and only if there exists an x_0 in $\mathcal{M}$ such that $\mathcal{M} - x_0$ is a hyperplane.

(3) (a) Show that an open subset of an LCS is connected if and only if it is arcwise connected. (b) If $\mathcal{X}$ is a real LCS and f is a real-valued continuous linear functional, show that $\mathcal{X} \backslash \ker f$ has the two components $\{x : f(x) < 0\}$ and $\{x : f(x) > 0\}$. (c) If $\mathcal{X}$ is a complex LCS and f is a continuous complex-valued linear functional, show that $\mathcal{X} \backslash \ker f$ is connected.

(4) Let $\mathcal{X}$ be a real LCS and prove the following. (a) The closure of an open half-space is a closed half-space, and the interior of a closed half-space is an open half-space. (b) If $A, B \subseteq \mathcal{X}$, then A and B are strictly separated if and only if there is a continuous linear functional $f : \mathcal{X} \to \mathbb{R}$ and a real number α such that $A \subseteq \{x : f(x) < \alpha\}$ and $B \subseteq \{x : f(x) > \alpha\}$. (c) If $A, B \subseteq \mathcal{X}$, then A and B are separated if and only if there is a continuous linear functional $f : \mathcal{X} \to \mathbb{R}$ and a real number α such that $A \subseteq \{x : f(x) \leq \alpha\}$ and $B \subseteq \{x : f(x) \geq \alpha\}$.

(5) Show that if $\mathcal{X}$ is an LCS and $f : \mathcal{X} \to \mathbb{R}$ is a continuous linear functional, then $f(A)$ is an open interval whenever A is an open convex subset of $\mathcal{X}$.

Chapter 8

Duality

In this chapter we will explore various topologies on a locally convex space $\mathcal{X}$ and its dual space $\mathcal{X}^*$, then we'll see some applications of these topologies that illustrate their power. In fact much of the utility of functional analysis for other parts of mathematics is the ability to rephrase a problem in a Banach space in terms of a problem for its dual space, where it is easier to solve. The reader might have noticed that two of the "three pillars" of the subject (HBT, OMT, PUB) are results that describe relations between a space and its dual.

8.1. Basics of duality

Once again we shift notation from the past.[1] When $\mathcal{X}$ is an LCS and $\mathcal{X}^*$ its dual space, the space of all continuous linear functionals on $\mathcal{X}$, for any x in $\mathcal{X}$ and x^* in $\mathcal{X}^*$ let

$$\langle x, x^* \rangle = \langle x^*, x \rangle = x^*(x)$$

Though the notation is inspired by the inner product in a Hilbert space, there are marked differences. To begin, it is irrelevant what occupies the first slot and what is in the second. On the other hand we do have something like the CBS inequality in that $|\langle x, x^* \rangle| \leq \|x\| \|x^*\|$.

---

[1] We have done this before and I am sure some readers find it disconcerting. The point is that notation is supposed to facilitate understanding. In this case, like the others where we have changed notation, I felt the initial choice was less complicated than the present one and therefore had an advantage when the concept was first introduced. When we first saw linear functionals, we emphasized that they were functions. Now we want to emphasize the interplay between a space and its dual. We have progressed and this new notation will promote a deeper understanding of duality.

The idea is that the elements of $\mathcal{X}$ and $\mathcal{X}^*$ play two roles. We are used to each x^* being a linear functional; that's how it's defined. But we have also looked at $\mathcal{X}^*$ as a vector space; so each x^* plays the role of vector. Similarly we now want to consider the elements x in $\mathcal{X}$ not only as vectors but also as linear functionals on $\mathcal{X}^*$. That is, each x in $\mathcal{X}$ corresponds to the linear functional on $\mathcal{X}^*$ defined by $x^* \mapsto \langle x, x^* \rangle$.

8.1.1. Definition. Let $\mathcal{X}$ be an LCS. For each x^* in $\mathcal{X}^*$ define the seminorm on $\mathcal{X}$,

$$p_{x^*}(x) = |\langle x, x^* \rangle|$$

The topology defined on $\mathcal{X}$ by the collection of seminorms $\{p_{x^*} : x^* \in \mathcal{X}^*\}$ is called the *weak topology*. Often this is denoted by $\sigma(\mathcal{X}, \mathcal{X}^*)$ (here the order of the slots is important) or by wk.

For each x in $\mathcal{X}$ define a seminorm on $\mathcal{X}^*$,

$$p_x(x^*) = |\langle x, x^* \rangle|$$

The topology defined on $\mathcal{X}^*$ by the collection of seminorms $\{p_x : x \in \mathcal{X}\}$ is called the *weak-star topology* and is denoted by $\sigma(\mathcal{X}^*, \mathcal{X})$ weak*, or wk*.

So a net $\{x_i\}$ converges weakly to 0 if and only if $\langle x_i, x^* \rangle \to 0$ for every x^* in $\mathcal{X}^*$. A subset U of $\mathcal{X}$ is weakly open if and only if for each x_0 in U there are $x_1^*, \ldots, x_m^*$ in $\mathcal{X}^*$ and $\epsilon_1, \ldots, \epsilon_m > 0$ such that $U \supseteq \bigcap_{k=1}^m \{x \in \mathcal{X} : |\langle x - x_0, x_k^* \rangle| < \epsilon_k\}$. We might point out that by replacing each x_k^* by $\epsilon_k^{-1} x_k^*$ we get that a necessary and sufficient condition for U to be weakly open is that for each x_0 in U there are $x_1^*, \ldots, x_m^*$ in $\mathcal{X}^*$ such that $U \supseteq \bigcap_{k=1}^m \{x \in \mathcal{X} : |\langle x - x_0, x_k^* \rangle| < 1\}$. (What are the analogous statements for the weak* topology?)

Observe that the given topology $\mathcal{T}$ on the LCS $\mathcal{X}$ contains the weak topology. In fact from what we just said in the last paragraph, if $U \in$ wk, $U \in \mathcal{T}$ since, by definition, each x^* is continuous in the original topology. Thus wk $\subseteq \mathcal{T}$ and the identity map $(\mathcal{X}, \mathcal{T}) \to (\mathcal{X}, \text{wk})$ is continuous. Usually the two topologies are different. Consider the following example.

8.1.2. Example. Let $\mathcal{H}$ be an infinite-dimensional Hilbert space with an orthonormal basis $\{e_n\}$. Since $\|h\|^2 = \sum_{n=1}^{\infty} |\langle h, e_n \rangle|^2$ and the dual of $\mathcal{H}$ is isomorphic to $\mathcal{H}$, we have that $e_n \to 0$ (wk). However $\|e_n\| = 1$ for all n, so the sequence $\{e_n\}$ does not converge to 0 in norm, the original topology on $\mathcal{H}$.

In spite of the differences between the weak and the original topologies, they share something in common. A remarkable and useful fact is that the two topologies have the same continuous linear functionals.

8.1.3. Theorem. *If $\mathcal{X}$ is an LCS, then $(\mathcal{X}, \text{wk})^* = \mathcal{X}^*$.*

Proof. If $\mathcal{T}$ is the topology on $\mathcal{X}$, we saw that wk $\subseteq \mathcal{T}$. Thus every f in $(\mathcal{X}, \text{wk})^*$ is in $\mathcal{X}^*$. The converse is trivial. ∎

8.1.4. Theorem. *If $\mathcal{X}$ is an LCS, then $(\mathcal{X}^*, \text{wk}^*)^* = \mathcal{X}$.*

Proof. It is clear that whenever $x \in \mathcal{X}$, $x^* \mapsto \langle x, x^* \rangle$ is a weak*-continuous linear functional. Now suppose $L : \mathcal{X}^* \to \mathbb{F}$ is a weak*-continuous linear functional. By definition there are $x_1, \ldots, x_n$ in $\mathcal{X}$ such that $|L(x^*)| \leq \sum_{k=1}^n |\langle x_k, x^* \rangle|$. But this implies that $\bigcap_{k=1}^n \ker x_k \subseteq \ker L$ and so there are $\alpha_1, \ldots, \alpha_n$ in $\mathbb{F}$ such that $L = \sum_{k=1}^n \alpha_k x_k$. (See Exercise 1.) That is, $L(x^*) = \sum_{k=1}^n \alpha_k \langle x_k, x^* \rangle = \langle x, x^* \rangle$, where $x = \sum_{k=1}^n \alpha_k x_k$. ∎

At the risk of being obscure, let's agree that unmodified topological statements about $\mathcal{X}$ mean that we are talking about the original topology, not the weak topology. So if we say a subset A is closed, we mean it is closed in the original topology. An amazing thing is that some topological statements are true simultaneously in both the original topology and the weak topology. We have seen one above; here is another.

8.1.5. Theorem. *If $\mathcal{X}$ is an LCS and A is a convex subset of $\mathcal{X}$, then the closure of A is the same as the weak closure.*

Proof. If $\mathcal{T}$ is the original topology on $\mathcal{X}$, we know that (wk) $\subseteq \mathcal{T}$; hence cl $A \subseteq$ wk-cl A. On the other hand, if $x \in \mathcal{X} \backslash$cl A, then Theorem 7.3.10 implies there is an x^* in $\mathcal{X}^*$, an α in $\mathbb{R}$, and an $\epsilon > 0$ such that Re $\langle a, x^* \rangle \leq \alpha < \alpha + \epsilon <$ Re $\langle x, x^* \rangle$ for every a in cl A. This implies that $A \subseteq B \equiv \{y \in \mathcal{X} : \text{Re} \langle y, x^* \rangle \leq \alpha\}$, a weakly closed set; hence wk-cl $A \subseteq B$. Since $x \notin B$, $x \in \mathcal{X} \backslash [\text{wk-cl } A]$ and we have proven the theorem. ∎

8.1.6. Corollary. *A convex subset of an LCS is closed if and only if it is weakly closed.*

The reader should be warned that when $\mathcal{X}$ is a normed space, the analogous statement in $\mathcal{X}^*$ to that in the preceding theorem is not true. There are convex sets in $\mathcal{X}^*$ that are norm closed but not weak*-closed. We'll see examples later.

Theorem 8.1.5 is a very powerful result, more so than its simple proof betrays. In fact that proof is simple because it uses a lot of material we have developed. To show just how strong it is, we offer the following illustration.

8.1.7. Proposition. *If X is compact and A is a convex subset of $C(X)$, then A is uniformly closed if and only if whenever $\{f_n\}$ is a bounded sequence in A such that there is an f in $C(X)$ with $f(x) = \lim_n f_n(x)$ for every x in X, then $f \in A$.*

Proof. If A satisfies the condition, then it is easy to see that it is uniformly closed. On the other hand if A is assumed norm closed and $\{f_n\}$ is a sequence in A satisfying the conditions stated there, then, by Proposition 6.7.6, $f_n \to f$ (wk). That is, f belongs to the weak closure of A. But by the preceding corollary, A is weakly closed so that $f \in A$. ∎

I doubt that the reader can find a direct proof of this last proposition. You might try counting the big results that are hidden within its proof.

8.1.8. Definition. If $\mathcal{X}$ is an LCS and $A \subseteq \mathcal{X}$, then the *polar* of A, denoted by A°, is defined by

$$A^\circ = \{x^* \in \mathcal{X}^* : |\langle a, x^* \rangle| \le 1 \text{ for all } a \text{ in } A\}$$

Analogously, if $B \subseteq \mathcal{X}^*$, the *prepolar* of B is the set

$$^\circ B = \{x \in \mathcal{X} : |\langle x, b^* \rangle| \le 1 \text{ for all } b^* \text{ in } B\}$$

If $A \subseteq \mathcal{X}$, then the *bipolar* of A is the set $^\circ(A^\circ)$. Similarly if $B \subseteq \mathcal{X}^*$, we call its bipolar the set $(^\circ B)^\circ$. If there is no chance of confusion we write these as $^\circ A^\circ$ and $^\circ B^\circ$.

8.1.9. Definition. If $\mathcal{X}$ is an LCS and $A \subseteq \mathcal{X}$, then the *annihilator* of A, denoted by $A^\perp$, is defined by

$$A^\perp = \{x^* \in \mathcal{X}^* : \langle a, x^* \rangle = 0 \text{ for all } a \text{ in } A\}$$

Analogously, if $B \subseteq \mathcal{X}^*$, the *preannihilator* of B is the set

$$^\perp B = \{x \in \mathcal{X} : \langle x, b^* \rangle = 0 \text{ for all } b^* \text{ in } B\}$$

If $A \subseteq \mathcal{X}$, then the *double annihilator* of A is the set $^\perp(A^\perp)$. If there is no chance of confusion we write this as $^\perp A^\perp$.

The author is aware of the reader's possible unease over all this. We have just had a high ratio of concepts to results and this is far from ideal. Now we have started flitting back and forth between an LCS and its dual. Nevertheless digesting these ideas and keeping them straight is important. We'll begin to develop some properties of polars and annihilators and this should make the reader more comfortable. But realize that it is exactly the ability to transfer problems in a normed space or LCS to a problem in the dual space that holds the key to power. At the end of this chapter we will, for example, give a proof of the famous Stone–Weierstrass Theorem using exactly such arguments. We also saw a bit of this power when we established the PUB, which we can rephrase using the current terminology by saying that a subset of a Banach space that is weakly bounded is norm bounded.

There is a strong connection between the annihilator of a set in an LCS and the orthogonal complement in a Hilbert space. Indeed since every continuous linear functional on a Hilbert space is represented by an inner product, the annihilator of a set in a Hilbert space is exactly its orthogonal complement.

For another example the reader should realize that if $\mathcal{X}$ is a normed space and $A = \text{ball}\,\mathcal{X}$, then $A^\circ = \text{ball}\,\mathcal{X}^*$; if $B = \text{ball}\,\mathcal{X}^*$, then $^\circ B = \text{ball}\,\mathcal{X}$. Also note that in this example the bipolar of A is A itself. We'll see an extension of this fact to the LCS setting shortly, but first let's look at the "calculus" of polars and annihilators.

8.1.10. Proposition. *If $\mathcal{X}$ is an LCS and A is a subset of $\mathcal{X}$, then the following hold.*

(a) A° *is a convex balanced* weak*-*closed subset of $\mathcal{X}^*$.*

(b) *If $A_1 \subseteq A$, then $A^\circ \subseteq A_1^\circ$.*

(c) *If α is a non-zero scalar, then $(\alpha A)^\circ = \alpha^{-1} A^\circ$.*

(d) $A \subseteq (^\circ A^\circ)$ *and* $A^\circ = (^\circ A^\circ)^\circ$.

Proof. The proof of the first three statements is an easy exercise as is the proof of the first part of (d), that $A \subseteq (^\circ A^\circ)$. Applying part (b) to this inclusion yields that $(^\circ A^\circ)^\circ \subseteq A^\circ$. To complete the proof we need only show that $A^\circ \subseteq (^\circ A^\circ)^\circ$. So let $x^* \in A^\circ$. If $x \in A$, then $|\langle x, x^* \rangle| \le 1$. This implies that $x \in [^\circ(A^\circ)] = {}^\circ A^\circ$ since x^* was arbitrarily chosen in A°. Contemplating this shows that this proves that $x^* \in (^\circ A^\circ)^\circ$. ∎

An analogous result holds for the prepolar of a set in $\mathcal{X}^*$ (Exercise 2).

8.1.11. Proposition. *If $\mathcal{X}$ is an LCS and A is a subset of $\mathcal{X}$, then the following hold.*

(a) $A^\perp$ *is a* weak*-*closed subspace of $\mathcal{X}^*$.*

(b) *If $A_1 \subseteq A$, then $A^\perp \subseteq A_1^\perp$.*

(c) $A \subseteq {}^\perp A^\perp$ *and* $A^\perp = (^\perp A^\perp)^\perp$.

(d) *If $\mathcal{M}$ is a linear manifold in $\mathcal{X}$, then $\mathcal{M}^\perp = \mathcal{M}^\circ$.*

Proof. The proof of the first two statements is an easy exercise and the proof of (c) is analogous to the proof of 8.1.10(d). To establish (d) observe that if $x^* \in \mathcal{M}^\circ$, then for every $t > 0$ and every x in $\mathcal{M}$, $1 \ge |\langle tx, x^* \rangle| = t|\langle x, x^* \rangle|$. So $t^{-1} \ge |\langle x, x^* \rangle|$ for all $t > 0$. Letting $t \to \infty$ we see that it must be that $x \in \mathcal{M}^\perp$. ∎

Again there are analogous results for the preannihilator (Exercise 2).

8.1.12. Theorem (Bipolar Theorem). *If $\mathcal{X}$ is an LCS and $A \subseteq \mathcal{X}$, then $^\circ A^\circ$ is the closed convex balanced hull of A; $^\perp A^\perp$ is the closed linear span of A.*

Proof. We only prove the statement for the bipolar. Let B be the closed convex balanced hull of A; that is, B is the intersection of all closed convex balanced sets containing A. Since $^{\circ}A^{\circ}$ is one such set, we have that $B \subseteq {^{\circ}A^{\circ}}$. Let $x_0 \in \mathcal{X} \backslash B$. By Theorem 7.3.10 there is an x^*, a real number α, and an $\epsilon > 0$ such that $\mathrm{Re}\,\langle b, x^* \rangle \leq \alpha < \alpha + \epsilon < \mathrm{Re}\,\langle x_0, x^* \rangle$ for every b in B. Since $0 \in B$, $\alpha > 0$; so if we replace x^* by $\alpha^{-1} x^*$, we have that there is a different $\epsilon > 0$ such that for every b in B, $\mathrm{Re}\,\langle b, x^* \rangle \leq 1 < 1 + \epsilon < \mathrm{Re}\,\langle x_0, x^* \rangle$. Now when $b \in B$, $e^{i\theta} b \in B$ for every θ, so this inequality implies $|\langle b, x^* \rangle| \leq 1 < 1 + \epsilon < \mathrm{Re}\,\langle x_0, x^* \rangle$ for all b in B. But this implies that $x^* \in B^{\circ}$ and hence $x_0 \notin ({^{\circ}A^{\circ}})$; that is, $x_0 \in \mathcal{X} \backslash {^{\circ}A^{\circ}}$ and the proof is complete. ∎

8.1.13. Corollary. *If $\mathcal{X}$ is an LCS and A is a closed convex balanced set, then A equals its bipolar.*

8.1.14. Corollary. *If $\mathcal{X}$ is an LCS and B is a subset of $\mathcal{X}^*$, then $^{\circ}(B^{\circ})$ is the weak*-closed convex balanced hull of B; $^{\perp}B^{\perp}$ is the weak*-closed linear span of B.*

Proof. Let $\mathcal{W}$ be the weak* topology on $\mathcal{X}^*$, and apply the Bipolar Theorem to the LCS $(\mathcal{X}^*, \mathcal{W})$. ∎

As the reader may have observed, so far the results in this section have all been for LCSs. Now we turn to a result for normed spaces that will have many consequences.

8.1.15. Theorem (Alaoglu's[2] Theorem). *If $\mathcal{X}$ is a normed space, then ball $\mathcal{X}^*$ is weak*-compact.*

Proof. For each x in $\mathcal{X}$, let $D_x = \{\alpha \in \mathbb{F} : |\alpha| \leq 1\}$, a copy of the closed unit ball in $\mathbb{F}$; put $D = \prod \{D_x : x \in \text{ball}\,\mathcal{X}\}$. By Tykhonov's Theorem, D is compact. The proof will be accomplished by showing that ball $\mathcal{X}^*$ with the weak* topology is homeomorphic to a closed subset of D. Note that we consider elements of D to be functions f from ball $\mathcal{X}$ into $\bigcup \{D_x : x \in \text{ball}\,\mathcal{X}\}$ such that $f(x) \in D_x$ for each x.

Define $\tau : \text{ball}\,\mathcal{X}^* \to D$ by $\tau(x^*)(x) = \langle x, x^* \rangle$. Since $|\tau(x^*)(x)| \leq 1$ for all x in ball $\mathcal{X}$, clearly $\tau(x) \in D$. To show that τ is injective, suppose $x_1^*, x_2^* \in \text{ball}\,\mathcal{X}^*$ and $\tau(x_1^*) = \tau(x_2^*)$. This implies $\langle x, x_1^* \rangle = \langle x_2^*, x \rangle$ for all x in ball $\mathcal{X}$, and therefore throughout $\mathcal{X}$.

τ is continuous. In fact let $\{x_i^*\}$ be a net in ball $\mathcal{X}^*$ that converges (wk*) to x^*. Then $\tau(x_i^*)(x) = \langle x, x_i^* \rangle \to \langle x, x^* \rangle = \tau(x^*)(x)$, so that $\tau(x_i^*) \to \tau(x^*)$.

---

[2]Leonidas Alaoglu was born in 1914 in Red Deer, Alberta, Canada, his parents being Greek immigrants. He received his doctorate from the University of Chicago in 1938. After a postdoctoral position at Harvard and a position at Purdue University, he went to work for the US Air Force. In 1953 he began working for the Lockheed corporation in Burbank, California, where he remained until his death in 1981. There is a yearly memorial lecture at Cal Tech named after him.

$\tau(\text{ball}\,\mathcal{X}^*)$ is closed in D. In fact, suppose $\{x_i^*\}$ is a net in ball $\mathcal{X}^*$ and $f \in D$ such that $\tau(x_i^*) \to f$ in D. Thus for each x in ball $\mathcal{X}$, $f(x) = \lim_i \tau(x_i^*)(x) = \lim_i \langle x, x_i^* \rangle$ in $\mathbb{F}$. Define $\tilde{f} : \mathcal{X} \to \mathbb{F}$ by $\tilde{f}(x) = t^{-1}f(tx)$, where t is a non-zero scalar such that $\|tx\| \leq 1$. First it must be shown that $\tilde{f}$ is well defined; that is, that the definition of $\tilde{f}(x)$ does not depend on the choice of t. This routine exercise involving nets is left to the reader, as is the proof that the resulting well defined function $\tilde{f} : \mathcal{X} \to \mathbb{F}$ is linear. It follows that $\tilde{f}$ has norm at most 1, since the fact that $f \in D$ implies that for all x in ball $\mathcal{X}$, $|\tilde{f}(x)| = |f(x)| \leq 1$. Therefore there is an x^* in ball $\mathcal{X}^*$ such that $\tilde{f} = x^*$; that is, $\tau(x^*) = f$ and so $\tau(\text{ball}\,\mathcal{X}^*)$ is closed in D.

Let $Z = \tau(\text{ball}\,\mathcal{X}^*)$, a compact subset of D. We need only show that $\tau^{-1} : Z \to \text{ball}\,\mathcal{X}^*$ is continuous. This is easy. If $\{\tau(x_i^*)\}$ is a net in Z that converges to $\tau(x^*)$, then the definition of the topology in D implies that for each x in ball $\mathcal{X}$, $\tau(x^*)(x) = \lim_i \tau(x_i^*)(x)$; that is $\langle x, x^* \rangle = \lim_i \langle x, x_i^* \rangle$ for all x in ball $\mathcal{X}$. But this is seen as equivalent to the condition that $x_i^* = \tau^{-1}(\tau(x_i^*)) \to x^*$ (wk*). ∎

A natural question is to ask when the weak or weak* topologies are metrizable. The answer is never unless the underlying normed space is finite-dimensional. However when we restrict ourselves to certain subsets we can sometimes get a positive answer. The next result is the most general one of this type.

8.1.16. Theorem. *If $\mathcal{X}$ is a Banach space, then* ball $\mathcal{X}^*$ *with its* weak*-*topology is metrizable if and only if $\mathcal{X}$ is separable.*

Proof. First assume that (ball $\mathcal{X}^*$, wk*) is metrizable and let $\{U_n\}$ be a sequence of relatively weak*-open sets in ball $\mathcal{X}^*$ such that $\bigcap_{n=1}^{\infty} U_n = (0)$. For each $n \geq 1$ there is a weak*-open set G_n in $\mathcal{X}^*$ such that $G_n \cap \text{ball}\,\mathcal{X}^* = U_n$. By definition there is a finite subset F_n of $\mathcal{X}$ such that $\{x^* : |\langle x, x^* \rangle| < 1$ for all x in $F_n\} \subseteq G_n$. Let $F = \bigcup_{n=1}^{\infty} F_n$, so that F is countable. We'll show that the linear span of F is dense in $\mathcal{X}$ by showing that $^\perp F^\perp = \mathcal{X}$; equivalently, we'll show that $F^\perp = (0)$. In fact, if $x^* \in F^\perp \cap \text{ball}\,\mathcal{X}^*$, then it follows that for each $n \geq 1$ and each x in F_n, $0 = |\langle x, x^* \rangle| < 1$ and so $x^* \in U_n$. Since $\bigcap_{n=1}^{\infty} U_n = (0)$, $x^* = 0$. It follows that the linear span of F with rational (or complex-rational) coefficients is both dense and countable.

Now assume that $\mathcal{X}$ is separable and let $\{x_n\}$ be a countable dense subset of ball $\mathcal{X}$. If $D_n = \{\alpha \in \mathbb{F} : |\alpha| \leq 1\}$ and $X = \prod_{n=1}^{\infty} D_n$, then X is a compact metric space. We'll show that (ball $\mathcal{X}^*$, wk*) is homeomorphic to a closed subset of X. Define $\tau : \text{ball}\,\mathcal{X}^* \to X$ by $\tau(x^*) = \{\langle x_n, x^* \rangle\}$. If $\{x_i^*\}$ is a net in ball $\mathcal{X}^*$ and $x_i^* \to x^*$ (wk*), then for each $n \geq 1$, $\langle x_n, x_i^* \rangle \to \langle x_n, x^* \rangle$ and this says that $\tau(x_i^*) \to \tau(x^*)$ in X; thus τ is continuous. By Alaoglu's

Theorem, ball $\mathcal{X}^*$ is weak*-compact and so $Z = \tau(\text{ball }\mathcal{X}^*)$ is a compact subset of X. We also have that τ is a closed map. So to show that τ is a homeomorphism onto Z we need only show that it is injective. But if $x^*, y^* \in \text{ball }\mathcal{X}^*$ and $\tau(x^*) = \tau(y^*)$, then $\langle x_n, x^* \rangle = \langle x_n, y^* \rangle$ for all $n \geq 1$. Since $\{x_n\}$ is dense in ball $\mathcal{X}^*$, $x^* = y^*$ and so τ is injective. ∎

We close this section with a strange result that will not be used again in this book but illustrates an oddity, raises a cautionary note about the weak topology, and underlines the necessity of becoming familiar with nets (§A.2).

8.1.17. Proposition. *If a sequence converges in the weak topology of ℓ^1, then it converges in norm.*

Proof. Here there is an advantage that is at least notational in thinking of the elements of ℓ^1 and ℓ^∞ as functions defined on $\mathbb{N}$ rather than sequences. Suppose $\{f_n\}$ is a sequence in ℓ^1 such that $f_n \to 0$ in $\sigma(\ell^1, \ell^\infty)$. For each $m \geq 1$ let

$$F_m = \{\phi \in \text{ball }\ell^\infty : |\langle f_n, \phi \rangle| \leq \epsilon/3 \text{ when } n \geq m\}$$

It is left to the reader to show that each F_m is weak*-closed in ball ℓ^∞. Because $f_n \to 0$ in $\sigma(\ell^1, \ell^\infty)$, ball $\ell^\infty = \bigcup_{m=1}^\infty F_m$. Now ball ℓ^∞ with the weak* topology is metrizable (8.1.16), so by the Baire Category Theorem (§A.1) there is an $m \geq 1$ such that F_m has non-empty interior in (ball ℓ^∞, wk*). It is routine (use Exercise 7 to give the details) that there is an integer $J \geq 1$, $\delta > 0$, and a ϕ_0 in F_m such that

$$S = \{\phi \in \text{ball }\ell^\infty : |\phi(j) - \phi_0(j)| < \delta \text{ for } 1 \leq j \leq J\} \subseteq F_m$$

Because $f_n \to 0$ weakly in ℓ^1, it follows that $\langle f_n, \phi_0 \rangle \to 0$ and $f_n(j) \to 0$ for all $j \geq 1$. Let $m_1 \geq m$ such that $|\langle f_n, \phi_0 \rangle| < \epsilon/3$ and $|f_n(j)| < (\epsilon/3)J$ whenever $1 \leq j \leq J$ and $n \geq m_1$. Fix $n \geq m_1$ and define ϕ in ℓ^∞ by $\phi(j) = \phi_0(j)$ when $1 \leq j \leq J$ and $\phi(j) = \text{sign}[f_n(j)]$ for $j > J$. It follows that $\phi \in S \subseteq F_m$. Hence

$$\|f_n\|_1 = \sum_{j=1}^J |f_n(j)| + \sum_{j=J+1}^\infty |f_n(j)|$$

$$= \sum_{j=1}^J |f_n(j)| + |\langle f_n, \phi - \phi_0 \rangle|$$

$$\leq \frac{\epsilon}{3} + |\langle f_n, \phi \rangle| + |\langle f_n, \phi_0 \rangle|$$

$$< \epsilon$$

∎

By Exercise 6 below, the weak and norm topologies on ℓ^1 are different; nevertheless they have the same convergent sequences. Moral: learn about nets if you work with weak and weak* topologies. The above proof and some additional references can be found in [**6**].

Exercises.

(1) Let $\mathcal{Y}$ be a vector space over $\mathbb{F}$ and let $f, f_1, \ldots, f_n$ be linear functionals defined on $\mathcal{Y}$. Show that $\bigcap_{k=1}^n \ker f_k \subseteq \ker f$ if and only if f is a linear combination of $f_1, \ldots, f_n$. (Note that we did this already (6.3.1) for the case $n = 1$.)

(2) State and prove a result for prepolars analogous to that of Proposition 8.1.10. Do the same for preannihilators.

(3) Let $\mathcal{X}$ be a complex LCS and let $\mathcal{X}_{\mathbb{R}}^*$ denote the continuous real-valued linear functionals on $\mathcal{X}$. Use the elements of $\mathcal{X}_{\mathbb{R}}^*$ to define the seminorms on $\mathcal{X}$ and show that the topology these seminorms define on $\mathcal{X}$ is the same as its weak topology.

(4) If $\mathcal{X}$ is an LCS, show that a subset A of $\mathcal{X}$ is bounded in the weak topology (that is, for every weakly open set U there is a scalar α with $\alpha A \subseteq U$) if and only if for every x^* in $\mathcal{X}^*$ there is a scalar α such that $\alpha x^* \in A^\circ$.

(5) If $\mathcal{X}$ is a normed space, show that the norm is lower semicontinuous in the weak topology; show that the norm on $\mathcal{X}^*$ is weak*-lower semicontinuous.

(6) If $\mathcal{X}$ is an infinite-dimensional normed space, show that the weak closure of $\{x \in \mathcal{X} : \|x\| = 1\}$ is all of ball $\mathcal{X}$.

(7) For ϕ, ψ in ℓ^∞, define

$$d(\phi, \psi) = \sum_{n=1}^{\infty} 2^{-n} |\phi(n) - \psi(n)|$$

Show that d is a metric and that the topology on ball ℓ^∞ defined by this metric d is the relative weak* topology.

(8) Let $B = $ ball $M[0,1]$ and for μ, ν in B, let

$$d(\mu, \nu) = \sum_{n=1}^{\infty} 2^{-n} \left| \int x^n \, d\mu \right|$$

Show that d is a metric on B that defines the weak* topology.

(9) If $\mathcal{X}$ is a Banach space such that $\mathcal{X}^*$ is separable, show that $\mathcal{X}$ is separable.

(10) It was shown in Theorem 1.4.6 that if X is a locally compact space
that is metrizable, then $C_0(X)$ is separable if and only if X is σ-
compact. Without using this result, prove that if X is compact and
$C(X)$ is separable, then X is a metric space. (Hint: Look at the
map $x \mapsto \delta_x$ of X into $M(X) = C(X)^*$ with the weak* topology
and show it is a homeomorphism of X onto its image.)

8.2. The dual of a quotient space and of a subspace

Here we want to examine the dual of a quotient and of a subspace of an LCS
and to relate the weak and weak* topologies to those of the original space.
Yes, this section is housekeeping: not particularly exciting, but necessary.
We want to be sure that there are no ambiguities when we discuss these
matters. For example, we'll show that if X is a Banach space and $M \leq X$,
then the dual space of X/M is naturally isomorphic to $M^\perp$. How does
the weak* topology on $(X/M)^*$ relate to the weak* topology on $M^\perp$ that
it inherits as a subspace of X^*?

When X is a normed space, we introduced the quotient of X by a closed
subspace M and defined the norm on X/M in (6.2.1). The first step in this
section is to discuss the seminorm on a quotient space induced by a seminorm
on the original space. The proof is left as an exercise as it parallels what
we did in the case of a normed space. Note that in §6.2 we insisted that
M be closed; with an LCS this becomes unnecessary. (Why?) In fact we
don't even need that X is an LCS in order to d.efine the seminorm on the
quotient space. The proofs of parts (a), (b), and (c) in the next proposition
follow the lines of the proofs of the corresponding parts of Theorem 6.2.2.

8.2.1. Proposition. *If p is a seminorm defined on a vector space X and
M is a linear manifold in X, then*

$$\widetilde{p}(x + M) = \inf\{p(x + y) : y \in M\}$$

*defines a seminorm $\widetilde{p}$ on X/M. If X is an LCS with its topology defined by
the family of seminorms $\mathcal{P}$ and $Q : X \to X/M$ is the quotient map, then
the following hold.*

(a) *The family of seminorms $\widetilde{\mathcal{P}} = \{\widetilde{p} : p \in \mathcal{P}\}$ defines the quotient topology
on X/M.*

(b) *A subset W of X/M is open if and only if $Q^{-1}(W)$ is open in X.*

(c) *If G is open in X, then $Q(G)$ is open in X/M.*

If X is an LCS and $x^* \in X^*$, then the restriction of x^* to M, $x^*|M$,
belongs to M^*. If X is a normed space, then $\|x^*|M\| \leq \|x^*\|$. Also the
HBT implies every continuous linear functional on M can be obtained as

the restriction $x^*|\mathcal{M}$ for some x^* in $\mathcal{X}^*$. This restriction is not unique. That is, there is another z^* in $\mathcal{X}^*$ with $z^*|\mathcal{M} = x^*|\mathcal{M}$; equivalently there are continuous linear functionals z^* such that $x^* - z^* \in \mathcal{M}^\perp$. If $x^* \in \mathcal{X}^*$ and $y^* \in \mathcal{M}^\perp$, then the restriction of $z^* = x^* + y^*$ to $\mathcal{M}$ defines the same element of $\mathcal{M}^*$ as does x^*. This implies that each element of the quotient space $\mathcal{X}^*/\mathcal{M}^\perp$ gives rise to a continuous linear functional on $\mathcal{M}$.

The next two theorems should be regarded as saying that all the good things that might happen relative to the dual of a quotient space and of a subspace, do happen – the exact opposite of Murphy's law. We begin with the dual of a subspace.

8.2.2. Theorem. *If $\mathcal{X}$ is an LCS, $\mathcal{M}$ is a linear manifold in $\mathcal{X}$, and $\rho : \mathcal{X}^*/\mathcal{M}^\perp \to \mathcal{M}^*$ is defined by $\rho(x^* + \mathcal{M}^\perp) = x^*|\mathcal{M}$, then ρ is a linear bijection. If $\mathcal{X}^*/\mathcal{M}^\perp$ has the quotient topology induced by $\sigma(\mathcal{X}^*, \mathcal{X})$ and $\mathcal{M}^*$ has its weak* topology, $\sigma(\mathcal{M}^*, \mathcal{M})$, then ρ is a homeomorphism. If $\mathcal{X}$ is a normed space and $\mathcal{M}$ is a closed subspace, then ρ is an isometry.*

Proof. We discussed before the statement of the theorem that ρ is a well defined map; it is straightforward to check that ρ is linear. To prove that ρ is a homeomorphism, let η^* denote the quotient topology on $\mathcal{X}^*/\mathcal{M}^\perp$ induced by $\sigma(\mathcal{X}^*, \mathcal{X})$. If $Q : \mathcal{X}^* \to \mathcal{X}^*/\mathcal{M}^\perp$ is the quotient map, we have that for y in $\mathcal{M}$, $Q^{-1}(\rho^{-1}\{y^* \in \mathcal{M}^* : |\langle y, y^* \rangle| < 1\}) = Q^{-1}\{x^* + \mathcal{M}^\perp : x^* \in \mathcal{X}^* \text{ and } |\langle y, x^* \rangle| < 1\} = \{x^* \in \mathcal{X}^* : |\langle y, x^* \rangle| < 1\} \in \sigma(\mathcal{X}^*, \mathcal{X})$. By Proposition 8.2.1(b), this implies that $\rho^{-1}\{y^* \in \mathcal{M}^* : |\langle y, y^* \rangle| < 1\}$ is open in the quotient topology of $\mathcal{X}^*/\mathcal{M}^\perp$, and so $\rho : (\mathcal{X}^*/\mathcal{M}^\perp, \eta^*) \to (\mathcal{M}^*, \mathrm{wk}^*)$ is continuous.

Let $x \in \mathcal{X}$ with $p_x(x^*) = |\langle x, x^* \rangle|$ the corresponding weak*-continuous seminorm on $\mathcal{X}^*$. Recall the definition of the seminorm $\widetilde{p_x}$ on $\mathcal{X}^*/\mathcal{M}^\perp$ above.

Claim. If $x \notin \mathcal{M}$, then $\widetilde{p_x} = 0$.

Let $\mathcal{Z}$ be the one-dimensional space spanned by the vector x. Since $x \notin \mathcal{M}$, $\mathcal{Z} \cap \mathcal{M} = (0)$. Fix x^* in $\mathcal{X}^*$ and define $f : \mathcal{M} + \mathcal{Z} \to \mathbb{F}$ by $f(y + \alpha x) = \langle y, x^* \rangle$ for all y in $\mathcal{M}$ and α in $\mathbb{F}$. Because $\mathcal{M}$ and $\mathcal{Z}$ are topologically complementary in $\mathcal{M} + \mathcal{Z}$ (6.6.1), f is continuous. By the HBT there is an x_1^* in $\mathcal{X}^*$ that extends f. Since $x^* - x_1^* \in \mathcal{M}^\perp$, $\widetilde{p_x}(x^* + \mathcal{M}^\perp) = \widetilde{p_x}(x_1^* + \mathcal{M}^\perp) \leq p_x(x_1^*) = |\langle x, x_1^* \rangle| = 0$, proving the claim.

Now let $\{x_i^* + \mathcal{M}^\perp\}$ be a net in $\mathcal{X}^*/\mathcal{M}^\perp$ such that $\rho(x_i^* + \mathcal{M}^\perp) = x_i^*|\mathcal{M} \to 0\,(\sigma(\mathcal{M}^*, \mathcal{M}))$. If $x \in \mathcal{X}$, we want to show that $\widetilde{p_x}(x_i^* + \mathcal{M}^\perp) \to 0$. According to the claim, we need only consider those x that are in $\mathcal{M}$. But when $x \in \mathcal{M}$, $\widetilde{p_x}(x_i^* + \mathcal{M}^\perp) \leq |\langle x, x_i^* \rangle| \to 0$. Thus $x_i^* + \mathcal{M}^\perp \to 0\,(\eta^*)$. This completes the proof that ρ is a homeomorphism.

It remains to show that ρ is an isometry when $\mathcal{X}$ is a normed space. If $x^* \in \mathcal{X}^*$ and $y^* \in \mathcal{M}^\perp$, then $\|x^*|\mathcal{M}\| = \|(x^* + y^*)|\mathcal{M}\| \leq \|x^* + y^*\|$; taking the infimum over all y^* in $\mathcal{M}^\perp$ we have that $\|\rho(x^* + \mathcal{M}^\perp)\| = \|x^*|\mathcal{M}\| \leq \|x^* + \mathcal{M}^\perp\|$. Now suppose that $f \in \mathcal{M}^*$. The HBT implies that there is an x^* in $\mathcal{X}^*$ such that $x^*|\mathcal{M} = f$ and $\|x^*\| = \|f\|$. Thus $\|\rho(x^* + \mathcal{M}^\perp)\| = \|f\| = \|x^*\| \geq \|x^* + \mathcal{M}^\perp\|$, completing the proof. ∎

8.2.3. Theorem. *Let $\mathcal{X}$ be an LCS, let $\mathcal{M} \leq \mathcal{X}$, and let $Q : \mathcal{X} \to \mathcal{X}/\mathcal{M}$ be the quotient map. The map $\rho : (\mathcal{X}/\mathcal{M})^* \to \mathcal{M}^\perp$ defined by $\rho(f) = f \circ Q$ is a linear bijection. If $(\mathcal{X}/\mathcal{M})^*$ is given its weak* topology, $\sigma((\mathcal{X}/\mathcal{M})^*, \mathcal{X}/\mathcal{M})$, and $\mathcal{M}^\perp$ is given the relative weak* topology from $\mathcal{X}^*$, $\sigma(\mathcal{X}^*, \mathcal{X})|\mathcal{M}^\perp$, then ρ is a homeomorphism. If $\mathcal{X}$ is a normed space and $\mathcal{M}$ is a closed subspace of $\mathcal{X}$, then ρ is an isometric isomorphism.*

Proof. It is easy to see that for each f in $(\mathcal{X}/\mathcal{M})^*$, $f \circ Q$ is not only a linear functional on $\mathcal{X}$ but actually annihilates $\mathcal{M}$ since $Q(x) = 0$ for every x in $\mathcal{M}$. It is also easy to see that ρ is linear, and, by definition, ρ is injective. Now to show that ρ maps into $\mathcal{X}^*$. If q is a continuous seminorm on $\mathcal{X}/\mathcal{M}$ such that $|f(x+\mathcal{M})| \leq q(x+\mathcal{M})$, then for all x in $\mathcal{X}$, $|\rho(f)(x)| = |f(Q(x))| \leq q(Q(x))$. But since Q is continuous, $q \circ Q$ is a continuous seminorm on $\mathcal{X}$ and hence $\rho(f) \in \mathcal{X}^*$.

The fact that ρ is surjective is immediate once we prove the following.

Claim. If $x^* \in \mathcal{M}^\perp \leq \mathcal{X}^*$ and p is a continuous seminorm on $\mathcal{X}$ such that $|\langle x, x^* \rangle| \leq p(x)$ for all x in $\mathcal{X}$, then there is an f in $(\mathcal{X}/\mathcal{M})^*$ such that $\rho(f) = x^*$ and $|f(x + \mathcal{M})| \leq \widetilde{p}(x + \mathcal{M})$.

Define $f : \mathcal{X}/\mathcal{M} \to \mathbb{F}$ by $f(x + \mathcal{M}) = \langle x, x^* \rangle$; this is well defined since $x^* \in \mathcal{M}^\perp$. Also for any y in $\mathcal{M}$ we have that $|f(x+\mathcal{M})| = |f(x+y+\mathcal{M})| = |\langle x+y, x^* \rangle| \leq p(x+y)$; taking the infimum over all such y shows $f \in (\mathcal{X}/\mathcal{M})^*$ and establishes the claim.

Now to show that ρ is a weak* homeomorphism. Let $w^* = \sigma(\mathcal{X}^*, \mathcal{X})$ and let $\sigma^* = \sigma((\mathcal{X}/\mathcal{M})^*, \mathcal{X}/\mathcal{M})$. If $\{f_i\}$ is a net in $(\mathcal{X}/\mathcal{M})^*$ and $f_i \to 0$ (σ^*), then for each x in $\mathcal{X}$ we have that $\langle x, \rho(f_i) \rangle = f_i(Q(x)) \to 0$; that is, $\rho(f_i) \to 0$ (w^*). Conversely if $\rho(f_i) \to 0$ (w^*), then for every x in $\mathcal{X}$, $f_i(x + \mathcal{M}) = \langle x, \rho(f_i) \rangle \to 0$ so that $f_i \to 0$ (σ^*).

Now assume $\mathcal{X}$ is a normed space and $\mathcal{M} \leq \mathcal{X}$. If $f \in (\mathcal{X}/\mathcal{M})^*$, $x^* = \rho(f)$ and, in the claim, we take $p(x) = \|x^*\|\|x\|$, then we get that $|f(x + \mathcal{M})| \leq \|x^*\|\|x + \mathcal{M}\|$ for all x in $\mathcal{X}$; hence $\|\rho(f)\| \leq \|x^*\|$. To get the equality of the norms let $\{x_n + \mathcal{M}\}$ be a sequence with $\|x_n + \mathcal{M}\| < 1$ and $|f(x_n+\mathcal{M})| \to \|f\|$. For each $n \geq 1$ there is a y_n in $\mathcal{M}$ with $\|x_n + y_m\| < 1$. Thus $\|\rho(f)\| \geq |\rho(f)(x_n + y_n)| = |f(x_n + \mathcal{M})| \to \|f\|$. ∎

Exercises.

(1) Let X be a compact space, let F be a closed subset of X, and let $\mathcal{M} = \{f \in C(X) : f(x) = 0$ for all x in $F\}$. Show that $\mathcal{M}^\perp = \{\mu \in M(X) : |\mu|(F) = 0\}$. What do the results in this section say about $(C(X)/\mathcal{M})^*$ and $\mathcal{M}^*$? See Exercise 6.2.3.

(2) Let $(X, \mathcal{A}, \mu)$ be a σ-finite measure space, $1 \leq p < \infty$, and let E be a measurable set. If $\mathcal{M} = \{f \in L^p(\mu) : f\chi_E = 0\}$ and $p^{-1} + q^{-1} = 1$, show that $\mathcal{M}^\perp = \{g \in L^q(\mu) : g\chi_{(X \setminus E)} = 0\}$. What do the results in this section say about $(L^p(\mu)/\mathcal{M})^*$ and $\mathcal{M}^*$?

8.3. Reflexive spaces

We know that when $\mathcal{X}$ is a normed space, its dual space $\mathcal{X}^*$ is a Banach space (1.5.6). Hence $\mathcal{X}^*$ also has a dual space, $(\mathcal{X}^*)^* \equiv \mathcal{X}^{**}$. But then $\mathcal{X}^{**}$ also has a dual space, $\mathcal{X}^{***}$. This continues. Does it ever stop? The answer is that sometimes it continues and sometimes it cycles back on itself (to be explained). In this section we'll look at the spaces where this recycling occurs, but first we introduce some notation that facilitates the discussion and marks a relevant phenomenon.

If $x \in \mathcal{X}$, define $\widehat{x} : \mathcal{X}^* \to \mathbb{F}$ by

$$\widehat{x}(x^*) = \langle x, x^* \rangle$$

for all x^* in $\mathcal{X}^*$. By Corollary 6.3.9, $\|x\| = \|\widehat{x}\|$. Thus the map $x \mapsto \widehat{x}$ is seen to be a linear isometry of $\mathcal{X}$ into $\mathcal{X}^{**}$. This map will be referred to as the *natural map* of $\mathcal{X}$ into its *second dual*.

8.3.1. Definition. A normed space $\mathcal{X}$ is said to be *reflexive* if the natural map into its second dual is surjective; equivalently, if $\mathcal{X}^{**} = \{\widehat{x} : x \in \mathcal{X}\}$.

Note that a reflexive normed space must be a Banach space since it is isometric to its double dual, which is always a Banach space. Also be aware that the requirement for reflexivity is not just that $\mathcal{X}$ and $\mathcal{X}^{**}$ be isometric but that the isometry be the natural map. In [20] there is an example of a non-reflexive Banach space that is isometrically isomorphic to its double dual but is not reflexive.

Rather than tracking the natural map $x \mapsto \widehat{x}$, we will assume that $\mathcal{X} \subseteq \mathcal{X}^{**}$.

8.3.2. Example. (a) If $1 < p < \infty$, $L^p(\mu)$ is reflexive. We know that the dual of $L^p(\mu)$ is $L^q(\mu)$, where $\frac{1}{p} + \frac{1}{q} = 1$ and the dual of $L^q(\mu)$ is $L^p(\mu)$. Establishing that $L^p(\mu)$ is reflexive amounts to examining these duality relations to see that the natural mapping is just the identity.

(b) Every Hilbert space is reflexive.

8.3.3. Example. c_0 is not reflexive. In fact $c_0^* = \ell^1$ and $c_0^{**} = (\ell^1)^* = \ell^\infty$. If we examine these duality relationships, then we conclude that the natural map from c_0 into its second dual is just the inclusion map $c_0 \hookrightarrow \ell^\infty$ and this certainly fails to be surjective. Another way to see this is to note that c_0 is separable, but ℓ^∞ is not (Exercise 1.3.12), so c_0 and ℓ^∞ cannot be isometrically equivalent.

Since $\mathcal{X}^{**}$ is the dual of $\mathcal{X}^*$, it has a weak* topology, $\sigma(\mathcal{X}^{**}, \mathcal{X}^*)$. By Alaoglu's Theorem ball $\mathcal{X}^{**}$ is $\sigma(\mathcal{X}^{**}, \mathcal{X}^*)$ compact. Note that since $\mathcal{X}$ is naturally embedded in $\mathcal{X}^{**}$, we can relativize the weak* topology of $\mathcal{X}^{**}$ to $\mathcal{X}$, but when we do this it is just the weak topology on $\mathcal{X}$, $\sigma(\mathcal{X}, \mathcal{X}^*)$. It is becoming clear that in the discussion that follows we better use the $\sigma(\cdot, \cdot)$ notation to designate the topologies rather than the words weak and weak*.

8.3.4. Proposition. *If $\mathcal{X}$ is a normed space, then* ball $\mathcal{X}$ *is* $\sigma(\mathcal{X}^{**}, \mathcal{X}^*)$ *dense in* ball $\mathcal{X}^{**}$.

Proof. Let σ denote the topology $\sigma(\mathcal{X}^{**}, \mathcal{X}^*)$ and let B be the σ closure of ball $\mathcal{X}$ in ball $\mathcal{X}^{**}$; since ball $\mathcal{X}^{**}$ is σ compact, $B \subseteq$ ball $\mathcal{X}^{**}$. If $x_0^{**} \in \mathcal{X}^{**} \backslash B$, then the HBT (here and later we will use "HBT" to refer not only to the theorem itself but to all its immediate consequences including the geometric consequences in §7.3) implies there is an x^* in $\mathcal{X}^*$, a real number α, and an $\epsilon > 0$ such that for all x in ball $\mathcal{X}$,

$$\operatorname{Re} \langle x, x^* \rangle < \alpha < \alpha + \epsilon < \operatorname{Re} \langle x^*, x_0^{**} \rangle$$

We use an argument we have seen before. The fact that $0 \in$ ball $\mathcal{X}$ means that $\alpha > 0$. So replacing x^* by $\alpha^{-1} x^*$ we get that there is an x^* and an $\epsilon > 0$ such that

$$\operatorname{Re} \langle x, x^* \rangle < 1 < 1 + \epsilon < \operatorname{Re} \langle x^*, x_0^{**} \rangle$$

for every x in ball $\mathcal{X}$. But $e^{i\theta} x \in$ ball $\mathcal{X}$ whenever $x \in$ ball $\mathcal{X}$ so we get that $|\langle x, x^* \rangle| < 1 < 1 + \epsilon < \operatorname{Re} \langle x^*, x_0^{**} \rangle$. This implies $\|x^*\| \leq 1$ so that $x^* \in$ ball $\mathcal{X}^*$. But then the inequality implies $x_0^{**} \notin$ ball $\mathcal{X}^{**}$. That is $\mathcal{X}^{**} \backslash B \subseteq \mathcal{X}^{**} \backslash$ ball $\mathcal{X}^{**}$ and the proof is complete. ∎

This next result is the main theorem on reflexive spaces.

8.3.5. Theorem. *If $\mathcal{X}$ is a Banach space, the following statements are equivalent.*

(a) $\mathcal{X}$ *is reflexive.*

(b) $\mathcal{X}^*$ *is reflexive.*

(c) $\sigma(\mathcal{X}^*, \mathcal{X}) = \sigma(\mathcal{X}^*, \mathcal{X}^{**})$.

(d) ball $\mathcal{X}$ *is* $\sigma(\mathcal{X}, \mathcal{X}^*)$ *(weakly) compact.*

Proof. (a) *implies* (c). This is clear since $\mathcal{X} = \mathcal{X}^{**}$.

(d) *implies* (a). Begin by observing that $\sigma(\mathcal{X}^{**}, \mathcal{X}^{*})|\mathcal{X} = \sigma(\mathcal{X}, \mathcal{X}^{*})$ for all Banach spaces. Condition (d) implies that ball $\mathcal{X}$ is $\sigma(\mathcal{X}^{**}, \mathcal{X}^{*})$ closed in ball $\mathcal{X}^{**}$. But the preceding proposition says that ball $\mathcal{X}$ is $\sigma(\mathcal{X}^{**}, \mathcal{X}^{*})$ dense in ball $\mathcal{X}^{**}$. Therefore ball $\mathcal{X}$ = ball $\mathcal{X}^{**}$ and so $\mathcal{X}$ is reflexive.

(c) *implies* (b). Alaoglu's Theorem implies that ball $\mathcal{X}^{*}$ is $\sigma(\mathcal{X}^{*}, \mathcal{X})$ compact. In light of (c), it is $\sigma(\mathcal{X}^{*}, \mathcal{X}^{**})$ compact. Since we know that (d) implies (a), if we apply this fact to $\mathcal{X}^{*}$ we get that $\mathcal{X}^{*}$ is reflexive.

(b) *implies* (a). ball $\mathcal{X}$ is norm closed in $\mathcal{X}^{**}$; therefore ball $\mathcal{X}$ is closed in the $\sigma(\mathcal{X}^{**}, \mathcal{X}^{***})$ topology of $\mathcal{X}^{**}$ (Corollary 8.1.6). But (b) holds so $\mathcal{X}^{*} = \mathcal{X}^{***}$. Hence this says that ball $\mathcal{X}$ is $\sigma(\mathcal{X}^{**}, \mathcal{X}^{*})$ closed in ball $\mathcal{X}^{**}$. But Proposition 8.3.4 says that ball $\mathcal{X}$ is $\sigma(\mathcal{X}^{**}, \mathcal{X}^{*})$ dense in ball $\mathcal{X}^{**}$. Therefore ball $\mathcal{X}$ = ball $\mathcal{X}^{**}$ and $\mathcal{X}$ is reflexive.

(a) *implies* (d). Alaoglu's Theorem implies that ball $\mathcal{X}^{**}$ is $\sigma(\mathcal{X}^{**}, \mathcal{X}^{*})$ compact. Since $\mathcal{X} = \mathcal{X}^{**}$, we have that ball $\mathcal{X}$ is weakly compact. ∎

8.3.6. Corollary. *If $\mathcal{X}$ is a reflexive Banach space and $\mathcal{M} \leq \mathcal{X}$, then $\mathcal{M}$ is reflexive.*

Proof. Since ball $\mathcal{M} = \mathcal{M} \cap$ ball $\mathcal{X}$, ball $\mathcal{M}$ is $\sigma(\mathcal{X}, \mathcal{X}^{*})$ compact. But $\sigma(\mathcal{M}, \mathcal{M}^{*})$ is the topology obtained by restricting $\sigma(\mathcal{X}, \mathcal{X}^{*})$ to $\mathcal{M}$ since, by the HBT, every continuous linear functional on $\mathcal{M}$ can be extended to $\mathcal{X}$. ∎

8.3.7. Corollary. *If $\mathcal{X}$ is a reflexive Banach space, $\mathcal{M} \leq \mathcal{X}$, and $x_0 \in \mathcal{X} \backslash \mathcal{M}$, then there is a point x in $\mathcal{M}$ such that $\|x_0 - x\| = \text{dist}(x_0, \mathcal{M})$.*

Proof. Let $d = \text{dist}(x_0, \mathcal{M})$. By Exercise 8.1.5, $x \mapsto \|x - x_0\|$ is weakly lower semicontinuous. According to the theorem above, norm bounded weakly closed sets are weakly compact. Hence $\mathcal{M} \cap \{x \in \mathcal{X} : \|x - x_0\| \leq 2d\}$ is weakly compact and lower semicontinuous functions attain their minimum on compact sets. ∎

Exercises.

(1) Show that $(\mathcal{X}^{*})^{**} = (\mathcal{X}^{**})^{*}$.

(2) Show that all finite-dimensional Banach spaces are reflexive and show that a Hilbert space is reflexive.

(3) Show that if $\mathcal{X}$ is reflexive and $\mathcal{M} \leq \mathcal{X}$, then $\mathcal{X}/\mathcal{M}$ is reflexive.

(4) If $\mathcal{M} \leq \mathcal{X}$, $\mathcal{M}$ is reflexive, and $\mathcal{X}/\mathcal{M}$ is reflexive, is $\mathcal{X}$ reflexive?

(5) Define $L : C[0,1] \to \mathbb{R}$ by $L(f) = \int_0^{\frac{1}{2}} f(t)dt - \int_{\frac{1}{2}}^1 f(t)dt$. Show that there is a function f in $C[0,1]$ such that the distance from f to

ker L is not attained. Thus $C[0,1]$ cannot be reflexive by Corollary 8.3.7.

(6) Show that for a compact metric space, $C(X)$ is reflexive if and only if X is finite.

(7) Define a sequence $\{x_n\}$ in a Banach space to be *weakly Cauchy* if for every x^* in $\mathcal{X}^*$ we have that $\{\langle x_n, x^* \rangle\}$ is a convergent sequence in $\mathbb{F}$. (a) Show that every weakly Cauchy sequence in a Banach space is norm bounded. (b) If $\mathcal{X}$ is a reflexive Banach space, show that every weakly Cauchy sequence in $\mathcal{X}$ has a weak limit point. (c) Let $f_n : [0,1] \to \mathbb{R}$ be defined by $f_n(t) = 1 - nt$ if $0 \le t \le \frac{1}{n}$ and $f_n(t) = 0$ for $\frac{1}{n} \le t \le 1$. Show that $\{f_n\}$ is a weakly Cauchy sequence in $C[0,1]$ that does not have a weak limit point in $C[0,1]$.

8.4. The Krein–Milman Theorem

8.4.1. Definition. If K is a convex subset of $\mathcal{X}$ and $a \in K$, then a is called an *extreme point* of K if there is no proper open line segment in K that contains a. Equivalently, if x and y are in K, $0 < t < 1$, and $a = tx + (1-t)y$, then $x = y = a$. Let $\operatorname{ext} K$ denote the set of extreme points of K.

8.4.2. Example. (a) Let $\mathcal{X} = \mathbb{R}$. If $K = [a,b]$, then $\operatorname{ext} K = \{a,b\}$. If $K = (a,b)$, then $\operatorname{ext} K = \emptyset$.

(b) If $\mathcal{X} = \mathbb{R}^2$ and $K = \{(x,y) : x^2 + y^2 \le 1\}$, then $\operatorname{ext} K = \{(x,y) : x^2 + y^2 = 1\}$. If $K = \{(x,y) : x \le 0\}$, then $\operatorname{ext} K = \emptyset$.

(c) If K is a convex polygon in $\mathbb{R}^2$ or a convex polyhedron in $\mathbb{R}^3$, then $\operatorname{ext} K$ is the set of vertices of K.

(d) If $\mathcal{X}$ is a normed space and $K = \operatorname{ball} \mathcal{X}$, then $\operatorname{ext} K \subseteq \{x : \|x\| = 1\}$.

(e) A generalization of part (d) is that $\operatorname{ext} K$ is always a subset of the topological boundary of K, a fact that is straightforward to verify.

(f) If $\mathcal{X} = L^1[0,1]$ and $K = \operatorname{ball} \mathcal{X}$, then $\operatorname{ext} K = \emptyset$. To see this let $f \in L^1[0,1]$ with $\|f\| = 1$ and choose x in $[0,1]$ with $\int_0^x |f(t)|dt = \frac{1}{2}$. Put $g(t) = 2f(t)$ for $0 \le t \le x$ and $g(t) = 0$ for $x \le t \le 1$; put $h(t) = 0$ for $0 \le t \le x$ and $h(t) = 2f(t)$ for $x \le t \le 1$. Then $\|g\| = \|h\| = 1$ and $f = \frac{1}{2}(g+h)$.

8.4.3. Proposition. *If K is a convex subset of the LCS $\mathcal{X}$ and $a \in K$, then the following statements are equivalent.*

(a) $a \in \operatorname{ext} K$.

(b) *If $x_0, x_1 \in \mathcal{X}$ and $a = \frac{1}{2}(x_1 + x_0)$, then either $x_0 \notin K$, $x_1 \notin K$, or $x_0 = x_1 = a$.*

(c) If $x_0, x_1 \in \mathcal{X}$, $0 < t < 1$, and $a = tx_1 + (1-t)x_0$, then either $x_0 \notin K$, $x_1 \notin K$, or $x_0 = x_1 = a$.

(d) If $x_1, \ldots, x_n \in K$ and $a \in \mathrm{co}\{x_1, \ldots, x_n\}$, then there is at least one j, $1 \le j \le n$, such that $a = x_j$.

(e) $K \backslash \{a\}$ is a convex set.

We leave the proof of the preceding proposition as Exercise 1.

8.4.4. Theorem (Krein[3]–Milman[4] Theorem). *If K is a non-empty convex compact subset of an LCS, then K is the closed convex hull of* ext K.

We might underline that a consequence of this theorem is that extreme points of K exist. In fact, the theorem says they exist in abundance.

Proof. According to Proposition 8.4.3(e), a is an extreme point precisely when $K \backslash \{a\}$ is a proper relatively open convex set; thus $K \backslash \{a\}$ is a maximal relatively open convex subset of K. We will begin the proof by showing the existence of such a set. Let $\mathcal{U}$ be the collection of all relatively open, non-empty convex subsets of K and order $\mathcal{U}$ by inclusion. Since $\mathcal{X}$ is locally convex, $\mathcal{U} \ne \emptyset$. Let $\mathcal{U}_0$ be a chain in $\mathcal{U}$ and put $U_0 = \bigcup \{U : U \in \mathcal{U}_0\}$. Clearly U_0 is open and it is straightforward to show that it is convex. Now if U_0 is not proper, then $\mathcal{U}_0$ is an open cover. By compactness, $\mathcal{U}_0$ has a finite subcover. But because $\mathcal{U}_0$ is a chain this implies there is one U in $\mathcal{U}_0$ that covers K, a contradiction to the fact that U must be proper. Thus $U_0 \in \mathcal{U}$, and therefore Zorn's Lemma implies there is a maximal element U in $\mathcal{U}$.

To show that $K \backslash U$ is a singleton is a bit harder than what we have done so far. For each x in K and $0 \le \lambda \le 1$, define $T_{x,\lambda} : K \to K$ by

----

[3]Mark Grigorievich Krein was born in 1907 in Kiev. He showed remarkable talent in mathematics and was attending research seminars in Kiev at the age of 14. He never completed his undergraduate degree, however, and at the age of 17 he ran away from home to Odessa. His talent was apparent and in spite of the lack of a degree he was admitted for graduate study at the university in 1926, where he completed his degree in 1929. He remained at the university as a faculty member and helped build it into one of the world's leading centers for functional analysis. In 1941, as the Nazis approached Odessa, he left and did not return until 1944. Soon afterwards he was dismissed by the Soviets for "Jewish nationalism." In fact the entirety of the functional analysis school in Odessa was closed down. Krein was never reinstated in spite of the efforts of other mathematicians, but he did obtain a chair at the Odessa Marine Engineering Institute and remained there until 1954 when he retired. Throughout his career he suffered from antisemitism. In spite of it all, he had a highly productive career and was most influential in the development of analysis. In addition to many books and papers, he had 51 doctoral students. He died in 1989 in Odessa.

[4]David Pinhusovich Milman was born in 1912 in Chechelnyk in the Ukraine. He received his doctorate in 1939 from Odessa State University under the direction of Mark Krein. He made many contributions to functional analysis and his two sons, Vitali and Pierre, both became mathematicians of note. In the 1970s he emigrated to Israel and became a professor at the University of Tel Aviv, where he died in 1982.

$T_{x,\lambda}(y) = \lambda y + (1-\lambda)x$. Note that $T_{x,\lambda}$ is continuous and $T_{x,\lambda}\left(\sum_{j=1}^{n} \alpha_j y_j\right) = \sum_{j=1}^{n} \alpha_j T_{x,\lambda}(y)$ when $\alpha_j \geq 0$ and $\sum_{j=1}^{n} \alpha_j = 1$. Thus whenever $x \in U$, $T_{x,\lambda}(U) \subseteq U$. Hence $U \subseteq T_{x,\lambda}^{-1}(U)$ and $T_{x,\lambda}^{-1}(U)$ is open and convex. Now take $y \in \mathrm{cl}\, U \backslash U$. It follows by Proposition 7.1.5 that for $0 \leq \lambda < 1$ and x in U, $T_{x,\lambda}(y) \in [x,y) \subseteq U$. Therefore $\mathrm{cl}\, U \subseteq T_{x,\lambda}^{-1}(U)$ when $0 \leq \lambda < 1$ and $x \in U$. By the maximality of U we have that $T_{x,\lambda}^{-1}(U) = K$. That is,

8.4.5 $T_{x,\lambda}(K) \subseteq U$ if $x \in U$ and $0 \leq \lambda < 1$

Claim. If V is an open convex subset of K, then either $V \cup U = U$ or $V \cup U = K$.

In fact (8.4.5) implies $V \cup U$ is convex so that the claim follows by the maximality of U in $\mathcal{U}$.

Suppose there are two distinct points a and b in $K \backslash U$ and let V be an open convex subset of K such that $a \in V$ and $b \notin V$. By the claim, $V \cup U = K$. But b does not belong to either V or U, so we have a contradiction. Therefore $K \backslash U$ is a singleton a and so $a \in \mathrm{ext}\, K$.

Now let E be the closed convex hull of $\mathrm{ext}\, K$ and assume there is a point x_0 in $K \backslash E$. By the HBT there is an x^* in $\mathcal{X}^*$, an α in $\mathbb{R}$, and an $\epsilon > 0$ with $\mathrm{Re}\, \langle x, x^* \rangle < \alpha < \alpha + \epsilon < \mathrm{Re}\, \langle x_0, x^* \rangle$ for every x in E. But $V = \{x \in K : \mathrm{Re}\, \langle x, x^* \rangle < \alpha\}$ is a relatively open convex subset of K that contains $\mathrm{ext}\, K$. Now $x_0 \notin V$, so V is proper. Hence there is a maximal element U in $\mathcal{U}$ that contains V. From what was done above, $K \backslash U = \{a\}$ for some a in $\mathrm{ext}\, K$. But then $a \in E \subseteq V \subseteq U$, a contradiction.   ∎

If $\mathcal{X}$ is a normed space, by Alaoglu's Theorem ball $\mathcal{X}^*$ is weak*-compact. Hence the Krein–Milman Theorem implies it is the weak*-closed convex hull of its extreme points. Since ball $L^1[0,1]$ has no extreme points (Example 8.4.2(f)), it is not the dual of a Banach space.

The next result is sometimes taken as part of the Krein–Milman Theorem.

8.4.6. Theorem. *If $\mathcal{X}$ is an LCS, K is a compact convex subset of $\mathcal{X}$, and F is a subset of K such that K is the closed convex hull of F, then $\mathrm{ext}\, K \subseteq \mathrm{cl}\, F$.*

Proof. Without loss of generality we assume that F is closed. Assume the theorem false and let $x_0 \in (\mathrm{ext}\, K) \backslash F$. So there is a continuous seminorm p on $\mathcal{X}$ such that $F \cap \{x : p(x - x_0) < 1\} = \emptyset$; put $U_0 = \{x : p(x) < \frac{1}{3}\}$. Thus $(x_0 + U_0) \cap (F + U_0) = \emptyset$ and so

$$x_0 \notin \mathrm{cl}\,(F + U_0)$$

The compactness of F implies there are $y_1, \ldots, y_n$ in F with $F \subseteq \bigcup_{j=1}^{n}(y_j + U_0)$. Let $K_j = \overline{\text{co}}[F \cap (y_j + U_0)]$; hence $K_j \subseteq y_j + \text{cl}\, U_0$. Since $K_1, \ldots, K_n$ are compact and convex, $\overline{\text{co}}(K_1 \cup \cdots \cup K_n) = \text{co}(K_1 \cup \cdots \cup K_n)$ (Exercise 10). Thus $K = \overline{\text{co}}(F) = \text{co}(K_1 \cup \cdots \cup K_n)$. Therefore there are non-negative real numbers $\alpha_1, \ldots, \alpha_n$ with $\sum_{j=1}^{n} \alpha_j = 1$ and there are points x_j in K_j with $x_0 = \sum_{j=1}^{n} \alpha_j x_j$. But x_0 is an extreme point of K so it must be that there is an x_j with $x_0 = x_j$; that is, $x_0 \in K_j \subseteq y_j + U_0 \subseteq \text{cl}\,(F + U_0)$, a contradiction. ∎

If K is a convex subset of a vector space $\mathcal{X}$ and $\mathcal{Y}$ is another vector space, say that a map $T : K \to \mathcal{Y}$ is *affine* if $T(\alpha a + \beta b) = \alpha T(a) + \beta T(b)$ whenever $a, b \in K$, $\alpha, \beta \geq 0$, and $\alpha + \beta = 1$. The maps $T_{x,\lambda}$ in the proof of the Krein–Milman Theorem are easily seen to be affine; also any linear map between vector spaces is affine when it is restricted to a convex subset of the domain. It is an easy consequence of the definitions that $T(K)$ is a convex subset of $\mathcal{Y}$. An affine map, however, does not necessarily take extreme points of its domain to extreme points of its image. In fact if T is the projection of the plane onto the x-axis and K is a closed disk, then there are many extreme points of K that are mapped onto non-extreme points of $T(K)$. We do have, however, the next result, which says that every extreme point of $T(K)$ is the image of an extreme point of the domain, provided K is compact and T is continuous.

8.4.7. Proposition. *If $\mathcal{X}$ and $\mathcal{Y}$ are LCSs, K is a compact convex subset of $\mathcal{X}$, and $T : K \to \mathcal{Y}$ is a continuous affine map, then $T(K)$ is a compact convex subset of $\mathcal{Y}$; if $y \in \text{ext}\, T(K)$, then there is an x in $\text{ext}\, K$ such that $T(x) = y$.*

Proof. We already have seen that $T(K)$ is convex and it is compact by the continuity of T. Now let $y \in \text{ext}\, T(K)$. It follows that $T^{-1}(y)$ is a compact convex subset of K; by the Krein–Milman Theorem there is an x in $\text{ext}\,[T^{-1}(y)]$. We now show that $x \in \text{ext}\, K$. In fact if $a, b \in K$ and $x = \frac{1}{2}(a + b)$, then $y = T(x) = \frac{1}{2}(T(a) + T(b))$. Since y is an extreme point of $T(K)$, $y = T(a) = T(b)$; that is, $a, b \in T^{-1}(y)$. The fact that $x \in \text{ext}\, T^{-1}(y)$ now implies that $x = a = b$ and so $x \in \text{ext}\, K$. ∎

Exercises.

(1) Prove Proposition 8.4.3.

(2) Show that ball c_0 has no extreme points.

(3) Show that the closed unit ball of $C_0(\mathbb{R})$ has no extreme points. Can you find a result, with its proof, that generalizes this result and the previous one.

(4) Let (X, μ) be a σ-finite measure space and show that the set of extreme points of the closed unit ball of $L^1(\mu)$ is $\{\mu(E)^{-1}\chi_E : E$ is a finite atom of $\mu\}$.

(5) Let (X, μ) be a σ-finite measure space and show that the set of extreme points of the closed unit ball of $L^\infty(\mu)$ is $\{f \in L^\infty(\mu) : |f(x)| = 1$ a.e.$[\mu]\}$.

(6) If $\mathcal{H}$ is a Hilbert space, show that the set of extreme point of the closed unit ball of $\mathcal{H}$ is $\{h \in \mathcal{H} : \|h\| = 1\}$. (Hint: Use the Parallelogram Law.)

(7) Show that if $(X, \mathcal{A}, \mu)$ is a σ-finite measure space and $1 < p < \infty$, then the set of extreme points of the closed unit ball of $L^p(\mu)$ is $\{f \in L^p(\mu)" \|f\|_p = 1\}$. (This extension of the preceding exercise may be difficult.)

(8) If X is a compact metric space, show that the set of extreme points of the closed unit ball of $C(X)$ is $\{f \in C(X) : |f| = 1\}$. How many extreme points does ball $C_{\mathbb{R}}[0, 1]$ have?

(9) Show that the closed unit ball of ℓ^1 is the norm closed convex hull of its extreme points even though it is not norm compact.

(10) If $\mathcal{X}$ is an LCS and $K_1, \ldots, K_n$ are compact convex sets in $\mathcal{X}$, show that $\overline{\mathrm{co}}(K_1 \cup \cdots \cup K_n) = \mathrm{co}(K_1 \cup \cdots \cup K_n)$.

(11) Let K be a convex set, $\mathcal{Y}$ a vector space, and $T : K \to \mathcal{Y}$ an affine map. If $y \in \mathrm{ext}\, T(K)$ and $x \in \mathrm{ext}\, T^{-1}(y)$, show that $x \in \mathrm{ext}\, K$. Give an example where there are points in $T^{-1}(y)$ that are not extreme points of K.

(12) If $\mathcal{H}$ is a Hilbert space and $T \in \mathcal{B}(\mathcal{H})$ such that either T or T^* is an isometry, show that T is an extreme point of the closed unit ball of ball $\mathcal{B}(\mathcal{H})$. (The converse of this is also true but the proof requires more operator theory than we have covered yet – rather, the proof I know needs more operator theory.)

(13) Find an example of a compact convex subset of an LCS whose set of extreme points is not closed. (There is such a set in $\mathbb{R}^3$, but it might be easier to find an example elsewhere.)

8.5. The Stone–Weierstrass Theorem

We continue the tradition that if $\alpha \in \mathbb{R}$, then $\bar{\alpha} = \alpha$; if $\alpha \in \mathbb{C}$, then $\bar{\alpha}$ is the complex conjugate. Similarly, if $f : X \to \mathbb{F}$, then $\bar{f}(x) = \overline{f(x)}$.

8.5.1. Theorem (The Stone[5]–Weierstrass[6] Theorem). *If X is compact and $\mathcal{A}$ is a closed subalgebra of $C(X)$, then $\mathcal{A} = C(X)$ if and only if the following conditions hold:*

(a) $1 \in \mathcal{A}$;

(b) *if $x, y \in X$ with $x \neq y$, then there is an f in $\mathcal{A}$ with $f(x) \neq f(y)$;*

(c) *if $f \in \mathcal{A}$, then $\bar{f} \in \mathcal{A}$.*

Note that if $C(X) = C(X)_{\mathbb{R}}$, condition (c) in the theorem is superfluous. In words the condition that appears in (b) is captured by saying that $\mathcal{A}$ *separates the points* of X. A bit of notation that is standard and will be used in this proof (and beyond) is that when μ is a finite measure on X and

---

[5]Marshall H. Stone was born in 1902 in New York. His father was Harlan Stone who, after time as the dean of the Columbia Law School, became a member of the US Supreme Court, including a term as its chief justice. Marshall Stone entered Harvard in 1919 intending to study law. He soon diverted to mathematics and received his doctorate in 1926 under the direction of David Birkhoff. Though he had brief appointments at Columbia and Yale, most of his early career was spent at Harvard. His initial work continued the direction it took under Birkhoff, but in 1929 he started working on hermitian operators. His American Mathematical Society book, *Linear Transformations in Hilbert space and Their Applications to Analysis*, became a classic. Indeed, a read of that book today shows how the arguments and clarity would easily lead to the conclusion that it is a contemporary monograph. During World War II he worked for the Navy and the War Department and in 1946 he left Harvard to become the chairman of the mathematics department at the University of Chicago. He himself said that this decision was arrived at because of "my conviction that the time was also ripe for a fundamental revision of graduate and undergraduate mathematical education." Indeed he transformed the department at Chicago. The number of theorems that bear his name is truly impressive. Besides the present theorem there is the Stone–Čech compactification, the Stone–von Neumann Theorem, the Stone Representation Theorem in Boolean algebra, and Stone's Theorem on one-parameter semigroups. He stepped down as chair at Chicago in 1952 and retired in 1968, but then went to the University of Massachusetts where he taught in various capacities until 1980. He loved to travel and on a trip to India in 1989 he died in Madras. He had 14 doctoral students.

[6]Karl Weierstrass was born in 1815 in Ostenfelde, Germany. His early schooling was rocky, though he exhibited greater than usual mathematical ability. The difficulty was that his father wanted him to pursue a career in finance, but his passion was for mathematics. He apparently at first reacted to this with a rebellious approach, neglecting all his studies and focusing on fencing and drinking. Nevertheless he studied mathematics on his own with extensive readings, even though he was supposed to follow a course in finance. After having left the university of Bonn without taking any examinations, he seems to have reached an understanding with his father and attended the academy at Münster with the intention of becoming a high school teacher. Here he came under the influence of Christoph Gudermann, a mathematician of note, and impressed his mentor with a paper on elliptic functions. In 1841 Weierstrass passed the exam to become a teacher, a career he followed for some years. In 1854 he published a paper on abelian functions and attracted considerable attention from the research world – sufficient for the University of Königsberg to give him an honorary doctorate, enabling him to launch his university career at Braunsberg. He obtained a chair at the University of Berlin in 1856, where he remained for the rest of his life. He had a profound influence on mathematics, setting new standards of rigor and fostering the careers of numerous mathematicians. He became known as the father of modern analysis. He was plagued by health problems that periodically surfaced and then ebbed. Starting in the early 1860s he lectured while seated and while a student assistant wrote on the board. During his last three years he was confined to a wheelchair and died of pneumonia in Berlin in 1897. He never married.

h is a bounded Borel function on X, then $h\mu$ is the measure on X whose value at a set E is $\int_E h d\mu$.

Proof. Recall the Riesz Representation Theorem that $C(X)^* = M(X)$. By the HBT, it suffices to show that $\mathcal{A}^\perp = (0)$. Assume this to be false. By Alaoglu's Theorem, ball $\mathcal{A}^\perp$ is weak*-compact and so by the Krein–Milman Theorem there is a measure μ in ext $[\text{ball }\mathcal{A}^\perp]$. Since $\mathcal{A}^\perp \neq (0)$, it must be that $\|\mu\| = 1$. Let $K = \operatorname{spt}\mu$, the support of μ. That is,

$$K = X \backslash \bigcup \{V : V \text{ is open and } |\mu|(V) = 0\}$$

It follows that $|\mu|(X \backslash K) = 0$. In fact, if L is a compact subset of $X \backslash K$, then $\{V : V \text{ is open and } |\mu|(V) = 0\}$ is an open cover of L and so there is a finite subcover; thus $|\mu|(L) = 0$ and the regularity of $|\mu|$ implies that $|\mu|(X \backslash K) = 0$. Also $\int f d\mu = \int_K f d\mu$ for every f in $C(X)$.

Claim. K is a singleton, $\{x_0\}$.

Note that once this claim is established we can rapidly conclude the proof. In fact, it follows that $\mu = \alpha \delta_{x_0}$ where α is a scalar with $|\alpha| = 1$. But since $1 \in \mathcal{A}$, $0 = \int 1 d\mu = \alpha$, contradicting the fact that $|\alpha| = 1$.

Suppose $x \in X$ and $x \neq x_0$; we'll show that $x \notin K$. By (b), there is a function f_1 in $\mathcal{A}$ with $f_1(x_0) \neq f_1(x) = \beta$. By (a), the constant function β belongs to $\mathcal{A}$, so $f_2 = f_1 - \beta \in \mathcal{A}$. By (c), $f_3 = |f_2|^2 = f_2 \overline{f_2} \in \mathcal{A}$. Note that $f_3(x) = 0 < f_3(x_0)$ and $f_3 \geq 0$. If $f = (1 + \|f_3\|)^{-1} f_3$, then $f \in \mathcal{A}$, $0 \leq f \leq 1$, $f(x) = 0$, and $0 < f(x_0) < 1$. Now if g is an arbitrary function in $\mathcal{A}$, then gf and $g(1-f)$ belong to $\mathcal{A}$. So the fact that $\mu \in \mathcal{A}^\perp$ implies that $0 = \int gf d\mu = \int g(1-f) d\mu$ for every g in $\mathcal{A}$. That is

$$f\mu \text{ and } (1-f)\mu \in \mathcal{A}^\perp$$

Put $\alpha = \|f\mu\| = \int f d|\mu|$. Since $f(x_0) > 0$, there is an open set V containing x_0 and an $\epsilon > 0$ such that $f(y) > \epsilon$ for all y in V. Since $V \cap \operatorname{spt}\mu \neq \emptyset$, $|\mu|(V) > 0$. Hence $\alpha = \int f d|\mu| \geq \int_V f d|\mu| \geq \epsilon |\mu|(V) > 0$. Similarly since $f(x_0) < 1$, we have that

$$0 < \alpha < 1$$

Finally we observe that $1 - \alpha = \int (1-f) d|\mu| = \|(1-f)\mu\|$.

Now we arrive at the fact that

$$\mu = \alpha \left[\frac{f\mu}{\|f\mu\|} \right] + (1-\alpha) \left[\frac{(1-f)\mu}{\|(1-f)\mu\|} \right]$$

But since μ is an extreme point of ball $(\mathcal{A}^\perp)$ and $0 < \alpha < 1$, we have that $\mu = \|f\mu\|^{-1} f\mu = \alpha^{-1} f\mu$. But this implies that $\alpha^{-1} f = 1$ a.e. $[|\mu|]$. Since f is continuous it follows that $f(y) = \alpha$ for all y in K. But $f(x) = 0 < \alpha$ so it must be that $x \notin K$. Since x was an arbitrary point in $X \backslash \{x_0\}$, it follows that $K = \{x_0\}$ and the proof is complete. ∎

The idea of proving this theorem using the Krein–Milman Theorem is due to [**10**].

With any good theorem it is important to test the hypothesis by finding examples that satisfy all the hypotheses save one and where the conclusion is invalid. See Exercise 1. Here is a result that partly does this but is much more.

8.5.2. Corollary. *If X is a compact space and $\mathcal{A}$ is a closed subalgebra of $C(X)$ that separates the points of X and is closed under complex conjugation, then either $\mathcal{A} = C(X)$ or there is a point x_0 in X such that $\mathcal{A} = \{f \in C(X) : f(x_0) = 0\}$.*

Proof. Identify $\mathbb{F}$ with the one-dimensional subspace of $C(X)$ consisting of the constant functions. Since $\mathbb{F}$ is one-dimensional, $\mathcal{A} + \mathbb{F}$ is a closed subspace of $C(X)$ (Corollary 6.2.3). It is easy to check that $\mathcal{A} + \mathbb{F}$ is an algebra and so $\mathcal{A} + \mathbb{F} = C(X)$ by the Stone–Weierstrass Theorem. If $\mathcal{A} \neq C(X)$, then $C(X)/\mathcal{A}$ is one-dimensional and therefore $\mathcal{A}^{\perp}$ is one-dimensional. Let $\mathcal{A}^{\perp} = \{\alpha\mu : \alpha \in \mathbb{F}\}$, where $\|\mu\| = 1$. But if $f \in \mathcal{A}$, then $f\mu \in \mathcal{A}^{\perp}$, so there is an α with $f\mu = \alpha\mu$. But since f is continuous this implies that $f = \alpha$ on spt μ. Since $\mathcal{A}$ separates the points of X this can only happen if the support of μ is a singleton $\{x_0\}$. It follows that $\mathcal{A}^{\perp} = \{\beta\delta_{x_0} : \beta \in \mathbb{F}\}$ and we have that $\mathcal{A} = {}^{\perp}(\mathcal{A}^{\perp}) = \{f \in C(X) : f(x_0) = 0\}$. ∎

We can use the preceding result to extend the Stone–Weierstrass Theorem to the locally compact setting.

8.5.3. Corollary. *If X is locally compact and $\mathcal{A}$ is a closed subalgebra of $C_0(X)$, then $\mathcal{A} = C_0(X)$ if and only if $\mathcal{A}$ satisfies the following:*

(a) *for each x in X there is a function f in $\mathcal{A}$ such that $f(x) \neq 0$;*

(b) *$\mathcal{A}$ separates the points of X;*

(c) *$\mathcal{A}$ is closed under complex conjugation.*

Proof. Let X_∞ be the one-point compactification of X and identify the functions f in $C(X_\infty)$ that vanish at the point at infinity with $C_0(X)$. So $\mathcal{A}$ is a closed subalgebra of $C(X_\infty)$. We know that $\mathcal{A} \neq C(X_\infty)$, so by the preceding corollary $\mathcal{A}$ is the algebra of all functions in $C(X_\infty)$ vanishing at some point. Using (a) we see that that point must be ∞, and so we get that $\mathcal{A} = C_0(X)$. ∎

There is an important generalization of the Stone–Weierstrass Theorem due to Errett Bishop that goes as follows. When X is compact and $\mathcal{A}$ is a closed subalgebra of $C(X)$ that contains the constants, say that a subset E of X is $\mathcal{A}$-antisymmetric if whenever $f \in \mathcal{A}$ and the restriction of f to E is real-valued, then f is constant on E. Bishop's Theorem states that if $g \in C(X)$ and for each $\mathcal{A}$-antisymmetric set E there is an f in $\mathcal{A}$ such that $g|E = f|E$, then $g \in \mathcal{A}$ [2]. It is not difficult to show that every set that is $\mathcal{A}$-antisymmetric is contained in a maximal $\mathcal{A}$-antisymetric set. Moreover the maximal $\mathcal{A}$-antisymetric sets are easily seen to be closed and pairwise disjoint. Thus if $\mathcal{S}$ is the collection of maximal $\mathcal{A}$-antisymetric sets, the conclusion of Bishop's generalization can be rephrased as $\mathcal{A} = \{f \in C(X) : f|S \in \mathcal{A}|S$ for every $S \in \mathcal{S}\}$. A proof following the lines of the proof given of the Stone–Weierstrass Theorem can be found in [16]. Also see the articles [4] and [29].

The next result has a proof that is close in spirit to the proof of the Stone–Weierstrass Theorem though it has little to do with the content of that result. A *probability measure* on $(X, \mathcal{A})$ is a positive measure μ on this space with $\mu(X) = 1$. We denote by $P(X)$ the set of all probability measures. Note that when X is a compact space, $P(X)$ is a convex set that is weak*-closed in ball $M(X)$ and hence is weak*-compact. Therefore $\text{ext } P(X) \neq \emptyset$. Before giving a characterization of the extreme points of $P(X)$ and ball $M(X)$, we need a lemma that could have been presented earlier.

8.5.4. Lemma. *If μ is a complex-valued measure on $(X, \mathcal{A})$ such that $\|\mu\| = \mu(X)$, then $\mu \geq 0$.*

Proof. First assume that μ is real-valued and consider the Hahn decomposition for μ, $X = E_+ \cup E_-$; so $\mu_{\pm} = \pm\mu_{E_{\pm}}$ are positive measures and $\mu = \mu_+ - \mu_-$. Note that $\mu_+(X) - \mu_-(X) = \mu(X) = \|\mu\| = \mu_+(X) + \mu_-(X)$; therefore $\mu_-(X) = 0$ and since $\mu_- \geq 0$, it follows that it is the zero measure.

Now assume that μ is complex-valued and write $\mu = \mu_1 + i\mu_2$. Since $\|\mu\| = \mu(X)$, we have that $\mu_2(X) = 0$ and $\|\mu_1\| \geq \mu_1(X) = \mu(X) = \|\mu\| \geq \|\mu_1\|$. Hence $\|\mu_1\| = \mu_1(X)$ and, by the first part of the proof, $\mu_1 \geq 0$. Suppose there is a set E in $\mathcal{A}$ with $\alpha = \mu_2(E) \neq 0$; so $\mu_2(X \backslash E) = -\alpha$. Thus $\|\mu\| = |\mu|(E) + |\mu|(X \backslash E) \geq |\mu(E)| + |\mu(X \backslash E)| = [\mu_1(E)^2 + \alpha^2]^{\frac{1}{2}} + [\mu_1(X \backslash E)^2 + \alpha^2]^{\frac{1}{2}} > \mu_1(X) = \|\mu\|$, a contradiction. $\blacksquare$

8.5.5. Theorem. *If X is a compact space, then*

$$\text{ext}\,[\text{ball}\,M(X)] = \{\alpha\delta_x : x \in X \text{ and } \alpha \in \mathbb{F} \text{ with } |\alpha| = 1\}$$

and

$$\text{ext}\,P(X) = \{\delta_x : x \in X\}$$

Proof. By Exercise 2 each $\delta_x \in \text{ext}\, P(X)$ and, as long as $|\alpha| = 1$, each $\alpha\delta_x \in \text{ext}\,[\text{ball}\, M(X)]$. We prove the converse.

The strategy in the proof is to show that each extreme point of $P(X)$ is in $\text{ext}\,[\text{ball}\, M(X)]$. Thus if we characterize the extreme points of ball $M(X)$ we will have established the second equality in the theorem. Assume that $\mu \in \text{ext}\, P(X)$ and there are measures ν_1, ν_2 in ball $M(X)$ with $\mu = \frac{1}{2}(\nu_1 + \nu_2)$. Since $1 = \|\mu\| \leq \frac{1}{2}(\|\nu_1\| + \|\nu_2\|) \leq 1$, it follows that $\|\nu_1\| = \|\nu_2\| = 1$. Since $\mu \geq 0$, $1 = \|\mu\| = \mu(X) = \frac{1}{2}[\nu_1(X) + \nu_2(X)]$. But $|\nu_1(X)| \leq 1$ and $|\nu_2(X)| \leq 1$; since 1 is an extreme point of $\text{cl}\,\mathbb{D}$, we have that $1 = \nu_j(X) = \|\nu_j\|$. Now Lemma 8.5.4 implies that both ν_1 and ν_2 are probability measures. Since μ is an extreme point of $P(X)$, we have that $\mu = \nu_1 = \nu_2$; hence $\mu \in \text{ext}\,[\text{ball}\, M(X)]$.

Now to show that every measure μ in $\text{ext}\,[\text{ball}\, M(X)]$ has the correct form. If K denotes the support of μ, then as in the proof of the Stone–Weierstrass Theorem we will show that K is a singleton and this will finish the proof. Assume that $x_0 \in K$ and assume there is an x in K different from x_0. Let U, V be open sets with x_0 in U, x in V, and $\text{cl}\, U \cap \text{cl}\, V = \emptyset$. By Urysohn's lemma there is a function f in $C(X)$ with $0 \leq f \leq 1$, $f(u) = 1$ on $\text{cl}\, U$, and $f(v) = 0$ on $\text{cl}\, V$. Consider the measures $f\mu$ and $(1 - f)\mu$. If $\alpha = \|f\mu\|$, then $\alpha = \int f\, d|\mu| \geq |\mu|(U) > 0$ since $U \cap K \neq \emptyset$. Similarly $1 - \alpha > 0$ since $V \cap K \neq \emptyset$. Thus $\alpha^{-1} f\mu$ and $(1 - \alpha)^{-1}(1 - f)\mu \in \text{ball}\, M(X)$ with $\mu = \alpha[\alpha^{-1} f\mu] + (1 - \alpha)[(1 - \alpha)^{-1}(1 - f)\mu]$; since μ is an extreme point we get that $\mu = \alpha^{-1} f\mu$. As in the proof of the Stone–Weierstrass Theorem this implies $f(y) = \alpha$ for all y in K. In particular, $f(x) = \alpha > 0$, but we know from the construction of f that $f(x) = 0$, a contradiction. Therefore $K = \{x_0\}$ and $\mu = \gamma\delta_{x_0}$. ∎

Exercises.

(1) In the Stone–Weierstrass Theorem take each possible pair of the conditions (a), (b), (c) and find an example of X and $\mathcal{A}$ such that $\mathcal{A}$ satisfies those two conditions but $\mathcal{A} \neq C(X)$.

(2) If X is a compact space and $x \in X$, show that δ_x is an extreme point of the set of probability measures on X and $\alpha\delta_x$ is an extreme point of ball $M(X)$ when $|\alpha| = 1$.

(3) Use the Stone–Weierstrass Theorem to give another proof of Proposition 4.7.3.

(4) For z a complex number, a polynomial in z and $\bar{z}$ is a finite linear combination of terms of the form $z^n \bar{z}^m$, where $n, m \geq 0$. If K is a compact subset of $\mathbb{C}$ show that the uniform closure of the polynomials in z and $\bar{z}$ is all of $C(K)$.

(5) If X is a compact subset of $\mathbb{C}^d$, show that the set of all polynomials in $z_1, \ldots, z_d, \bar{z}_1, \ldots, \bar{z}_d$ is uniformly dense in $C(X)$.

(6) Let (X, d) be a compact metric space, let $\{a_n\}$ be a countable dense subset of X, and for each $n \geq 1$ let $d_n : X \to \mathbb{R}$ be the continuous function $d_n(x) = d(a_n, x)$. Show that if $f \in C(X)$ and $\epsilon > 0$, then there is a function g that is the linear combination of the constant function and products of the functions $\{d_n\}$ such that $|f(x) - g(x)| < \epsilon$ for all x in X. Note that this shows that $C(X)$ is separable.

(7) If $\mathcal{A}$ is the uniformly closed subalgebra of $C_b(\mathbb{R})$ that is generated by $\sin x$ and $\cos x$, show that

$$\mathcal{A} = \{f \in C_b(\mathbb{R}) : f(x) = f(x + 2\pi) \text{ for all } x\}$$

(8) Let I be any non-empty set and consider the compact product space $X = [0, 1]^I$. For each i in I there is the coordinate projection $p_i : X \to [0, 1] \subseteq \mathbb{R}$ defined by $p_i(x) = x(i)$. Show that the algebra generated by $\{p_i : i \in I\}$ is uniformly dense in $C(X)$.

Chapter 9

Operators on a Banach Space

In this chapter we continue the study of linear transformations begun in Chapter 5. Though most of the material in that chapter concerned operators on a Hilbert space, §5.1 established some basic properties of linear transformations between Banach spaces. In this short chapter the emphasis will be on Banach spaces as we seek to extend some of the results on Hilbert space to this more general setting.

9.1. The adjoint

Here we will define the adjoint of a linear transformation between Banach spaces. Throughout the section, $\mathcal{X}$ and $\mathcal{Y}$ will denote Banach spaces. We start with the following result that will justify the definition of the adjoint.

9.1.1. Proposition. *If $\mathcal{X}$ and $\mathcal{Y}$ are Banach spaces and $T : \mathcal{X} \to \mathcal{Y}$ is a linear transformation, then the following statements are equivalent.*

(a) *T is bounded.*

(b) *If $y^* \in \mathcal{Y}^*$, then $x \mapsto \langle Tx, y^* \rangle$ is a continuous linear functional on $\mathcal{X}$.*

(c) *T is continuous if both $\mathcal{X}$ and $\mathcal{Y}$ have their weak topologies.*

Proof. If (a) holds, then $|\langle Tx, y^* \rangle| \leq \|T\|\|y^*\|\|x\|$, so $x \mapsto \langle Tx, y^* \rangle$ defines a bounded linear functional on $\mathcal{X}$. Now assume (b) is true and let's prove (c). An arbitrary defining seminorm for the weak topology on $\mathcal{Y}$ is p_{y^*}, where $y^* \in \mathcal{Y}^*$. For such a y^* let x^* be the element in $\mathcal{X}^*$ satisfying $\langle x, x^* \rangle = \langle Tx, y^* \rangle$ for all x as in (b). It follows that $p_{y^*}(Tx) = |\langle Tx, y^* \rangle| = p_{x^*}(x)$.

So if $\{x_i\}$ is a net in $\mathcal{X}$ and $x_i \to 0$ weakly, then $Tx_i \to 0$ weakly in $\mathcal{Y}$, proving (c). If we assume that (c) is true, then for any y^* in $\mathcal{Y}^*$ we have that $\sup\{|\langle Tx, y^*\rangle| : \|x\| \le 1\} < \infty$. That is, $T(\text{ball}\,\mathcal{X})$ is weakly bounded. By the PUB, $T(\text{ball}\,\mathcal{X})$ is norm bounded and so $T \in \mathcal{B}(\mathcal{X}, \mathcal{Y})$. ∎

If $T \in \mathcal{B}(\mathcal{X}, \mathcal{Y})$, the preceding proposition allows us to define the *adjoint* operator $T^* : \mathcal{Y}^* \to \mathcal{X}^*$ by

9.1.2 $\langle x, T^*(y^*)\rangle = \langle T(x), y^*\rangle$

for all x in $\mathcal{X}$ and y^* in $\mathcal{Y}^*$. The preceding proposition implies that the right-hand side of (9.1.2) defines a continuous linear functional on $\mathcal{X}$ and this equation says that we are denoting that linear functional by $T^*(y^*)$. This is quite similar to the way we defined the adjoint in a Hilbert space setting (5.4.4), but there is a small difference when $\mathbb{F} = \mathbb{C}$. When we identify the dual of a Hilbert space $\mathcal{H}$ with itself, for any h in $\mathcal{H}$ we define the linear functional L_h on $\mathcal{H}$ by $L_h(f) = \langle f, h\rangle$. We know from the Riesz Representation Theorem that the mapping $h \mapsto L_h$ is surjective, but it is not linear; it is conjugate-linear. That is, it is additive but for any scalar α, $L_{\alpha h} = \bar{\alpha} L_h$. This leads to small differences between the adjoint of an operator in the Hilbert space and Banach space settings as we will see in the next proposition, specifically in part (c). (Also see Exercise 1.)

9.1.3. Proposition. *Let $S, T \in \mathcal{B}(\mathcal{X}, \mathcal{Y})$ and let $\alpha, \beta \in \mathbb{F}$.*

(a) $T^* \in \mathcal{B}(\mathcal{Y}^*, \mathcal{X}^*)$ *and* $\|T^*\| = \|T\|$.

(b) $T^* : (\mathcal{Y}^*, \text{weak}^*) \to (\mathcal{X}^*, \text{weak}^*)$ *is continuous.*

(c) $(\alpha T + \beta S)^* = \alpha T^* + \beta S^*$.

(d) *If $\mathcal{Z}$ is a Banach space and $R \in \mathcal{B}(\mathcal{Y}, \mathcal{Z})$, then $(RT)^* = T^* R^*$.*

Proof. (a) If $y^* \in \text{ball}\,\mathcal{Y}^*$, then $|\langle x, T^*(y^*)\rangle| = |\langle Tx, y^*\rangle| \le \|T\|\|x\|$; thus $\|T^*\| \le \|T\|$ and so $T^* \in \mathcal{B}(\mathcal{Y}^*, \mathcal{X}^*)$. On the other hand,

$$\|Tx\| = \sup\{|\langle Tx, y^*\rangle| : y^* \in \text{ball}\,\mathcal{Y}^*\}$$
$$= \sup\{|\langle x, T^*(y^*)\rangle| : y^* \in \text{ball}\,\mathcal{Y}^*\}$$
$$\le \|x\|\|T^*\|$$

and so $\|T\| \le \|T^*\|$.

(b) A defining seminorm for the weak* topology on $\mathcal{X}^*$ is $p_{x^*}(x) = |\langle x, x^*\rangle|$, where $x \in \mathcal{X}$. But $p_x(T^* y^*) = p_{y^*}(Tx)$ and this shows (b).

The proofs of (c) and (d) are routine. ∎

Now that we have that $T^* : \mathcal{Y}^* \to \mathcal{X}^*$ is a bounded operator, we can define $T^{**} : \mathcal{X}^{**} \to \mathcal{Y}^{**}$ by

9.1.4 $$\langle T^{**}x^{**}, y^* \rangle = \langle x^{**}, T^*y^* \rangle$$

for all x^{**} in $\mathcal{X}^{**}$ and y^* in $\mathcal{Y}^*$. Now suppose that $x \in \mathcal{X}$ and consider x as an element of $\mathcal{X}^{**}$. So for each y^* in $\mathcal{Y}^*$ we have that $\langle T^{**}x, y^* \rangle = \langle x, T^*y^* \rangle = \langle Tx, y^* \rangle$. That is,

$$T^{**}|\mathcal{X} = T$$

We are in need of some examples.

9.1.5. Example. For a measure μ, a ϕ in $L^\infty(\mu)$, and $1 \le p < \infty$, we define the multiplication operator M_ϕ on $L^p(\mu)$ as in Example 5.1.2(c). Here $M_\phi^* : L^q(\mu) \to L^q(\mu)$ is none other than multiplication by ϕ; that is, $M_\phi^* = M_\phi$, but operating on a different space. Compare this to Example 5.4.5(a).

9.1.6. Example. If k and K are as in Proposition 5.1.9, then K^* is the integral operator on $L^q(\mu)$ with kernel $k^*(x, y) = k(y, x)$.

Recall that when X is a compact space the dual of $C(X)$ is the space of regular Borel measures on X, $M(X)$, with the total variation norm (4.3.8). We want to look at the adjoints of some operators between spaces of continuous functions and interpret what the adjoint does to a measure μ.

9.1.7. Example. If X and Y are compact spaces and $\tau : Y \to X$ is a continuous function, define $A : C(X) \to C(Y)$ by $Af = f \circ \tau$. It is easy to see that A is bounded; in fact, $|Af(y)| = |f(\tau(y))| \le \|f\|$, so $\|A\| \le 1$. On the other hand, $A1 = 1$ so we get that $\|A\| = 1$. It follows that for μ in $M(Y)$ and E a Borel subset of X,

$$(A^*\mu)(E) = \mu(\tau^{-1}(E))$$

See Exercise 3.

9.1.8. Theorem. *If $T \in \mathcal{B}(\mathcal{X}, \mathcal{Y})$, then*

$$\ker T^* = (\operatorname{ran} T)^\perp \quad and \quad \ker T = {}^\perp(\operatorname{ran} T^*)$$

The proof of this important result is a straightforward imitation of the proof of Theorem 5.4.10. Supplying the details of this proof would be a good place for the reader to become more comfortable with and fix the concept of the adjoint in his/her head.

9.1.9. Proposition. *If $T \in \mathcal{B}(\mathcal{X}, \mathcal{Y})$, then T is invertible if and only if T^* is invertible, in which case $(T^*)^{-1} = (T^{-1})^*$.*

Proof. We leave it to the reader to verify that the identity operator on a space $\mathcal{X}$, $1 = I$, has as its adjoint the identity on $\mathcal{X}^*$. Assume T is invertible and let $S = T^{-1}$. So $1 = ST$ and $1 = TS$; thus Proposition 9.1.3(d) implies $1 = T^*S^*$ and $1 = S^*T^*$. Thus we have that T^* is invertible and $(T^*)^{-1} = (T^{-1})^*$.

Now assume that T^* is invertible. By the Open Mapping Theorem there is a constant $c > 0$ such that $T^*(\text{ball } \mathcal{Y}^*) \supseteq c[\text{ball } \mathcal{X}^*]$. Thus for x in $\mathcal{X}$,

$$\|Tx\| = \sup\{|\langle Tx, y^*\rangle| : y^* \in \text{ball } \mathcal{Y}^*\}$$
$$= \sup\{|\langle x, T^*y^*\rangle| : y^* \in \text{ball } \mathcal{Y}^*\}$$
$$\geq \sup\{|\langle x, x^*\rangle| : \|x^*\| \leq c\}$$
$$= c^{-1}\|x\|$$

This tells us two things. First, that T is injective, and second, that $\operatorname{ran} T$ is closed (Lemma 5.6.13). But $\operatorname{ran} T$ is also dense since (9.1.8) $(\operatorname{ran} T)^\perp = \ker T^* = (0)$. Therefore T is bijective and hence invertible. ∎

The reader may have noticed that when discussing $C(X)$ for a compact space X, we have stopped specifying that X is a metric space. This certainly was the case when we saw the Stone–Weierstrass theorem, but it was occurring sporadically before that in examples and exercises. This continues. Partly this change is done because the results do not require metric spaces, though, as we pointed out, neither was that required in Chapter 2. The point then was that assuming X is a metric space simplified the arguments. That has ceased being the case, though in this more general setting we will be required to make arguments using nets rather than sequences. The author feels that is a bonus for the reader's education. Nevertheless, anyone who wishes to assume the hypothesis of metrizability is welcomed to do so and use arguments with sequences whenever the text uses nets. We start this in the next theorem.

9.1.10. Theorem (Banach–Stone Theorem). *If X and Y are compact spaces, then $A : C(X) \to C(Y)$ is a surjective isometry if and only if there is a homeomorphism $\tau : Y \to X$ and a continuous function $\alpha : Y \to \mathbb{F}$ with $|\alpha(y)| = 1$ for all y in Y such that*

$$Af(y) = \alpha(y)f(\tau(y))$$

for every f in $C(X)$.

Proof. It is easy to check that $A^* : M(Y) \to M(X)$ is also a surjective isometry. Hence $A^*(\text{ext}\,[\text{ball } M(Y)]) = \text{ext}\,[\text{ball } M(X)]$. By Theorem 8.5.5, for each y in Y there is a unique point $\tau(y)$ in X and a unique unimodular scalar $\alpha(y)$ such that

$$A^*(\delta_y) = \alpha(y)\delta_{\tau(y)}$$

We note that this defines two functions $\alpha : Y \to \mathbb{F}$ and $\tau : Y \to X$. Thus for any f in $C(X)$ and any y in Y we have that

9.1.11 $$(Af)(y) = \alpha(y)f(\tau(y))$$

This says that α and τ provide us with the desired formula; now we have to establish that these functions have the desired properties. The fact that α is continuous is easy since $\alpha = A(1) \in C(Y)$.

Claim. τ is continuous.

Suppose $\{y_i\}$ is a net in Y such that $y_i \to y$. Since X is compact, the net $\{\tau(y_i)\}$ has a cluster point x in X. Thus for each f in $C(X)$, $f(\tau(y_i)) \to_{\mathrm{cl}} f(x)$. But $Af(y_i) \to Af(y)$ since $Af \in C(Y)$. Using (9.1.11) and the fact that α is continuous we have that $f(\tau(y_i)) = \overline{\alpha(y_i)}Af(y_i) \to \overline{\alpha(y)}Af(y) = f(\tau(y))$. Thus $f(x) = f(\tau(y))$ for every f in $C(X)$. Therefore $x = \tau(y)$; that is, the net $\{\tau(y_i)\}$ in the compact set X has a unique cluster point, $\tau(y)$. This implies that $\tau(y_i) \to \tau(y)$ (A.2.6) and so the claim is established.

Because A is an isometry, τ must be surjective. In fact, if $\tau(Y)$ is a proper closed subset of X, then there is a non-zero f in $C(X)$ that vanishes on $\tau(Y)$. By (9.1.11) we get that $Af = 0$, contradicting the fact that A must be injective. Suppose y_1 and y_2 are two distinct points in Y; then $\overline{\alpha(y_1)}\delta_{y_1} \neq \overline{\alpha(y_2)}\delta_{y_2}$, no matter the values that α takes at these two points. Since A^* is injective, $\delta_{\tau(y_1)} = A^*[\overline{\alpha(y_1)}\delta_{y_1}] \neq A^*[\overline{\alpha(y_2)}\delta_{y_2}] = \delta_{\tau(y_2)}$, so that τ is one-to-one. Because τ is continuous and Y is compact, τ is also a closed map so that τ is a homeomorphism. ∎

Exercises.

(1) Let $1 \leq d < \infty$ and let the linear transformation $T : \mathbb{C}^d \to \mathbb{C}^d$ be represented by the matrix A. If $\mathbb{C}^d$ is considered as a Banach space, show that T^* is represented by the transpose of A. If $\mathbb{C}^d$ is considered as a Hilbert space, show that T^* is represented by the conjugate transpose of A.

(2) Verify the statements in Example 9.1.6.

(3) Verify the statements in Example 9.1.7. Also show that A is an isometry if and only if τ is surjective.

(4) Let $1 \leq p < \infty$ and define $S : \ell^p \to \ell^p$ by $S(\alpha_1, \alpha_2, \dots) = (0, \alpha_1, \alpha_2, \dots)$; compute S^*.

(5) Let $A \in \mathcal{B}(c_0)$, let $\{e_n\}$ be the standard basis in c_0, and put $\alpha_{mn} = (Ae_n)(m)$ for all $m, n \geq 1$. Prove the following. (a) For every $n \geq 1$, $\lim_m \alpha_{mn} = 0$. (b) $M = \sup_m \sum_{n=1}^{\infty} |\alpha_{mn}| < \infty$. (c) Conversely if $\{\alpha_{mn} : m, n \geq 1\}$ are scalars satisfying the conditions in (a) and

(b), then $(Ax)(m) = \sum_{n=1}^{\infty} \alpha_{mn}x(n)$ defines a bounded operator on c_0 with $\|A\| \le M$. (d) Calculate $A^* : \ell^1 \to \ell^1$ using the scalars $\{\alpha_{mn}\}$.

(6) Let $A \in \mathcal{B}(\ell^1)$, let $\{e_n\}$ be the standard basis in ℓ^1, and put $\alpha_{mn} = (Ae_n)(m)$ for all $m, n \ge 1$. Prove the following. (a) For every $m \ge 1$, $\sup_n |\alpha_{mn}| < \infty$. (b) $M = \sup_n \sum_{m=1}^{\infty} |\alpha_{mn}| < \infty$. (c) Conversely, if $\{\alpha_{mn} : m, n \ge 1\}$ are scalars satisfying (a) and (b), then $(Af)(m) = \sum_{n=1}^{\infty} \alpha_{mn}f(n)$ defines a bounded operator on ℓ^1 and $\|A\| = M$. (d) Calculate $A^* : \ell^\infty \to \ell^\infty$.

(7) If $B \in \mathcal{B}(\mathcal{Y}^*, \mathcal{X}^*)$, then there is an operator A in $\mathcal{B}(\mathcal{X}, \mathcal{Y})$ such that $B = A^*$ if and only if B is continuous when both $\mathcal{Y}^*$ and $\mathcal{X}^*$ have their weak* topologies.

9.2. Compact operators

Here we continue and amplify the study of compact linear transformations begun in §5.6. The first result extends part of Theorem 5.6.4 to the Banach space setting, though the proof is slightly more difficult.

9.2.1. Theorem (Schauder's[1] Theorem). *If $T \in \mathcal{B}(\mathcal{X}, \mathcal{Y})$, the T is compact if and only if T^* is compact.*

Proof. Assume T is compact and let $\{y_n^*\}$ be a sequence in ball $\mathcal{Y}^*$; we must show that $\{T^*y_n^*\}$ has a norm convergent subsequence. By Alaoglu's Theorem there is a y^* in ball $\mathcal{Y}^*$ such that $y_n^* \to_{\text{cl}} y^*$ (wk*). Since T^* is weak*-continuous, $T^*y_n^* \to_{\text{cl}} T^*y^*$ (wk*). We will show that T^*y^* is a norm cluster point of $\{T^*y_n^*\}$.

Let $\epsilon > 0$ and fix an $N \ge 1$. Since $T(\text{ball } \mathcal{X})$ has compact closure, it is totally bounded; thus there are vectors $x_1, \ldots, x_m$ in ball $\mathcal{X}$ such that $T(\text{ball } \mathcal{X}) \subseteq \bigcup_{k=1}^m \{y \in \mathcal{Y} : \|y - Tx_k\| < \epsilon/3\}$. Since $y_n^* \to_{\text{cl}} y^*$ (wk*), there is an $n \ge N$ such that $|\langle Tx_k, y_n^* - y^* \rangle| < \epsilon/3$ for $1 \le k \le m$. If $x \in \text{ball } \mathcal{X}$,

---

[1] Julius Schauder was born in 1899 in Lvov, which at the time was part of the Austrian-Hungarian Empire. During World War I he was drafted into the Austrian army and fought in Italy, where he was captured. When Austria collapsed he joined a Polish army formed in France. After the war Lvov once again became part of Poland and he completed his doctorate there in 1923, working under Steinhaus. With a fellowship from the Rockefeller Foundation he went to study in Paris and began his collaboration with Leray. His most famous work is a fixed point theorem named after him and the basic work on the Leray–Schauder degree, for which they received an international prize. He returned to the university in Lvov, which was occupied by the Soviets in 1939. In 1941 the German army entered the city. He died in 1943 at the hands of the Nazis in their systematic extermination of the Jews, though the exact circumstances of his death are not certain. His wife and daughter were sheltered by the resistance and for awhile hid in the sewers. Eventually they surrendered and were sent to the concentration camp in Lublin. His wife died there but his daughter, Eva, survived and after the war went to live with Schauder's brother in Italy.

choose k such that $\|Tx - Tx_k\| < \epsilon/3$. Thus

$$\begin{aligned}
|\langle x, T^*y^* - T^*y_n^*\rangle| &= |\langle Tx, y^* - y_n^*\rangle| \\
&\leq |\langle Tx - Tx_k, y^* - y_n^*\rangle| + |\langle Tx_k, y^* - y_n^*\rangle| \\
&< 2\|Tx - Tx_k\| + \epsilon/3 \\
&< \epsilon
\end{aligned}$$

Since x was an arbitrary element of ball $\mathcal{X}$ this implies $\|T^*y^* - T^*y_n^*\| < \epsilon$ so that T^* is compact.

For the converse, assume that T^* is compact and apply the first half of the proof to T^* to conclude that T^{**} is compact. Since $T = T^{**}|\mathcal{X}$, we have that T is compact. $\blacksquare$

Another part of Theorem 5.6.4 does not carry over from Hilbert space to Banach space: not every compact operator on a Banach space is the limit of finite rank operators. This was discussed after Theorem 5.6.4. However for certain Banach spaces it is true that every compact operator is the limit of finite rank ones. We start the process of verifying this for the space $C(X)$.

9.2.2. Definition. If X is a completely regular space and $\mathcal{F}$ is a subset of $C_b(X)$, then $\mathcal{F}$ is said to be *equicontinuous* if for every $\epsilon > 0$ and every point x_0, there is a neighborhood U of x_0 such that $|f(x) - f(x_0)| < \epsilon$ for all x in U and every f in $\mathcal{F}$.

Note that the fact that for each ϵ, x_0, and f there is such a neighborhood U is just the fact that f is continuous. The salient point here is that one open neighborhood U works for every function in the family $\mathcal{F}$. So every finite set in $C(X)$ is equicontinuous.

9.2.3. Theorem (Arzelà[2]–Ascoli[3] Theorem). *If X is compact, then a subset $\mathcal{F}$ of $C(X)$ is totally bounded if and only if $\mathcal{F}$ is bounded and equicontinuous.*

Proof. Assume that $\mathcal{F}$ is totally bounded and let $\epsilon > 0$. So there are $f_1, \ldots, f_n$ in $\mathcal{F}$ such that $\mathcal{F} \subseteq \bigcup_{k=1}^{n}\{f \in C(X) : \|f - f_k\| < \epsilon/3\}$. If $f \in \mathcal{F}$ and we choose f_k with $\|f - f_k\| < \epsilon/3$, then $\|f\| \leq \epsilon/3 + \max\{\|f_1\|, \ldots, \|f_n\|\}$ and we have that $\mathcal{F}$ is bounded. If $x_0 \in X$, let U be a neighborhood of x_0 such that for $1 \leq k \leq n$, $|f_k(x) - f_k(x_0)| < \epsilon/3$ when $x \in U$. Thus if $f \in \mathcal{F}$ and we choose f_k with $\|f - f_k\| < \epsilon/3$, then $|f(x) - f(x_0)| \leq |f(x) - f_k(x)| + |f_k(x) - f_k(x_0)| + |f_k(x_0) - f(x_0)| < \epsilon$. Hence $\mathcal{F}$ is equicontinuous.

---

[2]Cesare Arzelà was born in 1847 in La Spezia, Italy. He received his doctorate from the university in Pisa under the direction of Enrico Betti. He held positions at Florence and Palermo before he became a professor in Bologna in 1880. His most famous work is this result, where he established the condition as sufficient for compactness. He died in 1912 in La Spezia.

[3]Giulio Ascoli was born in Trieste in 1843. He received his degree from Pisa and became a professor at Milano in 1872. He had a distinguished career and this theorem is his most notable result. He died in 1896 in Milano.

Now assume that $\mathcal{F}$ is equicontinuous and $\mathcal{F} \subseteq \mathrm{ball}\, C(X)$. Fix $\epsilon > 0$ and for each x in X let U_x be an open neighborhood of x such that $|f(y) - f(x)| < \epsilon/3$ whenever $y \in U_x$. Thus $\{U_x : x \in X\}$ is an open cover of X and we can extract a finite subcover $\{U_{x_1}, \ldots, U_{x_n}\}$. Now choose $\alpha_1, \ldots, \alpha_m$ in $\mathbb{D}$ such that $\mathrm{cl}\,\mathbb{D} \subseteq \bigcup_{j=1}^{m} \{\alpha : |\alpha - \alpha_j| < \epsilon/6\}$. Note that the collection of all ordered n-tuples of elements from the set $\{\alpha_1, \ldots, \alpha_m\}$ is finite. We don't want to consider all such n-tuples, however, but only the set B of those ordered n-tuples $b = (\beta_1, \ldots, \beta_n)$ with $\beta_1, \ldots, \beta_n \in \{\alpha_1, \ldots, \alpha_m\}$ such that there is a f_b in $\mathcal{F}$ with $|f_b(x_j) - \beta_j| < \epsilon/6$. Since for each f in $\mathcal{F}$ we have that $f(X) \subseteq \mathrm{cl}\,\mathbb{D}$ we see that $B \neq \emptyset$. For each b in B fix one such function f_b in $\mathcal{F}$ so that $\{f_b : b \in B\}$ is a finite subset of $\mathcal{F}$.

Claim. $\mathcal{F} \subseteq \bigcup_{b \in B} \{f \in C(X) : \|f - f_b\| < \epsilon\}$, so that $\mathcal{F}$ is totally bounded.

If $f \in \mathcal{F}$, $\{f(x_1), \ldots, f(x_n)\} \subseteq \mathrm{cl}\,\mathbb{D}$ and so there is a $b = (\beta_1, \ldots, \beta_n)$ in B with $|f(x_k) - \beta_k| < \epsilon/6$ for $1 \leq k \leq n$. Thus $|f(x_k) - f_b(x_k)| < \epsilon/3$ for $1 \leq k \leq n$. For each x in X choose x_k with x in U_{x_k}. Hence $|f(x) - f_b(x)| \leq |f(x) - f(x_k)| + |f(x_k) - f_b(x_k)| + |f_b(x_k) - f_b(x)| < \epsilon$. Since x was arbitrary, the claim is established and the proof is complete. ∎

9.2.4. Corollary. *If X is compact and $\mathcal{F} \subseteq C(X)$, then $\mathcal{F}$ is compact if and only if $\mathcal{F}$ is closed, bounded, and equicontinuous.*

9.2.5. Theorem. *If X is compact, then $\mathcal{B}_{00}(C(X))$ is dense in $\mathcal{B}_0(C(X))$.*

Proof. Assume that T is a compact operator on $C(X)$. Fix $\epsilon > 0$. Since $T(\mathrm{ball}\, C(X))$ is totally bounded, it is equicontinuous. So for each x in X there is an open neighborhood U_x of x such that $|(Tf)(x) - (Tf)(y)| < \epsilon$ whenever $f \in \mathrm{ball}\, C(X)$ and $y \in U_x$. Let $x_1, \ldots, x_n \in X$ such that $X = \bigcup_{k=1}^{n} U_{x_k}$ and let $\phi_1, \ldots, \phi_n$ be a partition of unity for X subordinate to this open cover. Define $T_\epsilon : C(X) \to C(X)$ by

$$T_\epsilon f = \sum_{k=1}^{n} (Tf)(x_k)\phi_k$$

Clearly the range of T_ϵ is contained in the span of $\{\phi_1, \ldots, \phi_n\}$ and hence it has finite rank.

If $f \in \mathrm{ball}\, C(X)$ and x is any point in X, then

$$|(T_\epsilon f)(x) - (Tf)(x)| = \left| \sum_{k=1}^{n} [(Tf)(x_k) - (Tf)(x)]\phi_k(x) \right|$$

$$\leq \sum_{k=1}^{n} |(Tf)(x_k) - (Tf)(x)|\phi_k(x)$$

$$< \epsilon$$

since $|(Tf)(x_k) - (Tf)(x)| < \epsilon$ whenever $\phi_k(x) \neq 0$. Since f and x were arbitrary, $\|T_\epsilon - T\| < \epsilon$. ∎

Exercises.

(1) If $\mathcal{X}$ is reflexive and $T \in \mathcal{B}(\mathcal{X}, \mathcal{Y})$, show that $T(\text{ball } \mathcal{X})$ is closed.

(2) Let $(X, \mathcal{A}, \mu)$ be a measure space, $1 < p < \infty$, and $\frac{1}{p} + \frac{1}{q} = 1$. If k is an $\mathcal{A} \times \mathcal{A}$-measurable function on $X \times X$ such that

$$M = \left[\int \left(\int |k(x,y)|^p d\mu(x) \right)^{q/p} d\mu(y) \right]^{\frac{1}{q}} < \infty$$

and if $(Kf)(x) = \int k(x,y)f(y)d\mu(y)$, then K is a compact operator on $L^p(\mu)$ with $\|K\| \leq M$.

(3) If X is a compact space, μ is a positive Borel measure on X, $1 < p < \infty$, and $A \in \mathcal{B}(L^p(\mu), C(X))$, then $T : L^p(\mu) \to L^p(\mu)$ defined by $Tf = Af$ for every f in $L^p(\mu)$ is a compact operator.

(4) Let X be compact and let μ be a positive regular Borel measure on X. (a) If $f_1, \ldots, f_n, g_1, \ldots, g_n \in C(X)$ and $k(x,y) = \sum_{k=1}^n f_k(x)g_k(y)$, then $(Kf)(x) = \int k(x,y)f(y)d\mu(y)$ defines a finite rank operator. (b) If $k \in C(X \times X)$, show that $(Kf)(x) = \int k(x,y)f(y)d\mu(y)$ defines a compact operator on $C(X)$.

(5) If $\tau : [0,1] \to [0,1]$ is a continuous function and A is the operator defined on $C[0,1]$ by $Af = f \circ \tau$, give necessary and sufficient conditions on τ that A be compact.

(6) If $A \in \mathcal{B}(c_0)$ and (α_{mn}) is the corresponding matrix as in Exercise 9.1.5, give necessary and sufficient conditions on the matrix so that A be compact.

(7) If $A \in \mathcal{B}(\ell^1)$ and (α_{mn}) is the corresponding matrix as in Exercise 9.1.6, give necessary and sufficient conditions on the matrix so that A be compact.

(8) If (X, d) is a compact metric space and $\mathcal{F} \subseteq C(X)$, show that $\mathcal{F}$ is equicontinuous if and only if for every $\epsilon > 0$ there is a $\delta > 0$ such that when $d(x,y) < \delta$, then $|f(x) - f(y)| < \epsilon$ for every f in $\mathcal{F}$.

(9) If X is locally compact and $\mathcal{F} \subseteq C_0(X)$, show that $\mathcal{F}$ is totally bounded if and only if $\mathcal{F}$ is bounded, equicontinuous, and for every $\epsilon > 0$ there is a compact subset K of X such that for every f in $\mathcal{F}$ we have $\sup\{|f(x)| : x \in X \backslash K\} < \epsilon$.

(10) If X is locally compact, show that every compact operator on $C_0(X)$ is the limit of finite rank operators.

(11) Assume $\mathcal{X}$ is a Banach space with the property that there is a net $\{F_i\}$ of finite rank operators such that $\sup_i \|F_i\| < \infty$ and $\|F_i x - x\| \to 0$ for every x in $\mathcal{X}$. Show that for any compact operator T on $\mathcal{X}$, $\|TF_i - T\| \to 0$ and so there is a sequence of finite rank operators $\{T_n\}$ such that $\|T - T_n\| \to 0$.

(12) If $1 \leq p \leq \infty$ and $(X, \mathcal{A}, \mu)$ is a σ-finite measure space, show that there is a net of finite rank operators $\{F_i\}$ in $\mathcal{B}(L^p(\mu))$ as described in Exercise 11.

Chapter 10

Banach Algebras and Spectral Theory

The study of Banach algebras is a part of analysis that stands on its own with several good books on the subject and regular research conferences with this topic. We will skim the surface and focus only on certain aspects of the theory. However we will see enough of this to use it to explore the spectral theory of operators on a Banach space.

A sea change now occurs in this book: we will make everything complex. **Every vector space and every algebra considered from now on will be over the complex numbers.**

In fact we will shortly encounter a result that depends on having $\mathbb{C}$ as the underlying field and this result will be the cornerstone of the entire theory.

10.1. Elementary properties and examples

An *algebra* $\mathcal{A}$ over $\mathbb{C}$ is a vector space over $\mathbb{C}$ with a definition of multiplication and where all the usual associative and distributive laws are enforced: when $a, b, c \in \mathcal{A}$ and $\alpha \in \mathbb{C}$, $a(bc) = (ab)c$; $\alpha(ab) = (\alpha a)b = a(\alpha b)$; $a(b + c) = ab + ac$; $(b + c)a = ba + ca$. Equivalently it is a ring with the structure of a vector space over $\mathbb{C}$ and the usual associative laws hold that relate the two structures. Note that neither commutativity nor the existence of an identity is required.

10.1.1. Definition. A *Banach algebra* $\mathcal{A}$ is an algebra over $\mathbb{C}$ that also has a norm with respect to which $\mathcal{A}$ is a Banach space and such that

$$\|ab\| \leq \|a\|\|b\|$$

for all a, b in $\mathcal{A}$. If $\mathcal{A}$ has an identity element e, then it is required that $\|e\| = 1$.

It might be pointed out that both the inequality $\|ab\| \leq \|a\|\|b\|$ and the requirement $\|e\| = 1$ can be relaxed, but we won't go further with this. See Exercise 1. We might underline that the norm inequality in the definition guarantees that multiplication in $\mathcal{A}$ is continuous; that is, the map from $\mathcal{A} \times \mathcal{A} \rightarrow \mathcal{A}$ defined by $(a, b) \mapsto ab$ is continuous. Also note that if e is an identity and $\alpha \in \mathbb{C}$, then $\|\alpha e\| = |\alpha|$. For this reason when we have an identity we will identify the scalars as elements of $\mathcal{A}$. That is we will use α to denote the element αe in $\mathcal{A}$; in particular we write 1 for the identity element e.

10.1.2. Example. (a) If X is a compact space, then $C(X)$ is a Banach algebra, where we define the product of two functions pointwise. Here $C(X)$ is abelian and it has an identity (the constantly one function).

(b) If X is a locally compact space, then $C_0(X)$ is also a commutative Banach algebra, but it does not have an identity if X is not compact.

(c) If μ is a σ-finite measure on a set X, then $L^\infty(\mu)$ is an abelian Banach algebra with identity, where again multiplication is defined pointwise a.e. $[\mu]$.

(d) If $\mathcal{X}$ is a Banach space, then $\mathcal{B}(\mathcal{X})$ is a Banach algebra, where multiplication is defined by composition. In this case $\mathcal{B}(\mathcal{X})$ is not commutative when $\dim \mathcal{X} > 1$, but it does have an identity.

(e) If $\mathcal{X}$ is a Banach space, then $\mathcal{B}_0(\mathcal{X})$ is a Banach subalgebra of $\mathcal{B}(\mathcal{X})$, and it does not have an identity if $\dim \mathcal{X} = \infty$.

(f) $L^1(\mathbb{R})$ is a commutative Banach algebra where multiplication is defined as convolution (Proposition 4.11.3).

The next result shows that we can always adjoin an identity to a Banach algebra.

10.1.3. Proposition. *If $\mathcal{A}$ is a Banach algebra without an identity, let $\mathcal{A}_1 = \mathcal{A} \times \mathbb{C}$ and define algebraic operations and a norm on $\mathcal{A}_1$ as follows. For a, b in $\mathcal{A}$ and α, β in $\mathbb{C}$:*

(a) $(a, \alpha) + (b, \beta) = (a + b, \alpha + \beta)$;

(b) $\beta(a, \alpha) = (\beta a, \beta \alpha)$;

(c) $(a, \alpha)(b, \beta) = (ab + \alpha b + \beta a, \alpha \beta)$;

(d) $\|(a, \alpha)\| = \|a\| + |\alpha|$.

With these definitions $\mathcal{A}_1$ is a Banach algebra with identity $(0, 1)$ and $a \mapsto (a, 0)$ is an isometric isomorphism of $\mathcal{A}$ into $\mathcal{A}_1$ such that the image of $\mathcal{A}$ has codimension one in $\mathcal{A}_1$.

Moreover the algebra $\mathcal{A}_1$ is unique in the following sense. If $\mathcal{B}$ is a Banach algebra with identity such that there is an isometric isomorphism $\rho : \mathcal{A} \to \mathcal{B}$ with $\dim \mathcal{B}/\rho(\mathcal{A}) = 1$, then there is a surjective isometric isomorphism $\tau : \mathcal{A}_1 \to \mathcal{B}$ with $\tau(a,0) = \rho(a)$ for all a in $\mathcal{A}$.

Proof. Note that $\|(a,\alpha)(b,\beta)\| = \|ab + \alpha b + \beta a\| + |\alpha\beta| \leq \|a\|\|b\| + |\alpha|\|b\| + |\beta|\|\alpha\| + |\alpha||\beta| = \|(a,\alpha)\|\|(b,\beta)\|$. The rest of the proof is routine. ∎

The algebra $\mathcal{A}_1$ obtained in the preceding proposition is called the algebra obtained by *adjoining the identity* to $\mathcal{A}$. If X is locally compact and $\mathcal{A} = C_0(X)$ as in Example 10.1.2(b), then the algebra $\mathcal{A}_1$ can be identified with $C(X_\infty)$, where X_∞ is the one-point compactification of X. In fact $C_0(X)$ can be identified with $\{f \in C(X_\infty) : f(\infty) = 0\}$ and clearly this has codimension one in $C(X_\infty)$. Also if $\mathcal{X}$ is a Banach space, then the algebra obtained by adjoining an identity to $\mathcal{B}_0(\mathcal{X})$ is $\{T + \alpha \in \mathcal{B}(\mathcal{X}) : T \in \mathcal{B}_0(\mathcal{X}), \alpha \in \mathbb{C}\}$.

We end with the following result on direct sums of Banach algebras. Recall (Exercise 6.2.1) the definition of the direct sums of Banach spaces. The proof of this proposition is straightforward.

10.1.4. Proposition. *If $\{\mathcal{A}_n : n \geq 1\}$ are Banach algebras, then $\bigoplus_\infty \mathcal{A}_n$ and $\bigoplus_0 \mathcal{A}_n$ are Banach algebras when the algebraic operations are define coordinatewise.*

(a) $\bigoplus_\infty \mathcal{A}_n$ *(or $\bigoplus_0 \mathcal{A}_n$) is abelian if and only if each $\mathcal{A}_n$ is abelian.*

(b) $\bigoplus_\infty \mathcal{A}_n$ *has an identity if and only if each $\mathcal{A}_n$ has an identity.*

Exercises.

(1) Suppose that $\mathcal{A}$ is an algebra that is also a Banach space such that if $a \in \mathcal{A}$ the maps $x \mapsto ax$ and $x \mapsto xa$ are continuous. Let $\mathcal{A}_1 = \mathcal{A} \times \mathbb{C}$ have the algebraic operations as defined in Proposition 10.1.3 and for each a in $\mathcal{A}$ define $L_a : \mathcal{A}_1 \to \mathcal{A}_1$ by $L_a(x,\alpha) = (ax + \alpha a, 0)$. For each a in $\mathcal{A}$ define $|||a||| = \|L_a\|$. Show that $||| \cdot |||$ is a norm on $\mathcal{A}$ that is equivalent to the original norm and with this new norm $\mathcal{A}$ is a Banach algebra.

(2) Verify the statements made in Example 10.1.2.

(3) Prove Proposition 10.1.4.

(4) If $\mathcal{A} = \bigoplus_0 \mathcal{A}_n$, show that $\mathcal{A}$ does not have an identity and determine the algebra obtained by adjoining an identity to $\mathcal{A}$.

(5) If X and Y are topological spaces, let $X \oplus Y$ denote their disjoint union. If X and Y are compact, show that $C(X \oplus Y) = C(X) \oplus_\infty C(Y)$. If X and Y are locally compact, show that $C_0(X \oplus Y) = C_0(X) \oplus_\infty C_0(Y)$.

10.2. Ideals and quotients

If $\mathcal{A}$ is an algebra, a *left ideal* is a linear subspace $\mathcal{M}$ such that if $a \in \mathcal{A}$ and $x \in \mathcal{M}$, then $ax \in \mathcal{M}$. Similarly a *right ideal* is a linear subspace such that if $a \in \mathcal{A}$ and $x \in \mathcal{M}$, then $xa \in \mathcal{M}$. An *ideal* (or *bilateral ideal*) is a subspace $\mathcal{M}$ that is both a left and right ideal. An element a of an algebra $\mathcal{A}$ with identity is *left* (respectively, *right*) *invertible* if there is an element x in $\mathcal{A}$ such that $xa = 1$ (respectively, $ax = 1$). If a is both left and right invertible, we say that a is *invertible*. In this case there is a unique element a^{-1} with $1 = aa^{-1} = a^{-1}a$. (Look at Exercise 1 for an example of a left invertible element that is not invertible and such that the left inverse is not unique.)

We note that if $\mathcal{M}$ is a left ideal and contains a left invertible element a, then $\mathcal{M} = \mathcal{A}$. In fact, if $xa = 1$, then $1 \in \mathcal{M}$ and so for any b in $\mathcal{A}$, $b = b1 \in \mathcal{M}$. Similarly a right ideal that contains a right invertible element must be the entire algebra.

In this section we will examine more closely the invertible elements and the concept of an ideal for a Banach algebra and then explore the quotient of a Banach algebra by a closed bilateral ideal. We begin with the most basic fact in this circle of ideas.

10.2.1. Lemma. *If $\mathcal{A}$ is a Banach algebra with identity and $x \in \mathcal{A}$ with $\|1 - x\| < 1$, then x is invertible.*

Proof. Put $y = 1 - x$ and $r = \|y\|$; it follows from the Banach algebra inequality for the norm of a product that $\|y^n\| \le r^n$. Hence $\sum_{n=0}^{\infty} \|y^n\| < \infty$, so that $z = \sum_{n=0}^{\infty} y^n = \sum_{n=0}^{\infty} (1 - x)^n$ converges in $\mathcal{A}$ (Exercise 1.3.3). Also note that $y^n \to 0$. If $z_n = 1 + y + \cdots + y^n$, then $z_n(1 - y) = (1 + \cdots + y^n) - (y + \cdots + y^{n+1}) = 1 - y^{n+1} \to 1$. Since $z_n \to z$, this shows that $z(1 - y) = 1$. Similarly $(1 - y)z = 1$ so that $x = 1 - y$ is invertible with $x^{-1} = z$. ∎

10.2.2. Theorem. *If $\mathcal{A}$ is a Banach algebra with identity and G_ℓ, G_r, G denote the sets of left invertible, right invertible, and invertible elements in $\mathcal{A}$, then all three sets are open. Also the map $a \mapsto a^{-1}$ defined on G is continuous.*

Proof. Fix a_0 in G_ℓ and let b_0 be an element of $\mathcal{A}$ such that $b_0 a_0 = 1$. If $\|a - a_0\| < \|b_0\|^{-1}$, then $\|b_0 a - 1\| < 1$, and, by the preceding lemma, $b_0 a$ is invertible. If $b = (b_0 a)^{-1} b_0$, then $ba = 1$ and so $a \in G_\ell$. We have just shown that $B(a_0; \|b_0\|^{-1}) \subseteq G_\ell$; hence G_ℓ is open. Similarly G_r is open and therefore G is open since $G = G_\ell \cap G_r$.

Assume that $a_n \to 1$ and let $0 < \delta < 1$. When $\|a_n - 1\| < \delta$, the preceding lemma implies that a_n is invertible and $a_n^{-1} = [1 - (1 - a_n)]^{-1} = \sum_{k=0}^{\infty} (1 - a_n)^k$. Therefore $\|a_n^{-1} - 1\| = \left\|\sum_{k=1}^{\infty} (1 - a_n)^k\right\| \le \sum_{k=1}^{\infty} \|1 - a_n\|^k < \delta/(1 - \delta)$.

So for any $\epsilon > 0$ we can find a large N such that $\|a_n^{-1} - 1\| < \epsilon$ when $n \geq N$. That is, $a_n^{-1} \to 1$. This says that the map defined on G by $a \mapsto a^{-1}$ is continuous at 1.

If a is an arbitrary invertible element of $\mathcal{A}$ and $\{a_n\}$ is a sequence in G such that $a_n \to a$, then $a^{-1}a_n \to 1$. By the preceding paragraph $a_n^{-1}a \to 1$ and so $a_n^{-1} = (a_n^{-1}a)a^{-1} \to a^{-1}$. $\blacksquare$

The next result surfaced in the previous proofs and is worth recording.

10.2.3. Corollary. *Let $\mathcal{A}$ be a Banach algebra with identity.*

(a) *If $\|x - 1\| < 1$, then $x^{-1} = \sum_{n=0}^{\infty} (1 - x)^n$.*

(b) *If $b_0 a_0 = 1$ and $\|a - a_0\| < \|b_0\|^{-1}$, then a is left invertible.*

We say that $\mathcal{M}$ is a *maximal ideal* if it is a proper ideal and not contained in any properly larger ideal.

10.2.4. Example. If X is a compact space and for x_0 in X we set $\mathcal{M} = \{f \in C(X) : f(x_0) = 0\}$, then $\mathcal{M}$ is a maximal ideal in $C(X)$.

Also see Exercise 4.

10.2.5. Corollary. *If $\mathcal{A}$ is a Banach algebra with identity, then:*

(a) *the closure of a proper left, right, or bilateral ideal is a proper ideal of the same type;*

(b) *a maximal left, right, or bilateral ideal is closed.*

Proof. If G_ℓ is the set of left invertible elements and $\mathcal{M}$ is a proper left ideal, then $\mathcal{M} \cap G_\ell = \emptyset$. Since G_ℓ is open, $\mathrm{cl}\,\mathcal{M} \cap G_\ell = \emptyset$ and it is routine to verify that $\mathrm{cl}\,\mathcal{M}$ is a left ideal. A similar argument prevails for right and bilateral ideals.

If $\mathcal{M}$ is a maximal left ideal it is proper. Therefore $\mathrm{cl}\,\mathcal{M}$ is a proper left ideal that contains $\mathcal{M}$. Since $\mathcal{M}$ is maximal, $\mathcal{M} = \mathrm{cl}\,\mathcal{M}$. $\blacksquare$

The reader is asked to supply the details of the proof of the next result using Zorn's Lemma.

10.2.6. Proposition. *Every proper left, right, or bilateral ideal is contained in a maximal ideal of the same type.*

As usual, when we have an algebra and an ideal $\mathcal{M}$, the quotient $\mathcal{A}/\mathcal{M}$ is again an algebra. This carries over to the Banach algebra setting.

10.2.7. Proposition. *If $\mathcal{A}$ is a Banach algebra and $\mathcal{M}$ is a proper closed ideal and $\mathcal{A}/\mathcal{M}$ has the usual norm as a Banach space, then $\mathcal{A}/\mathcal{M}$ is also a Banach algebra. If $\mathcal{A}$ has an identity, so does $\mathcal{A}/\mathcal{M}$.*

Proof. We already know that $\mathcal{A}/\mathcal{M}$ is a Banach space. We only have to prove that if $a, b \in \mathcal{A}$, then $\|ab+\mathcal{M}\| = \|(a+\mathcal{M})(b+\mathcal{M})\| \leq \|a+\mathcal{M}\|\|b+\mathcal{M}\|$. Note that If $x, y \in \mathcal{M}$, then $(a + x)(b + y) = ab + (ay + xb + xy)$ and $ay + xb + xy \in \mathcal{M}$. Therefore $\|(a+\mathcal{M})(b+\mathcal{M})\| = \|ab+\mathcal{M}\| \leq \|ab + (ay + xb + xy)\| = \|(a + x)(b + y)\| \leq \|a + x\|\|b + y\|$. Taking the infimum first over all x in $\mathcal{M}$ and then over all y in $\mathcal{M}$ yields the desired inequality. ∎

Exercises.

(1) If S is the unilateral shift on ℓ^2 (5.3.3), show that S is left invertible in the algebra $\mathcal{B}(\ell^2)$ but not right invertible. Show that the left inverse for S is not unique.

(2) If $(X, \mathcal{A}, \mu)$ is a σ-finite measure space and I is a weak*-closed ideal in $L^\infty(\mu)$, show that there is a measurable set E such that $I = \chi_E L^\infty(\mu) = \{f \in L^\infty(\mu) : f(x) = 0 \text{ a.e.}[\mu] \text{ off } E\}$.

(3) Show that $M_d(\mathbb{C})$ has no proper ideals. How about $M_d(\mathbb{R})$?

(4) Find the minimal left ideals of $M_d(\mathbb{C})$. Find the maximal left ideals.

(5) Let $\mathcal{H}$ be a Hilbert space and L be a closed left ideal in $\mathcal{B}_0(\mathcal{H})$. (a) Show that if for every non-zero h in $\mathcal{H}$ there is a T in L with $Th \neq 0$, then $L = \mathcal{B}_0(\mathcal{H})$. (Hint: It might be helpful to note that if $T \in L$, then $T^*T \in L$.) (b) Characterize the maximal and minimal closed left ideals of $\mathcal{B}_0(\mathcal{H})$.

(6) Consider the algebra $L^1(\mathbb{R})$ where the multiplication is convolution and show that for any subset F of $\mathbb{R}$, $I_F = \{f \in L^1(\mathbb{R}) : \widehat{f}(t) = 0 \text{ for all } t \in F\}$ is an ideal of $L^1(\mathbb{R})$. Show that $I_F = I_{\text{cl } F}$ and that I_F is maximal if and only if F is a single point.

10.3. Analytic functions

We cannot proceed further without some knowledge of complex analysis. There are books on the subject, for example [**7**]. All that material is not necessary for what we need, which is rather meager but essential for us to continue to explore functional analysis. In this section we present a few results that will be required. If the reader has had a course in complex analysis, (s)he will not need this section and can proceed to the next.

The results here will be stated mostly without proof. The reader who has not seen this material should try to understand the statements of the results and take their veracity on faith until (s)he encounters a course in the subject. Throughout this section G will denote an open subset of the complex plane $\mathbb{C}$.

10.3.1. Definition. If $f : G \to \mathbb{C}$, then the derivative of f at a in G is defined as

$$f'(a) = \lim_{h \to 0} \frac{f(a+h) - f(a)}{h}$$

The function f is *analytic* on G if it is differentiable at each point of G. As in calculus if f has a derivative at each point of G this defines a new function $f' : G \to \mathbb{C}$ and we talk about its derivative.

One of the basic facts about analytic functions is that once we have the existence of one derivative on an open set, the function is infinitely differentiable. Given what we are used to in calculus, this is truly remarkable. Why does this happen? A little thought shows that the existence of a derivative of a function defined on an open subset of $\mathbb{C}$ is a far greater restriction than the same condition for a function defined on an open interval in $\mathbb{R}$. In fact there are basically only two ways for a real number h to approach 0 in $\mathbb{R}$, while in $\mathbb{C}$ there are uncountably many ways. So the existence of a derivative in $\mathbb{C}$ is a severe requirement and imposes many restrictions on the function such as infinite differentiability.

Often when we apply analytic functions to the study of Banach algebras, we will manufacture such functions defined on non-connected open sets. A typical situation is when we have two disjoint compact sets K_1 and K_2. We find disjoint open sets G_1 and G_2 such that $K_1 \subseteq G_1, K_2 \subseteq G_2$, and define $f : G_1 \cup G_2 \to \mathbb{C}$ to be the characteristic function of G_1. Because the open sets are disjoint, f is analytic.

The next result should not come as a surprise. But note that we require the underlying set G to be connected. If $G = G_1 \cup G_2$ and f are as in the preceding paragraph, then $f' \equiv 0$ but f is not constant.

10.3.2. Proposition. *If $f : G \to \mathbb{C}$ is analytic, G is connected, and $f'(z) = 0$ for all z in G, then f is constant.*

Obtaining examples of analytic functions is not difficult. Every polynomial in z is analytic, and the quotient of two analytic functions is analytic as long as we restrict to an open set where the denominator does not vanish. So a rational function $f(z) = p(z)/q(z)$ is analytic on the set $\{z \in \mathbb{C} : q(z) \neq 0\}$. Power series is another avenue to finding examples, but it is much more than that. Here is a bit of information about power series that may be known to some readers and not others.

10.3.3. Proposition. *Let $a \in \mathbb{C}$ and suppose $f(z) = \sum_{n=0}^{\infty} a_n (z-a)^n$ with $a_n \in \mathbb{C}$ for all $n \geq 0$. If the non-negative extended real number R is defined by*

$$\frac{1}{R} = \limsup_n |a_n|^{\frac{1}{n}}$$

and if $R > 0$, then:

(a) *if $|z - a| < R$ the series converges absolutely;*

(b) *if $|z - a| > R$ the terms in the series become unbounded and the series diverges;*

(c) *if $r < R$, the series converges uniformly for $|z - a| \leq r$.*

Moreover the number R is the unique number satisfying (a) *and* (b).

The number R defined above is called the *radius of convergence* of the power series.

10.3.4. Proposition. *If $f(z) = \sum_{n=0}^{\infty} a_n(z - a)^n$ has radius of convergence R, then f defines an analytic function on the disk $B(a; R)$ and*

$$f'(z) = \sum_{n=0}^{\infty} n a_n (z - a)^{n-1}$$

Conversely if $f : G \to \mathbb{C}$ is analytic and $B(a; r) \subseteq G$, then f has a power series representation $f(z) = \sum_{n=0}^{\infty} a_n(z - a)^n$ with radius of convergence $R \geq r$.

Notice that by differentiating the power series n times we get that the coefficients a_n can be captured from the function by

10.3.5
$$a_n = \frac{1}{n!} f^{(n)}(a)$$

The formula for the derivative in the last proposition says we can differentiate the power series term-by-term. We also have the definition of the exponential function by the power series

$$e^z = \exp z = \sum_{n=0}^{\infty} \frac{1}{n!} z^n$$

The reader can verify that the radius of convergence of this power series is $R = \infty$ and that the derivative of the exponential function is equal to the function itself. Other properties of the exponential function can also be derived. In particular $|e^{i\theta}| = 1$ for θ in $\mathbb{R}$. This can be seen by using the power series to show that $\overline{e^z} = e^{\bar{z}}$ so that

$$|e^{i\theta}|^2 = e^{i\theta} e^{-i\theta} = \left(\sum_{n=0}^{\infty} \frac{(i\theta)^n}{n!} \right) \left(\sum_{k=0}^{\infty} \frac{(-i\theta)^k}{k!} \right)$$

Now if you are adept at manipulating infinite series and recognizing classical formulas you can derive that the product of these two series is 1.

A *path* in G is a continuous function $\gamma : [a, b] \to G$. A *closed path* is a path $\gamma : [a, b] \to \mathbb{C}$ such that its initial and final points are the same; that is,

$\gamma(a) = \gamma(b)$. The *trace* of a path is defined as $\{\gamma\} = \{\gamma(t) : a \leq t \leq b\}$. This path is *rectifiable* if the function is of bounded variation. When $f : G \to \mathbb{C}$ is a continuous function, we define the integral of f along the rectifiable path γ as

$$\int_\gamma f = \int_\gamma f(z)dz = \int_a^b f(\gamma(t))d\gamma(t)$$

This integral is a Riemann–Stieltjes integral as in §1.1.

One of the basic results in complex analysis is Cauchy's Integral Formula. The reader can consult [**7**] for a thorough explication of this. Here we present a simple form of it that will be useful for our purposes.

10.3.6. Theorem (Cauchy's Integral Formula). *If* $f(z) = \sum_{n=0}^{\infty} a_n(z-a)^n$ *has radius of convergence* R, $0 < r < R$, *and* γ *is the curve* $\gamma(t) = a + r\exp(2\pi it)$ *for* $0 \leq t \leq 1$, *then for every* $n \geq 0$ *and* $|z - a| < r$

$$f^{(n)}(z) = \frac{n!}{2\pi i} \int_\gamma \frac{f(w)}{(w-z)^{n+1}} dw$$

The reader can verify that this holds when $z = a$ by an easy computation; this is what is used in the proof of the next result, which will be directly applied to Banach algebras.

10.3.7. Theorem (Liouville's[1] Theorem). *A bounded function that is analytic on the entire plane is constant.*

Proof. In light of Theorem 10.3.6 the proof is not difficult. Let $f : \mathbb{C} \to \mathbb{C}$ be analytic and suppose that $|f(z)| \leq M$ for all z. Fix a in $\mathbb{C}$ and let $r > 0$. If $\gamma(t) = a + r\exp(2\pi it)$ for $0 \leq t \leq 1$, then according to Cauchy's Integral

---

[1]Joseph Liouville was born in 1809 in Saint Omer, France near Calais. He entered the École Polytechnique in Paris in 1825. In 1831 he received his first academic appointment at the École Polytechnique as an assistant to Claude Mathieu. He was also lecturing at other institutions and spent some 35 hours a week giving classes. In 1836 Liouville founded *Journal de Mathématiques Pures et Appliquées*. His reputation soon achieved international stature and in 1838 he was appointed Professor of Analysis and Mechanics at the École Polytechnique. In addition to his mathematical life he was deeply involved in politics and was even elected to the Constitutional Assembly in 1848. He is best remembered in mathematics for his work on the Sturm–Liouville equations and on transcendental numbers, those that are not the root of any polynomial with rational coefficients. He was the first to construct an infinite collection of such numbers. In addition he had a major role in making the work of Galois known to the mathematics community. He died in Paris in 1882.

formula,

$$|f'(a)| = \left| \frac{1}{2\pi i} \int_\gamma \frac{f(z)}{(z-a)^2} dz \right|$$

$$= \frac{1}{2\pi} \left| \int_0^1 \frac{f(a + r\exp(2\pi it))}{(r\exp(2\pi it))^2} \gamma'(t) dt \right|$$

$$\leq \frac{1}{2\pi} \int_0^1 \frac{M}{|(r\exp(2\pi it))^2|} 2\pi r \, dt$$

$$= \frac{M}{r}$$

Now this holds whenever $r > 0$, so letting $r \to \infty$ shows that $f'(a) = 0$. Since a was arbitrary, f must be constant. ■

We would be remiss at this point if we did not point out that Liouville's Theorem can be used to prove the Fundamental Theorem of Algebra: every non-constant polynomial has a zero in $\mathbb{C}$. In fact assume there is a non-constant polynomial $p(z)$ with no zeros. Then $f(z) = [p(z)]^{-1}$ is a function that is analytic on $\mathbb{C}$. Because $p(z)$ is not constant, $|f(z)| \to 0$ as $z \to \infty$. (Why?) Hence f is bounded. By Liouville's Theorem, f must be constant. But this implies $p(z)$ is constant, a contradiction.

We conclude this section with a little lemma used later. It seems isolated and, here, an *ad hoc* result. However it is a special case of a topic called the winding number that is discussed in generality in [7].

10.3.8. Lemma. *Let $\lambda \in \mathbb{C}$, $r > 0$, and $\gamma(t) = \lambda + r\exp(2\pi it)$ for $0 \leq t \leq 1$. If $|\alpha - \lambda| \neq r$, then*

$$\int_\gamma \frac{1}{z - \alpha} dz = \begin{cases} 2\pi i & \text{if } |\alpha - \lambda| < r \\ 0 & \text{if } |\alpha - \lambda| > r \end{cases}$$

Proof. Assume $|\alpha - \lambda| < r$ and define $f : [0,1] \to \mathbb{C}$ by $f(s) = \int_\gamma (z - [s\alpha + (1-s)\lambda])^{-1} dz = \int_0^1 [\gamma(t) - (s\alpha + (1-s)\lambda)]^{-1} \gamma'(t) dt$. So $f(1)$ is the number we want to compute and $f(0)$ is easily computed to be $2\pi i$. We will show f is constant by showing that $f'(s) = 0$; this will give the result we want in this case. In fact differentiating under the integral sign we have that $f'(s) = -(\alpha - \lambda) \int_0^1 [\gamma(t) - (s\alpha + (1-s)\lambda)]^{-2} \gamma'(t) dt$. But we see that $g(t) = -[\gamma(t) - (s\alpha + (1-s)\lambda)]^{-1}$ is a primitive for this integrand. Since $g(1) = g(0)$ we have that $f'(s) = 0$.

Now assume that $|\alpha - \lambda| > r$ and define $f : [1, \infty) \to \mathbb{C}$ by $f(s) = \int_\gamma (z - s\alpha)^{-1} dz$. Again $f(1)$ is what we want and we can verify that $f(s) \to 0$ as $s \to \infty$; again it is routine to complete the proof by showing that $f'(s) = 0$. ■

10.4. The spectrum

It's time for another agreement.

For the remainder of the book when we consider a Banach algebra, we will always assume that it has an identity unless stated to the contrary.

When we want to consider an algebra without an identity, we will simply adjoin one, look at this new algebra, and then see whatever results we obtain say about the original algebra. In fact we will want to explore $\mathcal{B}_0(\mathcal{X})$ for a Banach space $\mathcal{X}$; in this case we simply assume it is contained in $\mathcal{B}(\mathcal{X})$.

10.4.1. Definition. If $\mathcal{A}$ is a Banach algebra with identity and $a \in \mathcal{A}$, then the *spectrum* of a is the set

$$\sigma(a) = \{\lambda \in \mathbb{C} : a - \lambda \text{ is not invertible}\}$$

The complement $\rho(a) = \mathbb{C}\backslash\sigma(a)$ is called the *resolvent set* of a. Similarly we can define the *left spectrum* $\sigma_\ell(a)$ and the *right spectrum* $\sigma_r(a)$.

10.4.2. Example. (a) If X is compact and $\mathcal{A} = C(X)$, then for each f in $C(X)$, $\sigma(f) = f(X)$. In fact this follows from the fact that a continuous function f in $C(X)$ is invertible if and only if $f(x) \neq 0$ for all x, in which case $f^{-1} = 1/f$. From here the fact that $\sigma(f) = f(X)$ is immediate.

(b) If $\mathcal{A} = M_d(\mathbb{C})$, the $d \times d$ matrices with entries in $\mathbb{C}$, then for each A in $\mathcal{A}$, $\sigma(A)$ is the set of eigenvalues of A. Note that this produces several interesting examples and furnishes us with several speed bumps less we get carried away and believe some things that are not always true. For example, if we abandoned our agreement made at the beginning of the chapter and allowed our Banach algebras to be over $\mathbb{R}$, then consider $\mathcal{A} = M_2(\mathbb{R})$ and

$$A = \begin{bmatrix} 0 & 1 \\ -1 & 0 \end{bmatrix}$$

It is easy to check that $\sigma(A) = \emptyset$ since $\det(A - \lambda) = \lambda^2 + 1$ and this has no roots in $\mathbb{R}$. However if we consider A as an element of $M_2(\mathbb{C})$, $\sigma(A) = \{i, -i\}$.

(c) If $\mathcal{X}$ is a Banach space and $A \in \mathcal{B}(\mathcal{X})$, then $\sigma(A) = \{\lambda \in \mathbb{C} : \ker(A - \lambda) \neq (0) \text{ or } \operatorname{ran}(A - \lambda) \neq \mathcal{X}\}$. In fact if $A \in \mathcal{B}(\mathcal{X})$ such that $A - \lambda$ is bijective, then $(A - \lambda)^{-1}$ is a well-defined linear transformation on $\mathcal{X}$; by the IMT, $(A - \lambda)^{-1}$ is bounded.

10.4.3. Example. If $\mathcal{H}$ is a Hilbert space and $T \in \mathcal{B}(\mathcal{H})$, then

$$\sigma_\ell(T) = \{\lambda \in \mathbb{C} : \inf\{\|(T - \lambda)h\| : \|h\| = 1\} = 0\}$$

To see this we show that T is left invertible if and only if $\inf\{\|Th\| : \|h\| = 1\} > 0$. If S is a bounded operator with $ST = 1$, then for $\|h\| = 1$, $1 = \|STh\| \leq \|S\|\|Th\|$ and so $\inf\{\|Th\| : \|h\| = 1\} \geq \|S\|^{-1}$. Conversely if

$\|Th\| \geq \epsilon$ when $\|h\| = 1$, then $\|Th\| \geq \epsilon\|h\|$ for all vectors h. This clearly shows that T is injective and Lemma 5.6.13 says that $\operatorname{ran} T$ is closed. By the IMT it follows that the map $T : \mathcal{H} \to \operatorname{ran} T$ has a bounded inverse $A : \operatorname{ran} T \to \mathcal{H}$. Now write $\mathcal{H} = \operatorname{ran} T \oplus (\operatorname{ran} T)^{\perp}$ and define $S : \mathcal{H} \to \mathcal{H}$ by $S(h + k) = Ah$ when $h \in \operatorname{ran} T$ and $k \in (\operatorname{ran} T)^{\perp}$. It follows that $ST = 1$. Also see Exercise 1.

In Example 10.4.2(b) we saw an element of a Banach algebra with empty spectrum. This was a consequence of having the underlying scalars the real numbers rather than $\mathbb{C}$. It is the next theorem that has caused us to depart from reality.

10.4.4. Theorem. *If $\mathcal{A}$ is a Banach algebra, then for each a in $\mathcal{A}$ the spectrum of a is a non-empty compact set in $\mathbb{C}$. Moreover if $|\lambda| > \|a\|$, then $\lambda \notin \sigma(a)$ and the function $z \mapsto (z - a)^{-1}$ is an $\mathcal{A}$-valued analytic function.*

Before starting the proof we need to say a few words about vector-valued analytic functions. The few words number 5: there are no big difficulties. Look once again at the definition of an analytic function in the preceding section, but now assume that $f : G \to \mathcal{X}$ for some Banach space $\mathcal{X}$. The formula for the derivative still makes sense, though convention says we should delete the h in the denominator and enter h^{-1} in the numerator on the left. Now proceed through the results of the last section with this extension in mind. The results hold almost verbatim except for small modifications to accommodate the fact that the analytic functions are Banach space-valued.

Another way to handle this extension of analytic functions is to let $x^* \in \mathcal{X}^*$ and consider $x^* \circ f : G \to \mathbb{C}$. If we apply the $\mathbb{C}$-version of a theorem to each $x^* \circ f$ we can usually obtain the vector-valued version. As an illustration, let's consider Liouville's Theorem. Suppose $f : \mathbb{C} \to \mathcal{X}$ is such that for every x^* in $\mathcal{X}^*$, $x^* \circ f$ is bounded. Then the scalar-valued version of Liouville's Theorem implies $x^* \circ f$ is constant. This implies that f is constant. In fact if $a \neq b$ in G and $f(a) \neq f(b)$, then the HBT implies there is an x^* in $\mathcal{X}^*$ such that $\langle f(a), x^* \rangle \neq \langle f(b), x^* \rangle$. This implies that $x^* \circ f$ is not constant, contradicting what was previously established. Let's also note that the PUB implies that $f : G \to \mathcal{X}$ is bounded if and only if $x^* \circ f$ is bounded for each x^* in $\mathcal{X}^*$.

We proceed with the proof of Theorem 10.4.4.

Proof. If $|\lambda| > \|a\|$, then $\lambda - a = \lambda(1 - \lambda^{-1}a)$ and $\|\lambda^{-1}a\| < 1$; by Lemma 10.2.1, $1 - \lambda^{-1}a$ is invertible and so $\lambda - a$ is invertible. This implies that $\sigma(a)$ is contained in the closed disk of radius $\|a\|$ and so $\sigma(a)$ is bounded. Now the map $\lambda \mapsto \lambda - a$ is continuous from $\mathbb{C}$ into $\mathcal{A}$ and the set G of invertible elements in $\mathcal{A}$ is open; since $\rho(a)$ is the inverse image of G under this map,

$\rho(a)$ is open. Therefore the spectrum is compact. We need to show that $\sigma(a) \neq \emptyset$ and this is where we need that the underlying field of scalars is $\mathbb{C}$ so we can employ the theory of analytic functions.

Let $F : \rho(a) \to \mathcal{A}$ be defined by $F(z) = (z - a)^{-1}$. Fix λ in the resolvent set; since $\rho(a)$ is open we can consider all the h with $|h|$ sufficiently small that $\lambda + h \in \rho(a)$. In the identity $x^{-1} - y^{-1} = x^{-1}(y - x)y^{-1}$, let $x = \lambda + h - a$ and $y = \lambda - a$ to obtain

$$\frac{F(\lambda + h) - F(\lambda)}{h} = \frac{(\lambda + h - a)^{-1} - (\lambda - a)^{-1}}{h}$$
$$= -(\lambda + h - a)^{-1}(\lambda - a)^{-1}$$

Since the inverse map is continuous on the set of invertible elements, we see that as $h \to 0$ the above quotient converges to $-(\lambda - a)^{-2}$; that is, F is differentiable at λ and $F'(\lambda) = -(\lambda - a)^{-2}$. Thus F is analytic. Assume that $\sigma(a) = \emptyset$. So F is an analytic function on all of $\mathbb{C}$. But we also have that $F(\lambda) \to 0$ as $\lambda \to \infty$. In fact we know from Corollary 10.2.3 that when $|\lambda| > \|a\|$, $F(\lambda) = \lambda^{-1}(1 - \lambda^{-1}a)^{-1} = \lambda^{-1} \sum_{n=0}^{\infty} \lambda^{-n}a^n$. Thus $\|F(\lambda)\| \leq |\lambda|^{-1}(1 - |\lambda|^{-1}\|a\|)^{-1} = (|\lambda| - \|a\|)^{-1}$ and so $\|F(\lambda)\| \to 0$ as $\lambda \to \infty$. Thus F is a bounded analytic function on all of $\mathbb{C}$, and Liouville's Theorem implies that F is constant. Since $F(z)$ converges to 0 as $z \to \infty$, this constant must be 0, something that is clearly false. Therefore we have a contradiction and $\sigma(a) \neq \emptyset$. ∎

Now that we know that the spectrum is always non-empty, we can proceed to develop some related concepts.

10.4.5. Definition. If $\mathcal{A}$ is a Banach algebra and $a \in \mathcal{A}$, then the *spectral radius* of a is defined as

$$r(a) = \max\{|\lambda| : \lambda \in \sigma(a)\}$$

Note that this definition makes sense since the spectrum is always non-empty. However it may be 0 even though $a \neq 0$ as we see by considering the algebra $M_2(\mathbb{C})$ and the element

$$A = \begin{bmatrix} 0 & 0 \\ 1 & 0 \end{bmatrix}$$

Here $\sigma(A) = \{0\}$ and so $r(A) = 0$.

10.4.6. Proposition. *If $a \in \mathcal{A}$, then*

$$r(a) = \lim_{n \to \infty} \|a^n\|^{\frac{1}{n}}$$

which limit always exists.

Proof. If $0 < |z| < \|a\|^{-1}$, then $z^{-1} - a = z^{-1}(1 - za)$ and $\|za\| < 1$; hence $z^{-1} - a$ is invertible and, by Corollary 10.2.3(a), $(z^{-1} - a)^{-1} = z \sum_{n=0}^{\infty} z^n a^n$. If R is the radius of convergence of this power series we have that $R \geq r(a)^{-1}$ and from (10.3.3)

$$R^{-1} = \limsup_n \|a^n\|^{\frac{1}{n}} \leq r(a)$$

(OK, we are using facts about power series where the coefficients come from a Banach space. It's not a big deal.) Now observe that for any complex number λ, $\lambda^n - a^n = (\lambda - a)(\lambda^{n-1} + \lambda^{n-2}a + \cdots + a^{n-1})$ and so it follows that if $\lambda - a$ is not invertible, then $\lambda^n - a^n$ is not invertible; that is, when $\lambda \in \sigma(a)$, $\lambda^n \in \sigma(a^n)$. So if $\lambda \in \sigma(a)$, $|\lambda|^n \leq \|a^n\|$ or $|\lambda| \leq \|a^n\|^{\frac{1}{n}}$. This implies that $r(a) \leq \|a^n\|^{\frac{1}{n}}$ for all $n \geq 1$. Therefore

$$r(a) \leq \liminf_n \|a^n\|^{\frac{1}{n}} \leq \limsup_n \|a^n\|^{\frac{1}{n}} \leq r(a)$$

Thus $\lim_n \|a^n\|^{\frac{1}{n}}$ exists and equals $r(a)$.   ∎

10.4.7. Proposition. *Let $\mathcal{A}$ be a Banach algebra and let $a \in \mathcal{A}$.*

(a) *If $\lambda \in \rho(a)$, then $\operatorname{dist}(\lambda, \sigma(a)) \geq \|(\lambda - a)^{-1}\|^{-1}$.*

(b) *If $\lambda, \mu \in \rho(a)$, then*

$$(\lambda - a)^{-1} - (\mu - a)^{-1} = (\mu - \lambda)(\lambda - a)^{-1}(\mu - a)^{-1}$$

The equation in part (b) of this proposition is called the *resolvent equation*.

Proof. The proof of (a) follows from Corollary 10.2.3 with a little manipulation. For part (b), if we let $x = \lambda - a$ and $y = \mu - a$ in the identity $x^{-1} - y^{-1} = x^{-1}(y - x)y^{-1}$, the resolvent equation is immediate.   ∎

We close by considering a question that may have already occurred to the reader. Suppose $\mathcal{A}$ and $\mathcal{B}$ are two Banach algebras with a common identity and $\mathcal{A} \subseteq \mathcal{B}$. If $a \in \mathcal{A}$, then we can look at the spectrum of a as an element of $\mathcal{A}$, $\sigma_{\mathcal{A}}(a)$, and as an element of $\mathcal{B}$, $\sigma_{\mathcal{B}}(a)$. For example let $\mathcal{B} = C(\partial \mathbb{D})$ and let $\mathcal{A}$ be the closure in $\mathcal{B}$ of all polynomials in the complex variable z. So $\mathcal{A}$ is the closed linear span of the functions $1, z, z^2, \ldots$. The function z is invertible in $\mathcal{B}$ so that $0 \notin \sigma_{\mathcal{B}}(z)$. In fact, according to Example 10.4.2(a), $\sigma_{\mathcal{B}}(z) = \partial \mathbb{D}$. However it is not invertible in $\mathcal{A}$, which is plausible since z^{-1} is not a polynomial. A proof of this, however, requires some complex analysis that won't be covered here; specifically it needs a result called the Maximum Modulus Theorem. Thus $0 \in \sigma_{\mathcal{A}}(z)$. (In fact the Maximum Modulus Theorem shows that $\sigma_{\mathcal{A}}(z) = \operatorname{cl} \mathbb{D}$.)

10.4.8. Proposition. *If $\mathcal{A}$ and $\mathcal{B}$ are Banach algebras with a common identity and $a \in \mathcal{A} \subseteq \mathcal{B}$, then $\sigma_{\mathcal{B}}(a) \subseteq \sigma_{\mathcal{A}}(a)$ and $\partial \sigma_{\mathcal{A}}(a) \subseteq \partial \sigma_{\mathcal{B}}(a)$.*

Proof. The first inclusion is the easiest: if $\lambda \in \mathbb{C} \backslash \sigma_{\mathcal{A}}(a)$, then $a - \lambda$ has an inverse in $\mathcal{A} \subseteq \mathcal{B}$, so that $\lambda \in \mathbb{C} \backslash \sigma_{\mathcal{B}}(a)$. Now assume that $\lambda \in \partial \sigma_{\mathcal{A}}(a)$. Since the first inclusion implies that $\operatorname{int} \sigma_{\mathcal{B}}(a) \subseteq \operatorname{int} \sigma_{\mathcal{A}}(a)$, we need only show that $\lambda \in \sigma_{\mathcal{B}}(a)$. Suppose this is not the case. So there is a b in $\mathcal{B}$ with $b(a - \lambda) = (a - \lambda)b = 1$. Let $\{\lambda_n\} \subseteq \mathbb{C} \backslash \sigma_{\mathcal{A}}(a)$ such that $\lambda_n \to \lambda$; hence $a - \lambda_n \to a - \lambda$. Now $(a - \lambda_n)^{-1} \in \mathcal{A} \subseteq \mathcal{B}$ and taking the inverse is a continuous function on the invertible elements of $\mathcal{B}$; hence $(a - \lambda_n)^{-1} \to b$. But $\mathcal{A}$ is complete and so $b \in \mathcal{A}$. Hence $\lambda \notin \sigma_{\mathcal{A}}(a)$, a contradiction. ∎

Let $\mathcal{A}$ be the subalgebra of $M_2(\mathbb{C})$ consisting of all matrices of the form

$$\begin{bmatrix} \alpha & 0 \\ 0 & 0 \end{bmatrix}$$

where $\alpha \in \mathbb{C}$, and let $\mathcal{B}$ be the algebra consisting of all 2×2 matrices of the form

$$\begin{bmatrix} \alpha & 0 \\ 0 & \beta \end{bmatrix}$$

where $\alpha, \beta \in \mathbb{C}$. Note that each of these algebras has an identity, $\mathcal{A} \subseteq \mathcal{B}$, but they do not have a common identity. So the preceding proposition does not apply.

Exercises.

(1) In Example 10.4.3 show that if we define the operator S by $S(h + k) = Ah + Bk$, where $B : (\operatorname{ran} T)^{\perp} \to \mathcal{H}$ is any bounded operator, then $ST = 1$. Hence if T is left invertible but not invertible, the left inverse is not unique. Prove that any left inverse of T must have this form.

(2) Let $(X, \mathcal{A}, \mu)$ be a σ-finite measure space and consider the Banach algebra $L^{\infty}(\mu)$. For ϕ in $L^{\infty}(\mu)$ show that the following statements are equivalent. (a) $\lambda \in \sigma(\phi)$. (b) $0 = \sup \{ \inf \{ |\phi(x) - \lambda| : x \in X \backslash E \} : E \in \mathcal{A}$ and $\mu(E) = 0 \}$. (c) If $\epsilon > 0$, $\mu(\{x : |\phi(x) - \lambda| < \epsilon\}) > 0$. (d) If ν is the measure defined on the Borel subsets of $\mathbb{C}$ by $\nu(\Delta) = \mu(\phi^{-1}(\Delta))$, then λ belongs to the support of ν. The spectrum $\sigma(\phi)$ is called the *essential range* of ϕ, a name that is justified by both parts (b) and (c).

(3) Show that if $\mathcal{A}$ is a Banach algebra, $\{a_n\} \subseteq \mathcal{A}$, $a_n \to a$, $\lambda_n \in \sigma(a_n)$, and $\lambda_n \to \lambda$ in $\mathbb{C}$, then $\lambda \in \sigma(a)$.

(4) If $\mathcal{A}$ is a Banach algebra, $a, b \in \mathcal{A}$, and $\lambda \neq 0$ with $\lambda - ab$ invertible, show that $\lambda - ba$ is invertible with $(\lambda - ba)^{-1} = \lambda^{-1} + \lambda^{-1}(\lambda - ab)^{-1}a$. Prove that $\sigma(ab) \cup \{0\} = \sigma(ba) \cup \{0\}$ and give an example such that $\sigma(ab) \neq \sigma(ba)$.

10.5. The spectrum of an operator

In this section we'll examine special properties of the spectrum of an operator on a Banach space, though the very first result includes some information on the spectrum of an operator on a Hilbert space; it once again underlines a distinction between these two settings when we examine the adjoint of an operator. The proof is left as an exercise. For any subset E of $\mathbb{C}$, let $E^* = \{\bar{z} : z \in E\}$.

10.5.1. Proposition. (a) *If $\mathcal{X}$ is a Banach space and $T \in \mathcal{B}(\mathcal{X})$, then $\sigma(T^*) = \sigma(T)$.*

(b) *If $\mathcal{H}$ is a Hilbert space and $T \in \mathcal{B}(\mathcal{H})$, then $\sigma(T^*) = \sigma(T)^*$.*

Throughout the section $\mathcal{X}$ will denote a Banach space over the complex numbers and $\mathcal{H}$ will be a Hilbert space. Recall (5.6.9) the definition of the point spectrum $\sigma_p(T)$ of an operator.

10.5.2. Definition. If $T \in \mathcal{B}(\mathcal{X})$, the *approximate point spectrum* is denoted by $\sigma_{ap}(T)$ and is defined by

$$\sigma_{ap}(T) = \{\lambda : \inf\{\|(T - \lambda)x\| : \|x\| = 1\} = 0\}$$

We note the obvious: $\sigma_p(T) \subseteq \sigma_{ap}(T)$. Also if $\lambda \in \sigma_{ap}(T)$, then there is a sequence of unit vectors $\{x_n\}$ such that $\|(T - \lambda)x_n\| \to 0$; hence the name "approximate" point spectrum.

10.5.3. Proposition. *If $T \in \mathcal{B}(\mathcal{X})$ and $\lambda \in \mathbb{C}$, then the following statements are equivalent.*

(a) $\lambda \notin \sigma_{ap}(T)$.

(b) *There is a constant $c > 0$ such that $\|(T - \lambda)x\| \geq c\|x\|$ for all x in $\mathcal{X}$.*

(c) $\ker(T - \lambda) = (0)$ *and* $\operatorname{ran}(T - \lambda)$ *is closed.*

Proof. By substituting $T - \lambda$ for T, we may assume that $\lambda = 0$.

(a) *implies* (b). If (b) is untrue, then for every positive integer n we can find a vector x_n with $\|Tx_n\| < n^{-1}\|x_n\|$. If we put $y_n = \|x_n\|^{-1}x_n$, then each y_n is a unit vector and $\|Ty_n\| \to 0$. Hence $0 \in \sigma_{ap}(T)$.

(b) *implies* (c). Assume (b) holds. Clearly $\ker T = (0)$ and Lemma 5.6.13 implies that $\operatorname{ran} T$ is closed.

(c) *implies* (a). Since $T : \mathcal{X} \to \operatorname{ran} T$ is a bounded bijection, the IMT implies there is a bounded operator $B : \operatorname{ran} T \to \mathcal{X}$ such that $BTx = x$ for every x in $\mathcal{X}$. Therefore $\|x\| \leq \|B\|\|Tx\|$ and we have that (c) holds with $c = \|B\|^{-1}$. ∎

10.5.4. Proposition. *For any T in $\mathcal{B}(\mathcal{X})$, the approximate point spectrum of T is closed.*

Proof. Suppose $\{\lambda_n\} \subseteq \sigma_{ap}(T)$ and $\lambda_n \to \lambda$. From the definition of $\sigma_{ap}(T)$ we have that for every $n \geq 1$ there is a unit vector x_n with $\|(T-\lambda_n)x_n\| < n^{-1}$. Therefore $\|(T-\lambda)x_n\| \leq \|(T-\lambda_n)x_n\| + \|(\lambda-\lambda_n)x_n\| \leq n^{-1} + |\lambda - \lambda_n|$. Hence $\lambda \in \sigma_{ap}(T)$. ∎

The reader may have realized that the preceding proof is similar to what was done in Example 10.4.3. In fact putting together what was obtained in that example with the preceding proposition we see that for a Hilbert space operator T, $\sigma_\ell(T) = \sigma_{ap}(T)$.

10.5.5. Proposition. *If $T \in \mathcal{B}(\mathcal{X})$, then $\partial\sigma(T) \subseteq \sigma_{ap}(T)$. In particular, $\sigma_{ap}(T) \neq \emptyset$.*

Proof. Let $\lambda \in \partial\sigma(T)$ and let $\{\lambda_n\}$ be a sequence in $\mathbb{C}\backslash\sigma(T)$ such that $\lambda_n \to \lambda$. We claim that $\|(T - \lambda_n)^{-1}\| \to \infty$. In fact if this is not the case, then by passing to a subsequence we can assume that there is a constant M with $\|(T-\lambda_n)^{-1}\| \leq M$ for all $n \geq 1$. But then $\|(T-\lambda)-(T-\lambda_n)\| = |\lambda-\lambda_n|$ and we can choose n sufficiently large that $\|(T - \lambda) - (T - \lambda_n)\| < M^{-1} \leq \|(T - \lambda_n)^{-1}\|^{-1}$. But then (10.2.3(b)) implies that $T - \lambda$ is invertible, a contradiction.

For each $n \geq 1$ choose a unit vector x_n with $\alpha_n = \|(T - \lambda_n)^{-1}x_n\| > \|(T - \lambda_n)^{-1}\| - n^{-1}$; so $\alpha_n \to \infty$. If $y_n = \alpha_n^{-1}(T - \lambda_n)^{-1}x_n$, then $\|y_n\| = 1$ and

$$(T - \lambda)y_n = (T - \lambda_n)y_n + (\lambda_n - \lambda)y_n$$
$$= \alpha_n^{-1}x_n + (\lambda_n - \lambda)y_n$$

Thus $\|(T - \lambda)y_n\| \leq \alpha_n^{-1} + |\lambda_n - \lambda|$, so that $\|(T - \lambda)y_n\| \to 0$ and we have established that $\lambda \in \sigma_{ap}(T)$. ∎

We extend the definition of the unilateral shift (5.3.3) from the setting of ℓ^2 to ℓ^p and derive its spectral properties.

10.5.6. Proposition. *If $1 \leq p \leq \infty$ and we define $S : \ell^p \to \ell^p$ by*

$$S(x_1, x_2, \dots) = (0, x_1, x_2, \dots)$$

then $\sigma(S) = \operatorname{cl}\mathbb{D}$, $\sigma_p(S) = \emptyset$, and $\sigma_{ap}(S) = \partial\mathbb{D}$. Moreover if $|\lambda| < 1$, then $\operatorname{ran}(S - \lambda)$ is closed and $\dim[\ell^p/\operatorname{ran}(S - \lambda)] = 1$.

Proof. We have many things to check, but none of them is hard. Let S_p denote the operator $S : \ell^p \to \ell^p$ defined in the proposition and define $T_p : \ell^p \to \ell^p$ by $T_p(x_1, x_2, \dots) = (x_2, x_3, \dots)$. We leave it to the reader to check that $T_p S_p = 1$ and for finite p, $T_q = S_p^*$, where $p^{-1} + q^{-1} = 1$. Hence for finite p, $\sigma(S_p) = \sigma(T_q)$.

Note that S_p is an isometry and so $\|S_p\| = 1$. Therefore $\sigma(S_p) \subseteq \text{cl}\,\mathbb{D}$. If $|\lambda| < 1$, put

$$x_\lambda = (1, \lambda, \lambda^2, \dots)$$

Clearly $x_\lambda \in \ell^\infty$ and for finite p, $\|x_\lambda\|_p^p = \sum_{n=0}^\infty (|\lambda|^p)^n = (1 - |\lambda|^p)^{-1} < \infty$, so that $x_\lambda \in \ell^p$. The reader can check that $T_p x_\lambda = \lambda x_\lambda$. Thus $\mathbb{D} \subseteq \sigma_p(T_p)$ for any value of p. For $1 \leq p < \infty$, $T_q = S_p^*$, so this implies that $\mathbb{D} \subseteq \sigma(S_p) = \sigma(T_q)$ when p is finite. But $S_\infty = T_1^*$, so we also have that $\mathbb{D} \subseteq \sigma(S_\infty)$. Thus $\sigma(S_p) = \text{cl}\,\mathbb{D}$ for $1 \leq p \leq \infty$.

If $\lambda \neq 0$ and $x \in \ell^p$, then $S_p x = \lambda x$ says that we must have $0 = \lambda x_1, x_1 = \lambda x_2, \dots$, which implies that $x_1 = 0, x_2 = 0$, and so on. So $\lambda \notin \sigma_p(S_p)$. Since S_p is an isometry we obtain that $\sigma_p(S_p) = \emptyset$ for $1 \leq p \leq \infty$.

If $|\lambda| < 1$ and $x \in \ell^p$, then $\|x\| = \|S_p x\| \leq \|(S_p - \lambda)x\| + |\lambda|\|x\|$, so that $\|(S - \lambda)x\| \geq (1 - |\lambda|)\|x\|$. Therefore $\lambda \notin \sigma_{ap}(S_p)$ and $\sigma_{ap}(S_p) \subseteq \partial\mathbb{D}$. But Proposition 10.5.5 shows the reverse inclusion, so we get that $\sigma_{ap}(S_p) = \partial\mathbb{D}$.

Claim. If $|\lambda| < 1$, then $\dim \ker(T_p - \lambda) = 1$ for $1 \leq p \leq \infty$.

In fact we already have that the vector $x_\lambda \in \ker(T_p - \lambda)$, so the dimension of the kernel is at least 1. If $T_p x = \lambda x$, then $(x_2, x_3, \dots) = (\lambda x_1, \lambda x_2, \dots)$. Thus for all $n \geq 1$, $x_{n+1} = \lambda x_n$. By iterating this equation we get that $x_{n+1} = \lambda^n x_1$; that is, $x = x_1 x_\lambda$. This establishes the claim.

If $|\lambda| < 1$, $1 \leq p < \infty$, and $1 = p^{-1} + q^{-1}$, then the claim together with the fact that for any operator A, $[\text{ran}\,A]^\perp = \ker A^*$ (Theorem 9.1.8) implies that $1 = \dim \ker(T_q - \lambda) = \dim[\text{ran}\,(S_p - \lambda)]^\perp = \dim[\ell^p/\text{ran}\,(S_p - \lambda)]$. Similarly $\dim[\ell^\infty/\text{ran}\,(S_\infty - \lambda)] = \dim\{^\perp[\text{ran}\,(S_\infty - \lambda)]\} = \dim \ker(T_1 - \lambda) = 1$. This completes the proof. ∎

This next corollary is a gathering together of facts about the operator T_p that appeared in the preceding proof. Clearly the operator T_p is the extension to ℓ^p of the backward shift defined on ℓ^2 in §5.4.

10.5.7. Corollary. *If $1 \leq p \leq \infty$ and $T : \ell^p \to \ell^p$ is defined by*

$$T(x_1, x_2, \dots) = (x_2, x_3, \dots)$$

then $\sigma(T) = \sigma_{ap}(T) = \text{cl}\,\mathbb{D}$ and $\sigma_p(T) = \mathbb{D}$. In fact for $|\lambda| < 1$ if $x_\lambda = (1, \lambda, \lambda^2, \dots)$, then $Tx_\lambda = \lambda x_\lambda$ and $\ker(T - \lambda)$ is spanned by x_λ.

The next example extends the idea of a diagonalizable operator from a Hilbert space to ℓ^p.

10.5.8. Example. If $\{\lambda_n\} \in \ell^\infty$ and $1 \leq p \leq \infty$, define $T : \ell^p \to \ell^p$ by $(Tx)(n) = \lambda_n x_n$. It is not hard to see that $\sigma(T) = \text{cl}\,\{\lambda_n\} = \sigma_{ap}(T)$ and $\sigma_p(T) = \{\lambda_n\}$.

We need to explore the behavior of an operator near an isolated point of the spectrum. This will be especially helpful in the next section when we examine the spectral properties of compact operators, but it is applicable in a much larger context. Actually what we do here is a special case of a more general approach for which we need more complex analysis than seems reasonable unless the reader has command of that subject. See §VII.4 and §VII.6 in [**8**] for the more general theory.

Let T be any bounded operator on a Banach space $\mathcal{X}$ and suppose λ is an isolated point of the spectrum. Let $r > 0$ such that $\operatorname{cl} B(\lambda; r) \cap \sigma(T) = \{\lambda\}$, and let γ be the circular path $\gamma(t) = \lambda + r \exp(2\pi i t)$, $0 \le t \le 1$. Examine the integral

10.5.9
$$E_\lambda = \frac{1}{2\pi i} \int_\gamma (z - T)^{-1} dz$$

Using the parameterization of this curve γ we have

$$E_\lambda = r \int_0^1 (r e^{2\pi i t} + \lambda - T)^{-1} e^{2\pi i t} dt$$

A typical Riemann sum approximating this integral is, after performing various cancellations, a finite sum of the form

10.5.10
$$E_\lambda \approx r \sum_k e^{2\pi i t_k} (t_k - t_{k-1})(r e^{2\pi i t_k} + \lambda - T)^{-1}$$

Looking at such an approximation carefully we see that E_λ is a bounded operator on $\mathcal{X}$.

The first thing we need to do is to show that the value of the integral (10.5.9) does not depend on the choice of radius of the circular path. (The reader who is conversant with complex analysis will recognize this as an easy application of Cauchy's Theorem.)

10.5.11. Lemma. *Let $T \in \mathcal{B}(\mathcal{X})$ and let λ be an isolated point of $\sigma(T)$. If $r, \rho > 0$ such that $\operatorname{cl} B(\lambda; r) \cap \sigma(T) = \{\lambda\} = \operatorname{cl} B(\lambda; \rho) \cap \sigma(T)$ and if γ_r and γ_ρ are the circular paths $\gamma_r(t) = \lambda + r \exp(2\pi i t)$ and $\gamma_\rho(t) = \lambda + \rho \exp(2\pi i t)$, then*

$$\frac{1}{2\pi i} \int_{\gamma_r} (z - T)^{-1} dz = \frac{1}{2\pi i} \int_{\gamma_\rho} (z - T)^{-1} dz$$

Therefore the definition of E_λ in (10.5.9) does not depend on the choice of radius r.

Proof. The proof is in the spirit of the proof of Lemma 10.3.8. Assume $r < \rho$ and for $0 \le s \le 1$ put $r_s = (1 - s)r + s\rho$. Let $\gamma_s(t) = \lambda + r_s \exp(2\pi i t)$ and define $F : [0, 1] \to \mathcal{B}(\mathcal{X})$ by

$$F(s) = \frac{1}{2\pi i} \int_{\gamma_s} (z - T)^{-1} dz = r_s \int_0^1 (r_s e^{2\pi i t} + \lambda - T)^{-1} e^{2\pi i t} dt$$

We want to show that F is constant and to accomplish this we show that $F'(s) = 0$. We have that

$$F'(s) = (\rho - r) \int_0^1 (r_s e^{2\pi it} + \lambda - T)^{-1} e^{2\pi it} dt$$

$$- (\rho - r) r_s \int_0^1 (r_s e^{2\pi it} + \lambda - T)^{-2} e^{4\pi it} dt$$

$$= (\rho - r) \int_0^1 \left[(r_s e^{2\pi it} + \lambda - T)^{-1} e^{2\pi it} - r_s (r_s e^{2\pi it} + \lambda - T)^{-2} e^{4\pi it} \right] dt$$

But the integrand in this last integral is seen to be equal to

$$\frac{d}{dt} \left[\frac{1}{2\pi i} \left[(r_s e^{2\pi it} + \lambda - T)^{-1} \right] \right]$$

That is, the integrand has a primitive and this primitive is periodic with period 1; therefore $F'(s) = 0$ and F is constant. The lemma follows. ■

What is actually at work here is that the family of curves $\{\gamma_s\}$ shows that the two circles are homotopic. It turns out that the integral is the same over any two curves homotopic in the open set $G \backslash \sigma(T)$. See [**7**].

10.5.12. Lemma. *The operator E_λ is an idempotent: $E_\lambda^2 = E_\lambda$.*

Proof. To show this we note that if we take a different radius $\rho > r$ such that $\operatorname{cl} B(\lambda; \rho) \cap \sigma(T) = \{\lambda\}$, then by the preceding lemma we also have that $E_\lambda = (2\pi i)^{-1} \int_\eta (\zeta - T)^{-1} d\zeta$, where $\eta(t) = \lambda + \rho \exp(2\pi it)$ for $0 \le t \le 1$. Using the resolvent equation to get that for $0 \le s, t \le 1$, $(\eta(s) - \gamma(t))(\gamma(t) - T)^{-1}(\eta(s) - T)^{-1} = (\gamma(t) - T)^{-1} - (\eta(s) - T)^{-1}$, we have that

$$E_\lambda^2 = \left[\frac{1}{2\pi i} \int_0^1 (\gamma(t) - T)^{-1} \gamma'(t) dt \right] \left[\frac{1}{2\pi i} \int_0^1 (\eta(s) - T)^{-1} \eta'(s) ds \right]$$

$$= -\frac{1}{4\pi^2} \int_0^1 \int_0^1 (\gamma(t) - T)^{-1} (\eta(s) - T)^{-1} \gamma'(t) \eta'(s) dt ds$$

$$= -\frac{1}{4\pi^2} \int_0^1 \int_0^1 \left[\frac{(\gamma(t) - T)^{-1} - (\eta(s) - T)^{-1}}{\eta(s) - \gamma(t)} \right] \gamma'(t) \eta'(s) dt ds$$

$$= -\frac{1}{4\pi^2} \int_0^1 \left[\int_0^1 \frac{1}{\eta(s) - \gamma(t)} \eta'(s) ds \right] (\gamma(t) - T)^{-1} \gamma'(t) dt$$

$$+ \frac{1}{4\pi^2} \int_0^1 \left[\int_0^1 \frac{1}{\eta(s) - \gamma(t)} \gamma'(t) dt \right] (\eta(s) - T)^{-1} \eta'(s) ds$$

$$= -\frac{1}{4\pi^2} \int_\gamma \left[\int_\eta \frac{1}{\zeta - z} d\zeta \right] (z - T)^{-1} dz$$

$$+ \frac{1}{4\pi^2} \int_\eta \left[\int_\gamma \frac{1}{\zeta - z} dz \right] (\zeta - T)^{-1} d\zeta$$

Now Lemma 10.3.8 implies $\int_\eta (\zeta - z)^{-1} d\zeta = 2\pi i$ since $|z - \lambda| < \rho$, and $\int_\gamma (\zeta - z)^{-1} dz = 0$ since $|\zeta - \lambda| > r$. Hence the above equation becomes

$$E_\lambda^2 = \frac{1}{2\pi i} \int_\gamma (z - T)^{-1} dz = E_\lambda$$

and we have proved the lemma. ∎

10.5.13. Proposition. *If $T \in \mathcal{B}(\mathcal{X})$ and λ is an isolated point in $\sigma(T)$, then the operator E_λ defined in (10.5.9) is an idempotent such that $AE_\lambda = E_\lambda A$ whenever $A \in \mathcal{B}(\mathcal{X})$ and $AT = TA$. If $\mathcal{X}_\lambda = E_\lambda \mathcal{X}$ then $\mathcal{X}_\lambda$ is a closed subspace such that:*

(a) $A\mathcal{X}_\lambda \subseteq \mathcal{X}_\lambda$ whenever $A \in \mathcal{B}(\mathcal{X})$ and $AT = TA$;

(b) $\sigma(T|\mathcal{X}_\lambda) = \{\lambda\}$ and the restriction of T to the complementary subspace $(1 - E_\lambda)\mathcal{X}$ has spectrum $\sigma(T)\backslash\{\lambda\}$.

Proof. We already established that E_λ is a bounded idempotent. Suppose that A is a bounded operator that commutes with T. So for any scalar α we have that $A(\alpha - T) = (\alpha - T)A$, from which it follows that for all α not in $\sigma(T)$, $A(\alpha - T)^{-1} = (\alpha - T)^{-1}A$. Therefore if we look at the definition of E_λ and approximate the defining integral by a Riemann sum as in (10.5.10) we see that A commutes with every such approximation to E_λ. Therefore $AE_\lambda = E_\lambda A$. From here (a) is immediate.

(b) Let $\mu \in \mathbb{C}$, $\mu \neq \lambda$ and choose $0 < r < |\mu - \lambda|$ such that $\operatorname{cl} B(\lambda; r) \cap \sigma(T) = \{\lambda\}$; put $\gamma(t) = \lambda + r \exp(2\pi i t)$, $0 \leq t \leq 1$. Note that the resolvent equation implies that for z on the circle γ, $(z - T)^{-1} - (z - \mu)^{-1} = (z - \mu)^{-1}(z - T)^{-1}(T - \mu)$. Now Lemma 10.3.8 implies that $\int_\gamma (z - \mu)^{-1} dz = 0$, so

$$E_\lambda = \frac{1}{2\pi i} \int_\gamma (z - T)^{-1} dz$$
$$= \frac{1}{2\pi i} \int_\gamma \left[(z - T)^{-1} - (z - \mu)^{-1} \right] dz$$
$$= \frac{1}{2\pi i} \int_\gamma (z - \mu)^{-1} (z - T)^{-1} (T - \mu) dz$$

If $x \in \mathcal{X}_\lambda$ and $(T - \mu)x = 0$, then

$$x = E_\lambda x = \frac{1}{2\pi i} \int_\gamma (z - \mu)^{-1} (z - T)^{-1} \left[(T - \mu)x \right] dz = 0$$

Hence $T|\mathcal{X}_\lambda - \mu$ is injective. If x is any vector in $\mathcal{X}_\lambda$, note that for every z on γ, $(z - T)^{-1}x \in \mathcal{X}_\lambda$ by part (a). If we approximate the next integral by a Riemann sum, we see that

$$y = \frac{1}{2\pi i} \int_\gamma (z - \mu)^{-1} \left[(z - T)^{-1}x\right] dz \in \mathcal{X}_\lambda$$

Now using the above alternative expression for E_λ, we see that

$$(T - \mu)y = (T - \mu)\left[\frac{1}{2\pi i} \int_\gamma (z - \mu)^{-1}\left[(z - T)^{-1}x\right] dz\right]$$
$$= \frac{1}{2\pi i} \int_\gamma (z - \mu)^{-1}(z - T)^{-1}(T - \mu)x\,dz$$
$$= E_\lambda x = x$$

and we have that $T|\mathcal{X}_\lambda - \mu$ is surjective. Hence $\mu \notin \sigma(T|\mathcal{X}_\lambda)$ whenever $\mu \neq \lambda$. Since the spectrum is always non-empty, we have that $\sigma(T|\mathcal{X}_\lambda) = \{\lambda\}$. Since $\mathcal{X} = \mathcal{X}_\lambda \oplus (1 - E_\lambda)\mathcal{X}$, we have that $T = T|\mathcal{X}_\lambda \oplus T|(1 - E_\lambda)\mathcal{X}$ and $\sigma(T) = \{\lambda\} \cup \sigma(T|(1 - E_\lambda)\mathcal{X})$. It is left to the reader to show that $\lambda \notin \sigma(T|(1 - E_\lambda)\mathcal{X})$. ∎

10.5.14. Example. A measurable function $k : [0, 1] \times [0, 1] \to \mathbb{C}$ is called a *Volterra*[2] *kernel* if it is bounded and $k(x, y) = 0$ when $x < y$. So k vanishes on the upper half of the unit square. If $1 \leq p \leq \infty$, define $V_k : L^p[0, 1] \to L^p[0, 1]$ by

$$V_k f(x) = \int_0^1 k(x, y)f(y)dy = \int_0^x k(x, y)f(y)dy$$

It follows that $V_k \in \mathcal{B}(L^p[0, 1])$ and $\|V_k\| \leq \|k\|_\infty$ (5.1.9). Note that if k is the characteristic function of the lower half of the unit square, then V_k is the Volterra operator discussed in Example 5.1.10.

If we have two Volterra kernels h and k, we define their product as

$$(hk)(x, y) = \int_0^1 h(x, t)k(t, y)dt = \int_y^x h(x, t)k(t, y)dt$$

The reader can check that hk is also a Volterra kernel, $\|hk\|_\infty \leq \|h\|_\infty \|k\|_\infty$, and $V_{hk} = V_h V_k$. (You might note the analogy with the multiplication of matrices and the composition of the linear transformations they represent.

---

[2]Vito Volterra was born in 1860 in Ancona in the Papal states of Italy on the Adriatic coast. At the age of 11 he began to study Legendre's Geometry and at 13 he took up the three-body problem. He pursued his doctorate in Pisa, where he studied under Betti and received the degree in 1882. The following year, after Betti's death, he became Professor of Mechanics at Pisa. In 1900 he moved to Rome. He pursued the study of functional analysis and partial differential equations and was led to a consideration of integral equations, the source of his present day fame. During World War I he joined the air force and promoted scientific collaboration among the allies. Volterra fought against the rise of fascism in Italy after the war, and in 1931 he refused to take an oath of allegiance to the Fascist Government. As a consequence he was forced to leave the University of Rome. He spent considerable time in the following years abroad but returned to Rome where he died in 1940.

Don't forget that an integral is a type of continuous sum. A matrix is a type of kernel on a finite purely atomic measure space and there is more than an analogy going on here.)

If k is a Volterra kernel, then

$$\sigma(V_k) = \{0\}$$

We will show this by using Proposition 10.4.6 to show that the spectral radius of V_k is 0. Start by observing that $V_k^n = V_{k^n}$, where the powers k^n are in the sense of the product of two Volterra kernels defined in the last paragraph. To show $r(V_k) = 0$, we need the following.

10.5.15. Claim. For all $n \geq 1$ and $y < x$, we have

$$|k^n(x, y)| \leq \frac{\|k\|_\infty^n}{(n-1)!}(x-y)^{n-1}$$

We use induction. The case $n = 1$ is clear, so assume (10.5.15) is valid for some $n \geq 1$. Thus

$$|k^{n+1}(x, y)| = \left| \int_y^x k(x, t)k^n(t, y)dt \right|$$

$$\leq \int_y^x |k(x, t)||k^n(t, y)|dt$$

$$\leq \|k\|_\infty \frac{\|k\|_\infty^n}{(n-1)!} \int_y^x (t-y)^{n-1}dt$$

$$= \frac{\|k\|_\infty^{n+1}}{n!}(x-y)^n$$

This establishes the claim.

From this claim we have that

$$\|V_k^n\| \leq \|k^n\|_\infty \leq \frac{\|k\|_\infty^n}{(n-1)!}$$

Taking n-th roots we have that

$$\|V_k^n\|^{\frac{1}{n}} \leq \frac{\|k\|_\infty}{[(n-1)!]^{\frac{1}{n}}} \to 0$$

We saw in Example 5.6.10 that for the Volterra operator there are no eigenvalues. In fact if $f \in L^p[0, 1]$ and $\int_0^x f(y)dy = 0$ for all x, then differentiation gives that $f = 0$. However for some Volterra kernels it is possible for 0 to be an eigenvalue. For example define $k(x, y)$ by

$$k(x, y) = \begin{cases} \chi_{(0, \frac{1}{2})}(y) & \text{when } y < x \\ 0 & \text{otherwise} \end{cases}$$

It follows that for every f in $L^p[0,1]$,

$$V_k f(x) = \begin{cases} \int_0^x f(y)dy & \text{if } x \leq \frac{1}{2} \\ \int_0^{\frac{1}{2}} f(y)dy & \text{if } x \geq \frac{1}{2} \end{cases}$$

So if f vanishes on $[0, \frac{1}{2}]$, $V_k f = 0$.

Exercises.

(1) Prove Proposition 10.5.1.

(2) If $\mathcal{X}$ is a Banach space and $T \in \mathcal{B}(\mathcal{X})$, show that $\sigma_\ell(T) = \sigma_r(T^*)$. What happens in the Hilbert space setting?

(3) Let K be any compact subset of $\mathbb{C}$ and show that there is an operator T on a separable Hilbert space with $\sigma(T) = K$.

(4) Let $1 \leq p \leq \infty$ and suppose that $0 < \lambda_1 \leq \lambda_2 \leq \cdots$ such that $r = \lim_n \lambda_n$ exists and is finite. If $T : \ell^p \to \ell^p$ is defined by $T(x_1, x_2, \dots) = (0, \lambda_1 x_1, \lambda_2 x_2, \dots)$, show that $\sigma(T) = \operatorname{cl} B(0; r)$, $\sigma_p(T) = \emptyset$, and $\sigma_{ap}(T) = \partial\sigma(T)$. If $|\lambda| < r$, show that $\operatorname{ran}(T - \lambda)$ is closed and has codimension 1.

(5) (a) Verify the statements made in Example 10.5.8. (b) What is the dimension of $\ker(T - \lambda_n)$? (c) For the operator T in Example 10.5.8, suppose λ is an isolated point of $\sigma(T)$ and determine the idempotent E_λ defined in (10.5.9). (Hint: How can a point λ in $\sigma(T)$ be isolated?)

(6) Verify the statements made in Example 10.5.14.

(7) If $\mathcal{H}$ is a Hilbert space and A is a hermitian operator on $\mathcal{H}$, show that $\sigma(A) = \sigma_{ap}(A)$. Is the same true when A is normal?

(8) Let $(X, \mathcal{A}, \mu)$ be a σ-finite measure space and let $1 \leq p \leq \infty$. For each ϕ in $L^\infty(\mu)$ define M_ϕ on $L^p(\mu)$ as in Example 5.1.2. Find $\sigma(M_\phi), \sigma_{ap}(M_\phi)$, and $\sigma_p(M_\phi)$. (Hint: Looking at Exercise 2 might prove helpful.)

10.6. The spectrum of a compact operator

This would be a good time to review §5.6, where we proved some results concerning eigenvalues of a compact operator on a Hilbert space that we now need in the Banach setting. We will restate these results here in the more general setting but refer to corresponding results in §5.6 for the proofs, which carry over to Banach spaces. We continue to assume that $\mathcal{X}$ will always denote a Banach space over the complex numbers.

10.6.1. Lemma. *If $T \in \mathcal{B}_0(\mathcal{X})$ and λ is a non-zero eigenvalue, then $\ker(T - \lambda)$ is finite-dimensional.*

For the proof see Proposition 5.6.11.

10.6.2. Lemma. *If $T \in \mathcal{B}_0(\mathcal{X})$, $\lambda \neq 0$, and $\lambda \in \sigma(T)$, then either $\lambda \in \sigma_p(T)$ or $\lambda \in \sigma_p(T^*)$.*

For the proof see Corollary 5.6.14.

Now for the fundamental result on the spectrum of a compact operator. The proof will use the preceding two lemmas and two more that will follow the statement.

10.6.3. Theorem. *If $\mathcal{X}$ is infinite-dimensional and $T \in \mathcal{B}_0(\mathcal{X})$, then one and only one of the following is true.*

(a) $\sigma(T) = \{0\}$.

(b) $\sigma(T) = \{0, \lambda_1, \ldots, \lambda_n\}$, *where each* $\lambda_k \neq 0$, $\lambda_k \in \sigma_p(T)$, *and* $\dim \ker(T - \lambda_k) < \infty$.

(c) $\sigma(T) = \{0, \lambda_1, \lambda_2, \ldots\}$, *where each* $\lambda_k \neq 0$, $\lambda_k \in \sigma_p(T)$, $\dim \ker(T - \lambda_k) < \infty$, *and* $\lambda_k \to 0$.

The first new lemma needed for the proof of this theorem is rather technical and is used in the proof of the next lemma

10.6.4. Lemma. *If $\mathcal{M} \leq \mathcal{N}$, $\mathcal{M} \neq \mathcal{N}$, and $\epsilon > 0$, then there is a y in $\mathcal{N}$ such that $\|y\| = 1$ and $\mathrm{dist}\,(y, \mathcal{M}) \geq 1 - \epsilon$.*

Proof. Before starting the proof, we might comment that this lemma is meant to get something close to what we almost automatically have in a Hilbert space setting. In fact if we had the same hypothesis in a Hilbert space, then we can find a unit vector y in $\mathcal{N}$ such that $y \perp \mathcal{M}$. In this case $\mathrm{dist}\,(y, \mathcal{M}) = 1$.

Put $\delta(y) = \mathrm{dist}\,(y, \mathcal{M}) = \inf\{\|y - x\| : x \in \mathcal{M}\}$. For any y_1 in $\mathcal{N}\backslash\mathcal{M}$, there is an x_0 in $\mathcal{M}$ such that $\delta(y_1) \leq \|x_0 - y_1\| \leq (1 + \epsilon)\delta(y_1)$. If we put $y_2 = y_1 - x_0$, then $(1 + \epsilon)\delta(y_2) = (1 + \epsilon)\inf\{\|y_2 - x\| : x \in \mathcal{M}\} = (1 + \epsilon)\inf\{\|y_1 - x_0 - x\| : x \in \mathcal{M}\} = (1 + \epsilon)\delta(y_1)$ since $x_0 \in \mathcal{M}$. Thus $(1 + \epsilon)\delta(y_2) > \|y_1 - x_0\| = \|y_2\|$. Let $y = \|y_2\|^{-1}y_2$. So for any x in $\mathcal{M}$ we have

$$\|y - x\| = \|y_2\|^{-1} \|y_2 - \|y_2\|x\|$$
$$> [1 + \delta(y_2)]^{-1} \|y_2 - \|y_2\|x\|$$
$$\geq (1 + \epsilon)^{-1}\delta(y_2)^{-1} \|y_2 - \|y_2\|x\|$$
$$\geq (1 + \epsilon)^{-1} > 1 - \epsilon$$

This completes the proof. ∎

10.6.5. Lemma. *If $T \in \mathcal{B}_0(\mathcal{X})$ and $\{\lambda_n\}$ is a sequence of distinct elements in $\sigma(T)$, then $\lambda_n \to 0$.*

Proof. For each $n \geq 1$ pick a non-zero vector x_n in $\ker(T - \lambda_n)$; put $\mathcal{M}_n = \bigvee\{x_1, \ldots, x_n\}$. Since the λ_n are distinct, $\dim \mathcal{M}_n = n$; hence $\mathcal{M}_n \leq \mathcal{M}_{n+1}$ and $\mathcal{M}_n \neq \mathcal{M}_{n+1}$. Taking $\mathcal{M}_0 = (0)$, the preceding lemma implies that for every $n \geq 1$, there is a y_n in $\mathcal{M}_n$ with $\|y_n\| = 1$ and $\mathrm{dist}\,(y_n, \mathcal{M}_{n-1}) > \frac{1}{2}$. If $y_n = \alpha_1 x_1 + \cdots + \alpha_n x_n$, then

$$(T - \lambda_n)y_n = \alpha_1(\lambda_1 - \lambda_n)x_1 + \cdots + (\lambda_{n-1} - \lambda_n)x_{n-1} \in \mathcal{M}_{n-1}$$

Therefore for $n > m$,

$$T(\lambda_n^{-1}y_n) - T(\lambda_m^{-1}y_m) = \lambda_n^{-1}(T - \lambda_n)y_n - \lambda_m^{-1}(T - \lambda_m)y_m + y_n - y_m$$

$$= y_n - [y_m + \lambda_m^{-1}(T - \lambda_m)y_m - \lambda_n^{-1}(T - \lambda_n)y_n]$$

Now the vector in the brackets in the last equation belongs to $\mathcal{M}_{n-1}$. Therefore $\|T(\lambda_n^{-1}y_n) - T(\lambda_m^{-1}y_m)\| \geq \mathrm{dist}\,(y_n, \mathcal{M}_{n-1}) > \frac{1}{2}$. Hence the sequence $\{T(\lambda_n^{-1}y_n)\}$ can have no convergent subsequence. But $\|\lambda_n^{-1}y_n\| = |\lambda_n|^{-1}$. If any subsequence of $\{\lambda_n^{-1}\}$ remains bounded, the fact that T is compact would imply that the corresponding subsequence of $\{T(\lambda_n^{-1}y_n)\}$ has a convergent subsequence. This contradiction implies $\lambda_n \to 0$. ∎

Proof of Theorem 10.6.3. We begin with the following.

Claim. If $\lambda \in \sigma(T)$ and $\lambda \neq 0$, then λ is an isolated point of $\sigma(T)$.

Suppose the claim is false; so there is a sequence $\{\lambda_n\}$ of distinct points in the spectrum of T such that $\lambda_n \to \lambda$. By Lemma 10.6.2 there is a subsequence that belongs to $\sigma_p(T)$ or there is a subsequence that belongs to $\sigma_p(T^*)$. In either case we have a sequence of distinct eigenvalues of a compact operator that converges to λ. According to Lemma 10.6.5, this sequence must converge to 0. Since we assumed $\lambda \neq 0$, we have a contradiction, establishing the claim.

Claim. If $\lambda \in \sigma(T)$ and $\lambda \neq 0$, then $\lambda \in \sigma_p(T)$ and $\dim \ker(T - \lambda) < \infty$.

By the first claim λ is an isolated point of $\sigma(T)$. We invoke Proposition 10.5.13 and define the idempotent E_λ (10.5.9). Let $\mathcal{X}_\lambda = E_\lambda \mathcal{X}$ and put $T_\lambda = T|\mathcal{X}_\lambda$. We know that $\sigma(T_\lambda) = \{\lambda\}$, so that T_λ is invertible. But T_λ is also compact since it is the restriction of a compact operator to an invariant subspace. The only way to make these two facts reconcilable is to have that $\dim \mathcal{X}_\lambda < \infty$. Since the spectrum of T_λ is the single point λ, $T_\lambda - \lambda$ is nilpotent and λ is an eigenvalue; it follows that $\lambda \in \sigma_p(T)$. Also $\ker(T - \lambda)$ is an invariant subspace for T and $T|\ker(T - \lambda)$ is compact and a non-zero multiple of the identity on this subspace; therefore $\dim \ker(T - \lambda) < \infty$. This proves the second claim.

To finish the proof, assume that $\mathcal{X}$ is infinite-dimensional. Since T is compact, it cannot be invertible; hence $0 \in \sigma(T)$. Assume the spectrum is infinite – otherwise we have, using the second claim, that either (a) or (b) holds. By the first claim all the non-zero elements of the spectrum are isolated points and so we have that $\sigma(T) = \{0, \lambda_1, \lambda_2, \dots\}$ and $\lambda_n \to 0$. By the second claim all these λ_n are eigenvalues of finite multiplicity. ∎

The Volterra operator V shows that Case (a) in Theorem 10.6.3 can happen. If we are given $\lambda_1, \dots, \lambda_n \neq 0$, define $D : \mathbb{C}^n \to \mathbb{C}^n$ by $D(z_1, \dots, z_n) = (\lambda_1 z_1, \dots, \lambda_n z_n)$ and put $T = V \oplus D$. It follows that T is compact on $\mathcal{X} = L^p(0, 1) \oplus \mathbb{C}^n$ and the spectrum of T satisfies Case (b). Note that by repeating any λ_k any finite number of times, we can arrange that $\dim(T - \lambda_k)$ has any finite dimension desirable. If $\{\lambda_n\}$ is a sequence of non-zero complex numbers converging to 0, define $D : \ell^p \to \ell^p$ by $T\{x_n\} = \{\lambda_n x_n\}$. Since $\lambda_n \to 0$, it follows that T is compact and we see that T satisfies Case (c). Again we can arrange that $\dim(T - \lambda_n)$ is any finite number we wish.

10.6.6. Theorem (The Fredholm[3] Alternative). *If $T \in \mathcal{B}_0(\mathcal{X})$, $\lambda \in \mathbb{C}$, and $\lambda \neq 0$, then $\operatorname{ran}(T - \lambda)$ is closed and $\dim \ker(T - \lambda) = \dim(T^* - \lambda) < \infty$.*

Proof. It suffices to assume that $\lambda \in \sigma(T)$. Put $\Delta = \sigma(T) \backslash \{\lambda\}$ and let $\mathcal{X}_\Delta = (1 - E_\lambda)\mathcal{X}$, $T_\lambda = T|\mathcal{X}_\lambda$, $T_\Delta = T|\mathcal{X}_\Delta$. From Proposition 10.5.13 we know that $\sigma(T_\lambda) = \{\lambda\}$ and $\sigma(T_\Delta) = \Delta$; so $T_\Delta - \lambda$ is invertible. Therefore $\operatorname{ran}(T - \lambda) = (T - \lambda)\mathcal{X}_\lambda + (T - \lambda)\mathcal{X}_\Delta = (T - \lambda)\mathcal{X}_\lambda + \mathcal{X}_\Delta$. Since $(T - \lambda)\mathcal{X}_\lambda$ is finite-dimensional, it follows from Corollary 6.2.3 that $\operatorname{ran}(T - \lambda)$ is closed.

We already know that $\dim \ker(T - \lambda) < \infty$. Note that

$$\mathcal{X}/\operatorname{ran}(T - \lambda) = (\mathcal{X}_\lambda + \mathcal{X}_\Delta)/(\operatorname{ran}(T_\lambda - \lambda) + \mathcal{X}_\Delta) \approx \mathcal{X}_\lambda/\operatorname{ran}(T_\lambda - \lambda)$$

Since $\mathcal{X}_\lambda$ is finite-dimensional, $\dim[\mathcal{X}/\operatorname{ran}(T - \lambda)] = \dim \mathcal{X}_\lambda - \dim \operatorname{ran}(T_\lambda - \lambda) = \dim \ker(T_\lambda - \lambda) = \dim \ker(T - \lambda)$ since $\ker(T - \lambda) \subseteq \mathcal{X}_\lambda$. But $[\mathcal{X}/\operatorname{ran}(T - \lambda)]^* = [\operatorname{ran}(T - \lambda)]^\perp (8.2.3) = \ker(T - \lambda)^*$ (9.1.8). Hence $\dim \ker(T - \lambda) = \dim(T^* - \lambda)$. ∎

---

[3] Erik Ivar Fredholm was born in 1866 in Stockholm to a wealthy, educated family. After receiving his baccalaureate in 1885, he entered the Royal Technological Institute in Stockholm and then transferred to the University of Uppsala, then the only university in Sweden that granted doctoral degrees. In spite of this he continued to study at Stockholm so as to work under the famous Swedish mathematician, Mittag-Leffler. He received the PhD in 1893 from Uppsala. In 1899 he journeyed to Paris to pursue his research in integral equations and potential theory. His work achieved instant recognition, attracting the attention of Hilbert. In 1906 he was appointed to a chair in mechanics and mathematical physics at Stockholm. His work in spectral theory, starting with integral operators but later generalized to abstract operators, has emblazoned his name in the history of mathematics with the Fredholm theory, of which the Fredholm Alternative is a special consequence. He had a productive career and died in 1927 in Stockholm.

10.6.7. Corollary. *If $T \in \mathcal{B}_0(\mathcal{X})$, $\lambda \in \mathbb{C}$, and $\lambda \neq 0$, then for every y in $\mathcal{X}$ there is an x such that $(T - \lambda)x = y$ if and only if the only vector x in $\mathcal{X}$ such that $(T - \lambda)x = 0$ is the vector $x = 0$.*

Most of the applications of the Fredholm Alternative are to integral equations. By the way, why does this theorem have the word "alternative" in its name? Look at Corollary 10.6.7 and rephrase it as follows: either $T - \lambda$ fails to be surjective or it fails to be injective.

Exercises.

(1) In Lemma 10.6.4 show that if the spaces $\mathcal{M}$ and $\mathcal{N}$ are finite-dimensional, then the unit vector y can be found with dist $(y, \mathcal{M}) = 1$.

(2) Show that for any T in $\mathcal{B}(\mathcal{X})$, eigenvectors for distinct eigenvalues are linearly independent.

(3) If $T \in \mathcal{B}_0(\mathcal{X})$ and p is a polynomial with $p(0) = 0$, show that $p(T) \in \mathcal{B}_0(\mathcal{X})$. What can you say about $\sigma(p(T))$?

10.7. Abelian Banach algebras

10.7.1. Theorem (Gelfand[4]–Mazur[5] Theorem). *If $\mathcal{A}$ is a Banach algebra with identity such that each non-zero element is invertible, then $\mathcal{A} = \mathbb{C}$. That is, $\mathcal{A} = \{\lambda 1 : \lambda \in \mathbb{C}\}$.*

Proof. If $a \in \mathcal{A}$, then $\sigma(a) \neq \emptyset$. But if $\lambda \in \sigma(a)$, then $a - \lambda$ does not have an inverse. By hypothesis, it must be that $a - \lambda = 0$. ∎

---

[4]Israel Moiseevich Gelfand was born in 1913 in Odessa. At the age of 16 he went to Moscow, where he took menial jobs such as a door keeper at the Lenin Library. He also began teaching various mathematics courses in the evening as well as attending lectures at Moscow University. In 1932 he was admitted as a research student and in 1935 he presented his thesis. He created the theory of commutative normed rings in 1938, which was the subject of his D.Sc. dissertation. In 1941 he was appointed a professor at Moscow State University. It was in the early 1940s that he began his collaboration with M. A. Naimark on what are now called C*-algebras (the next chapter). The range of Gelfand's mathematical interests is amazing and beyond the scope of this footnote to recount. He made contributions to various parts of algebra, analysis, and geometry and eventually gravitated toward biology and medicine. In 1990 he emigrated to the US and occupied a chair in the mathematics and biology departments at Rutgers. He had 23 doctoral students in his career and is acknowledged as one of the deepest mathematicians of the 20-th century. His awards include membership in the National Academy of Sciences and the Royal Society in Britain, the Steele Prize of the American Mathematical Society for Lifetime Achievement in 2005, and a MacArthur grant in 1994. He died in 2009 in New Brunswick, New Jersey.

[5]Stanisław Mazur was born in 1905 in Lvov, Poland. He entered the Polytechnic Institute in Lvov and became a student of Banach, receiving his doctorate in 1935 and his habilitation in 1938. He was an active member of the communist party. In 1948 he became a professor at the University of Warsaw and, due to his involvement with the communist party, was a high official in the science establishment. He made many contributions to functional analysis and had 21 doctoral students. He died in 1981 in Warsaw.

Note that we did not assume in this theorem that $\mathcal{A}$ is abelian, but the conclusion is that and more. An algebra $\mathcal{A}$ with the property that every non-zero element has an inverse is called a *division algebra*. If the division algebra is abelian, then, of course, it is a field.

This theorem enables us to proceed with the development of abelian Banach algebras. Recall Proposition 6.3.1 where we made a correspondence between hyperplanes in a Banach space and linear functionals. Here we make the analogous correspondence between maximal ideals in an abelian Banach algebra and continuous homomorphisms into $\mathbb{C}$. (Recall that in the case of an algebra, the homomorphism has to respect both vector addition and multiplication.)

10.7.2. Proposition. *If $\mathcal{A}$ is an abelian Banach algebra with identity and $\mathcal{M}$ is a maximal ideal of $\mathcal{A}$, then there is a continuous non-zero homomorphism $h : \mathcal{A} \to \mathbb{C}$ such that $\mathcal{M} = \ker h$. Conversely, if $h : \mathcal{A} \to \mathbb{C}$ is a non-zero homomorphism, then h is continuous and $\mathcal{M} = \ker h$ is a maximal ideal. The correspondence $h \mapsto \ker h$ between non-zero homomorphisms and maximal ideals is bijective.*

Proof. Let $\mathcal{M}$ be a maximal ideal; we know that $\mathcal{M}$ is closed (Corollary 10.2.5). Therefore $\mathcal{A}/\mathcal{M}$ is a Banach algebra with identity; let $\tau : \mathcal{A} \to \mathcal{A}/\mathcal{M}$ be the natural map. Suppose $a \in \mathcal{A}$ such that $\tau(a)$ is not 0 in the quotient algebra. Thus $\tau(a)(\mathcal{A}/\mathcal{M}) = \tau(a\mathcal{A})$ is a non-zero ideal in $\mathcal{A}/\mathcal{M}$. This implies $I = \tau^{-1}(\tau(a\mathcal{A}))$ is an ideal in $\mathcal{A}$ that properly contains $\mathcal{M}$. Hence $I = \mathcal{A}$, which implies $\tau(a)(\mathcal{A}/\mathcal{M}) = \tau(I) = \mathcal{A}/\mathcal{M}$. Therefore $\mathcal{A}/\mathcal{M}$ is a field and the Gelfand–Mazur Theorem implies $\mathcal{A}/\mathcal{M} = \mathbb{C}$. Define $\widetilde{h} : \mathcal{A}/\mathcal{M} \to \mathbb{C}$ by $\widetilde{h}(\lambda + \mathcal{M}) = \lambda$. It is easy to see that $\widetilde{h}$ is an isomorphism. Thus $h = \widetilde{h} \circ \tau$ is a homomorphism on $\mathcal{A}$ with $\ker h = \mathcal{M}$. Since h is a linear functional with closed kernel, it is continuous (6.3.2).

Now assume that h is a non-zero homomorphism on $\mathcal{A}$ and $\mathcal{M} = \ker h$. It is easy to see that $\mathcal{M}$ is an ideal and, since $\mathcal{A}/\mathcal{M} = \mathbb{C}$, $\mathcal{M}$ must be maximal. Therefore $\mathcal{M}$ is closed and h is continuous. If h_1 is another homomorphism on $\mathcal{A}$ with $\ker h_1 = \mathcal{M}$, then Proposition 6.3.1 implies there is a scalar α with $h_1 = \alpha h$. But $1 = h_1(1) = \alpha h(1) = \alpha$. ∎

10.7.3. Proposition. *If $\mathcal{A}$ is an abelian Banach algebra with identity and $h : \mathcal{A} \to \mathbb{C}$ is a non-zero homomorphism, then $\|h\| = 1$.*

Proof. Let $a \in \mathcal{A}$ and put $\lambda = h(a)$. Because $a - \lambda \in \ker h$, $a - \lambda$ cannot be invertible. Therefore $\lambda = h(a) \in \sigma(a)$. By Theorem 10.4.4 this implies $|h(a)| \leq \|a\|$; that is, $\|h\| \leq 1$. But $h(1) = 1$ if $h \neq 0$, so $\|h\| = 1$. ∎

In light of the preceding proposition we can consider the set Σ of all non-zero homomorphisms as a subset of ball $\mathcal{A}^*$.

10.7.4. Definition. If $\mathcal{A}$ is an abelian Banach algebra with identity, the *maximal ideal space* of $\mathcal{A}$ is the set Σ of non-zero homomorphisms of $\mathcal{A}$ into the complex numbers furnished with the relative weak* topology it has when considered as a subset of $\mathcal{A}^*$.

10.7.5. Theorem. *If $\mathcal{A}$ is an abelian Banach algebra with identity and Σ is its maximal ideal space, then Σ is compact and for any a in $\mathcal{A}$, $\sigma(a) = \{h(a) : h \in \Sigma\}$.*

Proof. In the preceding proposition we saw that $\sigma(a) \supseteq \{h(a) : h \in \Sigma\}$. On the other hand, if $\lambda \in \sigma(a)$, then $a - \lambda$ is not invertible so that $(a - \lambda)\mathcal{A}$ is a proper ideal in $\mathcal{A}$. Therefore there is a maximal ideal $\mathcal{M}$ that contains $(a - \lambda)\mathcal{A}$. By Proposition 10.7.2 there is an h in Σ with $\mathcal{M} = \ker h$; that is $\lambda = h(a)$. It remains to show that Σ is compact. But by Alaoglu's Theorem, to do this we need only show that Σ is a weak*-closed subset of ball $\mathcal{A}^*$. Let $\{h_i\}$ be a net in Σ and assume that $h_i \to L$ (wk*) in ball $\mathcal{A}^*$. Thus $L(ab) = \lim h_i(ab) = \lim h_i(a)h_i(b) = L(a)L(b)$. Since $L(1) = \lim h_i(1) = 1$, $L \neq 0$ and therefore $L \in \Sigma$. ∎

One example of an abelian Banach algebra with identity is $\mathcal{A} = C(X)$, where X is a compact space. If $x \in X$, then $\delta_x : C(X) \to \mathbb{C}$ defined by $\delta_x(f) = f(x)$ is a homomorphism.

10.7.6. Theorem. *If X is a compact space and Σ is the maximal ideal space of $C(X)$, then the map $x \mapsto \delta_x$ is a homeomorphism of X onto Σ.*

Proof. The hardest part of this proof is to show that if h is a non-zero homomorphism of $C(X)$ into $\mathbb{C}$, then there is a point x in X such that $h = \delta_x$. But the Riesz Representation Theorem implies there is a regular Borel measure μ on X with $h(f) = \int f d\mu$ for all f in $C(X)$. Let $x_0 \in \operatorname{spt} \mu$; we will show that $\operatorname{spt} \mu = \{x_0\}$. Once this is done it follows that there is a scalar α such that $\mu = \alpha \delta_{x_0}$. But then $\alpha = \int 1 d\mu = h(1) = 1$ and so $\mu = \delta_{x_0}$. Thus assume that $x \in X$ and $x \neq x_0$; we show that $x \notin \operatorname{spt} \mu$. Let U and V be two disjoint open sets with x_0 in U and x in V. Let $f_0, f \in C(X)$ such that $0 \leq f_0, f \leq 1$, $f_0(x_0) = 1$ and $f_0(y) = 0$ when $y \notin U$, $f(x) = 1$ and $f(y) = 0$ when $y \notin V$. Since $f_0 f = 0$, $0 = h(f_0 f) = h(f_0)h(f) = \left(\int f_0 d\mu\right)\left(\int f d\mu\right)$. Because $x_0 \in \operatorname{spt} \mu$, $\int f_0 d\mu \neq 0$; hence $\int f d\mu = 0$. But this implies that it must be that $x \notin \operatorname{spt} \mu$. Since x was arbitrary we have that $\operatorname{spt} \mu = \{x_0\}$ and so $\mu = \delta_{x_0}$.

What the preceding paragraph shows is that the map $\tau : X \to \Sigma$ defined by $\tau(x) = \delta_x$ is surjective. It is easily seen to be injective. (Verify!) Now assume that the net $\{x_i\}$ converges to x in X. Thus $f(x_i) \to f(x)$ for every f in $C(X)$. But this says that $\delta_{x_i} \to \delta_x$ in $\Sigma \subseteq C(X)^*$. Therefore τ is continuous. Since Σ is weak*-compact, τ is a closed map and therefore a homeomorphism. ∎

10.7.7. Definition. If $\mathcal{A}$ is an abelian Banach algebra with identity and $a \in \mathcal{A}$, define $\widehat{a} : \Sigma \to \mathbb{C}$ by

$$\widehat{a}(h) = h(a).$$

$\widehat{a}$ is called the *Gelfand transform* of a.

10.7.8. Theorem. *If $\mathcal{A}$ is an abelian Banach algebra with identity and maximal ideal space Σ, then for every a in $\mathcal{A}$, $\widehat{a} \in C(\Sigma)$ and $\|\widehat{a}\|_\infty \leq \|a\|$. The map $a \mapsto \widehat{a}$ is a continuous homomorphism of $\mathcal{A}$ into $C(\Sigma)$ of norm 1 and its kernel is*

$$\bigcap \{\mathcal{M} : \mathcal{M} \text{ is a maximal ideal of } \mathcal{A}\}$$

Moreover for each a in $\mathcal{A}$

$$\|\widehat{a}\|_\infty = \lim_{n \to \infty} \|a^n\|^{\frac{1}{n}}$$

Proof. If $a \in \mathcal{A}$, then the definition of the weak* topology on Σ implies $\widehat{a} \in C(\Sigma)$. Let $\gamma : \mathcal{A} \to C(\Sigma)$ be the map defined by $\gamma(a) = \widehat{a}$. Because each h in Σ is a homomorphism, it follows that γ is a homomorphism. Since $\Sigma \subseteq \operatorname{ball} \mathcal{A}^*$, it follows that $\|\gamma(a)\| \leq \|a\|$ for every a in $\mathcal{A}$; the fact that $\gamma(1) = 1$ implies $\|\gamma\| = 1$.

Note that $\|\widehat{a}\|_\infty$ is just the spectral radius of a, so that the formula for this norm is just a restatement of (10.4.6). It remains to determine $\ker \gamma$. But $a \in \ker \gamma$ if and only if $0 = \widehat{a}(h) = h(a)$ for each h in Σ. Since (10.7.2) says there is a bijective correspondence between maximal ideals in $\mathcal{A}$ and the kernels of homomorphisms, the designation of $\ker \gamma$ is automatic. ∎

The homomorphism $\gamma : \mathcal{A} \to C(\Sigma)$ defined by $\gamma(a) = \widehat{a}$ is also called the *Gelfand transform* of $\mathcal{A}$; $\ker \gamma$, as given in the theorem, is called the *radical* of $\mathcal{A}$. Note that if $\mathcal{A} = C(X)$ for a compact space X, the Gelfand transform $\gamma : C(X) \to C(\Sigma)$ is the identity map if Σ and X are identified.

A subset $\mathcal{S}$ of $\mathcal{A}$ is called a set of *generators* of $\mathcal{A}$ if $\mathcal{A}$ is the smallest Banach algebra with identity that contains $\mathcal{S}$. When $\mathcal{S}$ is a single element a, then we say that a is a generator of $\mathcal{A}$. It is straightforward that a is a generator of $\mathcal{A}$ if and only if $\{p(a) : p \text{ is a polynomial}\}$ is dense in $\mathcal{A}$. (Let's be clear that when we say a "polynomial" we mean a polynomial in the single (complex) variable z, not a polynomial $p(x, y)$ in two (real) variables.)

When X and Y are compact spaces and $\tau : Y \to X$ is a continuous map, then we define a contractive homomorphism $\tau^\# : C(X) \to C(Y)$ by $\tau^\#(f) = f \circ \tau$. Recall Example 9.1.7 and Exercise 9.1.3, where some of the properties of $\tau^{\#}$ were developed. In particular it was shown in the Banach–Stone Theorem (9.1.10) that when τ is a homeomorphism, $\tau^{\#}$ is a surjective isometry.

10.7.9. Proposition. *If A is an abelian Banach algebra with identity and A has a single generator a, then the map $\tau : \Sigma \to \sigma(a)$ defined by $\tau(h) = h(a) = \widehat{a}(h)$ is a homeomorphism; if $\gamma : A \to C(\Sigma)$ is the Gelfand transform and p is a polynomial, then $\gamma(p(a)) = \tau^{\#}(p)$.*

Proof. We already know that $\sigma(a) = \{h(a) : h \in \Sigma\}$, so τ does indeed map Σ onto $\sigma(a)$. If $\tau(h_1) = \tau(h_2)$, then $h_1(a) = h_2(a)$ and, by algebra, it follows that $h_1(p(a)) = h_2(p(a))$ for every polynomial $p(z)$. Since the polynomials in a are dense in A, $h_1 = h_2$; that is, τ is bijective. Also if $h_i \to h$ in Σ, then $\tau(h_i) = h_i(a) \to h(a) = \tau(h)$, so τ is continuous. Since Σ is compact, τ is a closed map and therefore a homeomorphism. ∎

10.7.10. Corollary. *If the Banach algebra A has two different generators a_1 and a_2, then $\sigma(a_1)$ and $\sigma(a_2)$ are homeomorphic subsets of $\mathbb{C}$.*

See Exercise 4 below.

Exercises.

(1) Let A be an abelian Banach algebra, but do not assume that A has an identity. (a) If $h : A \to \mathbb{C}$ is a homomorphism, show that h is continuous and $\|h\| \leq 1$. (Hint: Extend h to the algebra obtained by adjoining an identity to A (10.1.3).) (b) Let Σ denote the non-zero homomorphisms of A into $\mathbb{C}$. Show that the closure of Σ in the weak* topology of A^* is $\Sigma \cup \{0\}$. Conclude that Σ with the relative weak* topology is locally compact and its one-point compactification is homeomorphic to $\Sigma \cup \{0\}$. (c) Show that for every a in A, $\widehat{a} \in C_0(\Sigma)$. If $\gamma : A \to C_0(\Sigma)$ is defined by $\gamma(a) = \widehat{a}$, show that γ is a contractive homomorphism. (d) If X is a locally compact space and $A = C_0(X)$, show that $x \mapsto \delta_x$ is a homeomorphism of X onto Σ.

(2) If the finite set $\{a_1, \ldots, a_n\}$ generates the abelian Banach algebra with an identity A and Σ is the maximal ideal space of A, show that $\tau : \Sigma \to \mathbb{C}^n$ defined by $\tau(h) = (h(a_1), \ldots, h(a_n))$ is a homeomorphism of Σ onto a compact subset X of complex n-space.

(3) Let V be the Volterra operator on $L^2[0,1]$ (Example 5.1.10) and let A be the Banach subalgebra of $\mathcal{B}(L^2[0,1])$ generated by V and the identity operator. Determine the maximal ideal space of A and the radical of A.

(4) Show that $f(x) = \exp(\pi i x)$ is a generator of $C[0,1]$, but $g(x) = \exp(2\pi i x)$ is not.

(5) Show that $C(\partial \mathbb{D})$ does not have a single generator.

Chapter 11

C*-Algebras

C*-algebras are intimately connected with the algebra $\mathcal{B}(\mathcal{H})$ of all operators on a Hilbert space. $\mathcal{B}(\mathcal{H})$ is an example of one and, more significantly, an important theorem in the subject says that every C*-algebra is isomorphic to a subalgebra on $\mathcal{B}(\mathcal{H})$. (We will not see that result. A proof appears in [**8**], page 253.) In fact the treatment in this book only scratches the surface. A classic treatment is the book [**12**]. A complete modern treatment can be found in the four volume series by Kadison and Ringrose [**21**], [**22**], [**23**], [**24**].

11.1. Elementary properties and examples

If $\mathcal{A}$ is a Banach algebra, an *involution* on $\mathcal{A}$ is a map $a \mapsto a^*$ such that for a, b in $\mathcal{A}$ and α in $\mathbb{C}$: (i) $(a^*)^* = a$; (ii) $(ab)^* = b^*a^*$; (iii) $(\alpha a + b)^* = \overline{\alpha}a^* + b^*$. It is not hard to see that if $\mathcal{A}$ has an identity, then $1^* = 1$. In fact, $1^*a = (1^*a)^{**} = (a^*1)^* = (a^*)^* = a$, so that $1^* = 1$ by the uniqueness of the identity.

11.1.1. Definition. A C*-algebra is a Banach algebra $\mathcal{A}$ that has an involution such that $\|a^*a\| = \|a\|^2$ for every a in $\mathcal{A}$. This equation is often called the *C*-identity*.

11.1.2. Example. (a) If $\mathcal{H}$ is a Hilbert space, then $\mathcal{B}(\mathcal{H})$ is a C*-algebra. In fact Proposition 5.4.14 implies that for any T in $\mathcal{B}(\mathcal{H})$,

$$
\begin{aligned}
\|T^*T\| &= \sup\{|\langle T^*Th, h\rangle| : \|h\| = 1\} \\
&= \sup\{|\langle Th, Th\rangle| : \|h\| = 1\} \\
&= \sup\{\|Th\|^2 : \|h\| = 1\} = \|T\|^2
\end{aligned}
$$

(b) If $\mathcal{H}$ is a Hilbert space, then $\mathcal{B}_0(\mathcal{H})$ is a C*-algebra.

(c) If X is a compact space, then $C(X)$ is a C*-algebra where for f in $C(X)$, $f^*(x) = \overline{f(x)}$.

(d) If X is a locally compact space, then $C_0(X)$ is a C*-algebra.

(e) If $(X, \mathcal{A}, \mu)$ is a σ-finite measure space, then $L^\infty(\mu)$ is a C*-algebra when f^* is defined by $f^*(x) = \overline{f(x)}$.

11.1.3. Proposition. *If $\mathcal{A}$ is a C*-algebra, then $\|a^*\| = \|a\|$ for all a in $\mathcal{A}$.*

Proof. In fact, $\|a\|^2 = \|a^*a\| \leq \|a^*\|\|a\|$, so that $\|a\| \leq \|a^*\|$. Substituting a^* for a in this inequality gives that $\|a^*\| \leq \|(a^*)^*\| = \|a\|$. ∎

We now start the process of showing that if $\mathcal{A}$ is a C*-algebra without an identity, we can attach an identity and still have a C*-algebra. Of course this is reminiscent of Proposition 10.1.3, but we have the extra burden of showing that the norm we put on $\mathcal{A}_1 = \mathcal{A} + \mathbb{C}$ satisfies the C*-identity. The next result, which is clearly true when $\mathcal{A}$ has an identity, is the first step in that process.

11.1.4. Proposition. *If $\mathcal{A}$ is a C*-algebra and $a \in \mathcal{A}$, then*

$$\|a\| = \sup\{\|ax\| : x \in \text{ball}\,\mathcal{A}\}$$
$$= \sup\{\|xa\| : x \in \text{ball}\,\mathcal{A}\}$$

Proof. We can assume that $a \neq 0$. Put $\alpha = \sup\{\|ax\| : x \in \text{ball}\,\mathcal{A}\}$. Since $\|ax\| \leq \|a\|\|x\|$, it easily follows that $\alpha \leq \|a\|$. If we take $x = \|a\|^{-1}a^*$, then $x \in \text{ball}\,\mathcal{A}$ by the preceding proposition. Using this value of x in the definition of α we see that $\alpha = \|a\|$. The proof of the other equality is similar. ∎

We can give another interpretation of the last proposition. If $a \in \mathcal{A}$, define $L_a : \mathcal{A} \to \mathcal{A}$ by $L_a(x) = ax$; so L_a is a bounded operator on $\mathcal{A}$. This last proposition says $\|L_a\| = \|a\|$. The map $a \mapsto L_a$ is called the *left regular representation* of $\mathcal{A}$. It is easily checked that this is an isomorphism of algebras. (Similarly we can discuss the *right regular representation* of $\mathcal{A}$. However we will focus of the left hand version.) Since $\mathcal{A}$ is identified with $\{L_a : a \in \mathcal{A}\} \subseteq \mathcal{B}(\mathcal{A})$, we can look at the algebra $\mathcal{A}_1 = \{\alpha + L_a : a \in \mathcal{A}, \alpha \in \mathbb{C}\}$ with the norm it has as a subalgebra of $\mathcal{B}(\mathcal{A})$. This is how we adjoin the identity to $\mathcal{A}$, though we must define an involution on $\mathcal{A}_1$ and show that the norm satisfies the C*-identity. This is done in the next proposition.

If $\mathcal{A}$ and $\mathcal{C}$ are two C*-algebras, an algebra homomorphism $\nu : \mathcal{A} \to \mathcal{C}$ is called a *$*$-homomorphism* if it satisfies $\nu(a^*) = \nu(a)^*$ for all a in $\mathcal{A}$. Similarly we define a *$*$-isomorphism*. When $\mathcal{A}$ has an identity and ν is a

∗-homomorphism, it is not assumed that $\nu(1)$ is the identity of $\mathcal{C}$. When we come to proofs, however, this is easily circumvented as follows. Observe that $\mathcal{B} = \mathrm{cl}\, \nu(\mathcal{A})$ is a C*-algebra and that $\nu(1)$ must be an identity for $\mathcal{B}$. So in any argument we can assume that $\nu(1)$ is the identity of $\mathcal{C}$. When $\mathcal{A}$ does not have an identity, then we can extend ν to a ∗-homomorphism of the algebra obtained by adjoining the identity to $\mathcal{A}$, as we now see.

11.1.5. Proposition. *If $\mathcal{A}$ is a C*-algebra, then there is a C*-algebra $\mathcal{A}_1$ with an identity that contains $\mathcal{A}$ as an ideal. If $\mathcal{A}$ does not have an identity, then $\dim \mathcal{A}_1/\mathcal{A} = 1$. If $\mathcal{C}$ is a C*-algebra with identity and $\nu : \mathcal{A} \to \mathcal{C}$ is a ∗-homomorphism, then $\nu_1(a + \alpha) = \nu(a) + \alpha$ defines a ∗-homomorphism $\nu_1 : \mathcal{A}_1 \to \mathcal{C}$.*

Proof. We assume that $\mathcal{A}$ does not have an identity. Let $\mathcal{A}_1 = \{a + \alpha : a \in \mathcal{A} \text{ and } \alpha \in \mathbb{C}\}$. (This sum $a + \alpha$ can be considered as a formal sum or we could go through the process using the left regular representation of $\mathcal{A}$.) Define the algebraic operations in the obvious way and define an involution on $\mathcal{A}_1$ by $(a + \alpha)^* = a^* + \overline{\alpha}$; define a norm on $\mathcal{A}_1$ by

$$\|a + \alpha\| = \sup\{\|ax + \alpha x\| : x \in \mathrm{ball}\,\mathcal{A}\}$$

It is left to the reader to show that this norm makes $\mathcal{A}_1$ into a Banach algebra (see the proof of Proposition 10.1.3). We must show that this norm satisfies the C*-identity.

If $a + \alpha \in \mathcal{A}_1$ and $\epsilon > 0$, let $x \in \mathrm{ball}\,\mathcal{A}$ such that

$$\|a + \alpha\|^2 - \epsilon < \|ax + \alpha x\|^2 = \|(x^*a^* + \overline{\alpha}x^*)(ax + \alpha x)\|$$
$$= \|x^*(a+\alpha)^*(a+\alpha)x\| \le \|(a+\alpha)^*(a+\alpha)\|$$

Since ϵ was arbitrary we have that $\|a + \alpha\|^2 \le \|(a + \alpha)^*(a + \alpha)\|$. On the other hand, $\|(a+\alpha)^*(a+\alpha)\| \le \|(a+\alpha)^*\|\|a+\alpha\|$. So we need to prove that $\|(a + \alpha)^*\| \le \|a + \alpha\|$. But if $x, y \in \mathrm{ball}\,\mathcal{A}$, $\|y(a+\alpha)^*x\| = \|ya^*x + \overline{\alpha}yx\| = \|x^*ay^* + \alpha x^*y^*\| = \|x^*(a+\alpha)y^*\| \le \|a+\alpha\|$. Taking the supremum over all such x and y gives the desired inequality.

The proof of the statement about a ∗-homomorphism is routine. ∎

If $\mathcal{A}$ is a C*-algebra with identity and $a \in \mathcal{A}$, then $\sigma(a)$ is well defined; if $\mathcal{A}$ does not have an identity, then by the spectrum of a we mean $\sigma(a)$ when a is considered as an element of the algebra $\mathcal{A}_1$ obtained in the preceding proposition.

The next definition should have a familiar ring (5.8.9).

11.1.6. Definition. Let $\mathcal{A}$ be a C*-algebra and assume $a \in \mathcal{A}$. a is *hermitian* if $a^* = a$; a is *normal* if $a^*a = aa^*$. If $\mathcal{A}$ has an identity, then a is *unitary* if $a^*a = aa^* = 1$.

11.1.7. Proposition. *Let $\mathcal{A}$ be a C^*-algebra and let $a \in \mathcal{A}$.*

(a) *If a is invertible, then a^* is invertible and $(a^*)^{-1} = (a^{-1})^*$.*

(b) *$a = x + iy$, where x and y are hermitian elements in $\mathcal{A}$.*

(c) *If a is a unitary element, then $\|a\| = 1$.*

(d) *If $\mathcal{C}$ is a C^*-algebra and $\rho : \mathcal{A} \to \mathcal{C}$ is a $*$-homomorphism, then $\|\rho(a)\| \leq \|a\|$.*

(e) *If a is hermitian, then $\sigma(a) \subseteq \mathbb{R}$ and $\|a\| = r(a)$, its spectral radius.*

Proof. The proofs of (a), (b), and (c) are straightforward.

(e) Assume a is hermitian. So $\|a^2\| = \|a^*a\| = \|a\|^2$; by induction, $\|a^{2^n}\| = \|a\|^{2^n}$ for all $n \geq 1$. That is, $\|a^{2^n}\|^{1/2^n} = \|a\|$ for all $n \geq 1$. Since $r(a) = \lim_n \|a^n\|^{\frac{1}{n}}$, we have that $r(a) = \|a\|$.

Now let $\lambda \in \sigma(a)$, and write $\lambda = \alpha + it$ with $\alpha, t \in \mathbb{R}$. By replacing a with $a - \alpha$, we may assume that $it \in \sigma(a)$. If $s \in \mathbb{R}$, note that $i(s + t) = is + it \in \sigma(a + is)$ since $(a + is) - (is + it) = a - it$ is not invertible. Hence $|i(s+t)|^2 \leq r(a+is)^2 \leq \|a+is\|^2 = \|(a+is)^*(a+is)\| = \|(a-is)(a+is)\| = \|a^2 + s^2\| \leq \|a\|^2 + s^2$. Since $|i(s + t)|^2 = s^2 + 2st + t^2$, when we cancel s^2 we get

$$2st + t^2 \leq \|a\|^2$$

for all s in $\mathbb{R}$. The only way this can be true is if $t = 0$. Thus $\sigma(a) \subseteq \mathbb{R}$.

(d) We will assume that both $\mathcal{A}$ and $\mathcal{C}$ have an identity and that $\rho(1) = 1$. For any x in $\mathcal{A}$ it follows that $\sigma(\rho(x)) \subseteq \sigma(x)$. (Why?) Thus $r(\rho(x)) \leq r(x)$. Using part (e) this shows that $\|\rho(a)\|^2 = \|\rho(x^*x)\| = r(\rho(x^*x)) \leq r(x^*x) = \|x^*x\| = \|x\|^2$. $\blacksquare$

The reader might compare the next result with Proposition 10.4.8, which will be used in the proof.

11.1.8. Proposition. *If $\mathcal{A}$ and $\mathcal{B}$ are C^*-algebras with a common identity, $\mathcal{A} \subseteq \mathcal{B}$, and $a \in \mathcal{A}$, then $\sigma_{\mathcal{A}}(a) = \sigma_{\mathcal{B}}(a)$.*

Proof. By Proposition 10.4.8 we know that $\sigma_{\mathcal{B}}(a) \subseteq \sigma_{\mathcal{A}}(a)$ and $\partial\sigma_{\mathcal{A}}(a) \subseteq \partial\sigma_{\mathcal{B}}(a)$. First assume that $a = a^*$. By Proposition 11.1.7(e), $\sigma_{\mathcal{A}}(a), \sigma_{\mathcal{B}}(a) \subseteq \mathbb{R}$. Therefore they are equal to their boundaries in $\mathbb{C}$, and so we have that $\sigma_{\mathcal{A}}(a) = \sigma_{\mathcal{B}}(a)$. Now let a be arbitrary. It suffices to show that if a is invertible in $\mathcal{B}$, then it is invertible in $\mathcal{A}$. But if $b \in \mathcal{B}$ such that $ba = ab = 1$, then $(a^*a)(bb^*) = (bb^*)(a^*a) = 1$; so a^*a is invertible in $\mathcal{B}$. Since a^*a is hermitian, the first part of this proof says that it is invertible in $\mathcal{A}$. But since the inverse is unique, we have that $bb^* \in \mathcal{A}$. Therefore $b = b(b^*a^*) = (bb^*)a^* \in \mathcal{A}$. $\blacksquare$

Now we turn our attention to the basic attributes of homomorphisms of a C*-algebra into the complex numbers.

11.1.9. Proposition. *If $\mathcal{A}$ is a C*-algebra and $h : \mathcal{A} \to \mathbb{C}$ is a homomorphism, then the following hold.*

(a) *If $a = a^*$, then $h(a) \in \mathbb{R}$.*

(b) *$h(a^*) = \overline{h(a)}$.*

(c) *If $a = b^*b$ for some b in $\mathcal{A}$, then $h(a) \geq 0$.*

(d) *If $\mathcal{A}$ has an identity and u is a unitary in $\mathcal{A}$, then $|h(u)| = 1$.*

Proof. We can assume that $\mathcal{A}$ has an identity; otherwise attach an identity and extend h.

If $\lambda = h(a)$, then $a - \lambda \in \ker h$; so $a - \lambda$ is not invertible and $\lambda \in \sigma(a)$. Since a is hermitian, Proposition 11.1.7(e) implies $\lambda \in \mathbb{R}$, proving (a). For (b) write $a = x + iy$ with x and y hermitian. So $a^* = x - iy$ and $h(a^*) = h(x) - ih(y) = \overline{h(a)}$ since $h(x), h(y) \in \mathbb{R}$. It also follows that $h(b^*b) = h(b^*)h(b) = |h(b)|^2$, establishing (c). Part (d) is immediate from the definition of a unitary. ∎

Exercises.

(1) Check the details in Example 11.1.2.

(2) Let X be a locally compact space and for a C*-algebra $\mathcal{A}$ let $C_b(X, \mathcal{A})$ denote the set of all bounded continuous functions $f : X \to \mathcal{A}$; define $\|f\| = \sup\{\|f(x)\| : x \in X\}$ and $f^*(x) = f(x)^*$ for x in X. (a) Show that $C_b(X, \mathcal{A})$ is a C*-algebra. If $C_0(X, \mathcal{A})$ is the set of all f in $C_b(X, \mathcal{A})$ such that for every $\epsilon > 0$, $\{x \in X : \|f(x)\| \geq \epsilon\}$ is compact, show that $C_0(x, \mathcal{A})$ is a C*-algebra.

11.2. Abelian C*-algebras

The main result of this section is the first; it says that there is only one type of abelian C*-algebra, $C(X)$. Most of the work needed to establish this theorem has already been done.

11.2.1. Theorem. *If $\mathcal{A}$ is an abelian C*-algebra with identity, Σ is its maximal ideal space, and $\gamma : \mathcal{A} \to C(\Sigma)$ is the Gelfand transform, then γ is an isometric *-isomorphism of $\mathcal{A}$ onto $C(\Sigma)$.*

Proof. Recall from §10.7 the definition of the Gelfand transform $\gamma : \mathcal{A} \to C(\Sigma)$: $\gamma(x) = \widehat{x}$. We use $\|f\|_\infty$ to denote the norm of a function f in $C(\Sigma)$. Note that for x in $\mathcal{A}$, $\|\widehat{x}\|_\infty$ is the spectral radius of x; so Theorem 10.4.4 implies that $\|\widehat{x}\|_\infty \leq \|x\|$. On the other hand, when x is hermitian,

$\|\widehat{x}\|_\infty = \|x\|$ (11.1.7(e)). In particular, $\|x\|^2 = \|x^*x\| = \|\widehat{x^*x}\|_\infty = \|\widehat{x}\|_\infty^2$ for every x in $\mathcal{A}$, so γ is an isometry.

When $x \in \mathcal{A}$ and $h \in \Sigma$, $\widehat{x^*}(h) = h(x^*) = \overline{h(x)} = \overline{\widehat{x}(h)}$. Thus $\widehat{x^*} = \overline{\widehat{x}}$, or $\gamma(x^*) = \gamma(x)^*$ since the involution on $C(\Sigma)$ is defined by complex conjugation. This shows that γ is a $*$-homomorphism.

It remains to show that γ is surjective. Since γ is an isometry it has closed range, so it suffices to show that $\gamma(\mathcal{A})$ is dense in $C(\Sigma)$. To do this we will use the Stone–Weierstrass Theorem. We know that $\gamma(\mathcal{A})$ is a subalgebra of $C(\Sigma)$ that is closed under complex conjugation. Moreover $1 = \gamma(1)$, so it contains the constant functions. To show that $\gamma(\mathcal{A})$ separates the points and complete the proof, let h_1 and h_2 be distinct homomorphisms on $\mathcal{A}$. But distinctness here means there is an x in $\mathcal{A}$ with $h_1(x) \neq h_2(x)$; equivalently $\widehat{x}(h_1) \neq \widehat{x}(h_2)$, so that $\gamma(\mathcal{A})$ does indeed separate the points of Σ. ∎

Note that $L^\infty(\mu)$ is an abelian C*-algebra, so there is a compact space X such that it is $*$-isomorphic to $C(X)$. The spaces X that arise in this way are quite bizarre. As a start look at Exercise 8.

11.2.2. Corollary. *If $\mathcal{A}$ is a C*-algebra without an identity and Σ is its maximal ideal space, then the Gelfand transform $\gamma : \mathcal{A} \to C_0(\Sigma)$ is an isometric $*$-isomorphism of $\mathcal{A}$ onto $C_0(\Sigma)$.*

This follows from the theorem by adjoining an identity to $\mathcal{A}$ and observing that $\mathcal{A}_1$ is abelian. See Exercise 1.

To simplify matters, let's agree on the following.

Assumption. *All C*-algebras have an identity.*

The interested reader can try to extend results obtained below to algebras without an identity by adjoining one, using the results below, and carrying out the needed details to establish a result for the original algebra.

If $\mathcal{A}$ is a C*-algebra and $\mathcal{S} \subseteq \mathcal{A}$, let $C^*(\mathcal{S})$ be the C*-algebra generated by $\mathcal{S}$ and the identity. So $C^*(\mathcal{S})$ is the intersection of all the C*-algebras contained in $\mathcal{A}$ that contain $\mathcal{S}$ and the identity. When $\mathcal{S}$ consists of a single element a, we use $C^*(a)$ to denote the C*-algebra generated by $\{a\}$. Let $p(z, \overline{z})$ be a polynomial in the complex variable z and its conjugate $\overline{z}$. That is, $p(z, \overline{z})$ is the linear combination of terms of the form $z^n \overline{z}^m$, for $n, m \geq 0$. When $p(z, \overline{z})$ is such a polynomial and $a \in \mathcal{A}$, consider $p(a, a^*)$; so if $p(z, \overline{z}) = z^n \overline{z}^m$, $p(a, a^*) = a^n(a^*)^m$. It is easy to see that when a is normal, $C^*(a)$ is the closure of $\{p(a, a^*) : p(z, \overline{z})$ is a polynomial in z and $\overline{z}\}$ and is abelian. When a is not normal we must consider all polynomials in the non-commuting elements a and a^*.

The proof of the next result resembles that of Proposition 10.7.9 and will not be given.

11.2.3. Proposition. *If $\mathcal{A}$ is an abelian C^*-algebra with a single normal generator a, then the map $\tau : \Sigma \to \sigma(a)$ defined by $\tau(h) = h(a)$ is a homeomorphism; if $\gamma : \mathcal{A} \to C(\Sigma)$ is the Gelfand transform and $p(z, \overline{z})$ is a polynomial in z and $\overline{z}$, then $\gamma(p(a, a^*)) = \tau^\#(p)$.*

11.2.4. Theorem. *If a is a normal element in the C^*-algebra $\mathcal{A}$, then there is an isometric $*$-monomorphism $\rho : C(\sigma(a)) \to \mathcal{A}$ satisfying the following.*

(a) *$\rho[C(\sigma(a))] = C^*(a)$.*

(b) *$\rho(1) = 1$, $\rho(z) = a$, and $\rho(\overline{z}) = a^*$.*

Proof. Let $\gamma : C^*(a) \to C(\Sigma)$ be the Gelfand transform; so γ is an isometric $*$-isomorphism. Let $\tau : \Sigma \to \sigma(a)$ be the homeomorphism from the preceding proposition, and consider the induced map $\tau^\# : C(\sigma(a)) \to C(\Sigma)$ given by $\tau^\#(f) = f \circ \tau$. Because τ is a homeomorphism, $\tau^\#$ is an isometric $*$-isomorphism. Hence we can define $\rho : C(\sigma(a)) \to \mathcal{A}$ by

$$\rho(f) = \gamma^{-1}\left(\tau^\#(f)\right)$$

for every f in $C(\sigma(a))$. It readily follows that ρ is a $*$-monomorphism and $\rho(1) = 1$. Again using the properties of γ and $\tau^\#$ we have that $\rho(z) = a$ and $\rho(\overline{z}) = a^*$. ∎

When we have the situation of the preceding theorem, we let $f(a) = \rho(f)$ for every f in $C(\sigma(a))$. This mapping $f \mapsto f(a)$ is called the *functional calculus* for the normal element a. We get the following as a consequence of Theorem 11.2.4. The proof is transparent.

11.2.5. Corollary. (a) *If f is a real-valued function in $C(\sigma(a))$, then $f(a)$ is hermitian.*

(b) *If $|f(\lambda)| = 1$ for all λ in $\sigma(a)$, then $f(a)$ is a unitary.*

The next result is important even though the proof is easy. The reader should not be surprised at the ease of the proof since we have done a lot of work. In addition the proof is accomplished by hitting it with a big weapon, Theorem 11.2.4.

11.2.6. Theorem (Spectral Mapping Theorem). *If $\mathcal{A}$ is a C^*-algebra and a is a normal element of $\mathcal{A}$, then for every f in $C(\sigma(a))$*

$$\sigma(f(a)) = f(\sigma(a))$$

Proof. First realize that the definition of the spectrum is independent of the containing C*-algebra (11.1.8). Next if $\rho : C(\sigma(a)) \to C^*(a)$ is the isomorphism defined by the functional calculus, the fact that this is an isomorphism implies the equality of the spectrum. ∎

Exercises.

(1) Provide the details of the proof of Corollary 11.2.2.

(2) Prove the following converse to Proposition 11.2.3. If K is a compact subset of $\mathbb{C}$, then $C(K)$ has a single generator as a C*-algebra. (Contrast this with Exercise 10.7.5.)

(3) Formulate and prove a uniqueness statement for Theorem 11.2.4.

(4) For the functional calculus for a normal element of the C*-algebra $\mathcal{A}$, show that if $f(z) = \sum_{n=0}^{\infty} \alpha_n (z-\lambda)^n$ is a power series with radius of convergence R and $\sigma(a) \subseteq B(\lambda; R)$, then $f(a) = \sum_{n=0}^{\infty} \alpha_n (a - \lambda)^n$, where this series converges in the norm of $\mathcal{A}$.

(5) If $\mathcal{A}$ is a C*-algebra and a is a hermitian element of $\mathcal{A}$, show that e^{ia} is a unitary. (We are using the functional calculus to define e^{ia}.)

(6) Show that if $\mathcal{A}$ is a C*-algebra with generators $\{a_1, \ldots, a_n\}$, then there is a compact subset X of $\mathbb{C}^n$ and an isometric $*$-isomorphism $\rho : \mathcal{A} \to C(X)$ such that for $1 \le k \le n$, $\rho(z_k) = a_k$, where $z_k(\lambda_1, \ldots, \lambda_n) = \lambda_k$.

(7) If X is a compact space, show that X is totally disconnected if and only if $C(X)$ is the closed linear span of its projections. (A *projection* in a C*-algebra is a hermitian idempotent.)

(8) Show that if $(X, \mathcal{A}, \mu)$ is a σ-finite measure space, then the maximal ideal space of $L^\infty(\mu)$ is totally disconnected. (Use the preceding exercise.)

(9) This exercise uses Theorem 11.2.1 to prove the Spectral Theorem for compact hermitian operators on a Hilbert space from §5.7. Here $\mathcal{H}$ is a Hilbert space over $\mathbb{C}$ and T is a compact self-adjoint operator on $\mathcal{H}$. (a) Show that $C^*(T)$, the C*-algebra algebra generated by T and the identity, is abelian. (b) If Σ is the maximal ideal space of $C^*(T)$, prove that Σ is homeomorphic to $\sigma(T)$. Let $\sigma(T) = \{0, \lambda_1, \lambda_2, \ldots\}$. (c) Show that there is an isometric $*$-isomorphism $\rho : C^*(T) \to C(\sigma(T))$ such that $\rho(T) = x$, where $x : \sigma(T) \to \mathbb{R}$ is the identity function $x(t) = t$ for all t in $\sigma(T)$. (d) If χ_n is the characteristic function of $\{\lambda_n\}$, $n \ge 1$, show that $\chi_n \in C(\sigma(T))$ and $P_n = \rho^{-1}(\chi_n)$ is the orthogonal projection of $\mathcal{H}$ onto $\ker(T - \lambda_n)$. Show that P_n has finite rank. (e) Show that the series $\sum_{n=1}^{\infty} \lambda_n \chi_n$ converges uniformly on $\sigma(T)$ to the identity function x. (f) Now prove the Spectral Theorem as stated in §5.7.

(10) Let X be a compact space and fix a point x_0 in X. If $\mathcal{A}$ denotes the collection of all sequences $\{f_n\}$ of functions in $C(X)$ such that $\sup_n \|f_n\| < \infty$ and $\{f_n(x_0)\}$ is convergent, then $\mathcal{A}$ is an abelian C*-algebra. Determine its maximal ideal space.

11.3. Positive elements in a C*-algebra

There is an underlying order structure in a C*-algebra that plays a central role.

11.3.1. Definition. If $\mathcal{A}$ is a C*-algebra, then an element a in $\mathcal{A}$ is *positive* if a is hermitian and $\sigma(a) \subseteq [0, \infty)$. This is denoted by writing $a \geq 0$. The set of all positive elements in $\mathcal{A}$ is denoted by $\mathcal{A}_+$.

Yes, this differs from the definition of a positive operator given in (5.8.9); but we will shortly show that this definition, when applied to $\mathcal{B}(\mathcal{H})$, is equivalent to that one. Many examples of positive elements exist. For example a function f in $C(X)$ is a positive element if and only if it is a positive function; that is, $f(x) \geq 0$ for all x in X. We'll see more examples after we develop some of the theory. But note that if a is a normal element in a C*-algebra $\mathcal{A}$ and f is a positive function in $C(\sigma(a))$, then the Spectral Mapping Theorem (11.2.6) implies that $f(a)$ is a positive element of $\mathcal{A}$.

The next result is the first of two propositions that list equivalent formulations of positivity.

11.3.2. Proposition. *If $\mathcal{A}$ is a C*-algebra and $a \in \mathcal{A}$, then the following statements are equivalent.*

(a) $a \geq 0$.

(b) $a = a^*$ *and* $\|t - a\| \leq t$ *for all* $t \geq \|a\|$.

(c) $a = a^*$ *and* $\|t - a\| \leq t$ *for some* $t \geq \|a\|$.

Proof. (a) *implies* (b). By (a), $\sigma(a) \subseteq [0, \infty)$. Note that if $t \geq \|a\|$, then $t - s \leq t$ for all $s \geq 0$; so this holds for all s in $\sigma(a)$. If x denotes the function $x(s) = s$, then this says that in the C*-algebra $C(\sigma(a))$, $\|t - x\| \leq t$. Since the functional calculus of a is an isometric isomorphism, we have (b).

(c) *implies* (a). Since a is hermitian, $C^*(a)$ is abelian and we have the functional calculus. If $\sigma = \sigma(a) \subseteq \mathbb{R}$ and $t \geq \|a\| = \max\{|s| : s \in \sigma\}$, then (c) says that t exists with $|t - s| \leq t$. But if there is an s in σ that is negative, this is impossible. Hence we have that $\sigma \subseteq [0, \infty)$ and we have established (a). ∎

Below (11.3.6) we will see additional conditions equivalent to an element of a C*-algebra being positive, but first we need to lay some groundwork.

11.3.3. Proposition. *Let $\mathcal{A}$ be a C*-algebra.*

(a) $\mathcal{A}_+$ *is a closed cone. That is, it is a closed subset of $\mathcal{A}$ that satisfies:* $a + b \in \mathcal{A}_+$ *whenever* $a, b \in \mathcal{A}_+$ *and* $\alpha a \in \mathcal{A}_+$ *whenever* $a \in \mathcal{A}_+$ *and* $\alpha \in [0, \infty)$.

(b) *If* $a \in \mathcal{A}_+$ *and* $-a \in \mathcal{A}_+$, *then* $a = 0$.

Proof. Let $\{a_n\}$ be a sequence in $\mathcal{A}_+$ that converges to a. Clearly a is hermitian. For each $n \geq 1$ the preceding proposition implies $\|\|a_n\| - a_n\| \leq \|a_n\|$. Taking limits and using the same proposition we see that $a \in \mathcal{A}_+$. If $a \in \mathcal{A}_+$ and $\alpha \geq 0$, then αa is hermitian and $\sigma(\alpha a) = \alpha \sigma(a) \subseteq [0, \infty)$. If $a, b \in \mathcal{A}_+$, then

$$\|(\|a\| + \|b\|) - (a + b)\| \leq \|\|a\| - a\| + \|\|b\| - b\| \leq \|a\| + \|b\|$$

so that Proposition 11.3.2 shows that $a + b \in \mathcal{A}_+$.   ∎

The reader may have noticed from the preceding proofs that the functional calculus allows us to show that if something holds for continuous functions, it holds for normal elements of a C*-algebra. The next result illustrates this again.

11.3.4. Proposition. *If* a *is a hermitian element of the C*-algebra* $\mathcal{A}$, *then there are unique positive elements* u, v *in* $\mathcal{A}$ *such that* $a = u - v$ *and* $uv = vu = 0$.

Proof. Let σ denote the spectrum of a. So $\sigma \subseteq \mathbb{R}$ and we have the functional calculus for a. If $f(t) = \max\{t, 0\}$ and $g(t) = \max\{-t, 0\}$, then the Spectral Mapping Theorem implies that $u = f(a)$ and $v = g(a)$ are positive. Since $f(t) - g(t) = t$ for all t in σ, we have that $u - v = a$. Similarly $fg = 0$ on σ, so $uv = vu = 0$.

It remains to show uniqueness. Suppose u_1, v_1 are positive elements in $\mathcal{A}$ such that $a = u_1 - v_1$ and $u_1 v_1 = v_1 u_1 = 0$. It follows that $a^n = u_1^n + (-v_1)^n$ for every $n \geq 0$ and so $p(a) = p(u_1) + p(-v_1)$ for every polynomial. Let $\{p_n\}$ be a sequence of polynomials such that $p_n(t) \to f(t)$ uniformly for t in $\sigma(a) \cup \sigma(u_1) \cup \sigma(v_1)$, where $f(t) = \max\{t, 0\}$ as in the preceding paragraph. It follows that $p_n(a) \to u$, so that $u = f(u_1) + f(-v_1)$. Now $f(t) = t$ on $\sigma(u_1)$ and $f(t) = 0$ for t in $\sigma(-v_1)$; thus $f(u_1) = u_1$ and $f(-v_1) = 0$. Hence $u = u_1$. It is now easy algebra to show that $v = v_1$.   ∎

Recall that the above result was proved for the special case where $\mathcal{A} = \mathcal{B}_0(\mathcal{H})$ in Theorem 5.8.11, but with a different concept of positivity that will shortly be seen to be equivalent. The next result also appeared in this same context for $n = 2$ in Theorem 5.8.12.

11.3.5. Proposition. *If* a *is a positive element of the C*-algebra* $\mathcal{A}$ *and* $n \geq 1$, *then there is a unique positive element* b *in* $\mathcal{A}$ *such that* $a = b^n$.

Proof. Assume $a \geq 0$; so we have a functional calculus $f \mapsto f(a)$, where the function x is mapped to a. Since $\sigma(a) \subseteq [0, \infty)$, $x^{1/n} \in C(\sigma(a))$. If b

is the image of $x^{1/n}$ under the functional calculus, then $b = b^*$ and $b^n = a$. The proof of uniqueness is like that of Proposition 11.3.4. ∎

We are now in a position to give some additional formulations of positivity.

11.3.6. Proposition. *If a is an element of $\mathcal{A}$, then the following statements are equivalent.*

(a) *a is positive.*

(b) *There is an element b in $\mathcal{A}$ such that $a = b^*b$.*

(c) *There is a hermitian element b of $\mathcal{A}$ such that $a = b^2$.*

Proof. Proposition 11.3.5 shows that (a) implies (c). Clearly (c) implies (b). Now assume that $a = b^*b$ as in (b); so a is hermitian and we have the functional calculus. By Proposition 11.3.4 we can write $a = u - v$ with $u, v \geq 0$ and $uv = vu = 0$. We must show that $v = 0$.

If $bv = x + iy$, where x and y are hermitian, then $(bv)^*(bv) = (x - iy)(x + iy) = x^2 + y^2 + i(xy - yx)$. Looked at another way, $(bv)^*(bv) = vb^*bv = v(u - v)v = -v^3$. The Spectral Mapping Theorem implies $v^3 \geq 0$, so we have that $(bv)^*(bv) \leq 0$. Now Exercise 10.4.4 implies $\sigma((bv)^*(bv)) \cup \{0\} = \sigma((bv)(bv)^*) \cup \{0\}$; hence $(bv)(bv)^* \leq 0$. Again the alternative calculation yields $(bv)(bv)^* = x^2 + y^2 + i(yx - xy)$. Therefore

$$0 \geq (bv)^*(bv) + (bv)(bv)^* = 2(x^2 + y^2)$$

But the right hand side of this expression is in $\mathcal{A}_+$ by Proposition 11.3.3(a). By part (b) of that same proposition, we have that $x^2 + y^2 = 0$. From the above expression for $(bv)^*(bv)$, this implies $v^3 = 0$ and so $v = 0$. ∎

We can now show that the present definition of a positive element in a C^*-algebra is equivalent to the definition given for operators on a Hilbert space (5.8.9).

11.3.7. Proposition. *If $\mathcal{H}$ is a Hilbert space and $T \in \mathcal{B}(\mathcal{H})$, then $T \geq 0$ if and only if $\langle Th, h \rangle \geq 0$ for every h in $\mathcal{H}$.*

Proof. If $T \geq 0$, then, by the preceding proposition, there is an A in $\mathcal{B}(\mathcal{H})$ with $T = A^*A$. Therefore $\langle Th, h \rangle = \langle Ah, Ah \rangle \geq 0$. Now assume $\langle Th, h \rangle \geq 0$ for every h in $\mathcal{H}$. By Proposition 5.4.13, $T = T^*$; it remains to show that $\sigma(T) \subseteq [0, \infty)$. If $h \in \mathcal{H}$ and $\lambda < 0$, then

$$\|(T - \lambda)h\|^2 = \|Th\|^2 - 2\lambda\langle Th, h \rangle + \lambda^2\|h\|^2$$
$$\geq -2\lambda\langle Th, h \rangle + \lambda^2\|h\|^2$$
$$\geq \lambda^2\|h\|^2$$

By Proposition 10.5.3, $\lambda \notin \sigma_{ap}(T)$. From Example 10.4.3 we know that this means $T - \lambda$ is left invertible. But because $T - \lambda$ is hermitian, this means that $T - \lambda$ is invertible. (Why?) Thus $\lambda \notin \sigma(T)$ and so $T \geq 0$. ∎

We continue to concentrate on $\mathcal{B}(\mathcal{H})$ and will derive the polar decomposition of a bounded operator. This is the analogue of the factorization of a complex number λ as $\lambda = |\lambda| e^{i\theta}$. The correct analogue of the absolute value is easy enough. When $T \in \mathcal{B}(\mathcal{H})$, define

11.3.8 $$|T| = (T^*T)^{\frac{1}{2}}$$

We note the following agreeable property of the absolute value of an operator. If $h \in \mathcal{H}$, then $\|Th\|^2 = \langle Th, Th \rangle = \langle T^*Th, h \rangle = \langle |T|^2 h, h \rangle = \||T|h\|^2$; that is

$$\|Th\| = \||T|h\|$$

for all vectors h.

Finding the analog of $e^{i\theta}$ is a bit more involved. You might be tempted to say unitary or isometry, but this is inadequate (see Exercise 7).

11.3.9. Definition. A *partial isometry* on a Hilbert space $\mathcal{H}$ is an operator W in $\mathcal{B}(\mathcal{H})$ such that for h in $(\ker W)^\perp$, $\|Wh\| = \|h\|$. The space $(\ker W)^\perp$ is called the *initial space* for W and $\operatorname{ran} W$ is its *final space*.

So a partial isometry is an isometry from its initial space onto its final space. It is easy to manufacture examples of partial isometries. Just take two subspaces $\mathcal{M}$ and $\mathcal{N}$ of $\mathcal{H}$ having the same dimension, and define an isometry $V : \mathcal{M} \to \mathcal{N}$. Now define $W : \mathcal{H} \to \mathcal{H}$ by letting $Wh = Vh$ for h in $\mathcal{M}$ and $Wh = 0$ when $h \in \mathcal{M}^\perp$; the initial space of W is $\mathcal{M}$ and the final space is $\mathcal{N}$. The reader can see many properties of partial isometries in Exercises 8 through 11.

11.3.10. Theorem (Polar Decomposition). *If $T \in \mathcal{B}(\mathcal{H})$ then there is a partial isometry W with initial space $(\ker T)^\perp$ and final space $\operatorname{cl}(\operatorname{ran} T)$ such that $T = W|T|$. This decomposition is unique in the following sense: if $T = UP$ for a positive operator P and a partial isometry U with $\ker U = \ker P$, then $P = |T|$ and $U = W$.*

Proof. The goal is to define $W : \operatorname{ran} |T| \to \operatorname{ran} T$ such that $W(|T|h) = Th$ and then check that this is a partial isometry and has the advertised initial and final space. As noted after that definition of $|T|$, $\|Th\| = \||T|h\|$; so the proposed formula for W gives a well defined isometry from $\operatorname{ran} |T|$ onto $\operatorname{ran} T$. So we can extend W by continuity to $\operatorname{cl}[\operatorname{ran} |T|]$, and this gives an isometry from that space onto $\operatorname{cl}[\operatorname{ran} T]$; now define W to be 0 on $(\operatorname{ran} |T|)^\perp$ and we have a partial isometry satisfying $W|T| = T$. Also the final space of W is indeed $\operatorname{cl}(\operatorname{ran} T)$; we need to check the initial space.

We know that $\operatorname{ran} T^*$ is dense in $(\ker T)^\perp$ since $(\operatorname{ran} T^*)^\perp = \ker T$. Now if $h \in \operatorname{ran} T^*$, there is a g in $(\ker T^*)^\perp = \operatorname{cl}(\operatorname{ran} T)$ such that $T^* h = g$. Hence $\operatorname{ran}(T^*T)$ is dense in $\operatorname{cl}(\operatorname{ran} T^*) = (\ker T)^\perp$. But if $f \in \mathcal{H}$, $T^*Tf = |T|^2 f = |T|h$, where $h = |T|f$. That is, $\operatorname{ran}|T|$ is dense in $(\ker T)^\perp$ and so W has the desired initial space.

To establish uniqueness note that if $T = UP$, then $T^*T = PU^*UP$. If E is the projection onto the initial space of U, then $E = U^*U$ (Exercise 9). That is, E is the orthogonal projection onto $(\ker U)^\perp = (\ker P)^\perp = \operatorname{cl}(\operatorname{ran} P)$. Thus $T^*T = P^2$. By the uniqueness of the positive square root (11.3.5), $P = |T|$. For any vector h in $\mathcal{H}$, $U|T|h = UPh = Th = W|T|h$. That is, W and U agree on $\operatorname{ran}|T|$, a dense subset of their common initial space. Therefore $U = W$. ∎

Exercises.

(1) Complete the proof of Proposition 11.3.5.

(2) Let $(X, \mathcal{A}, \mu)$ be a σ-finite measure space. (a) If $\phi \in L^\infty(\mu)$, show that $M_\phi \geq 0$ if and only if $\phi(x) \geq 0$ a.e. $[\mu]$. (b) If $M_\phi \geq 0$, find $M_\phi^{1/n}$. (c) If M_ϕ is hermitian, find its positive and negative parts. (d) What is the polar decomposition of M_ϕ?

(3) Find an example of a positive operator on a Hilbert space that has a non-hermitian square root.

(4) If a is a hermitian element of the C*-algebra $\mathcal{A}$ and $a = u - v$ as in Proposition 11.3.4, show that $|a| = (a^*a)^{1/2} = u + v$.

(5) If $a, b \in \mathcal{A}_+$, $a \leq b$, and a is invertible, show that b is invertible and $0 \leq b^{-1} \leq a^{-1}$.

(6) Give an example of a C*-algebra $\mathcal{A}$ and positive elements a and b such that $a \leq b$, but $b^2 - a^2 \notin \mathcal{A}_+$.

(7) If $\mathcal{H}$ is an infinite-dimensional Hilbert space, let $\mathcal{M} \leq \mathcal{H}$ such that $\dim \mathcal{M}^\perp = 1$ and define T in $\mathcal{B}(\mathcal{H})$ as follows: on $\mathcal{M}$, T is an isometry with $\operatorname{ran} T = \mathcal{H}$ and $Th = 0$ for all h in $\mathcal{M}^\perp$. Show that there is no isometry W with $T = W|T|$.

(8) If $\mathcal{H}$ is a Hilbert space and $W \in \mathcal{B}(\mathcal{H})$, show that the following statements are equivalent. (a) W is a partial isometry. (b) W^* is a partial isometry. (c) W^*W is a projection. (d) WW^* is a projection. (e) $WW^*W = W$. (f) $W^*WW^* = W^*$.

(9) If W is a partial isometry, show that W^*W is the projection onto the initial space and WW^* is the projection onto the final space.

(10) If W_1 and W_2 are partial isometries, define $W_1 \preceq W_2$ to mean that $W_1^*W_1 \leq W_2^*W_2$, $W_1W_1^* \leq W_2W_2^*$, and $W_2h = W_1h$ when

h belongs to the initial space of W_1. (a) Show that this relation
defines a partial ordering on the set of all partial isometries. (b)
Show that W is a maximal partial isometry in this ordering if and
only if either W or W^* is an isometry.

(11) Using the terminology of the preceding exercise, show that the set of
extreme points of ball $\mathcal{B}(\mathcal{H})$ is the set of maximal partial isometries.

(12) Show that if $T \in \mathcal{B}(\mathcal{H})$, then there is a maximal partial isometry
W such that $T = W|T|$. Give a necessary and sufficient condition
that $T = W|T|$ for an isometry W. (Also see Exercise 7.)

(13) Find the polar decomposition of the following operators. (a) The
unilateral shift on ℓ^2. (b) M_ϕ on $L^2(\mu)$ for ϕ in $L^\infty(\mu)$. (c) The
unilateral weighted shift (Example 5.4.7).

(14) If T is a normal operator, show that the factors in its polar decomposition commute. Is the converse true?

(15) If $T \in \mathcal{B}(\mathcal{H})$ show that there is a positive operator P and a partial
isometry W such that $T = PW$. Discuss the uniqueness of P
and Q.

11.4. A functional calculus for normal operators

Now we depart from the abstract theory of C*-algebras and focus on $\mathcal{B}(\mathcal{H})$ as
we apply results from §11.2 to study normal operators on a Hilbert space. In
particular we will extend the functional calculus for normal operators beyond
what we obtained in Theorem 11.2.4 for normal elements of an abstract C*-
algebra.

Why should you be interested in normal operators? Since the answer to
such a question ultimately depends on an individual's orientation, their sense
of aesthetics, and their psychology, the answer for me is possibly not the
same as the answer for you. If applications are important to you, one could
say that normal operators, and certainly the unbounded hermitian ones,
are important in quantum mechanics. It is also true that as mathematical
objects they are a success story and therefore present a splendid opportunity
to discern how things are supposed to work out. For example, in the last
section of this book we present the multiplicity theory of normal operators.
This gives a necessary and sufficient condition for two normal operators on
a separable Hilbert space to be unitarily equivalent. (This was done for
compact normal operators in Theorem 5.8.2.) In any part of mathematics
the ultimate goal is to give a characterization of its equivalent objects. This
result ranks with the classification of finitely generated abelian groups as an
example of what we are supposed to be working toward.

Throughout this section, N is a normal operator on the Hilbert space $\mathcal{H}$.

Before going further let's establish a notational convenience. For sets K in the plane, let z denote the function defined on K whose value at z is z. That is, z is the identity function. We'll continue to also use z to denote a complex variable, so this introduces an ambiguity. The context of the usage, however, will remove that ambiguity. Similar comments apply to $\bar{z}$, the complex conjugate of z. As a consequence of the Stone–Weierstrass Theorem, it was established in Exercise 8.5.4 that for a compact subset K of $\mathbb{C}$ the algebra of all polynomials in z and $\bar{z}$ is dense in $C(K)$.

Given a normal operator on a Hilbert space $\mathcal{H}$, a particular compact subset of the complex plane that we are interested in is $\sigma(N)$. From §11.2 we have the functional calculus $f \mapsto f(N)$ from $C(\sigma(N))$ onto $C^*(N)$, in which the images of the functions $1, z, \bar{z}$ are $1, N, N^*$, respectively. We also have that this map is an isometric $*$-isomorphism. The first step in extending this functional calculus is to show how to associate a collection of measures with the normal operator.

11.4.1. Proposition. *If $e \in \mathcal{H}$, then there is a unique positive regular Borel measure μ_e on $\sigma(N)$ such that $\|\mu_e\| = \|e\|^2$ and*

$$\langle f(N)e, e \rangle = \int_{\sigma(N)} f \, d\mu_e$$

for all f in $C(\sigma(N))$.

Proof. If we let $L(f) = \langle f(N)e, e \rangle$, then we see that $L : C(\sigma(N)) \to \mathbb{C}$ is a linear functional. Moreover if $f \in C(\sigma(N))_+$, then $f(N) \geq 0$ and so $L(f) \geq 0$ by Theorem 11.3.7. That is, L is a positive linear functional on $C(\sigma(N))$ and so μ_e exists by the Riesz Representation Theorem. ∎

We fix the notation of the preceding proposition. If $e \in \mathcal{H}$, let $\mathcal{H}_e = \bigvee\{N^n N^{*m}e : n, m \geq 0\}$. Of course both the measure μ_e and the space $\mathcal{H}_e$ depend on N as well as the vector e, but in the notation we'll suppress this part of the dependence. So $\mathcal{H}_e$ is the smallest subspace of $\mathcal{H}$ that contains e and is invariant for both N and N^*. Anytime a subspace $\mathcal{M}$ is invariant for both an operator T and its adjoint T^*, we say that $\mathcal{M}$ is a reducing subspace for T. (See §5.5.) The reason for this terminology is that it follows that $\mathcal{M}^\perp$ is also invariant for both T and T^* (Proposition 5.5.9). Therefore the study of T is reduced to the study of the two operators $T|\mathcal{M}$ and $T|\mathcal{M}^\perp$, which are possibly easier to handle. For e in $\mathcal{H}$ let $N_e = N|\mathcal{H}_e$. We leave it to the reader to verify that N_e is also a normal operator, and $\sigma(N_e) \subseteq \sigma(N)$.

If K is any compact subset of $\mathbb{C}$ and μ is a positive regular Borel measure on K, denote by N_μ the operator $N_\mu : L^2(\mu) \to L^2(\mu)$ defined by

11.4.2 $(N_\mu f)(z) = zf(z)$

for all z in K. So the operator N_μ defined above is the operator M_z defined on $L^2(\mu)$ just as we define the operator M_ϕ as multiplication by the function ϕ in $L^\infty(\mu)$.

11.4.3. Theorem. *If N is a normal operator on $\mathcal{H}$ and $e \in \mathcal{H}$, then the operator N_e on $\mathcal{H}_e$ is unitarily equivalent to N_{μ_e} on $L^2(\mu_e)$. That is, there is a unitary operator $U : L^2(\mu_e) \to \mathcal{H}_e$ such that*

$$UN_{\mu_e} = N_e U$$

In fact, for any f in $C(\sigma(N))$ we have that $Uf(N_{\mu_e}) = f(N_e)U$.

Proof. Recall that $C(\sigma(N))$ is dense in $L^2(\mu_e)$ (Theorem 2.4.26 and Exercise 2.7.4). For all f in $C(\sigma(N))$, define $Uf = f(N)e$ in $\mathcal{H}_e$. We observe that

$$\|Uf\|^2 = \|f(N)e\|^2 = \langle f(N)e, f(N)e \rangle = \langle f(N)^* f(N)e, e \rangle = \int |f|^2 d\mu_e$$

by Proposition 11.4.1. This says that U is an isometry on a dense manifold in $L^2(\mu_e)$, and so U extends to an isometry from $L^2(\mu_e)$ into $\mathcal{H}_e$. Also note that when $m, n \geq 0$,

$$U(z^n \bar{z}^m) = N^n N^{*m} e$$

so the range of U includes a dense linear manifold in $\mathcal{H}_e$; therefore U is a unitary. This last equation also shows that $UN_{\mu_e}(z^n \bar{z}^m) = U(z^{n+1} \bar{z}^m) = N^{n+1} N^{*m} e = N_e U(z^n \bar{z}^m)$.

If $f, g \in C(\sigma(N))$, $Uf(N_{\mu_e})g = U(fg) = f(N)g(N)e = f(N_e)Ug$. Since $C(\sigma(N))$ is dense in $L^2(\mu_e)$, the proof is complete. ∎

11.4.4. Definition. If $T \in \mathcal{B}(\mathcal{H})$, say that a vector e is a *star-cyclic vector* for T if there is no proper subspace of $\mathcal{H}$ that contains e and is reducing for T. The operator T is a *star-cyclic operator* if it has a star-cyclic vector.

So a normal operator N is star-cyclic if and only if there is a vector e such that the space $\mathcal{H}_e$ defined above is all of $\mathcal{H}$. We observe that for any compactly supported regular Borel measure μ on $\mathbb{C}$, the vector 1 is star-cyclic for N_μ since the polynomials in z and $\bar{z}$ are dense in $L^2(\mu)$. Combining this observation with the preceding theorem we obtain the following.

11.4.5. Corollary. *A normal operator is star-cyclic if and only if it is unitarily equivalent to N_μ for some compactly supported regular Borel measure μ on $\mathbb{C}$.*

In general, if μ and ν are two measures defined on the same σ-algebra of subsets of X, say that μ and ν are *mutually absolutely continuous* when they have the same sets of measure 0; that is, $\mu \ll \nu$ and $\nu \ll \mu$. We will denote this by writing $[\mu] = [\nu]$. In the literature this relation is often called the *equivalence* of the measures.

11.4.6. Theorem. *If μ and ν are two positive regular Borel measures that are compactly supported on $\mathbb{C}$, then N_μ and N_ν are unitarily equivalent if and only if $[\mu] = [\nu]$.*

Proof. Assume that $[\mu] = [\nu]$. So by the Radon–Nikodym Theorem there are positive functions g in $L^1(\mu)$ and h in $L^1(\nu)$ such that $\nu = g\mu, \mu = h\nu$. It follows that $g(z)h(z) = 1$ a.e. $[\mu]$ and a.e. $[\nu]$. Define $U : L^2(\mu) \to L^2(\nu)$ by $Uf(z) = \sqrt{h(z)}f(z)$. Thus $\|Uf\|^2 = \int |f(z)|^2 h(z)d\nu(z) = \int |f(z)|^2 d\mu(z) = \|f\|^2$. Thus U is an isometry. Similarly if $V : L^2(\nu) \to L^2(\mu)$ is defined by $Vf(z) = \sqrt{g(z)}f(z)$, then V is an isometry. Because $gh = 1$, it follows that $V = U^{-1} = U^*$. Thus U is a unitary. Moreover $UN_\mu f = U(zf) = \sqrt{h}zf = N_\nu Uf$, so that $N_\nu = UN_\mu U^*$.

Now assume that there is a unitary $U : L^2(\mu) \to L^2(\nu)$ such that $N_\nu = UN_\mu U^*$. Let $h = U(1)$ and put $\sigma = \sigma(N_\mu) = \sigma(N_\nu)$. It follows that for any polynomial $p(z, \bar{z})$, $p(N_\nu, N_\nu^*)U = Up(N_\mu, N_\mu^*)$; by taking uniform limits we have $f(N_\nu)U = Uf(N_\mu)$ for every f in $C(\sigma)$. (Put this in your memory banks.) Hence for f in $C(\sigma)$, $\int f d\mu = \langle f(N_\mu)1, 1 \rangle = \langle Uf(N_\mu)1, U1 \rangle = \langle f(N_\nu)U1, U1 \rangle = \int f|h|^2 d\nu$. Since this holds for every continuous function on the support of the two measures, $\mu = |h|^2\nu$. Similarly if $g = U^*1$, $\nu = |g|^2\mu$. This says that μ and ν are mutually absolutely continuous. ∎

From here on we will seldom use the term "regular Borel measure," as all the measures we treat will be of this type. The preceding theorem hints that absolute continuity of measures will be important. The next result strengthens this assertion.

11.4.7. Proposition. *If $h \in \mathcal{H}_e$, then $\mu_h \ll \mu_e$. Conversely, if ν is a positive measure such that $\nu \ll \mu_e$, then there is a vector h in $\mathcal{H}_e$ such that $\nu = \mu_h$.*

Proof. Let $U : L^2(\mu_e) \to \mathcal{H}_e$ be the unitary obtained in Theorem 11.4.3 and let $g \in L^2(\mu_e)$ such that $Ug = h$. Then for any f in $C(\sigma(N))$,

$$\int f d\mu_h = \langle f(N)h, h \rangle = \langle f(N_e)h, h \rangle$$

$$= \langle f(N_e)Ug, Ug \rangle = \langle Uf(N_{\mu_e})g, Ug \rangle$$

$$= \langle f(N_{\mu_e})g, g \rangle$$

$$= \int f|g|^2 d\mu_e$$

Since this holds for every f in $C(\sigma(N))$, $\mu_h = |g|^2\mu_e$ and so $\mu_h \ll \mu_e$.

For the converse, the Radon–Nikodym Theorem implies there is a non-negative function ψ in $L^1(\mu_e)$ such that $\nu = \psi\mu_e$. Let $g \in L^2(\mu_e)$ such that $\psi = |g|^2$ and put $h = Ug$. By an argument similar to that used for the first half of the proof (Verify!), it follows that $\nu = \mu_h$. ∎

11.4.8. Proposition. (a) *If $\{e_n\}$ is an orthonormal basis for $\mathcal{H}$, and $h = \sum_{n=1}^{\infty} \alpha_n e_n$, then $\mu_h = \sum_{n=1}^{\infty} |\alpha_n|^2 \mu_{e_n}$.*

(b) *There is a vector e in $\mathcal{H}$ such that for every h in $\mathcal{H}$, $\mu_h \ll \mu_e$.*

Proof. (a) First observe that Proposition 11.4.1 implies that $\|\mu_{e_n}\| = \|e_n\|^2 = 1$ for each $n \geq 1$. Hence $\nu = \sum_{n=1}^{\infty} |\alpha_n|^2 \mu_{e_n}$ is a finite measure of total variation $\|h\|^2$. For f in $C(\sigma(N))$, we have that

$$\int f d\mu_h = \langle f(N)e, e \rangle = \sum_{n=1}^{\infty} |\alpha_n|^2 \langle f(N)e_n, e_n \rangle = \sum_{n=1}^{\infty} |\alpha_n|^2 \int f d\mu_{e_n} = \int f d\nu$$

Since the function f is arbitrary, we have equality of the measures ν and μ_h.

(b) Let $\{e_n\}$ be an orthonormal basis for $\mathcal{H}$ and put $e = \sum_{n=1}^{\infty} \lambda_n e_n$ for any square summable sequence $\{\lambda_n\}$ such that $\lambda_n \neq 0$ for all n. If $h = \sum_{n=1}^{\infty} \alpha_n e_n$, then $|\alpha_n|^2 \mu_{e_n} \ll |\lambda_n|^2 \mu_{e_n}$ for all $n \geq 1$, so that part (a) implies $\mu_h \ll \mu_e$. ∎

11.4.9. Definition. If N is a normal operator, a vector e in $\mathcal{H}$ is called a *maximal vector* for N if $\mu_h \ll \mu_e$ for every h in $\mathcal{H}$. A measure μ on $\sigma(N)$ is called a *scalar-valued spectral measure* for N if there is a maximal vector e with $[\mu] = [\mu_e]$; equivalently, if $\mu_h \ll \mu$ for every h in $\mathcal{H}$.

So the preceding proposition says maximal vectors exist. Of course they are not unique as the proof of (11.4.8) amply demonstrates. In fact the introduction of the concept of a scalar-valued spectral measure recognizes this. If e and f are two maximal vectors for the normal operator N, they are related at the level of the measures μ_e and μ_f, but what other relationship they may have is not clear. By Theorem 11.4.6 this condition is equivalent to the requirement that $N|\mathcal{H}_e \cong N|\mathcal{H}_f$. We also note that if μ is a scalar-valued spectral measure for N, then $\sigma(N)$ is the support of μ.

11.4.10. Proposition. *If N is a normal operator on $\mathcal{H}$ and $g, h \in \mathcal{H}$, then there is a complex-valued measure ν on $\sigma(N)$ such that $\nu \ll \mu_g$ and*

11.4.11
$$\langle f(N)g, h \rangle = \int f \, d\nu$$

for all f in $C(\sigma(N))$. Conversely, if $e \in \mathcal{H}$ and ν is a complex measure on $\sigma(N)$ such that $\nu \ll \mu_e$, then there are vectors g, h in $\mathcal{H}_e$ such that (11.4.11) holds for all f in $C(\sigma(N))$ and $\mathcal{H}_g = \mathcal{H}_e$.

Proof. First observe that $f \mapsto \langle f(N)g, h \rangle$ defines a bounded linear functional on $C(\sigma(N))$; therefore there is a complex measure ν on $\sigma(N)$ with $\langle f(N)g, h \rangle = \int f \, d\nu$ for every f in $C(\sigma(N))$. Also

$$\left| \int f \, d\nu \right|^2 = |\langle f(N)g, h \rangle|^2$$
$$\leq \|f(N)g\|^2 \|h\|^2$$
$$= \|h\|^2 \langle f(N)^* f(N)g, g \rangle$$
$$= \|h\|^2 \int |f|^2 \, d\mu_g$$

For any compact subset K of $\sigma(N)$, let $\{f_n\}$ be a sequence of continuous functions on $\sigma(N)$ such that $\chi_K \leq f_n \leq 1$ and $f_n(z) \to \chi_K(z)$ for all z in $\sigma(N)$. So $\nu(K) = \lim_n \int f_n \, d\nu$ and $\int |f_n|^2 d\mu_g \to \mu_g(K)$. (Why?) Hence $|\nu(K)| \leq \|h\|^2 \mu_g(K)$ for all compact sets K. Now suppose Δ is a Borel subset of $\sigma(N)$ such that $\mu_g(\Delta) = 0$; it follows that $\nu(K) = 0$ for every compact set contained in Δ. Therefore $\nu(\Delta) = 0$ and we have that $\nu \ll \mu_g$.

For the converse let ψ be the Radon–Nikodym derivative of ν with respect to μ_e; so $\nu = \psi \mu_e$. Put $\Delta = \{z \in \sigma(N_e) : \psi(z) \neq 0\}$. Let $U : L^2(\mu_e) \to \mathcal{H}_e$ be the unitary obtained in (11.4.3) such that $U N_{\mu_e} = N_e U$. Define a function g_1 on $\sigma(N_e)$ by $g_1(z) = \sqrt{|\psi(z)|}$ for z in Δ and $g_1(z) = 1$ when $z \notin \Delta$. Define another function by letting $h_1(z) = \overline{\psi(z)}/\sqrt{|\psi(z)|}$ for z in Δ and $h_1(z) = 0$ when $z \notin \Delta$. Note that $g_1, h_1 \in L^2(\mu_e)$ and $\psi = g_1 \overline{h_1}$. Let $g = U g_1, h = U h_1$. If $f \in C(\sigma(N))$, then $\int f \, d\nu = \int f g_1 \overline{h_1} d\mu_e = \langle f(N)g, h \rangle$. The fact that $\mathcal{H}_g = \mathcal{H}_e$ follows by Exercise 2 and the fact that $g_1(z) > 0$ a.e. $[\mu_e]$. ∎

The reader may have noticed that there is a small discrepancy between the first part of the preceding proposition and the converse; see Exercise 3. The next result is phrased for arbitrary compact metric spaces rather than compact subsets of the plane.

11.4.12. Proposition. *If X is a compact metric space and μ is a positive measure on X, then for every ϕ in $L^\infty(\mu)$ there is a sequence $\{f_n\}$ in $C(X)$ such that $\|f_n\| \le \|\phi\|_\infty$ and $f_n(x) \to \phi(x)$ a.e. $[\mu]$. Consequently, ball $C(X)$ is weak*-dense in ball $L^\infty(\mu)$.*

Proof. Let $\mathcal{M}$ denote the set of all functions ϕ in $L^\infty(\mu)$ that satisfy the conclusion of the theorem. It is easy to see that $\mathcal{M}$ is a linear manifold. We need an equivalent formulation of membership in $\mathcal{M}$ that involves a metric as this will be easier to manipulate than pointwise convergence a.e. $[\mu]$. Yes, it does involve an unusual blend – an L^1-condition applied to bounded functions.

Claim. $\phi \in \mathcal{M}$ if and only if there is a sequence $\{f_n\}$ in $C(X)$ with $\|f_n\| \le \|\phi\|_\infty$ and $\int |f_n - \phi| d\mu \to 0$.

In fact if $\phi \in \mathcal{M}$, let $\{f_n\}$ be a sequence in $C(X)$ such that $\|f_n\| \le \|\phi\|_\infty$ and $f_n(x) \to \phi(x)$ a.e. $[\mu]$. So $|f_n(x) - \phi(x)| \le 2\|\phi\|_\infty$ and the DCT implies that $\int |f_n - \phi| d\mu \to 0$. Conversely, if $\int |f_n - \phi| d\mu \to 0$ for a sequence $\{f_n\}$ in $C(X)$ with $\|f_n\| \le \|\phi\|_\infty$, then there is a subsequence $\{f_{n_k}\}$ such that $f_{n_k}(x) \to \phi(x)$ a.e. $[\mu]$ (2.5.5).

Claim. $\mathcal{M}$ is norm closed in $L^\infty(\mu)$.

Let ϕ be in the norm closure of $\mathcal{M}$ with $\|\phi\|_\infty = 1$ and suppose $\{\phi_n\}$ is a sequence in ball $\mathcal{M}$ such that $\|\phi - \phi_n\|_\infty < n^{-1}$. By the first claim, for each $n \ge 1$ there is an f_n in $C(X)$ with $\|f_n\| \le \|\phi_n\|_\infty \le 1$ such that $\int |f_n - \phi_n| d\mu < n^{-1}$. Thus $\int |f_n - \phi| d\mu < 2n^{-1}$, so that $\phi \in \mathcal{M}$ by the first claim.

Recall that the simple functions are norm dense in $L^\infty(\mu)$ (Theorem 2.7.12). (Also see Theorem 2.4.26.) So in light of the second claim and the fact that $\mathcal{M}$ is a linear subspace, to prove the proposition it suffices to prove that the characteristic functions belong to $\mathcal{M}$. First assume that K is a compact set. There is a sequence $\{f_n\}$ in $C(X)$ with $\chi_K \le f_n \le 1$ such that $f_n(x) \to \chi_K(x)$; this says that $\chi_K \in \mathcal{M}$. If E is any Borel set, then we can find a sequence of compact sets $\{K_n\}$ such that $\mu(E \backslash K_n) \to 0$. Since $\chi_{K_n} \in \mathcal{M}$, there is a continuous function f_n with $\|f_n\| \le 1$ and $\int |f_n - \chi_{K_n}| d\mu < n^{-1}$. Since $\int |\chi_E - \chi_{K_n}| d\mu = \mu(E \backslash K_n)$, we obtain that $\chi_E \in \mathcal{M}$.

The fact that ball $C(X)$ is weak*-dense in ball $L^\infty(\mu)$ is an easy consequence of the DCT. ∎

The reader might compare the preceding result with Proposition 8.3.4. The pertinent difference is that in that proposition we get that ball $\mathcal{X}$ is weak*-dense in ball $\mathcal{X}^{**}$, but in the preceding proposition $L^\infty(\mu)$ is not the second dual of $C(X)$.

11.4.13. Proposition. *If $\mathcal{H}$ is a Hilbert space and $\{T_n\}$ is a uniformly bounded sequence in $\mathcal{B}(\mathcal{H})$ such that $\lim_n \langle T_n e, h \rangle$ exists for every e, h in $\mathcal{H}$, then there is a bounded operator T on $\mathcal{H}$ such that $\|T\| \leq \sup_n \|T_n\|$ and $\langle Te, h \rangle = \lim_n \langle T_n e, h \rangle$ for all e, h in $\mathcal{H}$.*

Proof. Note that $u(e, h) = \lim_n \langle T_n e, h \rangle$ defines a bounded sesquilinear form and so the existence of T is a consequence of Theorem 5.4.2. ∎

There is a topology involved with the last proposition, though we will not explore this here. For each pair of vectors e, h in $\mathcal{H}$ we can define a seminorm on $\mathcal{B}(\mathcal{H})$ by $p_{e,h}(T) = |\langle Te, h \rangle|$. The locally convex topology defined by all such seminorms is called the *weak operator topology*. This plays an important role in the study of operators on a Hilbert space, but we don't need it here. The interested reader can consult [**8**].

We are now in a position to extend the functional calculus for a normal operator. The reader should do Exercise 10.4.2 if (s)he has not already done this. In particular that exercise contains the definition of the essential range of a function in $L^\infty(\mu)$.

11.4.14. Theorem. *If N is a normal operator and μ is a scalar-valued spectral measure, then there is a mapping $\rho : L^\infty(\mu) \to \mathcal{B}(\mathcal{H})$ having the following properties.*

(a) *ρ is an isometry and a $*$-isomorphism onto its image.*

(b) *$\rho(f) = f(N)$ for every continuous function f on $\sigma(N)$.*

(c) *If $\{\phi_i\}$ is a net in $L^\infty(\mu)$, then $\phi_i \to 0$ (wk*) in $L^\infty(\mu)$ if and only if for all vectors g, h in $\mathcal{H}$ we have that $\langle \rho(\phi_i)g, h \rangle \to 0$.*

(d) *For every ϕ in $L^\infty(\mu)$, $\sigma(\rho(\phi))$ is the essential range of ϕ.*

Also ρ is unique in the sense that if $\gamma : L^\infty(\mu) \to \mathcal{B}(\mathcal{H})$ is an isometry that is a $*$-isomorphism onto its image such that $\gamma(f) = f(N)$ for every f in $C(\sigma(N))$ and γ satisfies the same continuity condition as stated in (c), then $\gamma = \rho$.

Proof. Let's begin by observing that a net $\{\phi_i\}$ in $L^\infty(\mu)$ converges (wk*) to ϕ if and only if $\int \phi_i d\nu \to \int \phi d\nu$ for every measure ν that is absolutely continuous with respect to μ. In fact by the Radon–Nikodym Theorem we can identify $L^1(\mu)$ with the space of all measures absolutely continuous with respect to μ.

To define ρ, let $\phi \in L^\infty(\mu)$. Proposition 11.4.12 guarantees the existence of a sequence $\{f_n\}$ in $C(X)$ such that $\|f_n\| \leq \|\phi\|_\infty$ and $f_n(x) \to \phi(x)$ a.e. $[\mu]$. Hence $f_n \to \phi$ (wk*) in $L^\infty(\mu)$. Now let $e, h \in \mathcal{H}$. By Proposition 11.4.10 there is a complex measure $\nu \ll \mu_e$ such that $\langle f(N)e, h \rangle = \int f d\nu$

for all f in $C(\sigma(N))$. Since $\mu_e \ll \mu$, $\langle f_n(N)e, h \rangle = \int f_n d\nu \to \int \phi d\nu$. According to Proposition 11.4.13 there is a bounded operator T on $\mathcal{H}$ such that $\langle Te, h \rangle = \lim_n \langle f_n(N)e, h \rangle$ for all e, h in $\mathcal{H}$. We denote this operator by $T = \rho(\phi)$. It must be checked that the definition of $\rho(\phi)$ does not depend on the choice of the sequence $\{f_n\}$, and this is left to the reader. Note that we automatically have that (b) is true.

(c) Let e be a maximal vector so that $[\mu] = [\mu_e]$, and let $U : L^2(\mu_e) \to \mathcal{H}_e$ be the unitary such that $UN_{\mu_e} = N_e U$. Assume $\phi_i \to 0$ (wk*) in $L^\infty(\mu) = L^\infty(\mu_e)$. So if $g, h \in \mathcal{H}$, Proposition 11.4.10 implies there is a measure ν with $\nu \ll \mu$ such that $\langle f(N)g, h \rangle = \int f d\nu$ for all f in $C(\sigma(N))$. It follows that $\langle \rho(\phi)g, h \rangle = \int \phi d\nu$ for ϕ in $L^\infty(\mu)$, and so $\langle \rho(\phi_i)g, h \rangle \to 0$.

Now assume that for all vectors g, h in $\mathcal{H}$ we have that $\langle \rho(\phi_i)g, h \rangle \to 0$. As before we have that Proposition 11.4.10 implies that if $\nu \ll \mu$, then there are vectors g, h in $\mathcal{H}$ such that $\langle \rho(\phi)g, h \rangle = \int \phi d\nu$ for all ϕ in $L^\infty(\mu)$. Thus $\int \phi_i d\nu = \langle \rho(\phi_i)g, h \rangle \to 0$, so that $\phi_i \to 0$ in the weak* topology of $L^\infty(\mu)$.

Claim. ρ is a *-homomorphism.

The fact that ρ is linear is routine; we prove that it is multiplicative. Let $\phi, \psi \in L^\infty(\mu)$, fix vectors g, h in $\mathcal{H}$, and suppose that $\{f_n\}$ is a uniformly bounded sequence such that $f_n \to \phi$ a.e. $[\mu]$. So $f_n \to \phi$ (wk*). For any u in $C(X)$ we also have that $\|f_n u\| \le \|\phi u\|_\infty$ and $f_n u \to \phi u$ a.e. $[\mu]$. Thus $f_n u \to \phi u$ (wk*) in $L^\infty(\mu)$, and so $\langle \rho(f_n u)g, h \rangle \to \langle \rho(\phi u)g, h \rangle$. But by (b), $\rho(f_n u) = \rho(f_n)\rho(u)$, so $\langle \rho(f_n u)g, h \rangle = \langle \rho(f_n)[\rho(u)g], h \rangle \to \langle \rho(\phi)[\rho(u)g], h \rangle$. Therefore we have that

$$\rho(\phi u) = \rho(\phi)\rho(u)$$

whenever $\phi \in L^\infty(\mu)$ and $u \in C(X)$. Now let $\{u_n\}$ be a uniformly bounded sequence in ball $C(X)$ such that $u_n \to \psi$ a.e. $[\mu]$. It follows that $\phi u_n \to \phi \psi$ (wk*) in $L^\infty(\mu)$, and so by (c)

$$\langle \rho(\phi \psi)g, h \rangle = \lim_n \langle \rho(\phi u_n)g, h \rangle = \lim_n \langle \rho(\phi)\rho(u_n)g, h \rangle = \langle \rho(\phi)\rho(\psi)g, h \rangle$$

Thus ρ is multiplicative.

The proof that $\rho(\overline{\phi}) = \rho(\phi)^*$ is straightforward.

Claim. ρ is an isometry.

Since in the definition of $\rho(\phi)$ we can take the sequence $\{f_n\}$ to satisfy $\|f_n\| \le \|\phi\|_\infty$ (11.4.12) and we know that $\|f_n(N)\| = \|f_n\|$, it follows that $\|\rho(\phi)\| \le \|\phi\|_\infty$. If $\epsilon > 0$, let $g \in L^1(\mu)$ with $\int |g|d\mu = 1$ such that $\|\phi\|_\infty - \epsilon < \left| \int \phi g d\mu \right|$. Let e be a maximal vector with $[\mu] = [\mu_e]$; so there is a $\widetilde{g}$ in $L^1(\mu_e)$ with $\int |\widetilde{g}|d\mu_e = 1$ such that $g\mu = \widetilde{g}\mu_e$. Let $g_1, g_2 \in L^2(\mu_e)$ such that $\widetilde{g} = g_1 \overline{g_2}$ and $1 = \|g_1\|_2 = \|g_2\|_2$ (How do we get this last condition?). By (11.4.3) there is a unitary $U : L^2(\mu_e) \to \mathcal{H}_e$ such that $U N_{\mu_e} = N_e U$; let $U g_j = h_j \in \mathcal{H}_e$, $j = 1, 2$. Note that since U preserves norms, $\|h_1\| = \|h_2\| = 1$. So for any f in $C(\sigma(N))$

$$\int f g d\mu = \int f \widetilde{g} d\mu_e = \int f g_1 \overline{g_2} d\mu_e = \langle f(N_e) h_1, h_2 \rangle$$

It follows that $\int \phi g d\mu = \langle \rho(\phi) h_1, h_2 \rangle$. Thus

$$\|\phi\|_\infty - \epsilon < \left| \int \phi g d\mu \right| = |\langle \rho(\phi) h_1, h_2 \rangle| \le \|\rho(\phi)\|$$

This proves the claim and, therefore, finishes the proof of (a).

(d) Since ρ is a $*$-isomorphism between two C*-algebras, it preserves spectra; therefore (d) follows from Exercise 10.4.2.

Now to prove uniqueness. Let $\phi \in L^\infty(\mu)$ and again invoke Proposition 11.4.12 to find a sequence $\{f_n\}$ in $C(\sigma(N))$ such that $\|f_n\| \le \|\phi\|_\infty$ and $f_n(x) \to \phi(x)$ a.e. $[\mu]$. It follows from (c) and the assumption on γ that for all g, h in $\mathcal{H}$ we have that $\langle \rho(\phi)g, h \rangle = \lim_n \langle f_n(N)g, h \rangle = \langle \gamma(\phi)g, h \rangle$. That is, $\rho(\phi) = \gamma(\phi)$. ∎

Note that part (d) in Theorem 11.4.14 is a Spectral Mapping Theorem since the spectrum of ϕ in the algebra $L^\infty(\mu)$ is exactly its essential range (10.4.2).

We define $\phi(N) = \rho(\phi)$ for the map ρ in the preceding theorem. This gives us the promised extension of the functional calculus. A lacking ingredient in fully understanding this functional calculus is to identify the range of ρ. In fact, it can be shown that this is the closure of $C^*(N)$ in the weak operator topology mentioned above and also $\operatorname{ran} \rho$ is the set of all operators T in $\mathcal{B}(\mathcal{H})$ that commute with every operator that commutes with N. See [**8**].

11.4.15. Example. If μ is a positive measure on a compact subset of $\mathbb{C}$ and $N = N_\mu$ on $L^2(\mu)$, then μ is a scalar-valued spectral measure for N and for each ϕ in $L^\infty(\mu)$, $\phi(N) = M_\phi$. In fact this last statement follows by the uniqueness part of the last theorem.

We close with an equivalent and useful condition for a vector to be a maximal vector for a normal operator.

11.4.16. Proposition. *If N is a normal operator on $\mathcal{H}$ with μ as a scalar-valued spectral measure and $e \in \mathcal{H}$, then e is a maximal vector if and only if $\phi \in L^\infty(\mu)$ and $\phi(N)e = 0$ implies that $\phi = 0$.*

Proof. Assume e is a maximal vector; so we can take $\mu = \mu_e$. If $\phi(N)e = 0$, then $0 = \|\phi(N)e\|^2 = \langle \phi(N)^* \phi(N)e, e \rangle = \int |\phi|^2 d\mu_e$; hence $\phi = 0$ in $L^\infty(\mu)$.

For the converse assume e is not a maximal vector; so there is a Borel set Δ with $\mu(\Delta) > 0$ and $\mu_e(\Delta) = 0$. Taking $\phi = \chi_\Delta$, we have that $\phi \neq 0$ in $L^\infty(\mu)$, but $\|\phi(N)e\|^2 = \int |\phi|^2 d\mu_e = \mu_e(\Delta) = 0$. ∎

Also see Exercise 7 below.

Exercises.

(1) If $\{h_n\}$ is a sequence in $\mathcal{H}$ and $h_n \to h$, is there anything we can say about the sequence of measures $\{\mu_{h_n}\}$?

(2) Show that a function f in $L^2(\mu)$ is a star-cyclic vector for N_μ if and only if $|f(z)| > 0$ a.e. $[\mu]$.

(3) In the converse part of Proposition 11.4.10, if the Radon–Nikodym derivative ψ of the measure ν with respect to μ does not belong to $L^2(\mu)$, show that there is no vector h in $\mathcal{H}_e$ such that $\int f d\nu = \langle f(N)e, h \rangle$ for all f in $C(\sigma(N))$.

(4) If N is a diagonalizable normal operator, determine its scalar-valued spectral measure. Give a necessary and sufficient condition on a vector for it to be a maximal vector.

(5) If X is a compact metric space and μ and ν are two positive measures on X, let $N = N_\mu \oplus N_\nu$. Give an example of a maximal vector for N.

(6) Let λ be Lebesgue measure on $[0, 1]$ and let $\mu = \sum_{n=1}^\infty 2^{-n} \delta_{a_n}$, where $\{a_n\}$ are the rational numbers in the unit interval. Show that there is a sequence of continuous functions $\{f_n\}$ on the unit interval such that $f_n \to 1$ (wk*) in $L^\infty(\lambda)$ and $f_n \to 0$ (wk*) in $L^\infty(\mu)$. (Hint: What is the relationship between $L^\infty(\lambda + \mu)$, $L^\infty(\lambda)$, and $L^\infty(\mu)$?)

(7) Improve Proposition 11.4.16 by showing that e is a maximal vector for N if and only if $f = 0$ when $f \in C(\sigma(N))$ and $f(N)e = 0$.

11.5. The commutant of a normal operator

Here we examine the nature of operators that commute with a normal operator. When $\mathcal{A}$ is any collection of operators on a Hilbert space $\mathcal{H}$, $\mathcal{A}' = \{T \in \mathcal{B}(\mathcal{H}) : TA = AT \text{ for all } A \in \mathcal{A}\}$ is called the *commutant* of $\mathcal{A}$. It is not difficult to see that the commutant of any set of operators is an algebra that contains the identity.

We begin with an algebra of normal operators.

11.5.1. Theorem. *If $(X, \mathcal{A}, \mu)$ is a σ-finite measure space and $\mathcal{A}_\mu = \{M_\phi : \phi \in L^\infty(\mu)\}$, then $\mathcal{A}_\mu' = \mathcal{A}_\mu$.*

Proof. Clearly $\mathcal{A}_\mu \subseteq \mathcal{A}'_\mu$, so fix a T in $\mathcal{A}'_\mu$.

Case 1: $\mu(X) < \infty$. In this case, $1 \in L^2(\mu)$ so we have that $T(1) = \phi \in L^2(\mu)$. If $\psi \in L^\infty(\mu)$, then $\psi \in L^2(\mu)$ and $T(\psi) = TM_\psi(1) = M_\psi T(1) = \phi\psi$. If we set $\Delta_n = \{x \in X : |\phi(x)| \geq n\}$, then taking $\psi = \chi_{\Delta_n}$ in the preceding equation we get

$$\|T\|^2 \mu(\Delta_n) = \|T\|^2 \|\chi_{\Delta_n}\|^2 \geq \|T(\chi_{\Delta_n})\|^2$$
$$= \|\phi\chi_{\Delta_n}\|^2 = \int_{\Delta_n} |\phi|^2 d\mu \geq n^2 \mu(\Delta_n)$$

Since T is bounded, this says that $\mu(\Delta_n) = 0$ for all n sufficiently large. Hence $\phi \in L^\infty(\mu)$, $T = M_\phi$, and so $\|T\| = \|\phi\|_\infty$ (5.1.8).

Case 2. $\mu(X) = \infty$. Suppose $\Delta \in \mathcal{A}$ has $\mu(\Delta) < \infty$ and consider $L^2(\mu|\Delta) = \{f \in L^2(\mu) : f(x) = 0 \text{ off } \Delta\}$. For f in $L^2(\mu|\Delta)$, $Tf = TM_{\chi_\Delta}f = M_{\chi_\Delta}T(f) \in L^2(\mu|\Delta)$. So we can define the operator $T_\Delta = T|L^2(\mu|\Delta)$ and we see that by Case 1 there is a function ϕ_Δ in $L^\infty(\mu|\Delta)$ such that $T_\Delta f = \phi_\Delta f$ for all f in $L^2(\mu|\Delta)$. Suppose we have two sets of finite measure, Δ and Σ. It is straightforward to show that $\phi_\Delta|\Delta \cap \Sigma = \phi_\Sigma|\Delta \cap \Sigma$. It also follows that $\|T\| \geq \|T_\Delta\| \geq \|\phi_\Delta\|_\infty$ for every measurable set of finite measure Δ.

Now we use the fact that the measure space is σ-finite and write $X = \bigcup_{n=1}^\infty \Delta_n$ with $\mu(\Delta_n) < \infty$ for each $n \geq 1$. By the argument just made, if we define ϕ on X by setting $\phi(x) = \phi_{\Delta_n}(x)$ when $x \in \Delta_n$, then ϕ is a well defined measurable function on X. Since $\|\phi_{\Delta_n}\|_\infty \leq \|T\|$ for every $n \geq 1$, we have that $\phi \in L^\infty(\mu)$. It is routine to verify that $T = M_\phi$. ∎

11.5.2. Theorem (Fuglede[1]–Putnam[2] Theorem). *If N and M are normal operators on $\mathcal{H}$ and $\mathcal{K}$ and $B : \mathcal{K} \to \mathcal{H}$ is a bounded operator such that $NB = BM$, then $N^*B = BM^*$.*

Proof. From the hypothesis it follows that $p(N)B = Bp(M)$ for any polynomial $p(z)$. For a fixed z in $\mathbb{C}$, $\exp(i\bar{z}N)$ and $\exp(i\bar{z}M)$ are limits of the

---

[1] Bent Fuglede was born in 1925 in Copenhagen. He received his first degree in 1948 at the University of Copenhagen then studied in the US until 1951, after which he returned to Denmark. In 1954 he joined the Matematisk Institut, University of Copenhagen, where he held various positions; in 1965 he became a professor of mathematics at the University of Copenhagen, where he stayed until his retirement in 1992. He made many contributions to several parts of analysis, received several honors, and had six doctoral students.

[2] Calvin R. Putnam was born in 1924 in Baltimore and received all his degrees from Johns Hopkins University, including the doctorate in 1948. After two more years on the staff at Johns Hopkins and a year at the Institute for Advanced Study at Princeton, he joined the faculty at Purdue University in 1951. He remained there until his retirement in 1992. During that time he published an impressive number of papers and had six doctoral students. He was influential in the development of operator theory and I had the pleasure of collaborating with him on a paper. He died in Lafayette, Indiana in 2008.

same sequence of polynomials in N and M, respectively. Hence

$$\exp(i\bar{z}N)B = B\exp(i\bar{z}M)$$

for all z in $\mathbb{C}$. By Exercise 1 this implies that for all z,

$$f(z) \equiv e^{-izN^*}Be^{izM^*}$$
$$= e^{-izN^*}e^{-i\bar{z}N}Be^{i\bar{z}M}e^{izM^*}$$
$$= e^{-i(zN^*+\bar{z}N)}Be^{i(\bar{z}M+zM^*)}$$

Now f is seen to be an analytic function on the entire plane. Also the fact that $zN^* + \bar{z}N$ and $zM^* + \bar{z}M$ are hermitian implies that $e^{-i(zN^*+\bar{z}N)}$ and $e^{i(\bar{z}M+zM^*)}$ are unitary (Exercise 11.2.5). Therefore $\|f(z)\| \leq \|B\|$ for all z. By Liouville's Theorem (10.3.7), f is constant. Therefore $0 = f'(z) = -iN^*e^{-izN^*}Be^{izM^*} + ie^{-izN^*}BM^*e^{izM^*}$. Evaluating when $z = 0$ gives that $0 = -iN^*B + iBM^*$, whence the theorem. ∎

11.5.3. Corollary. *If μ is a compactly supported measure in $\mathbb{C}$, then*

$$\{N_\mu\}' = \{M_\phi : \phi \in L^\infty(\mu)\}$$

Proof. If $T \in \{N_\mu\}'$, then the Fuglede–Putnam Theorem implies we also have $TN_\mu{}^* = N_\mu{}^*T$. Using algebra we get that $TM_\phi = M_\phi T$ when ϕ is a polynomial in z and $\bar{z}$. Taking the uniform and then weak* limits produces the result. ∎

11.5.4. Proposition. *Let N and M be normal operators on $\mathcal{H}$ and $\mathcal{K}$. If $T : \mathcal{H} \to \mathcal{K}$ is a bounded operator such that $TN = MT$, then the following hold.*

(a) $\mathrm{cl}\,(\mathrm{ran}\,T)$ *is a reducing subspace for M.*

(b) $\ker T$ *is a reducing subspace for N.*

(c) *If $N_1 = N|(\ker T)^\perp$ and $M_1 = M|\mathrm{cl}\,(\mathrm{ran}\,T)$, then $N_1 \cong M_1$.*

Proof. (a) If $h \in \mathcal{H}$, then $MTh = TNh \in \mathrm{ran}\,T$, so $\mathrm{cl}\,(\mathrm{ran}\,T)$ is invariant under M. By the Fuglede–Putnam Theorem, $TN^* = M^*T$, so the same type of argument shows that $\mathrm{cl}\,(\mathrm{ran}\,T)$ is also invariant for M^*. The proof of (b) is similar.

(c) Using parts (a) and (b), we see that to prove (c) it suffices to assume that $(\ker T)^\perp = \mathcal{H}$ and $\mathrm{cl}\,(\mathrm{ran}\,T) = \mathcal{K}$ and show that $N \cong M$. Consider the polar decomposition $T = UA$, with $A \geq 0$ (11.3.10). Because of the assumption that T is injective with dense range, $U : \mathcal{H} \to \mathcal{K}$ is a unitary. Since $TN = MT$ and $TN^* = M^*T$, we also have $T^*M = NT^*$ and $T^*M^* = N^*T^*$. Thus $(T^*T)N = T^*MT = N(T^*T)$; so $A = (T^*T)^{\frac{1}{2}} \in \{N\}'$. Therefore $MUA = MT = TN = UAN = UNA$; that is, $MU = UN$ on $\mathrm{ran}\,A$.

But $\ker A = \ker T = (0)$, so $\mathrm{cl}\,(\mathrm{ran}\,A) = (\ker T)^\perp = \mathcal{H}$, and so $MU = UN$; that is, $M \cong N$. ∎

11.5.5. Corollary. *Two normal operators are similar if and only if they are unitarily equivalent.*

Exercises.

(1) Show that when A and B are commuting operators,
$$\exp(A + B) = (\exp A)(\exp B)$$

(2) If μ is a compactly supported measure on the complex plane, what is $\{N_\mu \oplus N_\mu\}'$?

(3) If $\mathcal{A}$ is a subalgebra of $\mathcal{B}(\mathcal{H})$, show that $\mathcal{A}$ is a maximal abelian subalgebra of $\mathcal{B}(\mathcal{H})$ if and only if $\mathcal{A} = \mathcal{A}'$.

(4) Find the commutant of a diagonal normal operator on a separable Hilbert space.

11.6. Multiplicity theory

In this section we develop a complete set of unitary (or isomorphism) invariants for a normal operator. That is, to each normal operator N on a separable Hilbert space, we associate a collection $\mathcal{M}(N)$ of objects such that two normal operators N_1 and N_2 are unitarily equivalent if and only if $\mathcal{M}(N_1) = \mathcal{M}(N_2)$. There aren't many mathematical concepts for which we can do this. If we consider $d \times d$ matrices and the notion of isomorphism is similarity, then canonical Jordan forms furnish the invariants. If we consider finitely generated abelian groups, then it is possible to attach to each such group a finite ordered set of positive integers such that two of these groups are isomorphic if and only if they generate the same set of integers. Normal operators furnish another example of such a classification.

We maintain the notation of the preceding sections. In particular for a normal operator N and any vector e in $\mathcal{H}$, $\mathcal{H}_e$ is the smallest reducing subspace for N containing the vector e. Here is the theorem we want to prove.

11.6.1. Theorem. *For any normal operator N there is a sequence of unit vectors $\{e_n\}$ such that the following hold.*

(a) $\mathcal{H}_{e_n} \perp \mathcal{H}_{e_m}$ *for $n \neq m$.*

(b) $\mathcal{H} = \bigoplus_{n=1}^\infty \mathcal{H}_{e_n}$.

(c) $\mu_{e_{n+1}} \ll \mu_{e_n}$ *for all $n \geq 1$.*

Once we have this theorem, here is what we do. If $n \geq 1$, let $N_n = N|\mathcal{H}_{e_n} \cong N_{\mu_{e_n}}$. By (b),

$$N = \bigoplus_{n=1}^{\infty} N|\mathcal{H}_{e_n} \cong \bigoplus_{n=1}^{\infty} N_{\mu_{e_n}}$$

Now consider the collection of all sequences $([\mu_1], [\mu_2], \dots)$, where $\mu_1, \mu_2, \dots$ are measures with compact support and with $\mu_{n+1} \ll \mu_n$ for all $n \geq 1$. Notice that the objects in the sequence are not the measures but the sets of all measures that have the same sets of measure 0 as the μ_n. So the requirement is that $[\mu_{n+1}] \subseteq [\mu_n]$ for all $n \geq 1$. We will see shortly that this is a complete set of unitary invariants for normal operators. First we need some preliminary work, starting with a lemma that is a "cancellation" process for unitary equivalence.

11.6.2. Lemma. *If $N, A,$ and B are normal operators with N star-cyclic and $N \oplus A \cong N \oplus B$, then $A \cong B$.*

Proof. The proof is not difficult but it is cumbersome and requires careful bookkeeping. Assume N, A, B operate on $\mathcal{K}, \mathcal{H}_A,$ and $\mathcal{H}_B$, respectively. Let $U : \mathcal{K} \oplus \mathcal{H}_A \to \mathcal{K} \oplus \mathcal{H}_B$ be a unitary such that $U(N \oplus A)U^{-1} = N \oplus B$. We can write U as a matrix with operator entries:

$$U = \begin{bmatrix} U_{11} & U_{12} \\ U_{21} & U_{22} \end{bmatrix}$$

where $U_{11} : \mathcal{K} \to \mathcal{K}$, $U_{12} : \mathcal{H}_A \to \mathcal{K}$, $U_{21} : \mathcal{K} \to \mathcal{H}_B$, $U_{22} : \mathcal{H}_B \to \mathcal{H}_B$. Expressing $N \oplus A$ and $N \oplus B$ as matrices we have

$$\begin{bmatrix} N & 0 \\ 0 & A \end{bmatrix} \text{ and } \begin{bmatrix} N & 0 \\ 0 & B \end{bmatrix}$$

The equation $U(N \oplus A) = (N \oplus B)U$ becomes

11.6.3 $$\begin{bmatrix} U_{11}N & U_{12}A \\ U_{21}N & U_{22}A \end{bmatrix} = \begin{bmatrix} NU_{11} & NU_{12} \\ BU_{21} & BU_{22} \end{bmatrix}$$

Similarly $U(N \oplus A)^* = (N \oplus B)^*U$ becomes

11.6.4 $$\begin{bmatrix} U_{11}N^* & U_{12}A^* \\ U_{21}N^* & U_{22}A^* \end{bmatrix} = \begin{bmatrix} N^*U_{11} & N^*U_{12} \\ B^*U_{21} & B^*U_{22} \end{bmatrix}$$

(We will refer to the resulting equations obtained by setting equal the corresponding entries of these matrices as $(11.6.3)_{ij}$ and $(11.6.4)_{ij}$ for $i, j = 1, 2$.)

Looking at the equations $UU^* = 1$ and $U^*U = 1$ and performing the required matrix multiplication we get

11.6.5 $$\begin{cases} \text{(a)} & U_{11}^*U_{12} + U_{21}^*U_{22} = 0 \quad \text{on } \mathcal{H}_A \\ \text{(b)} & U_{11}U_{11}^* + U_{12}U_{12}^* = 1 \quad \text{on } \mathcal{K} \\ \text{(c)} & U_{21}U_{11}^* + U_{22}U_{12}^* = 0 \quad \text{on } \mathcal{K} \end{cases}$$

We apply Proposition 11.5.4 to $(11.6.3)_{22}$ to conclude that $(\ker U_{22})^{\perp}$ reduces A, $\mathrm{cl}\,(\mathrm{ran}\,U_{22}) = (\ker U_{22})^{\perp}$ reduces B, and

11.6.6 $$A|(\ker U_{22})^{\perp} \cong B|(\ker U_{22})^{\perp}$$

What about $A|(\ker U_{22})$ and $B|(\ker U_{22})$? Note that if these two restrictions are unitarily equivalent, we are done. Let's see what happens.

Let $h \in \ker U_{22} \subseteq \mathcal{H}_A$; so using the matrix representation of U we see that $U(0 \oplus h) = U_{12}h \oplus 0$. Since U is an isometry it follows that U_{12} maps $\ker U_{22}$ isometrically onto a closed subspace of $\mathcal{K}$; put $\mathcal{M}_1 = U_{12}(\ker U_{22}) \leq \mathcal{K}$. Now $(11.6.3)_{12}$ and $(11.6.4)_{12}$ and the fact that $\ker U_{22}$ reduces A imply that $\mathcal{M}_1$ reduces N. Thus the unitary $U_{12}|\ker U_{22}$ implements the equivalence

11.6.7 $$A|\ker U_{22} \cong N|\mathcal{M}_1$$

Similarly, U_{21}^* maps $\ker U_{22} = (\mathrm{ran}\,U_{22})^{\perp}$ isometrically onto $\mathcal{M}_2 = U_{21}^*(\ker U_{22}^*)$, $\mathcal{M}_2$ reduces N, and

11.6.8 $$B|\ker U_{22}^* \cong N|\mathcal{M}_2$$

Note that if $\mathcal{M}_1 = \mathcal{M}_2$, then (11.6.6), (11.6.7), and (11.6.8) imply $A \cong B$ and the proof is complete. Can we hope for this?

If $h \in \ker U_{22}$, then (11.6.5(a)) implies that $U_{11}^* U_{12}h = 0$; thus $\mathcal{M}_1 = U_{12}(\ker U_{22}) \subseteq \ker U_{11}^*$. On the other hand, if $g \in \ker U_{11}^*$, then (11.6.5(b)) implies that $g = (U_{11}U_{11}^* + U_{12}U_{12}^*)g = U_{12}U_{12}^*g$. But by (11.6.5(c)), $U_{22}U_{12}^*g = 0$, so $U_{12}^*g \in \ker U_{22}$. Hence $g \in U_{12}(\ker U_{22})$. Therefore

$$\mathcal{M}_1 = \ker U_{11}^*$$

Similarly

$$\mathcal{M}_2 = \ker U_{11}$$

The extremely careful reader will have noticed that so far we have not used the hypothesis that N is star-cyclic. But $(11.6.3)_{11}$ implies $U_{11} \in \{N\}'$. Since N is star-cyclic, Theorem 11.4.3 and Corollary 11.5.3 imply that U_{11} is normal. Therefore (5.4.17), $\ker U_{11} = \ker U_{11}^*$; that is, $\mathcal{M}_1 = \mathcal{M}_2$. ∎

The preceding lemma is not true without the hypothesis that N is star-cyclic. For example, let N be the 0 operator, let A be the identity, and let $B = 0 \oplus 1$, all living on infinite-dimensional spaces.

The next two lemmas will enable us to prove Theorem 11.6.1. The first of these is just a special case of the second, and its only purpose is to enable us to prove the second.

11.6.9. Lemma. *Let μ be a measure with compact support K in $\mathbb{C}$, let Δ be a Borel subset of K with $\mu(\Delta) > 0$, and put $\nu = \mu|\Delta$. If $N = N_\mu \oplus N_\nu$ and*

$g \oplus h \in L^2(\mu) \oplus L^2(\nu)$ such that $|h(z)| > 0$ a.e. $[\nu]$, then there is an f in $L^2(\mu)$ such that $f \oplus h$ is a maximal vector for N and $g \oplus h \in \mathrm{cl}\,[C^*(N)(f \oplus h)]$.

Proof. Define $f(z) = g(z)$ if $z \in \Delta$ and $f(z) = 1$ otherwise. Put $\mathcal{M} = \mathrm{cl}\,[C^*(N)(f \oplus h)] = \mathrm{cl}\,\{\phi f \oplus \phi h : \phi \in L^\infty(\mu)\}$, and put $\Delta' = K \backslash \Delta$. Note that for any ϕ in $L^\infty(\mu)$, $\phi\chi_{\Delta'} \oplus 0 = \phi\chi_{\Delta'}(f \oplus h) \in \mathcal{M}$. Hence $L^2(\mu|\Delta') \oplus (0) \subseteq \mathcal{M}$, and so $(1-g)\chi_{\Delta'} \oplus 0 \in \mathcal{M}$. Therefore $g \oplus h = f \oplus h - (1-g)\chi_{\Delta'} \oplus 0 \in \mathcal{M}$ as desired.

To see that $f \oplus h$ is maximal for N we use Proposition 11.4.16. Let $\phi \in L^\infty(\mu)$ such that $\phi(N)(f \oplus h) = \phi f \oplus \phi h = 0$; thus $\phi(z)f(z) = \phi(z)h(z) = 0$ a.e. $[\mu]$. Since $|h(z)| > 0$ a.e. $[\mu]$ on Δ, $\phi(z) = 0$ a.e. $[\mu]$ on Δ. But $f(z) = 1$ on Δ', so we have that $\phi = 0$. ∎

11.6.10. Lemma. *If N is a normal operator on $\mathcal{H}$ and $k \in \mathcal{H}$, then there is a maximal vector e such that $k \in \mathcal{H}_e$.*

Proof. The idea of this proof is to reduce the general case to that covered by the preceding lemma. So let e_0 be any maximal vector, put $\mu = \mu_{e_0}$, and let $\mathcal{G} = \mathrm{cl}\,\{\phi(N)e_0 : \phi \in L^\infty(\mu)\} = \mathcal{H}_{e_0}$. Write $k = g_1 + h_1$, where $g_1 \in \mathcal{G}$ and $h_1 \in \mathcal{G}^\perp$. Put $\mathcal{L} = \mathcal{H}_{h_1}$ and $\eta = \mu_{h_1}$.

Since $\eta \ll \mu$, there is a Borel set Δ such that $[\eta] = [\mu|\Delta]$. Also $N|\mathcal{G} \cong N_\mu$ and $N|\mathcal{L} \cong N_\eta \cong N_\nu$, where $\nu = \mu|\Delta$. Let $U : \mathcal{G} \oplus \mathcal{L} \to L^2(\mu) \oplus L^2(\nu)$ be a unitary such that $U(N|\mathcal{G} \oplus N|\mathcal{L})U^{-1} = N_\mu \oplus N_\nu$. We note that $U(g_1 \oplus h_1) = g \oplus h \in L^2(\mu) \oplus L^2(\nu)$, where g and h satisfy the hypothesis of Lemma 11.6.9. If $f \in L^2(\mu)$ as in that lemma and $e = U^{-1}(f \oplus h)$, then e is the sought after maximal vector. ∎

Proof of Theorem 11.6.1. Fix an orthonormal basis $\{f_n\}$ for $\mathcal{H}$ with f_1 a maximal vector for N; this will act as a type of control to ensure that when we have picked the sequence $\{e_n\}$ we satisfy (b). Let $e_1 = f_1$ and let f_2' be the orthogonal projection of f_2 onto $\mathcal{H}_{e_1}^\perp$. By Lemma 11.6.10 there is a maximal vector e_2 for $N_2 = N|\mathcal{H}_{e_1}^\perp$ such that $\|e_2\| = 1$ and $f_2' \in \mathrm{cl}\,\{\phi(N_2) : \phi \in L^\infty(\mu_{e_2})\}$. Note that $\mathrm{cl}\,\{\phi(N_2)e_2 : \phi \in L^\infty(\mu_{e_2})\} = \mathcal{H}_{e_2}$, $\mathcal{H}_{e_2} \perp \mathcal{H}_{e_1}$, $\{f_1, f_2\} \subseteq \mathcal{H}_{e_1} \oplus \mathcal{H}_{e_2}$, and $\mu_{e_2} \ll \mu_{e_1}$. Continue by induction. The fact that at each stage we have $\{f_1, \ldots, f_n\} \subseteq \bigoplus_{j=1}^n \mathcal{H}_{e_j}$, guarantees that part (b) is satisfied. Parts (a) and (c) follow from the induction process. ∎

Using Theorem 11.6.1 we can now carry out the plan that was outlined after we stated that result. First we present an easy result that will be useful here and later.

11.6.11. Proposition. *If $N = N_{\mu_1} \oplus N_{\mu_2} \oplus \cdots$ with $\mu_{n+1} \ll \mu_n$ for all $n \geq 1$, then:*

(a) *μ_1 is a scalar-valued spectral measure for N.*

(b) *If $\phi \in L^\infty(\mu_1)$, then $\phi(N) = \phi(N_{\mu_1}) \oplus \phi(N_{\mu_2}) \oplus \cdots = M_\phi \oplus M_\phi \oplus \cdots$.*

Proof. If $h = h_1 \oplus h_2 \oplus \cdots \in L^2(\mu_1) \oplus L^2(\mu_2) \oplus \cdots$, then $\mu_h = \sum_{n=1}^\infty \mu_{h_n}$. By (11.4.7), each $\mu_{h_n} \ll \mu_n \ll \mu_1$, so $\mu_h \ll \mu_1$. By definition, μ_1 is a scalar-valued spectral measure. The proof of (b) is an easy exercise. ∎

11.6.12. Theorem. (a) *If N is a normal operator on a separable Hilbert space, then there is a possibly finite sequence of measures $\{\mu_n\}$ on $\sigma(N)$ such that $\mu_{n+1} \ll \mu_n$ for all $n \geq 1$ and*

$$N \cong N_{\mu_1} \oplus N_{\mu_2} \oplus \cdots$$

(b) *If N and $\{\mu_n\}$ are as in (a) and $M \cong \bigoplus_{n=1}^\infty N_{\nu_n}$, where $\nu_{n+1} \ll \nu_n$ for all $n \geq 1$, then $N \cong M$ if and only if $[\mu_n] = [\nu_n]$ for each n.*

Proof. Part (a) follows from Theorem 11.6.1 and Theorem 11.4.3. The main task is to prove (b).

If $[\mu_n] = [\nu_n]$ for all $n \geq 1$, then the fact that $N \cong M$ follows readily from Theorem 11.4.6. Now assume that $N \cong M$. The first step is to show that $[\mu_1] = [\nu_1]$. This is best done by dropping the notation of the theorem and becoming more abstract. So let N and M operate on the spaces $\mathcal{H}$ and $\mathcal{K}$, let μ and ν be scalar-valued spectral measures for N and M, respectively, and let $U : \mathcal{H} \to \mathcal{K}$ be a unitary such that $MU = UN$. We want to show that $[\mu] = [\nu]$. Let e be a maximal vector for N and assume that $\mu = \mu_e$. It readily follows that $\sigma(N) = \sigma(M)$ and that $p(M, M^*)U = Up(N, N^*)$ for every polynomial $p(z, \bar{z})$. By taking uniform limits we have that $f(M)U = Uf(N)$ for every f in $C(\sigma(N))$. Letting $k = Ue$, we have that for each f in $C(\sigma(N))$, $\int f d\mu = \langle f(N)e, e \rangle = \langle Uf(N)e, Ue \rangle = \langle f(M)k, k \rangle = \int f d\nu_k$, where ν_k is the measure for M defined by k. (The consistent notation would be μ_k rather than ν_k, but this seems confusing.) Hence, by the uniqueness part of the Riesz Representation Theorem, $\mu = \nu_k \ll \nu$ and so $[\mu] \subseteq [\nu]$. If we reverse the roles of N and M we get the other inclusion. Therefore we have that $[\mu] = [\nu]$ as desired.

Since μ_1 and ν_1 are the scalar-valued spectral measures for N and M, we have that $[\mu_1] = [\nu_1]$ by the preceding paragraph. Hence $N_{\mu_1} \cong N_{\nu_1}$ by Theorem 11.4.6. By Lemma 11.6.2, $N_{\mu_2} \oplus N_{\mu_3} \oplus \cdots \cong N_{\nu_2} \oplus N_{\nu_3} \oplus \cdots$. Continuing by induction we have the desired conclusion. ∎

11.6.13. Example. Let N be a normal matrix in $M_d(\mathbb{C})$ and let $\lambda_1, \ldots, \lambda_d$ be its eigenvalues repeated according to their multiplicity. Let $\lambda_1^1, \ldots, \lambda_{n_1}^1$ be the distinct eigenvalues and let μ_1 be the sum of unit point masses at these distinct eigenvalues. Now let $\lambda_1^2, \ldots, \lambda_{n_2}^2$ be the distinct eigenvalues of N that have multiplicity 2; let μ_2 be the sum of unit point masses at these eigenvalues. Continue. This produces atomic measures $\mu_1, \ldots, \mu_p$ with $\mu_p \ll \cdots \ll \mu_1$ and $N \cong N_{\mu_1} \oplus \cdots \oplus N_{\mu_p}$ as in Theorem 11.6.12.

Now we give an alternative complete unitary invariant for normal operators that is the generalization of what was given in Theorem 5.9.3 for compact normal operators. Adopt the notation of Theorem 11.6.12 but assume $\|\mu_n\| = 2^{-n}$ for all n. Put $\mu = \mu_1$, and let E_1 be a Borel set having full μ measure. For each $n \geq 2$ let E_n be a Borel set such that $[\mu_n] = [\mu_{n-1}|E_n]$. Since $\mu_{n+1} \ll \mu_n$ for each n, we may assume that $E_{n+1} \subseteq E_n$ for each n; put $E_\infty = \bigcap_{n=1}^\infty E_n$. Now put $\Delta_n = E_n \backslash E_{n+1}$. The way to think of these sets Δ_n is that on Δ_n the scalar-valued spectral measure μ "appears" n times: once from μ_1, once from μ_2, and so on until we reach μ_n and then $\mu_m(\Delta_n) = 0$ for all $m > n$. The set Δ_∞ is where μ appears infinitely often. Of course it may be that some of the sets Δ_n are invisible to μ. For example, if $[\mu_2] = [\mu_3]$, then $[\mu_3] = [\mu_2|E_3] = [\mu|E_2]$, and so $\mu(\Delta_3) = 0$. We now define a *multiplicity function* $m_N : \mathbb{C} \to \{0, \infty\} \cup \mathbb{N}$ by setting $m_N(z) = n$ when $z \in \Delta_n$ and 0 otherwise. We will see shortly that the pair (μ, m_N) is a complete unitary invariant for normal operators. But before stating the formal result it is helpful to make an observation using the functional calculus.

For any Borel set Δ consider the operator $\chi_\Delta(N)$. Using the fact that $\phi \mapsto \phi(N)$ is a *-isomorphism from $L^\infty(\mu)$ onto its image in $\mathcal{B}(\mathcal{H})$ it follows that $\chi(N)$ is a projection and $N\chi_\Delta(N) = \chi_\Delta(N)N$. Therefore the range of this projection, $\chi_\Delta(N)\mathcal{H}$, is a reducing subspace for N. The reader can examine Exercise 2 for more on this projection and the function $\Delta \mapsto \chi_\Delta(N)$, but for now it is simply an observation we will make use of.

11.6.14. Theorem. *If N is a normal operator on a separable Hilbert space $\mathcal{H}$ with scalar-valued spectral measure μ, then there is a Borel function $m_N : \mathbb{C} \to \{0, \infty\} \cup \mathbb{N}$ such that the following hold:*

(a) $\mu(\{z : m_N(z) = 0\}) = 0.$

(b) *If $1 \leq n \leq \infty$ and $\Delta_n = \{z : m_N(z) = n\}$, then*

11.6.15 $N|[\chi_{\Delta_n}(N)\mathcal{H}] \cong N_{\mu|\Delta_n} \oplus \cdots \oplus N_{\mu|\Delta_n} (n \text{ times}).$

If M is another normal operator with scalar-valued spectral measure ν and function m_M that satisfies the corresponding conditions (a) *and* (b), *then $N \cong M$ if and only if $[\mu] = [\nu]$ and $m_N = m_M$ a.e. $[\mu]$.*

Proof. The function m_N is defined before the statement of the theorem. Part (a) holds by construction. To establish (b) we unwind the construction of the sets Δ_n. As before the statement of the theorem, we put $\mu = \mu_1$. Note that $[\mu|E_n] = [\mu_n]$ and $E_n = \Delta_\infty \cup \bigcup_{k=n}^\infty \Delta_k$. From Proposition 11.6.11 we have that

$$\chi_{\Delta_n}(N) \cong \chi_{\Delta_n}(N_{\mu_1}) \oplus \chi_{\Delta_n}(N_{\mu_2}) \oplus \cdots$$

Now for $k > n$, $\mu_k(\Delta_n) = 0$, so $\chi_{\Delta_n}(N_{\mu_k}) = 0$. When $k \leq n$, $\chi_{\Delta_n}(N_{\mu_k}) = M_{\chi_{\Delta_n}}$ on $L^2(\mu_k)$; thus $\chi_{\Delta_n}(N_{\mu_k})L^2(\mu_k) = L^2(\mu_k|\Delta_n)$ when $k \leq n$. Therefore

$$\chi_{\Delta_n}(N)\mathcal{H} \cong L^2(\mu_1|\Delta_n) \oplus \cdots \oplus L^2(\mu_n|\Delta_n)$$
$$\cong L^2(\mu|\Delta_n) \oplus \cdots \oplus L^2(\mu|\Delta_n)$$

This establishes (b) when $n < \infty$. The case of an infinite n is similarly proven.

Now assume that the operator M is given with scalar-valued spectral measure ν and the function m_M. If N and M are unitarily equivalent, then $[\mu] = [\nu]$ and the above construction of the sets $\{\Delta_n\}$ and the function m_N applied to M will show that $m_N = m_M$ a.e. $[\mu]$.

Conversely suppose $[\mu] = [\nu]$ and $m_N = m_M$ a.e. $[\mu]$. With a suitable modification we can assume that $m_N = m_M$ everywhere. For $1 \leq n \leq \infty$, let $\Delta_n = \{z : m_M(z) = n\} = \{z : m_N(z) = n\}$. If $\mathcal{K}$ is the space on which M operates, then, by part (a), for M, $\mathcal{K} = \chi_{\Delta_\infty}(M)\mathcal{K} \oplus \bigoplus_{n=1}^{\infty} \chi_{\Delta_n}(M)\mathcal{K}$. Since $[\mu] = [\nu]$, $N_{\mu|\Delta_n} \cong N_{\nu|\Delta_n}$ for $1 \leq n \leq \infty$. Therefore by property (b) for M and N,

$$M|\mathcal{K}_n \cong N_{\nu|\Delta_n} \oplus \cdots \oplus N_{\nu|\Delta_n} \ (n \text{ times})$$
$$\cong N_{\mu|\Delta_n} \oplus \cdots \oplus N_{\mu|\Delta_n} \ (n \text{ times})$$
$$\cong N|\left[\chi_{\Delta_n}(N)\mathcal{H}\right]$$

Therefore

$$M = M|\mathcal{K}_\infty \oplus \bigoplus_{n=1}^{\infty} M|\mathcal{K}_n$$
$$\cong N|\left[\chi_{\Delta_\infty}(N)\mathcal{H}\right] \oplus \bigoplus_{n=1}^{\infty} N|\left[\chi_{\Delta_n}(N)\mathcal{H}\right]$$
$$= N \qquad \blacksquare$$

So we see that on the space $\chi_{\Delta_n}(N)\mathcal{H}$ in the preceding theorem, the action of the operator N is repeated n times; that is, it has multiplicity n there. In Example 11.6.13 the set Δ_n consists of all the eigenvalues λ such that $\dim \ker(N - \lambda) = n$.

So Theorem 11.6.14 says that two normal operators are unitarily equivalent if and only if they have mutually absolutely continuous scalar-valued spectral measures and the same multiplicity function. The reader might compare the preceding theorem with Theorem 5.8.2.

There is another representation of normal operators we should discuss before concluding this chapter. Recall that if $(X, \mathcal{A}, \mu)$ is a σ-finite measure space and $\phi \in L^\infty(\mu)$, then M_ϕ is a normal operator. We prove a converse of this for normal operators on a separable Hilbert space.

11.6.16. Theorem. *If N is a normal operator on a separable Hilbert space $\mathcal{H}$, then there is a σ-finite measure space $(X, \mathcal{A}, \mu)$ such that $L^2(\mu)$ is separable and there is a function ϕ in $L^\infty(\mu)$ such that $N \cong M_\phi$.*

Proof. Let's start with the representation of the N obtained in the previous theorem: let ν be the scalar-valued spectral measure for N and let m_N be its multiplicity function. For $1 \le n \le \infty$ put $\Delta_n = \{z : m_N(z) = n\}$ and let X_n be the disjoint union of n copies of Δ_n, denoted by $\Delta_n^1, \ldots, \Delta_n^n$. For example, we could take $\Delta_n^j = \Delta_n \times \{j\} \subseteq \mathbb{C}^2$. We note that the sets $\{X_n : 1 \le n \le \infty\}$ are pairwise disjoint. Put $X = X_\infty \cup \bigcup_{n=1}^\infty X_n$ and define $\mathcal{A}$ to be the collection of all subsets E of X such that for $1 \le n \le \infty$, $E \cap X_n = \bigcup_{j=1}^n E_n^j$ with E_n^j a Borel subset of Δ_n^j. It is easy to show that $\mathcal{A}$ is a σ-algebra of subsets of X. Now we define a measure μ_n on X_n by letting $\mu_n(\bigcup_{j=1}^n E_n^j) = \sum_{j=1}^n \nu(E_n^j)$ when E_n^j is a Borel subset of Δ_n^j. The measure μ is defined on $(X, \mathcal{A})$ by letting $\mu(E) = \mu_\infty(E \cap X_\infty) + \sum_{n=1}^\infty \mu_n(E \cap X_n)$. Again it is easy to verify that $(X, \mathcal{A}, \mu)$ is a σ-finite measure space.

Now define $\phi : X \to \mathbb{C}$ as follows: if we did as we suggested and set $\Delta_n^j = \Delta_n \times \{j\}$, then $\phi(z, j) = z$ when $(z, j) \in \Delta_n \times \{j\}$. It is straightforward that $\phi \in L^\infty(\mu)$.

Now we show that $N \cong M_\phi$ on $L^\infty(\mu)$. In fact we know that

$$N \cong N|\left[\chi_{\Delta_\infty}(N)\mathcal{H}\right] \oplus \bigoplus_{n=1}^\infty N|\left[\chi_{\Delta_n}(N)\mathcal{H}\right]$$

and for $1 \le n \le \infty$ we have (11.6.15). We also have that

$$M_\phi \cong M_\phi|L^2(\mu_\infty) \oplus \bigoplus_{n=1}^\infty M_\phi|L^2(\mu_n)$$

So we will be done if we can show that for $1 \le n \le \infty$,

$$M_\phi|L^2(\mu_n) \cong N_{\mu|\Delta_n} \oplus \cdots \oplus N_{\mu|\Delta_n} \; (n \text{ times}).$$

But

$$M_\phi|L^2(\mu_n) \cong \bigoplus_{j=1}^n M_\phi|\left[L^2(\mu_n|\Delta_n^j)\right]$$

and it is immediately seen from the definition of ϕ that

$$N_{\mu|\Delta_n} \cong M_\phi|\left[L^2(\mu_n|\Delta_n^j)\right]$$

concluding the proof. ∎

The reader might compare the preceding theorem with Theorem 5.8.2.

A natural question arises about calculating the unitary invariants for the normal operator M_ϕ. This can be complicated. As a little thought will reveal the scalar-valued spectral measure is $E \mapsto \mu\big(\phi^{-1}(E)\big)$. The complication comes in defining the multiplicity function $m = m_{M_\phi}$. The intuition is to define $m(z)$ to be the number of points in $\phi^{-1}(z)$. However ϕ is only defined a.e. $[\mu]$ and so the number of points in such a set becomes intolerably ambiguous. For a resolution of this, we refer the reader to the paper [1] which everyone who has reached this far in this book can now readily comprehend.

Exercises.

(1) Let μ and ν be two compactly supported measures on $\mathbb{C}$, and find the decomposition of $N = N_\mu \oplus N_\nu$ as in Theorem 11.6.12. (Hint: The Lebesgue–Radon–Nikodym Theorem might be useful.)

(2) If M and N are two normal operators and $\{\nu_n\}$ and $\{\mu_n\}$ are the sequences of measures obtained in Theorem 11.6.12 for M and N, respectively, what is the corresponding sequence of measures for $M \oplus N$?

(3) Give all the details of the proof of Theorem 11.6.16.

(4) Suppose that μ is a compactly supported measure on $\mathbb{C}$ and $m : \mathbb{C} \to \mathbb{N} \cup \{\infty\}$ is a Borel function such that $\mu(\{z : m(z) = 0\}) = 0$. Find a normal operator N such that μ is its scalar-valued spectral measure and m is its multiplicity function.

(5) Let μ and ν be two compactly supported measures on $\mathbb{C}$, and find the scalar-valued spectral measure of $N = N_\mu \oplus N_\nu$ and its multiplicity function m_N as in Theorem 11.6.14.

(6) If M and N are two normal operators with scalar-valued spectral measures ν and μ and multiplicity functions m_M and m_N, what is the scalar-valued spectral measure and multiplicity function of $M \oplus N$?

(7) Let $\{\mu_n\}$ be a sequence of compactly supported measures on $\mathbb{C}$ such that $\mu_{n+1} \ll \mu_n$ for all n and suppose that M is a normal operator whose scalar-valued spectral measure ν is absolutely continuous with respect to each μ_n. Show that

$$N_{\mu_1} \oplus N_{\mu_2} \oplus \cdots \cong M \oplus N_{\mu_1} \oplus N_{\mu_2} \oplus \cdots$$

Appendix

A.1. Baire Category Theorem

A.1.1. Theorem (Baire[1] Category Theorem). *If (X, d) is a complete metric space and $\{U_n\}$ is a sequence of open subsets of X each of which is dense, then $\bigcap_{n=1}^{\infty} U_n$ is dense.*

Proof. To show that $\bigcap_{n=1}^{\infty} U_n$ is dense it suffices to show that if G is a non-empty open subset of X, then $G \cap \bigcap_{n=1}^{\infty} U_n \neq \emptyset$. (Why?) Since U_1 is dense, there is an open disk $B(x_1; r_1)$ such that $B(x_1; r_1) \subseteq G \cap U_1$. This is the first step in an induction argument that establishes that for each $n \geq 2$ there is an open disk $B(x_n; r_n)$ with $r_n < n^{-1}$ such that

$$\operatorname{cl} B(x_n; r_n) \subseteq B(x_{n-1}; r_{n-1}) \cap U_n$$

---

[1] René-Louis Baire was born in Paris in 1874. His father was a tailor and his family was poor. However he won a scholarship that enabled him to receive an excellent education and he distinguished himself from the start. Soon he gained admission to the prestigious École Normale Supérieure. After this education and further study in mathematics, he obtained a position as a professor at a lycée and began the research that led him to the introduction of his classification of functions of a real variable. He was awarded a scholarship that enabled him to study in Italy where he met Volterra. His dissertation on discontinuous functions earned him a doctorate in 1899, and in 1901 he secured a position on the faculty at the University of Montpelier. Throughout his life he suffered from poor health, though he continued his research throughout. In 1907 he was promoted to a professorship at Dijon, but his health interfered with his teaching duties. In 1914 he requested a leave and traveled to Lausanne, Switzerland. While there World War I broke out and he was unable to return to France. Over time he developed a dislike of the younger Lebesgue and felt that his own work was unappreciated. His health deteriorated, depression ensued, and he spent the rest of his life on the shores of Lac Léman in Switzerland. It was there that he received the Chevalier de la Legion d'Honneur and in 1922 he was elected to the Académie des Sciences. He published significant works on number theory and functions. He died 1932 in Chambéry near Geneva.

Observe that this implies that when $n > N$,

A.1.2 $\operatorname{cl} B(x_n; r_n) \subseteq B(x_N; r_N) \cap U_N \subseteq G \cap U_N$

Thus for $n, m \geq N$, $d(x_n, x_m) < 2N^{-1}$ and so $\{x_n\}$ is a Cauchy sequence. Since X is complete, there is an x in X such that $x_n \to x$. The same observation (A.1.2) reveals that when $n > m > N$, $x \in \operatorname{cl} B(x_n; r_n) \subseteq B(x_m; r_m) \subseteq G \cap U_N$. Since N was arbitrary, $x \in G \cap \bigcap_{N=1}^{\infty} U_N$ and we have established density. ∎

The next result is the form of the theorem that will be used most often and will also be referred to as the Baire Category Theorem.

A.1.3. Corollary. *If (X, d) is a complete metric space and $\{F_n\}$ is a sequence of closed subsets such that $X = \bigcup_n F_n$, there is an n such that* $\operatorname{int} F_n \neq \emptyset$.

Proof. Suppose $\operatorname{int} F_n = \emptyset$ for each $n \geq 1$ and put $U_n = X \backslash F_n$. It follows that each U_n is open and dense, so the theorem implies that $\bigcap_{n=1}^{\infty} U_n$ is dense. But the hypothesis of this corollary implies that this intersection is empty, a dramatic contradiction. ∎

We might mention that the Baire Category Theorem is also valid for a locally compact space. In fact the proof of this is similar to the one given above where we use the finite intersection property for compact sets in place of the completeness of the metric space.

A.2. Nets

Here we give a sequence of definitions and propositions that lay out the theory of nets, which is used in this book especially when the weak and weak* topologies are discussed in Chapter 7. We give few proofs; the reader is asked to supply those that are missing. Two references where the reader will find all this material and more are [**26**] and [**13**].

A *directed set* is a set I together with a partial ordering $\leq$ with the property that if $i, j \in I$, then there is a k in I with $i, j \leq k$. A *net* in a topological space X is a pair (x, I), where I is a directed set and x is a function from I into X. We will often (usually) write the net as $\{x_i : i \in I\}$ or $\{x_i\}$ if the directed set I is understood. Note that every sequence is a net. Also note that a directed set has the property that if $j_1, \dots, j_n \in I$, then there is an i in I with $i \geq j_k$ for $1 \leq k \leq n$.

A.2.1. Example. (a) Let $\mathcal{E}$ be a set and order $2^{\mathcal{E}}$ by inclusion. That is, if $E, F \in 2^{\mathcal{E}}$, say that $E \leq F$ precisely when $E \subseteq F$. Then $2^{\mathcal{E}}$ is a directed set.

(b) Again if $\mathcal{E}$ is a set, $\mathcal{F}$ is the collection of all finite subsets of $\mathcal{E}$, and $\mathcal{F}$ is ordered by inclusion, then $\mathcal{F}$ is a directed set.

(c) If $\mathcal{E}$ and $\mathcal{F}$ are as in (b) and $x : \mathcal{E} \to \mathbb{F}$ is a function, then $\{\sum_{e \in F} x(e) : F \in \mathcal{F}\}$ is a net. Technically we could define $\tilde{x} : \mathcal{F} \to \mathbb{F}$ by $\tilde{x}(F) = \sum_{e \in F} x(e)$ and $(\tilde{x}, \mathcal{F})$ is a net.

(d) Let X be a topological spaces and for some x_0 in X, let $\mathcal{U}$ denote the collection of all open subsets of X that contain x_0. It is easy to see that $\mathcal{U}$ is a directed set under reverse inclusion; that is, if $U, V \in \mathcal{U}$, declare $U \geq V$ to mean that $U \subseteq V$. If for each U in $\mathcal{U}$, we pick a point x_U in U, then $\{x_U : U \in \mathcal{U}\}$ is a net.

A.2.2. Definition. If $\{x_i\}$ is a net in a topological space X, say that the net *converges* to x, in symbols $x_i \to x$, if for every open set G containing x there is an i_0 in I such that $x_i \in G$ for all $i \geq i_0$. Say that $\{x_i\}$ *clusters* at x, in symbols, $x_i \to_{\mathrm{cl}} x$, if for every open set G containing x and for every j in I there is an $i \geq j$ with x_i in G.

It is easy to see that if we have a sequence, this definition of convergence is the same as the concept of a convergent sequence. Similarly a sequence clusters at a point x in this sense if and only if it clusters in the sense of sequences. The reader might note that the net $\{x_U\}$ defined in Example A.2.1(d) converges to x_0.

A.2.3. Proposition. (a) *If a net in a topological spaces converges to x, then it clusters at x.*

(b) *A net in a Hausdorff space can converge to only one point.*

Be cautioned that though there is a concept of a subnet and a net that clusters at a point has a subnet that converges to it, the definition is not the obvious one. The interested reader can consult the references for the definition. We won't use the idea of a subnet. I have found that I can usually avoid it and have decided to live my life that way.

A.2.4. Theorem. *Let X and Y be topological spaces.*

(a) *If $f : X \to Y$ and $x \in X$, then f is continuous at x if and only if whenever $x_i \to x$ in X, $f(x_i) \to f(x)$ in Y. If f is continuous at x and $x_i \to_{\mathrm{cl}} x$, then $f(x_i) \to_{\mathrm{cl}} f(x)$.*

(b) *A subset F of X is closed if for every net $\{x_i\}$ in F that converges to a point x, we have that $x \in F$.*

A.2.5. Theorem. *A topological space X is compact if and only if every net in X has a cluster point.*

Proof. Suppose X is compact and $\{x_i\}$ is a net in X. For every j in I let F_j be the closure of $\{x_i : j \geq i\}$. If $j_1, \ldots, j_n \in I$, there is an $i \geq j_k$ for $1 \leq k \leq n$; hence $x_i \in \bigcap_{k=1}^n F_{j_k}$. So $\{F_i : i \in I\}$ has the finite intersection property; since X is compact there is an x in $\bigcap_i F_i$. If G is an open set containing $x \in G$ and $i_0 \in I$, then $x \in F_{i_0}$ so that $G \cap \{x_i : i \geq i_0\} \neq \emptyset$. That is, there is an $i \geq i_0$ with $x_i \in G$. By definition, this says that $x_i \to_{\mathrm{cl}} x$.

Now assume X is not compact and let $\mathcal{G}$ be an open cover of X without a finite subcover. Let I be the set of all finite subsets of $\mathcal{G}$ and order I by inclusion as in Example A.2.1(b). By assumption, for every $i = \{G_1, \ldots, G_n\}$ in I there is a point x_i in X such that $x_i \notin \bigcup_{k=1}^n G_k$; $\{x_i\}$ is a net in X. This net does not have a cluster point. In fact if there is a cluster point x, then there is a G in $\mathcal{G}$ that contains x. But $\{G\} \in I$ and so there is an $i = \{G, G_1, \ldots, G_n\} \geq \{G\}$ with $x_i \in G$. But by definition $x_i \notin G \cup \bigcup_{k=1}^n G_k$, a contradiction. $\blacksquare$

The next theorem will be very useful.

A.2.6. Theorem. *If X is a compact space and $\{x_i : i \in I\}$ is a net in X with a unique cluster point x, then $\{x_i\}$ converges to x.*

Proof. Fix a proper open set G containing x. If it were the case that the net does not converge to x, then for every i_0 in I, there is an i in I with x_i in $X \backslash G$. This says that $J = \{j \in I : x_j \in X \backslash G\} \neq \emptyset$. In fact this also says that J is directed. Indeed, if $j_1, j_2 \in J$, let $i_0 \in I$ such that $i_0 \geq j_1, j_2$. By what we have said, there is a j in J with $j \geq i_0 \geq j_1, j_2$. Thus $\{x_j : j \in J\}$ is a net in $X \backslash G$. Since $X \backslash G$ is compact, there is a y in $X \backslash G$ such that $\{x_j : j \in J\} \to_{\mathrm{cl}} y$. The reader can check that this implies that y is a cluster point of the original net. Since $y \neq x$, we have a contradiction. $\blacksquare$

Bibliography

[1] M. B. Abrahamse and T. L. Kriete [1973], "The spectral multiplicity of a multiplication operator," *Indiana Math. J.*, **22**, pp. 845–857.

[2] E. Bishop [1961], "A generalization of the Stone–Weierstrass Theorem," *Pacfic J. Math.* **11**, pp. 777–783.

[3] J. K. Brooks [1971], "The Lebesgue Decomposition Theorem for Measures," *Amer. Math. Monthly* **78**, pp. 660–662.

[4] R. B. Burckel [1984], "Bishop's Stone-Weierstrass theorem," *Amer. Math. Monthly* **91**, pp. 22–32.

[5] L. Carleson [1966], "On the convergence and growth of partial sums of Fourier series," *Acta Mathematica* **116**, pp. 135–157.

[6] J. B. Conway [1969], "The inadequacy of sequences," *Amer. Math. Monthly* **76**, pp. 68 – 69.

[7] J. B. Conway [1978], *Functions of One Complex Variable*, Springer-Verlag, New York.

[8] J. B. Conway [1990], *A Course in Functional Analysis*, Springer-Verlag, New York.

[9] M. M. Day [1958], *Normed linear spaces*, Springer-Verlag, Berlin.

[10] L. de Branges [1959], "The Stone–Weierstrass theorem," *Proc. Amer. Math. Soc.* **10**, pp. 822–824.

[11] J. Diestel [1984], *Sequences and Series in Banach Spaces*, Springer-Verlag.

[12] J. Dixmier [1964], *Les C*-Algèbras et leurs Représentations*, Gautiers-Villars, Paris.

[13] J.. Dugundji [1967], *Topology*, Allyn and Bacon.

[14] Per Enflo [1973], "A counterexample to the approximation problem in Banach spaces," *Acta Mathematica* **130**, pp. 309–317.

[15] G. B. Folland [1999], *Real Analysis: Modern Techniques and Their Applications*, Wiley (Hoboken).

[16] I. Glicksberg [1962], "Measures orthogonal to algebras and sets of antisymmetry," *Trans. Amer. Math. Soc.* **105**, pp. 415–435 .

[17] S. Grabiner [1986], "The Tietze extension theorem and the open mapping theorem," *Amer. Math. Monthly* **93**, pp. 190–191.

[18] E. Hewitt and K. Stromberg [1975], *Real and Abstract Analysis*, Springer-Verlag, New York.

[19] R. A. Hunt [1967], "On the convergence of Fourier series, orthogonal expansions and their continuous analogies," pp. 235–255, *Proc. Conf. at Edwardsville, Ill*, Southern Illinois Univ. Press, Carbondale.

[20] R. C. James [1951], "A non-reflexive Banach space isometric with its second conjugate space," *Proc. Nat. Acad. Sci. USA* **37**, pp. 174–177.

[21] R. V. Kadison and J. R. Ringrose [1997a], *Fundamentals of the Theory of Operator Algebras. Volume I: Elementary Theory*, Amer. Math. Soc., Providence.

[22] R. V. Kadison and J. R. Ringrose [1997b], *Fundamentals of the Theory of Operator Algebras. Volume II: Advanced Theory*, Amer. Math. Soc, Providence.

[23] R. V. Kadison and J. R. Ringrose [1998a], *Fundamentals of the Theory of Operator Algebras. Volume III*, Amer. Math. Soc, Providence.

[24] R. V. Kadison and J. R. Ringrose [1998b], *Fundamentals of the Theory of Operator Algebras. Volume IV*, Amer. Math. Soc, Providence.

[25] J. L. Kelley [1966], "Decomposition and representation theorems in measure theory," *Math. Ann.* **163**, pp. 89–94.

[26] J. L. Kelley [2008], *General Topology*, Ishi Press.

[27] H. Kestelman [1971], "Mappings with Non-Vanishing Jacobian," *Amer. Math. Monthly* **78**, pp. 662–663.

[28] J. Lindenstrauss and L. Tzafriri [1971], "On complemented subspaces problem," *Israel J. Math.* **5**, pp. 153–156.

[29] T. J. Ransford [1984], "A short elementary proof of the Bishop–Stone–Weierstrass theorem," *Math. Proc. Camb. Phil. Soc.* **96**, pp. 309–311.

[30] W. Rudin [1991], *Functional Analysis*, McGraw-Hill, New York.

[31] M. E. Taylor [2006], *Measure Theory and Integration*, Amer. Math. Soc., Providence.

[32] D. E. Varberg [1965], "On absolutely continuous functions," *Amer. Math. Monthly* **72**, pp. 831–841.

[33] A. Villani [1984], "On Lusin's condition for the inverse function," *Rendiconti Circolo Mat. Palermo* **33**, pp. 331–333.

List of Symbols

Index